100 FAMILIES OF FLOWERING PLANTS

100 FAMILIES OF FLOWERING PLANTS

MICHAEL HICKEY & CLIVE KING

FOREWORD BY S.M. WALTERS
Director of the University Botanic Garden, Cambridge

CAMBRIDGE UNIVERSITY PRESS

Cambridge
London New York New Rochelle
Melbourne Sydney

Published by the Press Syndicate of the University of Cambridge
The Pitt Building, Trumpington Street, Cambridge CB2 1RP
32 East 57th Street, New York, NY 10022, USA
296 Beaconsfield Parade, Middle Park, Melbourne 3206, Australia

First published 1981

Set by Ward Partnership and printed in Great Britain
at the Alden Press, Oxford

British Library cataloguing in publication data

Hickey, Michael
One hundred families of flowering plants.
1. Angiosperms – Dictionaries
I. Title II. King, Clive John
582'.13'03 QK495.A1 79-42670

ISBN 0 521 23283 X hard covers
ISBN 0 521 29891 1 paperback

To our wives, Robin and Margaret

CONTENTS

Contents

Contents

Contents

FOREWORD

I am very pleased to see this book completed, for several reasons. Firstly, I really believe that it will provide teachers and students of Angiosperm systematics with a body of information which it is difficult, if not impossible, to find in the existing literature, and I look forward to using it in my own teaching. Secondly, it represents what seems to me an ideal collaboration between two 'Cambridge Botanic Garden men' – if I may so describe the authors – whose talents are impressively complementary, and such collaboration is the natural outcome of a Garden which has a long tradition of practical horticulture allied to scientific scholarship. Thirdly, and perhaps most importantly, I see the book as a small but significant contribution to the very necessary task of interesting a much wider public in the classification of the Flowering Plants, by encouraging people to look carefully for themselves, as Michael Hickey has done, to see the remarkable detail of floral structure on which the classification is based. Careful observation lies at the root of all worthwhile scientific enquiry, and biology is *par excellence* the science of careful observation.

Having witnessed the various stages in the production, I can record my warm admiration of the authors' tenacity of purpose and meticulous attention to detail, qualities which have in my opinion produced a far more valuable text than might at first sight be realised. Descriptions and statements in the literature have been checked against the living plant – a process all too rarely undertaken – and much detailed clarification has resulted. Undoubtedly errors and ambiguities remain, but they must be greatly reduced by the scrupulous checking which the authors have carried out. Generations of students will, I hope, test the value of the book, and come to appreciate its worth.

S.M. Walters
Director
University Botanic Garden
Cambridge

PREFACE

According to various authorities there are some 300–400 families of flowering plants in the world today. To understand fully the relationships between these families and to appreciate the diversity of their floral structure would require research into a wider range of plants than time and available facilities normally permit. However, a basic knowledge can be obtained by the study of a selected group of families, and therefore we decided to restrict the number dealt with in this book to one hundred. We thought that this number would be sufficiently large to enable students at schools and universities to gain a reasonable general knowledge of this aspect of botany.

In selecting the hundred families the following points were borne in mind: (1) The range of families should be sufficiently wide to illustrate adequately the similarities and differences existing in floral structure. (2) The typical representative(s) of a family should, as far as possible, be readily available, either as members of the native flora, or as introduced plants commonly grown in gardens or glasshouses.

In limiting ourselves to a maximum of one hundred families we have had to omit many which are both large and important, especially some which are prominent in tropical or subtropical regions.

In general, the descriptions of the families have been based on information given in Willis's *Dictionary of the Flowering Plants and Ferns*, revised by H.K. Airy Shaw, and published by Cambridge University Press in 1973. In cases where long-established families have been divided up, however, this course has not always been followed, the broader view of the family being considered to be of more value to the students for whom this book is designed. The families are arranged in the same order as that of the second edition of *The Identification of Flowering Plant Families* by P.H. Davis and J. Cullen, published by Cambridge University Press in 1979.

The text for each family is in two parts. The first part treats the family as a whole and gives its world distribution, general characteristics, principal economic and ornamental plants, and a classification which includes the mention of some of the larger or more important genera together with their distribution and the number of species they contain. The second part is devoted to the detailed description of a plant chosen as a typical representative of the family – in a few cases it was considered that more than one plant was necessary to show the variation in floral structure existing within the family. This part gives the distribution of the plant, its vegetative characteristics, floral formula, details of the flower and inflorescence, pollination mechanism, and suggestions for alternatives if the plant described proves to be unobtainable.

The pen and ink drawings which show, in detail, the floral structure of the plant(s) representing each family are an essential part of the book, and all of them have been made from living material. The drawings are accompanied by extended captions which in most cases include measurements of the floral parts. Since many of the drawings are larger than life-size in order to show more clearly the parts concerned, it was felt desirable to provide these measurements so that students, when examining a flower or fruit, should have a clear mental picture of the actual size of the part in question. It must be emphasised that the measurements were taken from the specimens used for the drawings, and that care was taken to ensure that these were typical in form. However, variation within a species or developmental factors may mean that a student will sometimes find that a certain part does not agree exactly with the description or dimensions given. The abbreviations L.S. and T.S. (for longitudinal and transverse sections respectively) have been found convenient to indicate what is seen when a part has been cut lengthwise or across at any point. Their use has therefore not been restricted to describing thin slices of material such as those normally used in microscopic work.

It is hoped that this book will be useful, not only to students, but also to professional botanists and interested amateurs of natural history and horticulture.

We would be grateful if users of the book would notify us of any errors or omissions that may come to their notice.

M.H.
C.J.K.

ACKNOWLEDGEMENTS

We would like to express our appreciation and gratitude to Dr S.M. Walters, Director of the University Botanic Garden, Cambridge, for acting as adviser and for his encouragement in the writing and illustrating of this book. We would also like to thank him and his staff for providing much of the plant material required for the illustrations. A generous amount of material was also supplied by Mr Philip Butler, Curator of the Birmingham Botanical Garden, and by J.K. Barry, R. and C. Draper, W.G. Hennessy, M. Newland and E.W. Pymont. To all these we give our thanks. We are grateful also to J. and H. Blackwood, R.E. Edwards, R. Gregory, T.V. and J. Ireland, S.W. Smith (formerly Head of Science, St Paul's College, Cheltenham), A. Tandy, D. Tullis and Dr P.F. Yeo (Taxonomist, University Botanic Garden, Cambridge) for giving us information and assistance in various ways. We are greatly indebted to Dr Margaret Bradshaw of the University of Durham for her valuable comments on the Introduction, and to Mrs Ann Hill for her sustained effort in producing the typescript. Finally, we wish to thank Dr Alan Winter and others at the Cambridge University Press for their advice and help throughout the production of this book.

SIGNS AND ABBREVIATIONS

Note. Throughout the text the numbers given in parentheses after the genera listed for each family indicate the number of species known throughout the world.

Sign	Meaning
♂	male
♀	female
⚥	hermaphrodite
∞	indefinite number
×	hybrid
( )	united
A	androecium
C	corolla
$\overline{G}$	gynoecium (ovary inferior)
$\underline{G}$	gynoecium (ovary superior)
K	calyx
P	perianth
L.S.	longitudinal section
T.S.	transverse section

Abbreviation	Meaning
c.	(circa) about, approximately
cm	centimetre(s)
m	metre(s)
mm	millimetre(s)
2-merous	dimerous
3-merous	trimerous
4-merous	tetramerous
5-merous	pentamerous
6-merous	hexamerous
adj.	adjective
plur.	plural

A measurement given without qualification refers to length. Two measurements connected by × indicate length followed by width. Further measurements in parentheses indicate exceptional cases outside the normal range.

FLORAL FORMULA

Some of the above signs and abbreviations occur in the Floral Formula, which is a convenient form of 'shorthand' for representing the structure of a flower. The letters K, C, A, and G are used to indicate the whorls of floral parts, beginning with the outermost whorl and working inwards towards the centre of the flower. Where there is no separation into calyx and corolla the letter P is used in place of K and C.

Each letter is followed by one or more figures showing the number of parts comprising each whorl, e.g., A5 indicates an androecium consisting of 5 stamens, and A5+5 shows that there are 2 whorls of stamens with 5 in each whorl. Where the number of parts is large and imprecise the sign '∞' is used. If the number of parts in a whorl is variable, a dash joining 2 figures indicates the range of variation, e.g., A12–20 (in *Reseda lutea*) means that from 12 to 20 stamens may be found in a flower of this species.

The parts forming a whorl are sometimes connate, e.g., G(3) denotes a gynoecium of 3 united carpels. (A bracket may be placed above the letters concerned to indicate that 2 whorls are joined together, e.g., $\overset{\frown}{\mathrm{C}(5)\ \mathrm{A}5}$, but this practice has not been adopted here.)

The position of the ovary is shown by a line above or below the letter G, representing an inferior or superior ovary respectively. In rare instances of dioecious plants lacking a perianth, e.g., female flowers in *Betula pendula*, it is not possible to show the relationship of the ovary to the other floral parts and the line is therefore omitted.

INTRODUCTION

CLASSIFICATION AND EVOLUTION

The classification of the flowering plants or Angiosperms is based upon comparative study of the structure of their flowers and fruits. Thus, one of the most obvious differences between plants, namely whether they are trees, shrubs or herbs, is not a character used in the main classification into different families, and indeed many important families of flowering plants, such as Rosaceae and Compositae, contain both woody and herbaceous representatives. But it is significant, and also useful in identification, that certain families, such as those comprising the catkin-bearing group or Amentiferae, are wholly or very largely composed of woody plants, whilst others are entirely herbaceous.

It is generally assumed that the change from woody to herbaceous forms, and vice versa, has occurred many times within different evolutionary lines of flowering plants. Many botanists also hold the view that the primitive Angiosperm was a dicotyledon with woody stems and large, terminal flowers like the modern *Magnolia*, but this theory has not been proved to the satisfaction of all. Neither is it certain how the monocotyledons originated, though it may be that the common ancestor of both monocotyledons and dicotyledons was some shrubby Angiosperm with primitive floral structure belonging to a group which is now completely extinct. Although the form of the earliest flowering plants is still unknown, it seems clear from the fossil record that they arose in the early Cretaceous period (120 million years ago) and that by the end of that period (80–90 million years ago) they had ousted the conifers and cycads from their position of dominance and established themselves in their stead as the characteristic form of land plant, a place they have continued to hold to the present day.

THE FLOWERING PLANT

With very few exceptions, growth in the flowering plants is maintained underground by a **root** system and above ground by a **shoot** system. The shoot system consists of one or more stems bearing leaves which are arranged spirally or in pairs, or more rarely in whorls of three or more, along each stem. The point from which a leaf arises is called a **node**, and the portion of the stem between two nodes is described as an **internode**. The angle formed by a leaf and its parent stem is known as an **axil**, and buds appearing in this position are called **axillary**. A bud may also be found at the apex of the stem, in which case it is described as **terminal**. Each bud is, in fact, a shoot with very short internodes and immature leaves. Some buds do not develop further but, if they do so, the internodes lengthen and the leaves enlarge, forming a typical shoot.

A flower may be regarded as a bud which is modified for the purpose of reproduction of the parent plant. The outermost leaves have changed little, retaining their leaf-like appearance and protective function. The inner leaves, on the other hand, have in many cases undergone considerable modification in shape and colour, and have become highly specialised in order to perform a variety of functions necessary for the successful propagation of the plant concerned. These modified leaves are now referred to as floral parts.

In the more primitive families such as Magnoliaceae, the floral parts are many or **indefinite** in number and are arranged on an elongated axis. In more advanced families there is a reduction in the number of parts which then arise from a much shortened axis. In either case, the production of a flower means that growth of the stem concerned is now limited, although it may continue to elongate sufficiently in order to allow the flower-cluster or **inflorescence** to develop fully.

Growth of a stem by apical extension is termed **monopodial** and results in an inflorescence known as a **raceme.** In contrast to this, the main stem may cease to lengthen owing to a flower being formed at its apex. If this occurs, side or **lateral** branches may arise from buds below this flower. This is called **sympodial** growth and results in an inflorescence known as a **cyme.** Determining whether an inflorescence is racemose or cymose may sometimes be difficult but they are usually distinguishable by the order of flower development. In a raceme the youngest flower is situated at the apex of the stem, while in a cyme the oldest flower occupies this position. Grouping the flowers together into an inflorescence, of whatever kind, renders them more conspicuous and is an important aid to pollination.

The term inflorescence is often used solely for a cluster of individual flowers, but, correctly employed, it includes the stem from which the flowers arise which is termed the **peduncle.** Sometimes, however, the flowers are solitary on a stem as in *Galanthus*, Snowdrop. If an individual flower is unstalked it is said to be sessile, but if stalked this stalk is known as a **pedicel.** There may often be scale or leaf-like structures arising from the peduncle or pedicel. These are termed **bracts** or **bracteoles**, and are usually simple in shape compared with the foliage leaves. Where bracteoles are present in dicotyledons, there are usually two situated opposite each other on the pedicel, but only a single bracteole may be present in monocotyledons. In many inflorescences each pedicel arises from the axil of a bract on the peduncle. In the Compositae and Dipsacaceae the inflorescence is condensed to a head of sessile flowers known as a **capitulum** which is borne on the thickened and flattened apex of the peduncle. The bracts of these flowers are crowded into one or more whorls round the capitulum, and together form an **involucre.**

FLORAL STRUCTURE

The flower is a specialised structure evolved for the process of seed formation, which is normally the result of a sexual union. Some plants, however, have the ability to form seeds without the aid of fertilisation. This form of reproduction, which is found in several common genera of Rosaceae (e.g., *Alchemilla*, *Rosa* and *Rubus*) and Compositae (e.g., *Hieracium* and *Taraxacum*) is known as **apomixis** (more strictly **agamospermy**).

The floral axis on which the flower-parts are arranged is termed the **receptacle** or **torus.** The parts are divided into accessory organs which are non-reproductive,

Fig. I. Aestivation.

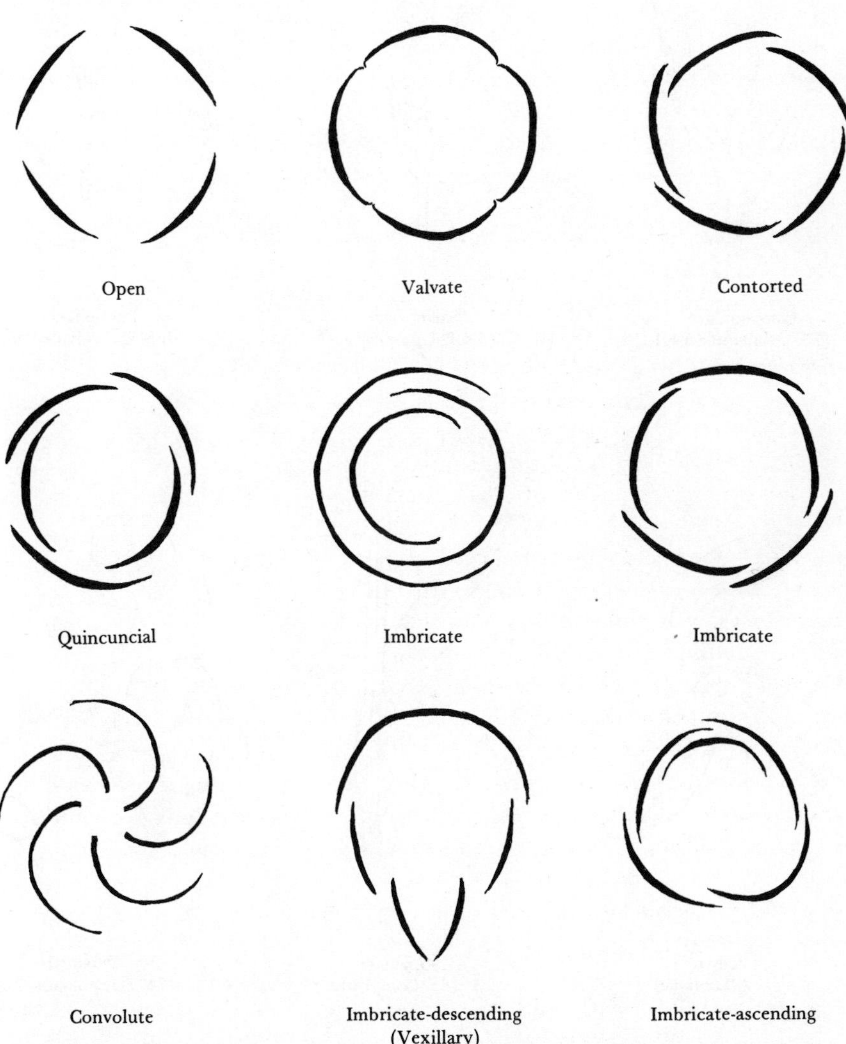

Fig. II. Corolla forms and special structures.

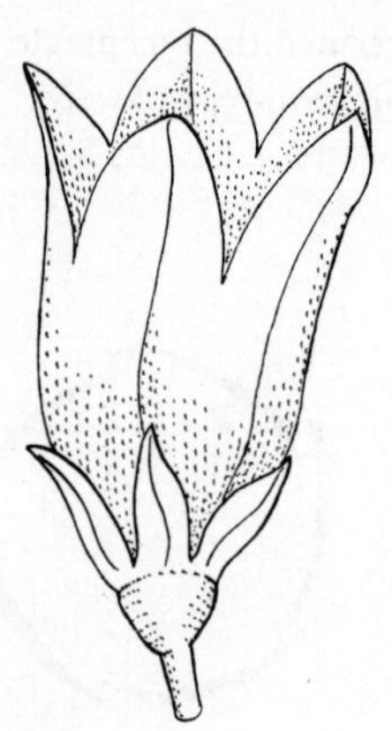

Campanulate
(78. Campanulaceae)

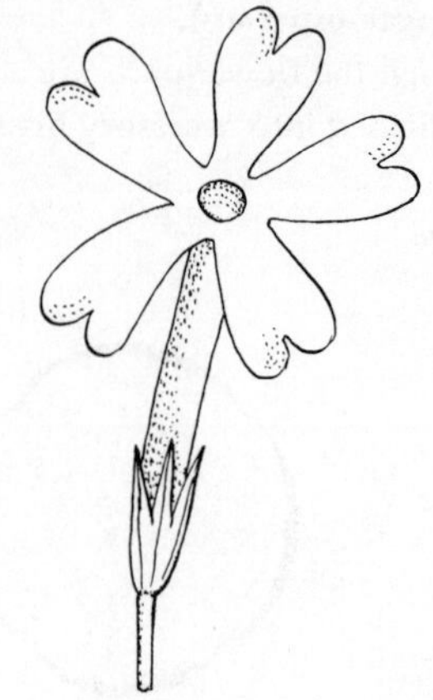

Salverform
(69. Polemoniaceae)

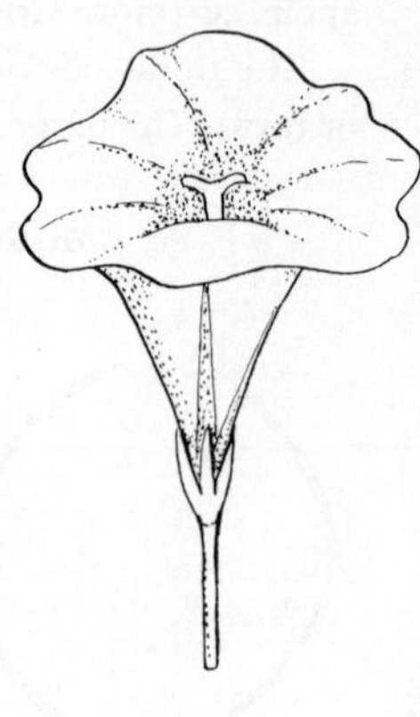

Funnelform
(68. Convolvulaceae)

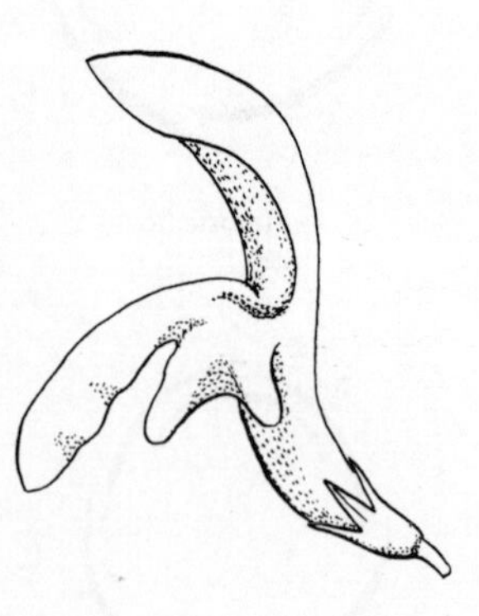

Labiate
(72. Labiatae)

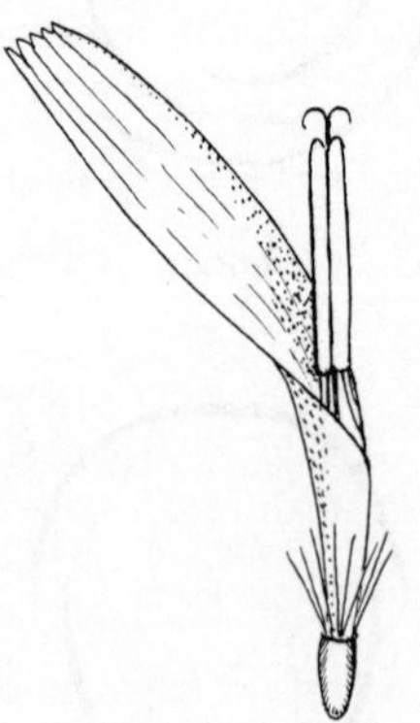

Ligulate
(83. Compositae)

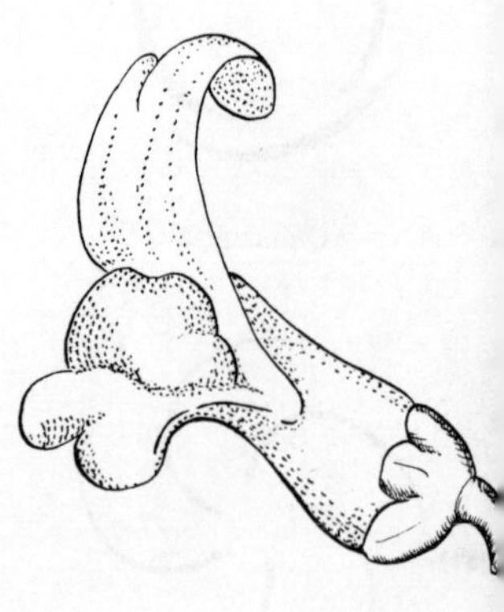

Personate
(74. Scrophulariaceae)

Fig. II. *continued*

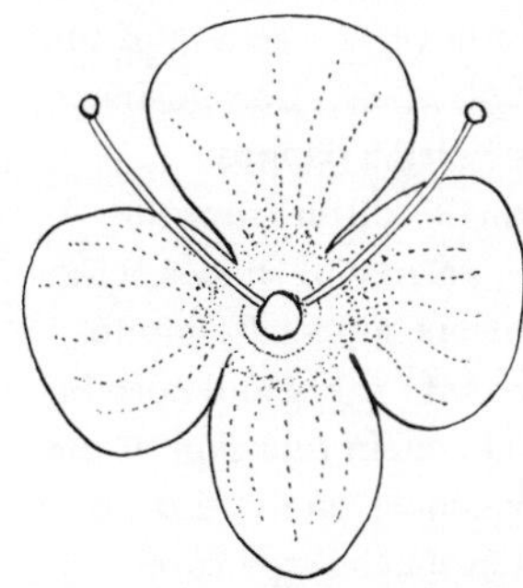

Rotate
(74. Scrophulariaceae–*Veronica*)

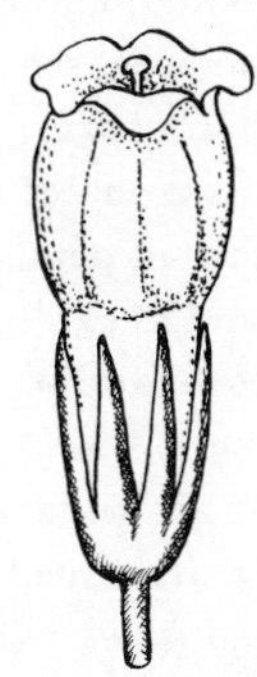

Tubular
(70. Boraginaceae–*Symphytum*)

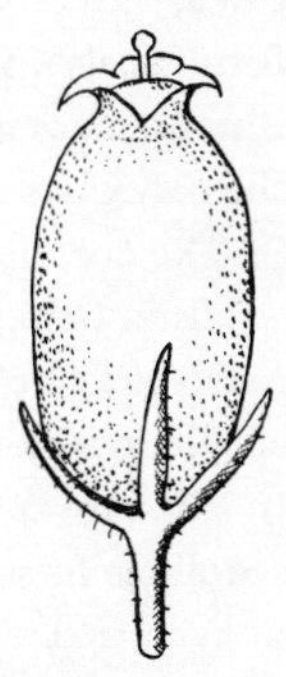

Urceolate
(33. Ericaceae–*Erica*)

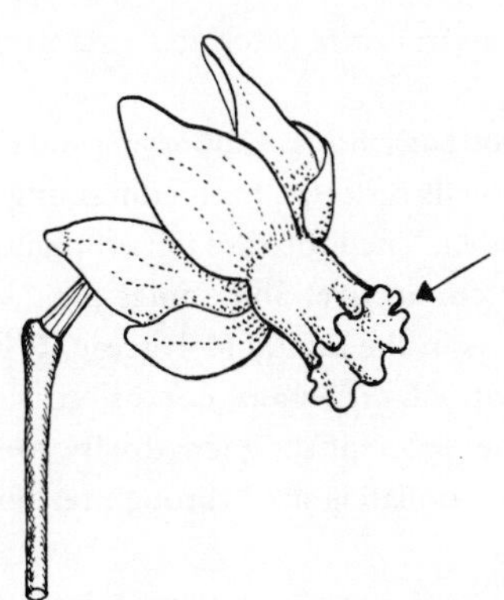

Corona
(98. Amaryllidaceae–*Narcissus*)

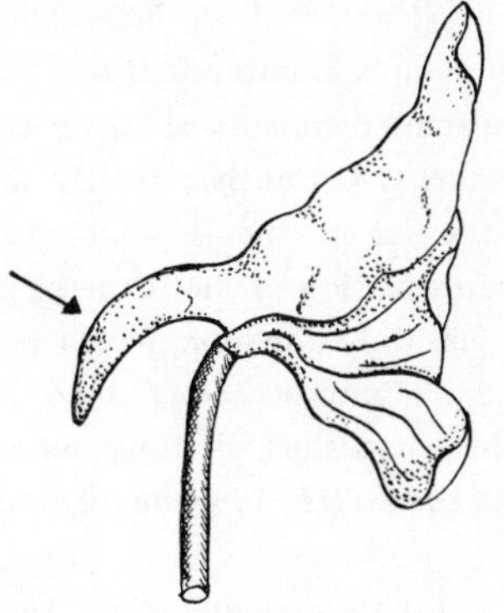

Spur
(5. Ranunculaceae–*Delphinium*)

and essential organs which are reproductive. In some families, e.g., Salicaceae, one or more of the accessory organs may be absent.

ACCESSORY ORGANS

The principal accessory organs are the **calyx** and the **corolla**, collectively known as the **perianth.** The calyx consists of a number of leaf-like structures, usually green but sometimes coloured, known as **sepals.** These may be united to form a **calyx-tube.** Coloured or petaloid sepals are found in certain genera of the Ranunculaceae, e.g., *Caltha* and *Helleborus.* In some families, e.g., Compositae, the calyx has been modified into a group of hairs and is called a **pappus.**

The corolla consists of a number of **petals** which normally differ from the sepals in being larger, coloured, and in exhibiting a wider variety of shapes. The petals may be united to form a **corolla-tube.** In some flowers, especially monocotyledons, e.g., *Tulipa,* Tulip, there is no differentiation into calyx and corolla. In this case the perianth-segments are known as **tepals.** The main function of the corolla is to attract pollinating agents, usually insects, but more rarely birds, bats or even snails. In animal-pollinated flowers there are often special structures which secrete a sugary solution known as **nectar,** the nectar-secreting organ being termed a **nectary.** The form of the nectary may vary considerably between one plant and another.

ESSENTIAL ORGANS

The male organs are known collectively as the **androecium** and the female organs as the **gynoecium.**

The Androecium

Structure. Each individual unit of the androecium is termed a **stamen** if functional or a **staminode** if non-functional.

The stamen consists of two main parts, the **anther,** and its supporting stalk, the **filament.** The anther usually comprises two cells or lobes, each containing two pollen-sacs – **thecae** – which house the pollen. The lobes are joined together by a continuation of the filament known as the **connective.** Sometimes the anther has only one lobe, with two pollen-sacs, as in the family Malvaceae. It is then termed **monothecous,** in contrast to the normal, **dithecous,** condition.

Dehiscence is usually **longitudinal,** each anther-lobe splitting lengthwise to liberate the pollen. In some cases, e.g., Ericaceae, pollen is shed through terminal

Fig. III (opposite). Top – half-flower of dicotyledon, e.g., *Geranium*: 1, branched stigma; 2, style; 3, anther; 4, filament; 5, free stamen; 6, free petal of 5-petalous corolla; 7, superior ovary of 5 united carpels (1 ovule per carpel); 8, free sepal of 5-sepalous calyx; 9, receptacle; 10, pedicel; Bottom – half-flower of monocotyledon, e.g., *Leucojum*: 1, outer perianth-segment; 2, inner perianth-segment; 3, stigma; 4, style; 5, anther; 6, filament; 7, stamen; 8, receptacle; 9, inferior ovary containing ovules; 10, pedicel.

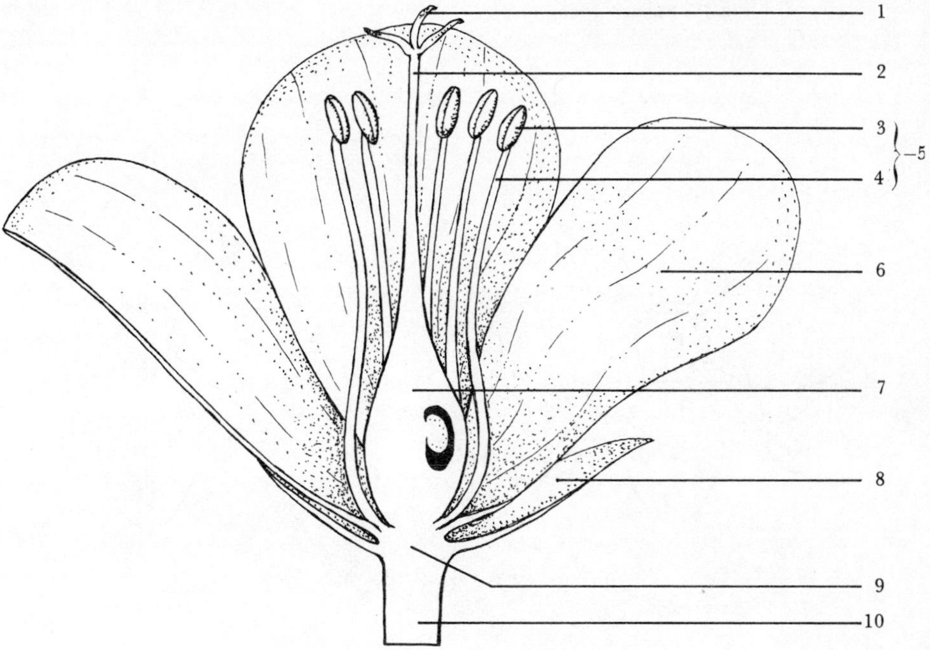

1
2
3
4
5
6
7
8
9
10

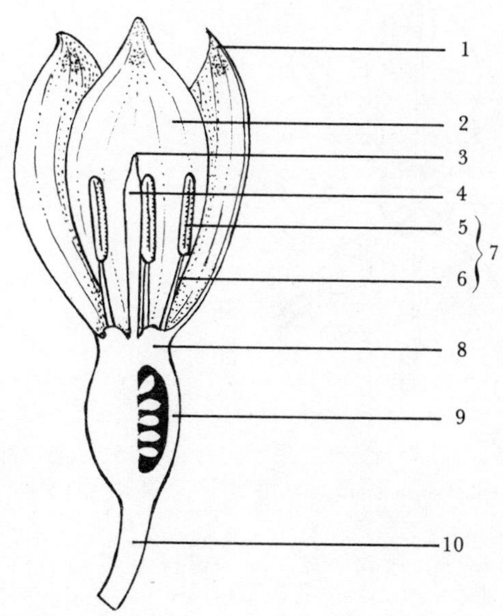
1
2
3
4
5
6
7
8
9
10

Fig. IV. Stamen arrangement.

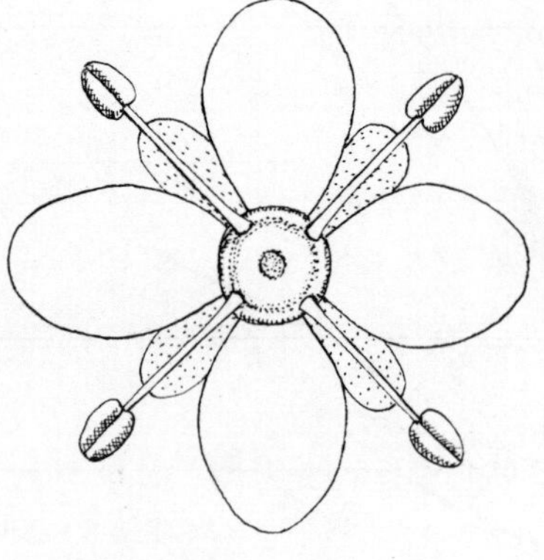

Antisepalous

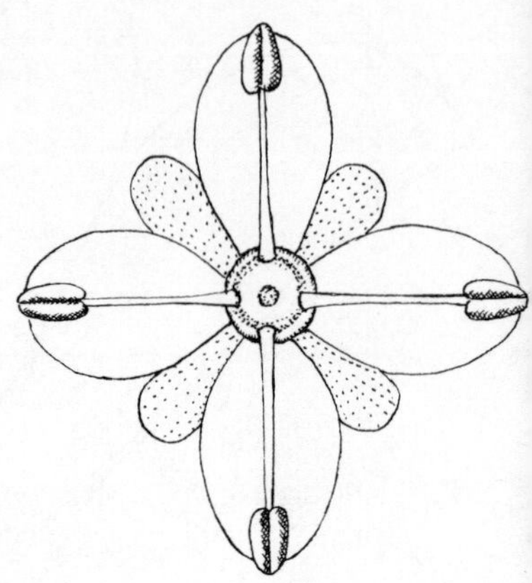

Antipetalous

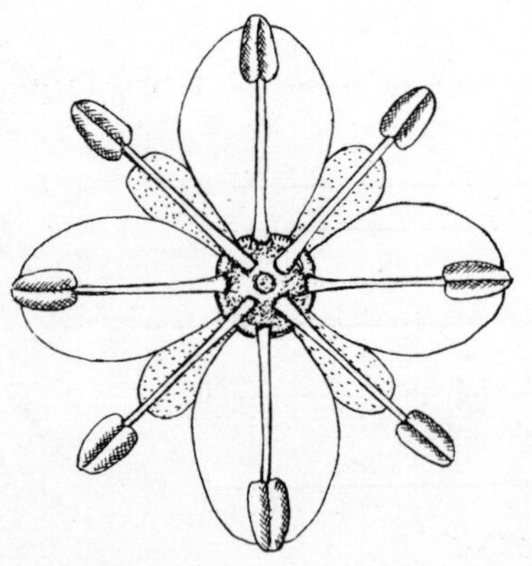

Obdiplostemonous

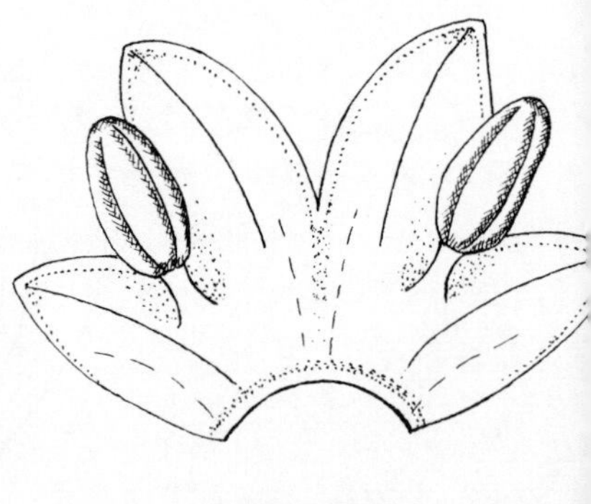

Epipetalous

Fiv. IV. *continued*

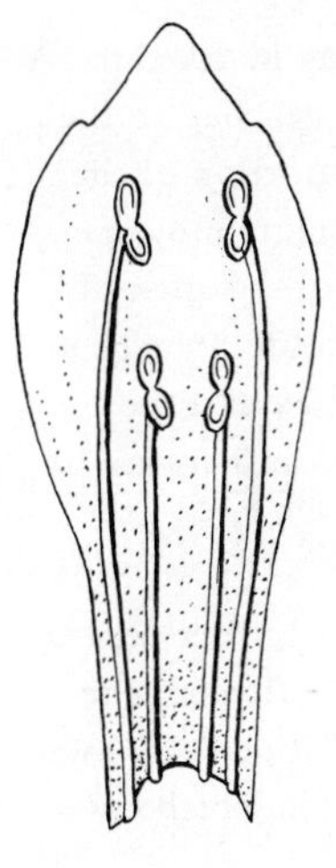

Didynamous
(72. Labiatae)

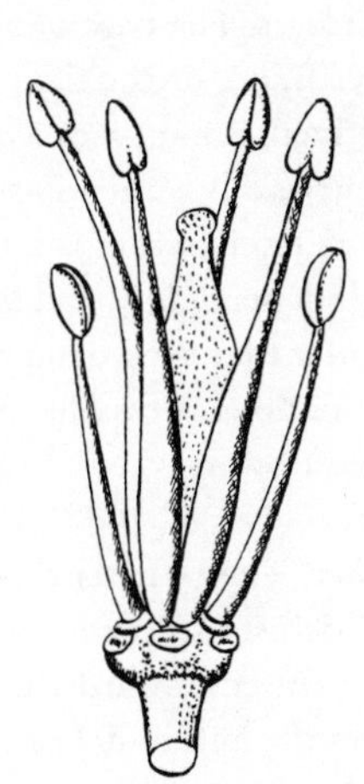

Tetradynamous
(31. Cruciferae)

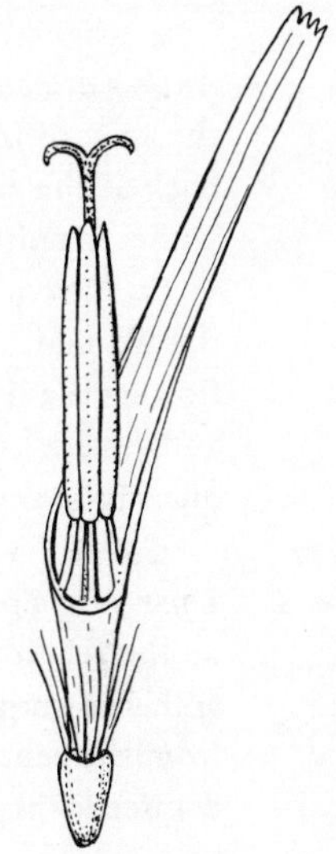

Syngenesious
(83. Compositae)

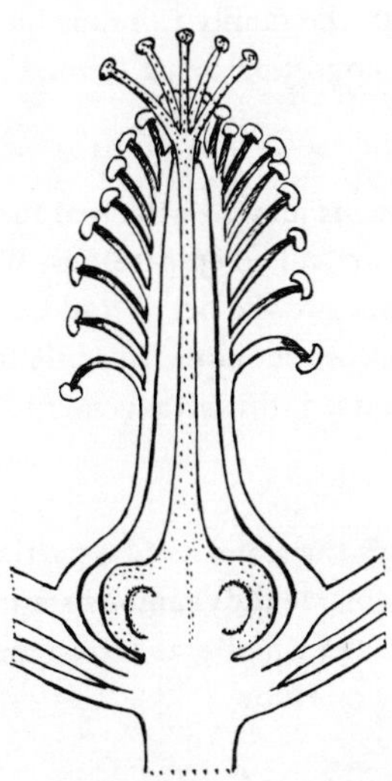

Monadelphous (21. Malvaceae)

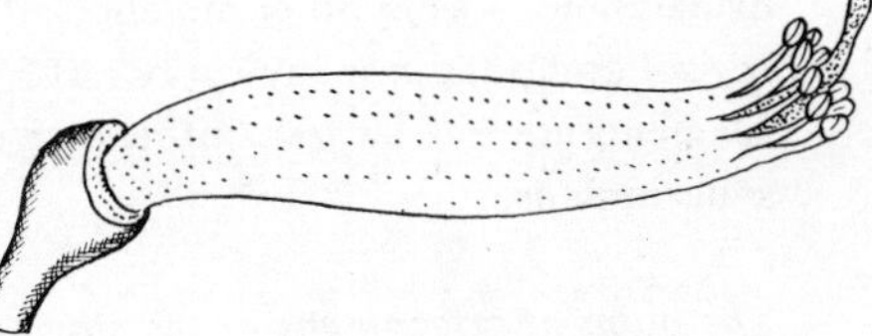

Monadelphous (39. Leguminosae)

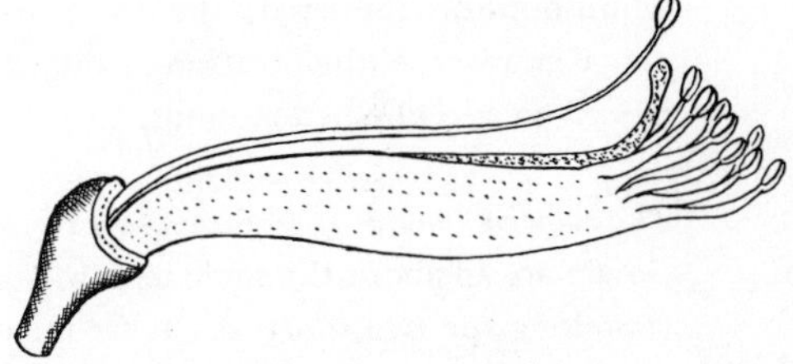

Diadelphous (39. Leguminosae)

pores, this being called **poricidal** dehiscence. In other cases, e.g., Lauraceae, dehiscence is valvate, the pollen being shed when a flap-like valve opens upwards.

Anther attachment. (i) Fixed anthers. The base of the anther may be fixed to the apex of the filament, i.e., **basifixed**, or the filament may be attached to the back of the anther and be **dorsifixed.** Sometimes a considerable portion of the anther is united with the filament and it is then incapable of separate movement.

(ii) Moveable anthers. If a dorsifixed anther is attached only to the apex of the filament, it is able to move independently and is termed **versatile.** In certain flowers the filaments are so slender that they bend over at the apex and the anthers are then described as **pendulous.** Versatile and pendulous anthers are frequently found in wind-pollinated flowers.

The relative position of the anther. The term used to describe the anther depends on the relationship of its line of dehiscence with the centre of the flower. The anther is known as **introrse** if it dehisces inwards, and **extrorse** if it dehisces away from the centre. In some families the anther dehisces at the side, in which case dehiscence is described as **lateral.**

The fusing of the stamens. The stamens may be free from each other or they may be united in some way. Their filaments may all be united to form a tube round the gynoecium as in the Malvaceae and some of the Leguminosae, in which case they are termed **monadelphous.** They may be in two groups – **diadelphous** – as in other members of the Leguminosae, or they may form several groups as in many species of *Hypericum*. In the family Compositae the filaments are free but the anthers are united, this condition being termed **syngenesious.**

The point of attachment of the stamens. The stamens may be attached to the receptacle or to the petals. In the latter case they are called **epipetalous.** Where the stamens are in two whorls, the inner whorl opposite the petals and the outer whorl opposite the sepals, the flower is termed **diplostemonous.** In some families, e.g., Geraniaceae, the position of the whorls is reversed, this arrangement being described as **obdiplostemonous.**

The relative length of the stamens. In most families the stamens of a particular flower are all about the same length, but in the Labiatae **didynamous** stamens (two long and two short) occur frequently, and in the Cruciferae **tetradynamous** stamens (four long and two short) are the typical condition.

The Gynoecium

The Carpel. The basic unit of the gynoecium is the **carpel** which at maturity forms the true fruit enclosing the seed. The carpel is made up of several parts and is situated above, below or on the same plane as the other floral organs. The basal part of the carpel is known as the **ovary.** This encloses and protects the unfertilised 'seeds' or **ovules.** The cavity within the ovary which contains the ovules is termed the **loculus.** A stalk, the **funicle**, connects each ovule to a particular part of the ovary known as the **placenta.** The gynoecium may consist of free carpels, in which case it is termed **apocarpous.** *Ranunculus repens*, Creeping Buttercup, has several free carpels, while *Vicia faba*, Broad Bean, has only one.

Fig. V. Anther forms: 1, anther-cell (theca); 2, connective; 3, filament.

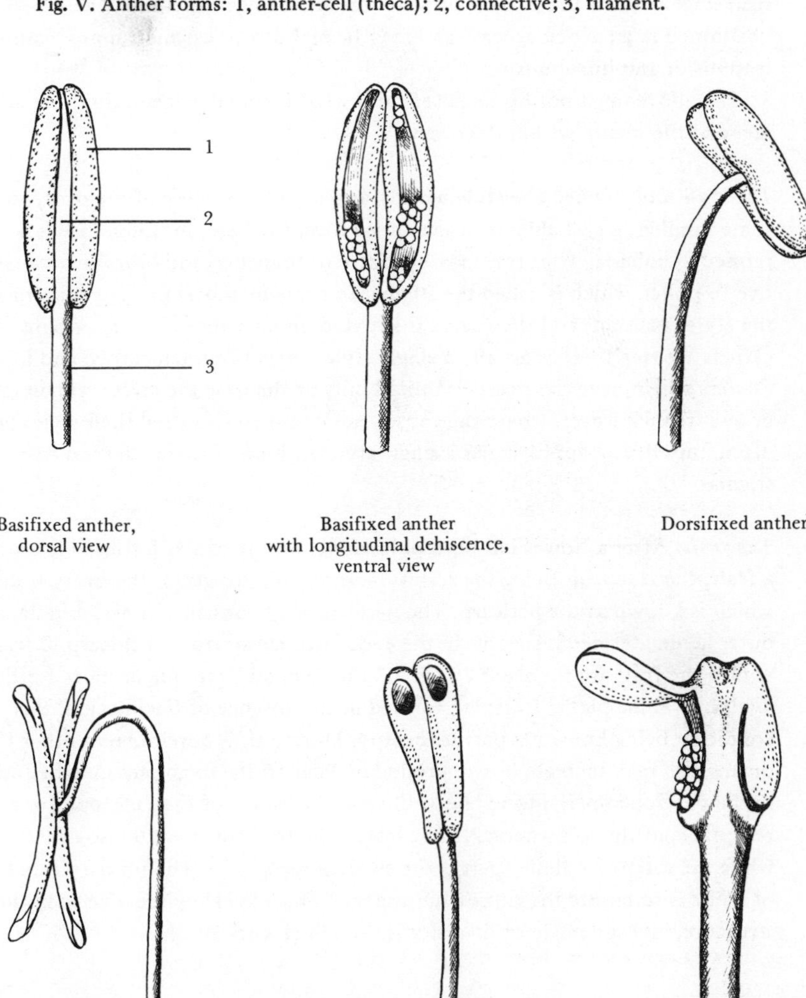

In many flowers the carpels are united or **syncarpous** as in *Lycopersicon esculentum*, Tomato.

The ovule and its orientation. The ovule is made up of central tissue known as the **nucellus** in which the embryo sac develops and, surrounding the nucellus, the **integuments** which form the outer covering of the ovule. There is a minute opening in the integuments known as the **micropyle**, which at this stage exposes a portion of the nucellus and contributes to the process of fertilisation and at the seed stage permits the passage of water essential to germination. The ovule is positioned in a particular way, and may be **orthotropous, anatropous, campylotropous** or **amphitropous.**

The above must not be confused with what is called **placentation**, the arrangement of the ovules within the ovary.

The style and stigma. The **style** usually arises from the apex of the ovary, but in some families, e.g., Labiatae, it may arise from the base, in which case it is termed **gynobasic.** The style may be entire or branched and bears an area receptive to pollen which is called the **stigma.** In some members of the Papaveraceae the style is absent, and the stigma is situated directly on the ovary. Where the carpels are free there is usually a single style present on each carpel, and in flowers which have their carpels united only at the base the styles remain more or less free. However, where the carpels are completely united their styles are often united also, appearing as a single style with one fused or several free stigmas.

The fruit. After a flower has been pollinated and the ovules fertilised it produces a fruit, the true fruit being the result of secondary growth of the ovary wall which is known as the **pericarp.** The pericarp may consist of inner, middle, and outer layers, termed respectively the **endocarp, mesocarp** and **epicarp.** Enclosed within the fruit are the seeds which have developed from one or more fertilised ovules. In some plants fruits are formed in the absence of fertilisation, this condition being known as **parthenocarpy.** During fruit development other floral organs also may increase in size producing false fruits. Examples may be found in the developed perianth of the Mulberry, the bracts of the Pineapple, and the receptacle of the Strawberry. In the latter, the true fruits are the so-called 'pips', while the attractive fleshy part is the swollen receptacle. The primary function of fruits is to ensure the successful dispersal of seeds. They are diverse in their structure, and the range of diversity is used in classification.

Fig. VI. Ovule forms: 1, micropyle; 2, chalaza; 3, nucellus with embryo; 4, integuments; 5, funicle.

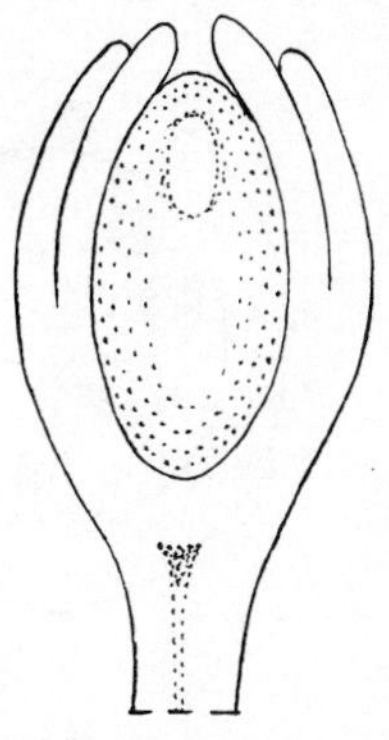

Orthotropous

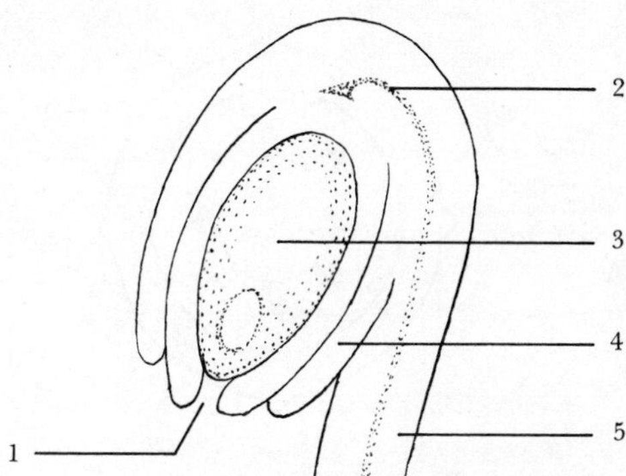

Anatropous

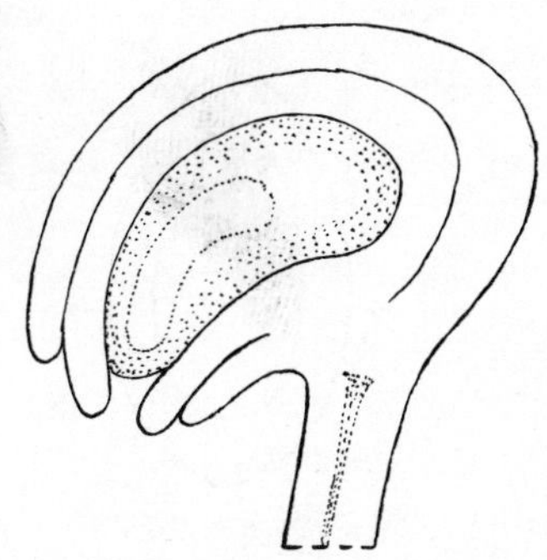

Campylotropous

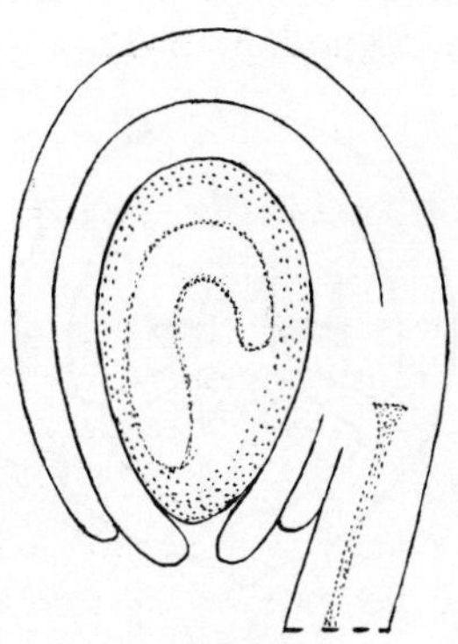

Amphitropous

Fig. VII. Placentation.

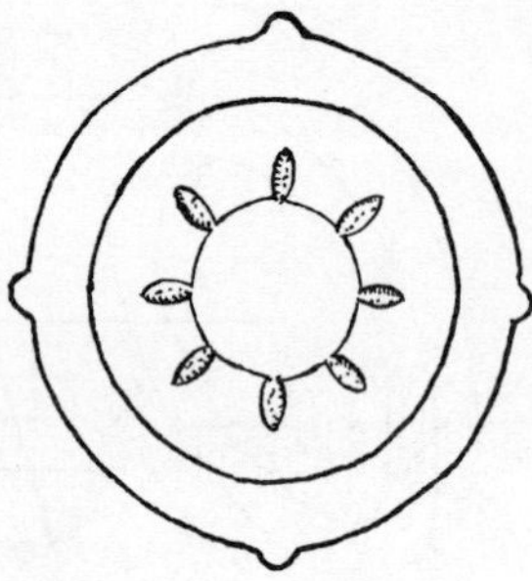

Free-central

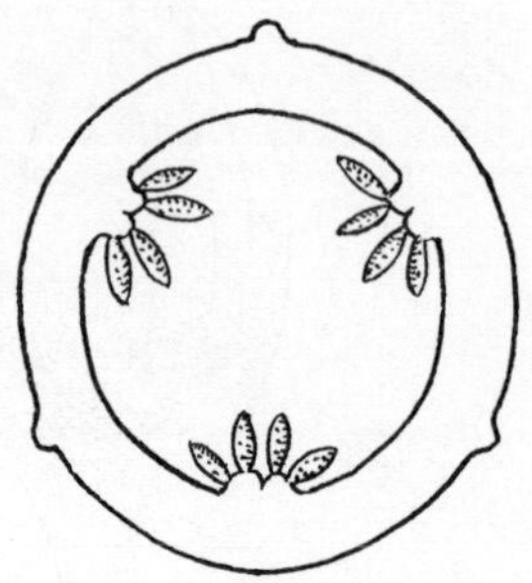

Parietal

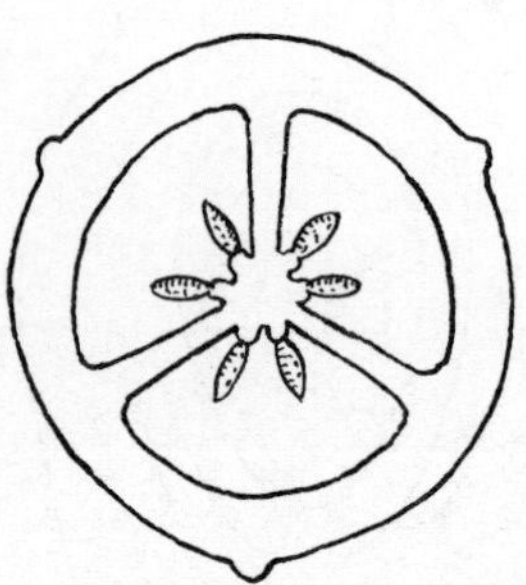

Axile

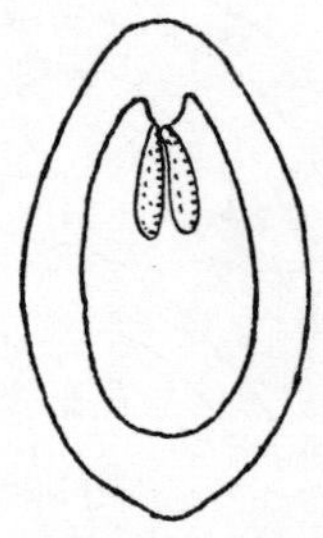

Marginal

Fig. VIII. Position of ovary.

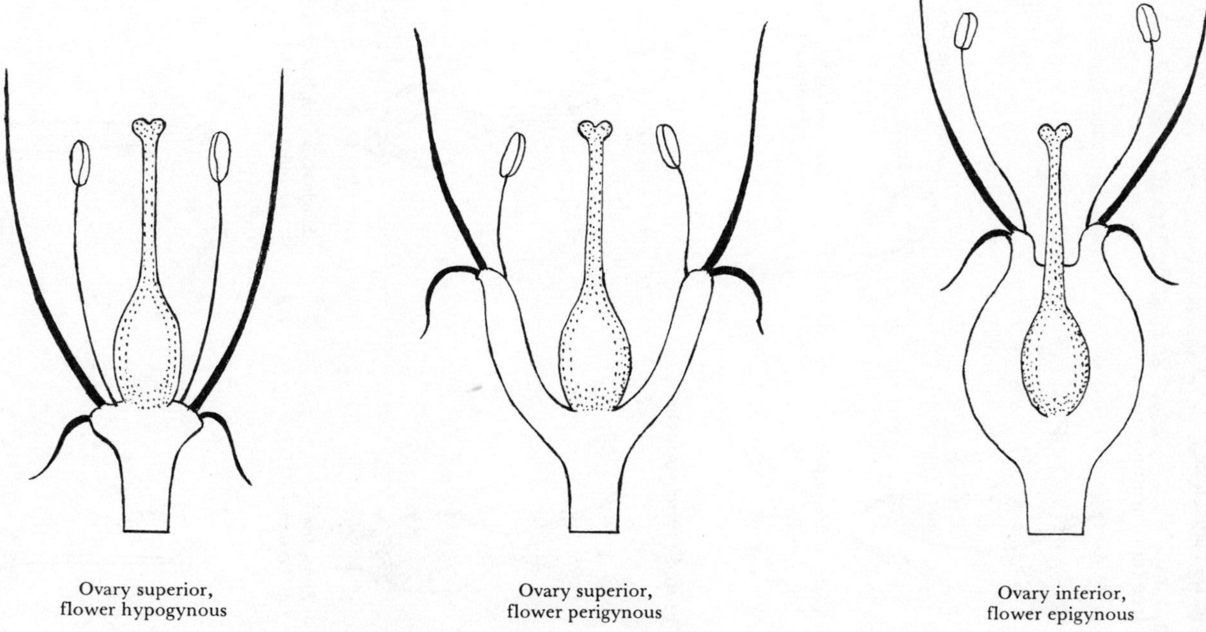

Fig. IX. Carpel forms.

Carpels free (apocarpous)

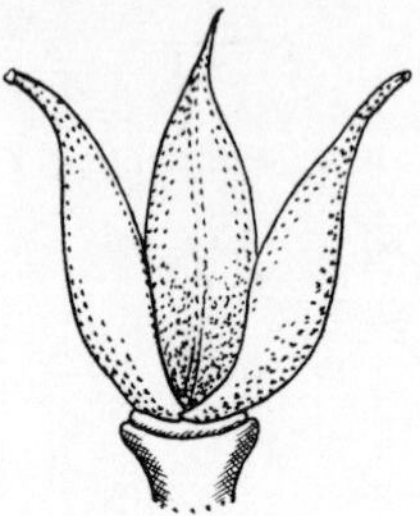

Several free carpels
(5. Ranunculaceae)

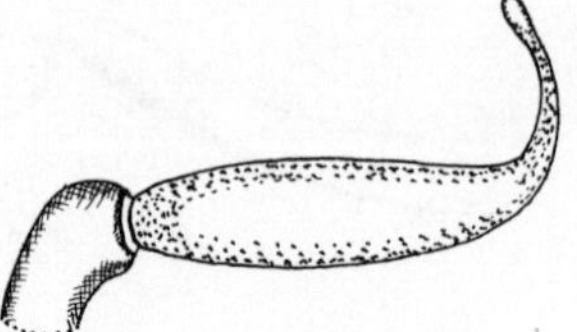

One free carpel
(39. Leguminosae)

Carpels united (syncarpous)

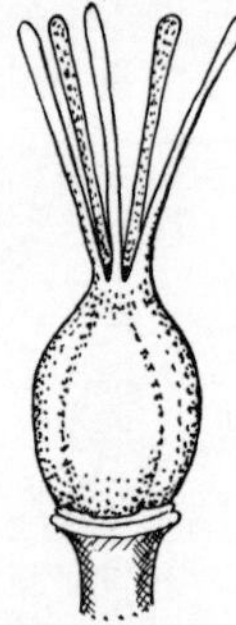

Styles free
(56. Linaceae)

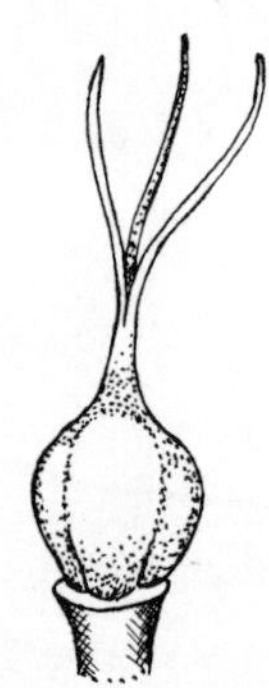

Single style with branches
(88. Cyperaceae)

Single style with capitate stigma
(34. Primulaceae)

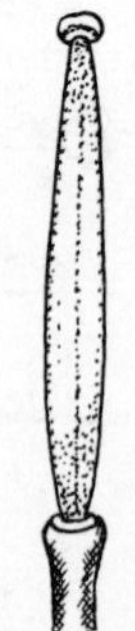

Style absent and stigma sessile
(31. Cruciferae)

1 Magnoliaceae Juss.

Magnolia family

12 genera and 230 species

Distribution. Temperate and tropical E. Asia and America.

General characteristics. Trees or shrubs with alternate, simple, stipulate leaves. Oil passages in parenchyma. Stipules large, enclosing young growth, deciduous, leaving annular scar round node. Flowers terminal or axillary, usually solitary and hermaphrodite, with deciduous spathe-like bracts. Perianth-segments free, arranged in whorls, usually all petaloid. Stamens hypogynous, numerous, arranged spirally. Ovary superior, the free or rarely united carpels usually numerous, and arranged spirally on the elongated receptacle. Ovules 1—several, with usually marginal placentation. Fruit apocarpous (follicle or samara-like) or a woody or fleshy syncarp. Seed with abundant endosperm.

Economic and ornamental plants. Liriodendron tulipfera, Tulip Tree, is commercially important in N. America for its timber, known as Canary Whitewood. Some species of *Magnolia*, *Manglietia* and *Michelia* are also a source of wood. Many species of *Magnolia* are planted for decoration in suitably sheltered situations and a number of hybrids and cultivars have been raised. *Liriodendron* is also grown for its flowers which it produces in abundance at maturity.

Classification. The following classification, based on that of Dandy (1), divides the family into 2 tribes:

1. Magnolieae (anthers introrse or dehiscing laterally; fruiting carpels not samaroid; testa free from endocarp, externally arilloid).
   Magnolia (80) Himalaya to Japan and Indonesia; eastern N. America to Venezuela, W. Indies.
   Michelia (50) tropical Asia, China.
   Talauma (50) E. Himalaya to S.E. Asia; Mexico to tropical S. America, W. Indies.
   Manglietia (30) S.E. Asia to Sumatra.
2. Liriodendreae (anthers extrorse; fruiting carpels samaroid; testa adherent to endocarp, not arilloid).
   Liriodendron one species in eastern N. America and 1 in China.

MAGNOLIA × SOULANGEANA Soul.

Distribution. *M.* × *soulangeana* is a hybrid between 2 Chinese species, *M. denudata*, which has white flowers, and *M. liliflora*, which has the perianth-segments purple on the outside and white on the inside. It was raised in the early nineteenth century by Soulange-Bodin at Fromont, near Paris, and is probably the most popular *Magnolia* for general planting, being tolerant of most soils and little affected by atmospheric pollution.

Vegetative characteristics. A large shrub with several wide-spreading stems, or a low tree, with ovate to obovate leaves tapering rather abruptly to a point.

Floral formula. P9 A∞ $\underline{G}\infty$

Flower and inflorescence. The hermaphrodite, actinomorphic, terminal flowers begin to appear on the naked shoots in April and May, but continue to develop until the plant is in full leaf in early June. The petaloid perianth-segments are usually white with a rose-purple base. Selected forms of the hybrid, tending towards one or other parent, are given an additional (cultivar) name. The flower colour of these cultivars, which varies from pure white in 'Alba Superba' to a rich rosy red in 'Rustica Rubra', gives an indication of the range of variation which may be expected in a plant of hybrid origin.

Pollination. Observations made on certain species of *Magnolia* suggest that beetles play a part in the pollination of this genus. These insects find shelter in the centre of the flower where the 3 innermost perianth-segments arch over the reproductive organs. When the anthers dehisce, the pollen falls on to the beetles, and at the same time the perianth-segments open wide. No longer sheltered, the insects fly off to a flower in the earlier, female stage and transfer the pollen to the receptive stigmas.

Alternative flowers for study. In the absence of *M.* × *soulangeana*, any other member of the genus will prove an acceptable substitute. Unfortunately *M.* × *soulangeana* is shy to fruit, but many other species fruit more abundantly. One of these is *M. tripetala*, the Umbrella Tree, a hardy N. American species which produces strongly scented, creamy white flowers in May and June simultaneously with the leaves, to be followed later in the year by attractive clusters of red fruits.

Fig. 1 Magnoliaceae, *Magnolia* × *soulangeana*

A L.S. of flower with several of the perianth-segments removed to reveal the elongated receptacle which bears numerous stamens and numerous free carpels. The perianth-segments are imbricated and quickly fade and fall. All the floral parts are arranged in a spiral manner.
Perianth-segments (shorter in outer whorl): 9–10 × 2.4–3.3 cm

B One of the many stamens. The anthers are introrse, 2-celled and dehisce longitudinally.
Stamens: 10–11 mm

C Close-up of the lower portion of the elongated receptacle, showing a few of the numerous carpels with their curved styles. The round scars indicate the position of the stamens which were removed in order to allow the carpels to be seen more clearly. Below these are larger, horizontal scars where 2 of the perianth-segments were attached.

D L.S. of the elongated receptacle or androgynophore, showing how the individual carpels are attached to the upper part and the stamens to the lower portion.
Androgynophore (at anthesis): 23 mm

E L.S. of a single carpel inserted on the upper portion of the androgynophore. The carpel has a curved style with a long stigmatic surface on one side.
Style: 3–4 mm Loculus of ovary: 2 × 1 mm

F T.S. of upper portion of androgynophore and some of the carpels. Three of these have been cut to expose the anatropous ovules attached to the ovary wall. In sections, the spiral arrangement of the carpels is clearly visible.
T.S. of androgynophore (at anthesis): 5–6 mm

G The aggregation of fruits which have developed on the androgynophore. Each fruit consists of a single follicle which dehisces along the dorsal suture. The follicle contains one or two large seeds which eventually hang from it on a thread-like structure formed by the unravelling of the spiral vessels of the funicle. The seed has a small embryo but a well-developed endosperm.
Group of fruits: 8 cm Seed: 12 × 10 mm

Fig. 1

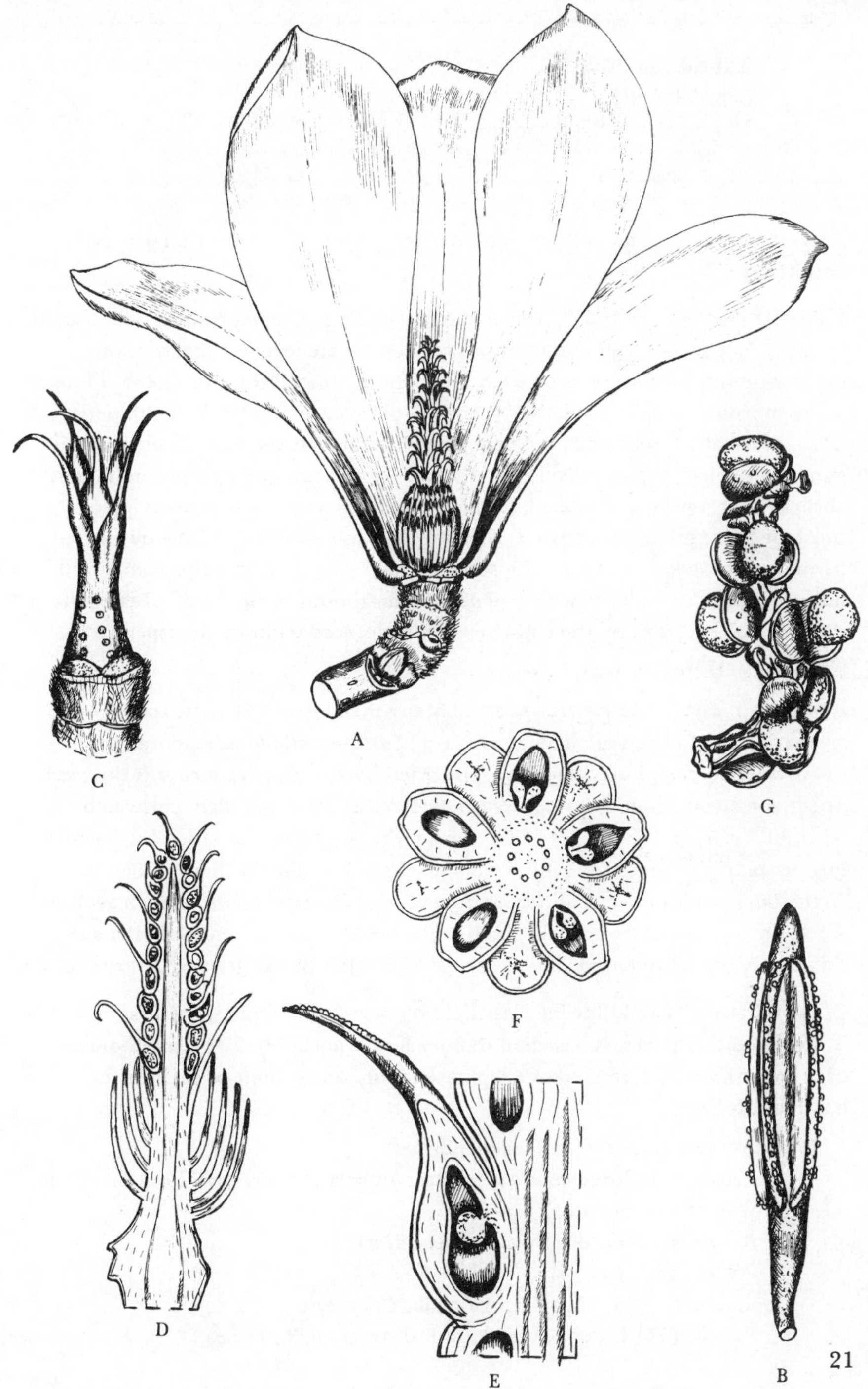

2 Lauraceae Juss.

Laurel family

32 genera and 2500 species

Distribution. Tropics and subtropics; chief centres of distribution S.E. Asia and Brazil.

General characteristics. Trees or shrubs (sometimes monoecious) with usually leathery, evergreen, alternate or opposite, exstipulate leaves. Numerous oil-cavities present. Inflorescence racemose, cymose, umbelliform or mixed. Flowers actinomorphic, usually 3-merous, hermaphrodite or unisexual. Perianth perigynous, of usually 6 segments, in 2 whorls of 3, united at the base. Stamens perigynous or epigynous, usually 12, in 4 whorls of 3, some reduced to staminodes; anthers 2- or 4-celled, opening by valves, usually introrse, but in many cases those of the third whorl extrorse. Axis more or less concave and the ovary free from it at the sides. Ovary usually superior, of 1 carpel, or more probably of 3 united carpels, unilocular, with 1 pendent, anatropous ovule. Fruit a berry, often more or less enclosed by the cup-like receptacle; seed without endosperm; embryo straight.

Economic and ornamental plants. Many members of the Lauraceae contain aromatic oils of economic importance, e.g., *Sassafras albidum*, sassafras, and *Cinnamomum camphora*, camphor. The dried bark of *C. zeylanicum* is the source of the spice cinnamon. *Persea americana*, Avocado Pear, is widely cultivated for its edible fruit, and the leaves of *Laurus nobilis* (see below) are used for flavouring. Some of the larger trees are commercially important for their timber, e.g., *Umbellularia californica*, Californian Laurel, and *Ocotea bullata*, known as Stinkwood on account of its pungent smell when freshly cut. Ornamental plants in this family are represented almost exclusively in the British Isles by *Laurus nobilis*.

Classification. The following classification into 2 subfamilies, the first of which is divided into 5 tribes, is based on that by Kostermans (2). The Cassythoideae have sometimes been treated as a separate family on account of their different habit and foliage.

I. Lauroideae (arborescent; leaves normal).

1. Perseeae (inflorescence paniculate; umbels of flowers without involucre; fruit without cupule).
   Beilschmiedia (200) tropics, Australasia.
   Persea (150) tropics.
   Endiandra (80) Malaysia, Australia, Polynesia.
   Phoebe (70) Indomalaysia, tropical America, W. Indies.

2. Cinnamomeae (as Perseeae, but fruit base embedded in a cupule).
   Ocotea (300–400) tropical and subtropical America, tropical and S. Africa.
   Cinnamomum (250) E. Asia, Indomalaysia.
3. Laureae (umbels of flowers surrounded by involucre of large, decussate, persistent bracts).
   Litsea (400) S. and E. Asia, Australasia, America.
   Lindera (100) Himalaya to E. and S.E. Asia, N. America.
   Laurus (2) Mediterranean region, Canary Islands, Madeira.
4. Cryptocaryeae (as Cinnamomeae; ovary superior, but fruit completely enclosed in the accrescent perianth-tube).
   Cryptocarya (200–250) tropics and subtropics.
5. Hypodaphnideae (as Cryptocaryeae, but ovary inferior)
   Hypodaphnis (1) tropical W. Africa, *H. zenkeri*.

II. Cassythoideae (parasitic twiners without proper leaves).
   Cassytha (20) Old World tropics.

LAURUS NOBILIS L.

Sweet Bay, Bay Laurel, Poet's Laurel

Distribution. Native in the Mediterranean region and naturalised elsewhere. It is cultivated in gardens as a shrub or tree, but is perhaps more frequently grown as a clipped specimen in tubs or other containers.

Vegetative characteristics. A dioecious, aromatic, evergreen shrub or tree from 2 to 20 m in height with blackish bark. The leaves have short petioles and are alternate, dark green, glossy and glabrous, lanceolate or oblong-lanceolate in shape and acute or acuminate at the apex; the margins are sometimes undulate.

Floral formula. Male: P(4) A8–12
Female: P(4) A(staminodes) 2–4 $\underline{G}(3)$

Flower and inflorescence. The actinomorphic, more or less inconspicuous, yellowish green, unisexual flowers appear in late April and form subsessile umbels in the axils of the leaves. The unisexuality of the flowers is due to abortion, as is shown by the presence of staminodes in the female flower.

Pollination. Bees, wasps and ichneumon flies have been observed visiting the flowers in search of pollen, which is readily available from the prominent anthers.

Alternative flowers for study. *L. nobilis* is often cultivated in the British Isles and is probably the most easily obtainable representative of the family. The strongly aromatic, evergreen shrub or tree *Umbellularia californica*, Californian Laurel, which is occasionally found in parks or large gardens, has hermaphrodite flowers with a perianth of 6 segments and 4-valved anthers. *Persea americana* has a similar floral structure, but in this species the flowers are followed by large, edible fruits sold commercially as 'Avocado Pears'.

Fig. 2 Lauraceae, *Laurus nobilis*

A Two young male inflorescences arising from the axils of the leaves.
Male flower-bud: 5 mm in diameter

B L.S. of a male flower exposing 3 of the 8–12 stamens which are arranged in whorls. The 4-lobed perianth is subtended by an involucre of bracts.
Male flower: 8 mm in diameter Perianth-segments: 5 x 2–3 mm

C A single stamen detached from the flower. The basifixed, introrse anthers are 2-celled, each cell dehiscing by a flap-like valve which opens upwards. Near to the base of the filament are 2 glandular outgrowths. The proportions of the stamens and the shape of the glands are somewhat variable.
Filaments: 2–2.5 mm Anthers: 1.75–2 mm

D A rare abnormality observed in one of the male flowers showing partly the characteristics of a perianth-segment and partly those of a half-stamen.
Abnormal segment: 4 mm Anther: 1.5 mm

E A young female inflorescence subtended by 2 bracts.
Female flower-bud: 5 x 2.5 mm Bract: 3.25 x 3 mm

F A female flower with one of the perianth-segments removed to reveal two of the 4 staminodes usually found opposite the segments in female flowers.
Perianth-segments: 3 x 2 mm

G Detail of a single staminode from a female flower attached to a part of the receptacle.
Staminode: 3 x 1.25 mm

H L.S. of the superior ovary bearing a short style and unbranched stigma. The unilocular ovary encloses a solitary, pendulous, anatropous ovule attached to an apical placenta. On each side of the ovary is a staminode and a perianth-segment.
Ovary: 2 x 1.25 mm

I Two of the shiny, green fruits that form the infructescence, showing the well-developed receptacle. The berries turn black as they ripen.
Fruit: 13 x 10 mm

Fig. 2

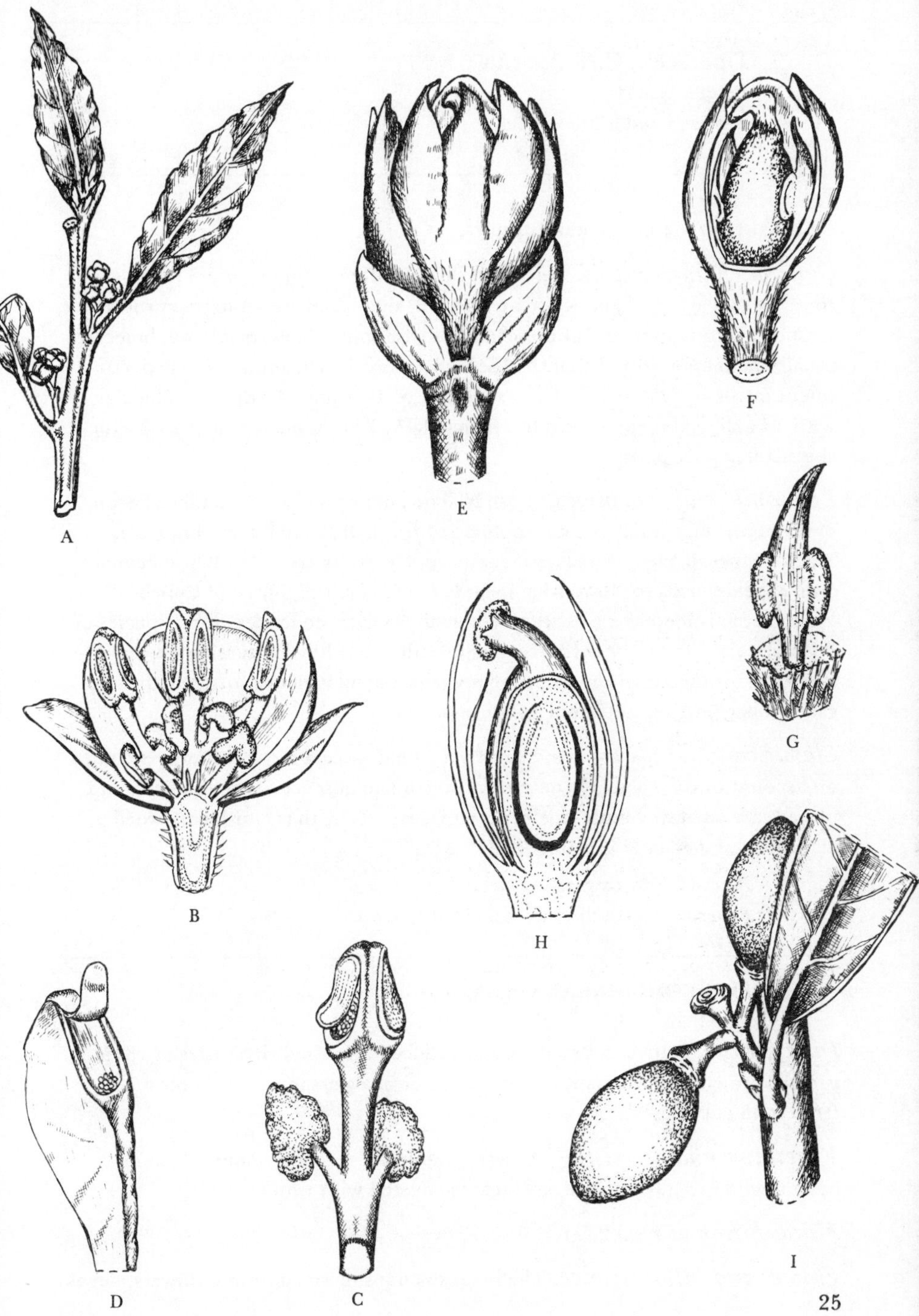

3 Piperaceae C.A. Agardh

Pepper family

8 genera and 3100 species

Distribution. Tropics and subtropics.

General characteristics. Herbs and shrubs, often climbing, more rarely small trees. Leaves alternate, opposite or whorled, simple, entire, sometimes succulent, with or without stipules. Inflorescence a dense spike. Flowers minute, bracteate, usually hermaphrodite. Perianth absent. Stamens 1–10; anthers 2-celled, confluent into 1 in *Peperomia*. Ovary superior, of 1–5 united carpels, unilocular, with 1 basal, orthotropous ovule; stigmas 1–5. Fruit a small drupe; seed with abundant, mealy perisperm.

Economic and ornamental plants. The only genus in this family of economic importance is *Piper*. The whole dried fruits of *P. nigrum* are known as Black Pepper; if the outer layer is removed, the fruits are called White Pepper. Other species used for flavouring include *P. cubeba*, the source of Cubebs. *P. betle*, Betel, is used as a masticatory in many Asiatic countries. Some species of *Piper* are also grown for their ornamental value, but horticulturally the genus *Peperomia* is the more important, many species and varieties (often with variegated foliage) being cultivated as house plants.

Classification. Peperomia and some very small genera are sometimes excluded on account of differences in anatomy, pollen and absence of stipules, etc. and placed in a separate family, the Peperomiaceae. Here, the traditional, broader view of the Piperaceae is maintained.

Piper (2000) tropics.
Peperomia (1000) tropics and subtropics.

PEPEROMIA INCANA A. Dietr.

Distribution. Native in Brazil, and introduced into the British Isles in 1815. It is often grown as a pot plant for the house or in a glasshouse and roots readily from stem cuttings.

Vegetative characteristics. A fleshy subshrub, reaching about 30 cm in height, with broadly cordate, stiff leaves covered with white hairs.

Floral formula. P0 A2 $\underline{G}1$

Flower and inflorescence. The inconspicuous, hermaphrodite flowers, devoid of a perianth, are borne in long spikes, usually in February, though flowering can occur at any time of the year in cultivation. As the flowers are minute and are deeply sunk into the fleshy axis of the inflorescence, there is some difficulty in seeing the floral parts.

Pollination. Little is known about the pollination of *Peperomia* or other genera in the Piperaceae. However, in spite of the outward similarity of the minute-flowered, spicate inflorescence to wind-pollinated families such as Betulaceae and Corylaceae, the Piperaceae are considered to be closely related to the insect-pollinated Magnoliaceae, and it is not unlikely that both families are entomophilous. The flowers are protogynous, which encourages cross-pollination, and they are not specialised, so that various types of insects may be involved. Further points in favour of insect-pollination are that the pollen is of a glutinous nature, and not easily removed by wind or rain, and that at least in the case of *Peperomia resediflora*, the flowers are scented.

Alternative flowers for study. In the absence of *P. incana*, the ease with which the flowers can be studied is a most important factor when choosing an alternative species. In this respect, *P. magnoliifolia*, *P. sandersii* and *P. caperata* are suitable and readily available as house plants. Observation of several species will show minor differences in the form of the stamens and stigma, including the relative position of the stigmatic papillae. The closely related genus *Piper* has spikes of usually unisexual flowers, in which case the plants may be monoecious or dioecious. Species of *Piper* normally have 2 or 4 carpels and distinctly 2-celled anthers, in contrast to *Peperomia* species which have a single carpel and confluent anther-cells.

Fig. 3 Piperaceae, *Peperomia incana*

A The terminal, spicate inflorescence, which develops as a continuation of the shoot and is composed of numerous, closely set flowers. Below the inflorescence are 2 of the alternate, fleshy, white-hairy leaves.
Inflorescence: 6.5 cm Larger leaf: 4 cm

B A portion of the flower-spike, showing the arrangement of the flowers, each in the axil of a prominent bract.

C Anterior view of the apetalous flower subtended by a more or less orbicular floral bract. Above the bract can be seen the ovary with its tufted stigma. The 2 stamens, one of which is beginning to dehisce, are situated at the sides.

D T.S. of the fleshy axis of the inflorescence taken just below the middle. The section shows the minute, apetalous flowers at different stages of development embedded in the tissue of the axis.
Axis of inflorescence: *c.* 2 mm in diameter

E Part of a T.S. of the inflorescence, showing a single flower at an early stage. At the centre is the ovary with its tufted stigma. On each side are the stamens, with confluent anther-cells, which arise from the base of the ovary.
Ovary: 0.3 × 0.2 mm Anthers: 0.3–0.5 mm across lobes

F L.S. of a single flower, showing the position of one of the stamens in relation to the rest of the flower (indicated by a dotted outline). The anther consists of 2 confluent cells.
Anthers: 0.4–0.5 mm

G The anther at dehiscence, with yellow pollen grains.

H L.S. of a single flower showing the position of the ovary, which is unilocular and contains a single, basal, orthotropous ovule. The stigma is composed of a tuft of papillae. After fertilisation, the ovary develops into a small drupaceous fruit. Below the flower is the prominent floral bract.
Ovary: 0.5 × 0.4 mm Bract: 0.4 mm

Fig. 3

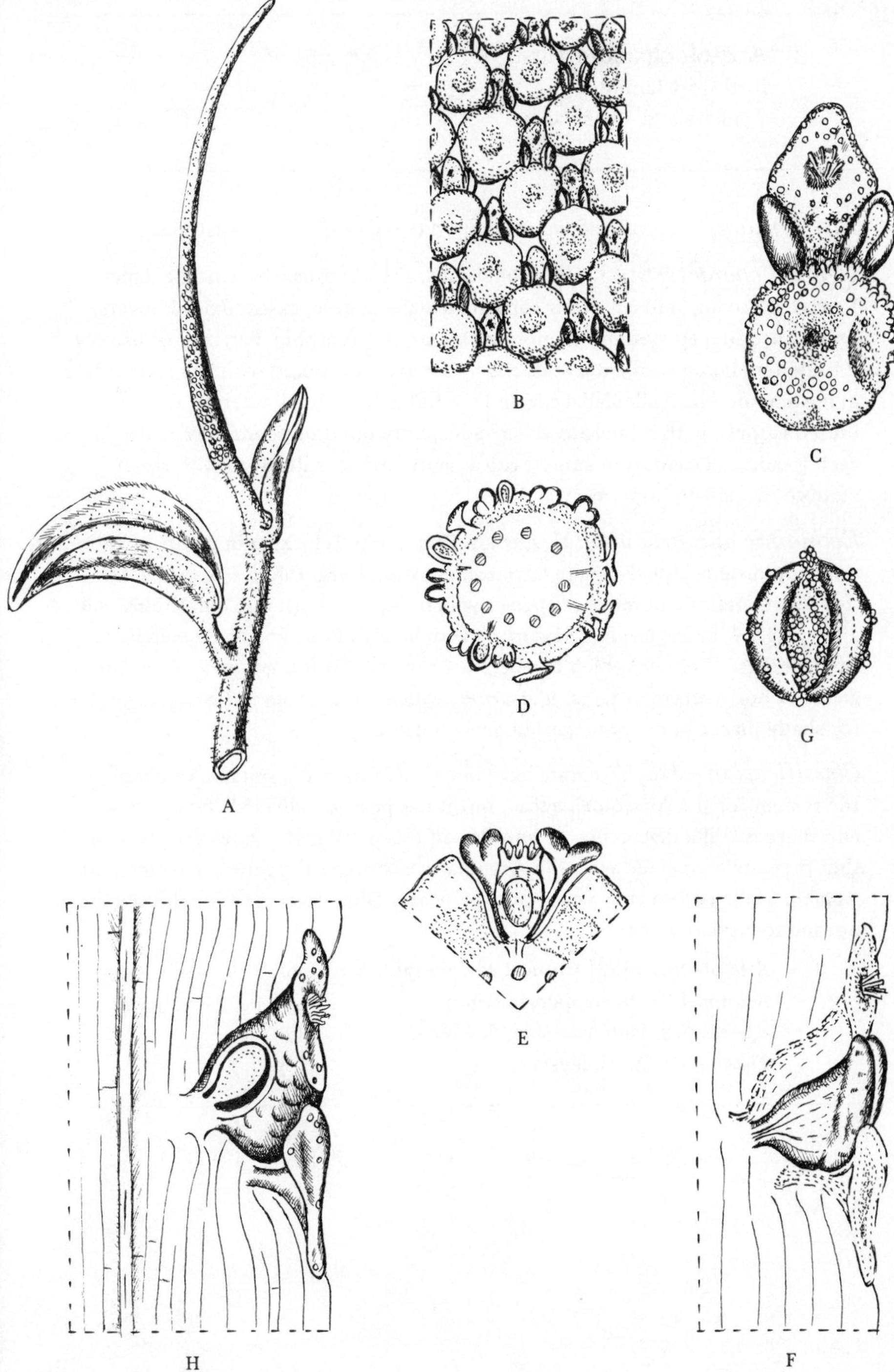

4 Aristolochiaceae Juss.

Birthwort family

7 genera and 444 species

Distribution. Tropical and warm temperate regions, except Australia.

General characteristics. Herbs or shrubs, the latter usually twining lianes. Leaves alternate, stalked, often cordate, usually simple, exstipulate. Flowers hermaphrodite, epigynous, actinomorphic or zygomorphic. Perianth of usually 3 united, petaloid segments. Stamens 6–36, free, or united with the style into a gynostemium (cf. Asclepiadaceae and Orchidaceae). Ovary inferior, of 4–6 united carpels, with numerous anatropous, horizontal or pendulous ovules in each loculus. Placentation axile. Fruit a septicidal capsule. Seed with small embryo, in rich endosperm.

Economic and ornamental plants. This family is lacking in plants of economic importance, but 2 genera have some horticultural value. The genus *Aristolochia* contains a number of species which require glasshouse conditions, and a few, e.g., *A. macrophylla* (*A. durior*) Dutchman's Pipe, and *A. clematitis*, Birthwort (see below), which can be grown outside in this country. The other genus of horticultural interest is *Asarum*, which includes several species suitable for shady places in the rock garden or woodland.

Classification. Many attempts have been made to find a suitable position in the system for the Aristolochiaceae, but it has proved a difficult family to place, and there is still a divergence of opinion on this matter. It is however probable that they are connected with the Annonaceae through the genera *Thottea* and *Apama*. Some authors have even put them near Dioscoreaceae though they are not monocotyledonous.

Aristolochia (350) tropical and temperate regions.
Asarum (70) N. temperate region.
Apama (12) Indomalaysia, S. China.
Thottea (9) W. Malaysia

ARISTOLOCHIA CLEMATITIS L.

Birthwort

Distribution. Probably native in E. and S.E. Europe. It was formerly cultivated as a medicinal plant and has now become naturalised throughout most of Europe, including the eastern and southern parts of England.

Vegetative characteristics. A glabrous, foetid, perennial herb with a long, creeping rhizome and broadly ovate-cordate leaves.

Floral formula. P(3) A(6) $\overline{\text{G}}$(6)

Flower and inflorescence. Axillary clusters of zygomorphic, hermaphrodite flowers on short pedicels appear from June to September on the erect stems. The perianth has a yellow tube and a brownish limb.

Pollination. Small flies are attracted by the foetid smell of the protogynous flowers. On entering the flower, the flies crawl down the perianth-tube. The sides of the tube have long, downwardly-directed hairs that become denser towards the base (see Fig. 4.B) and prevent the insects crawling out again. If they are carrying pollen from another flower, they dust the ripening stigma with it in their vain attempts to escape. A day or so later, the anthers dehisce, covering the flies with pollen. At the same time the hairs within the perianth wither, liberating the captive insects, which emerge and transport the pollen to the mature stigma of another flower. (cf. Araceae)

Alternative flowers for study. The genus *Aristolochia* contains many species with very much larger flowers, some of which are grown in glasshouses, e.g., *A. elegans* and *A. grandiflora*, and these would be very suitable alternatives. Apart from *A. clematitis*, the only member of the family found wild in Britain is *Asarum europaeum*, Asarabacca, a plant with actinomorphic flowers appearing from May to August which is restricted to a few localities, mainly in the southern half of the country. As mentioned above, several species of *Asarum* are cultivated as rock garden or woodland plants, and an interesting comparison may be made between flowers of this genus and those of *Aristolochia*.

Fig. 4 Aristolochiaceae, *Aristolochia clematitis*

A A single, zygomorphic, hermaphrodite flower with an inferior ovary. The perianth, formed from 3 united, petaloid sepals, has a swollen, globose base or utricle. Above this, it narrows into a curved tube, finally opening out into an entire limb which is sometimes as long as the tube.
Entire flower: 20–30 mm Ovary: *c.* 5 mm
Perianth-tube: *c.* 14 mm

B L.S. of the utricle, which encloses the 6 stamens adnate to the stylar column. The sessile, 2-celled anthers are extrorse and dehisce longitudinally. Above the anthers are lobes which have evolved from modified anther connectives and have assumed stigmatic functions. The whole column is known as a gynostemium.
Gynostemium: 1.5 × 1.5 mm

C L.S. of the inferior ovary and gynostemium. Several anatropous ovules are attached to the axile placenta.
Ovary (at anthesis): 4.5–5 mm

D T.S. of ovary with axile placentation. The ovary is composed of 6 fused carpels with one loculus per carpel.
Ovary (at anthesis): 1.25 mm in diameter

E The complete fruit, a comparatively large, septicidal capsule, which opens lengthwise to liberate the seeds.
Fruit: 25 × 24 mm

F A single, flattened, triangular seed containing a small embryo and copious endosperm.
Seed: 10 × 9 mm

Fig. 4

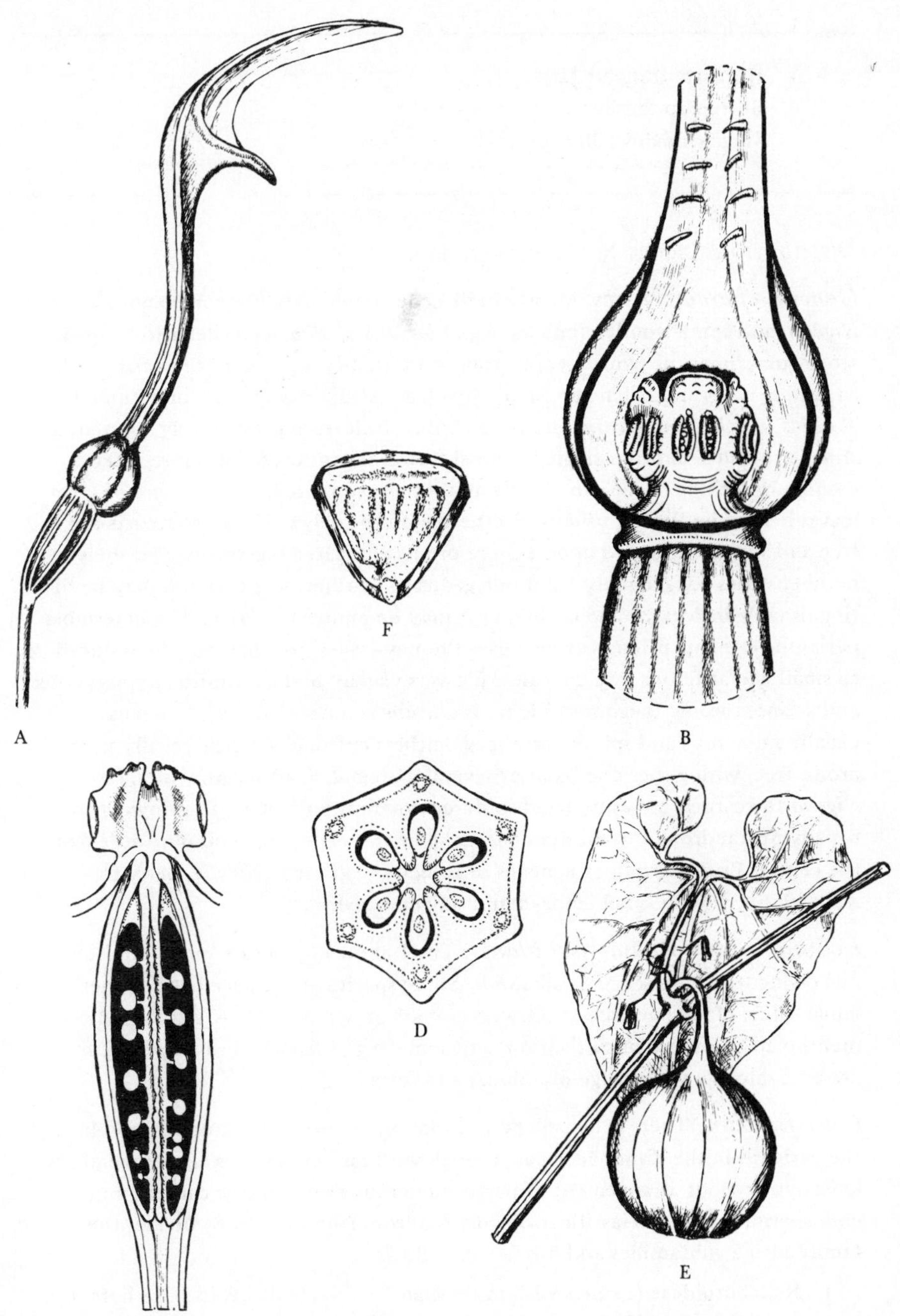

5 Ranunculaceae Juss.

Buttercup family

50 genera and 1900 species

Distribution. Chiefly N. temperate region.

General characteristics. Mostly herbaceous perennials, but some annuals, e.g., *Nigella*, and some woody climbers, e.g., *Clematis vitalba*, Traveller's Joy. Rootstock sometimes tuberous. Leaves usually alternate (opposite in *Clematis*) with sheathing bases and often very much divided; usually exstipulate, but stipulate in *Trollius*, *Caltha*, *Thalictrum* and *Ranunculus*. Inflorescence variously formed; in *Anemone* and *Eranthis* a single terminal flower is produced, but more often a cymose branching occurs. In *Nigella* and *Anemone* there is an involucre of green leaves below the flower, usually alternate with the calyx. The flower-parts are free and spirally arranged upon a more or less elongated receptacle, and their number varies considerably from one genus to another. The perianth may be distinguishable into calyx and corolla or it may be undifferentiated. The innermost perianth-segments often bear nectaries ('honey-leaves') or they may be reduced to small nectariferous sacs or scales. Flowers usually hermaphrodite, hypogynous and actinomorphic (zygomorphic in *Aconitum* and *Delphinium*). Stamens usually numerous and spirally arranged, anthers extrorse. Carpels usually numerous, free, with either one basal or several marginal, anatropous ovules. (In *Nigella* the carpels are united and in *Actaea* there is only one.) The flowers are usually protandrous and the stamens, as their anthers open, bend outwards from the centre. Fruit a group of achenes or follicles (capsule in *Nigella*, berry in *Actaea*). Seeds with small embryo and oily endosperm.

Economic and ornamental plants. The Ranunculaceae are mostly poisonous owing to the presence of alkaloids. Some species are or have been of medicinal value. Many have showy flowers and are grown in gardens. All the genera mentioned below are in cultivation, and some, e.g., *Clematis* and *Delphinium*, are available in a wide range of colours and forms.

Classification. There is considerable variation in vegetative structure and in the perianth in the Ranunculaceae, though with rare exceptions all the members have hypogynous arrangement of parts, numerous stamens, free carpels, and endospermic seeds. A classification adapted from Hutchinson (3) divides the family into 2 subfamilies and 5 tribes as follows:

I. Helleboroideae (carpels with more than 1 ovule; fruit a follicle or berry).
  1. Helleboreae (flowers actinomorphic).
     Aquilegia (100) N. temperate region.

Caltha (32) arctic and N. temperate regions, New Zealand, temperate S. America.

Trollius (25) N. temperate and arctic regions.

Helleborus (20) Europe, Mediterranean region to Caucasus, 1 species in W. China.

Nigella (20) Europe, Mediterranean region to central Asia.

2. Delphinieae (flowers zygomorphic).

Aconitum (300) N. temperate region.

Delphinium (250) N. temperate region.

II. Ranunculoideae (carpels with 1 ovule; fruit a group of achenes, very rarely a berry).

3. Ranunculeae (leaves alternate, flowers not subtended by involucre of leaves, calyx usually caducous, corolla usually present).

Ranunculus (400) cosmopolitan.

Thalictrum (150) N. temperate region, tropical S. America, tropical and S. Africa.

4. Anemoneae (leaves alternate, flowers subtended by involucre of leaves, calyx usually persistent and coloured, corolla absent).

Anemone (150) cosmopolitan.

5. Clematideae (leaves opposite, calyx coloured, corolla absent).

Clematis (250) cosmopolitan.

DELPHINIUM AMBIGUUM L.

(D. AJACIS auct., CONSOLIDA AMBIGUA (L.) Ball & Heyw.)

Larkspur

Distribution. Native in the Mediterranean region, but locally naturalised elsewhere in Europe, including the British Isles. It seeds freely, and often escapes from gardens where it is grown as an ornamental plant.

Vegetative characteristics. A pubescent, annual herb, with an erect, simple or branched stem reaching a height of 1 m. The lower leaves are deeply palmately cut and have long petioles, the upper leaves are similar in shape but are more or less sessile.

Floral formula. P5+(2) A∞ $\underline{G}$1

Flower and inflorescence. The bright blue, zygomorphic, hermaphrodite flowers are borne in 4–16-flowered racemes in June and July. Occasionally the flowers may be pink or white. They have 5 outer perianth-segments and 2 united inner ones bearing nectaries, the 'honey-leaves'. The pedicels, at anthesis, are usually at such an angle with the stems that the flowers are horizontal, but later they bend upwards so that the follicles stand erect. Double forms, with perianth-segments of varying shapes and sometimes with more than one carpel, occur in cultivation.

Pollination. Only long-tongued insects, mainly bumble-bees, can reach the nectar that is concealed in a spur formed by the 2 united honey-leaves. The flowers are protandrous, and an insect visiting the flower in its male stage becomes dusted with pollen from the mature stamens. After shedding their pollen, the stamens wither and bend downwards, making room for the maturing style. A pollen-carrying insect, which visits the flower in its female stage, transfers the pollen to the receptive stigma and cross-pollination is accomplished.

Alternative flowers for study. Other suitable species of *Delphinium* which are often cultivated include *D. orientalis*, *D. consolida*, *D. grandiflorum* and *D. elatum*. The first 2 species are annuals like *D. ambiguum* and are sometimes placed with it in the genus *Consolida*, *D. consolida* then being called *C. regalis*. The other 2 are perennials, which differ further in having 4 free honey-leaves (the 2 upper ones spurred), stamens in 8 spirally arranged series and 3–5 follicles. Comparison should be made with the other genera mentioned in the Classification section which are readily available in the native British flora or in cultivation.

Fig. 5.1 Ranunculaceae, *Delphinium ambiguum*

A A zygomorphic, hermaphrodite flower in its natural position. 4 of the 5 perianth-segments (sometimes referred to as petaloid sepals) can be seen. The upper (posterior) perianth-segment is prolonged at its base into a spur. The 2 united honey-leaves, visible at the centre of the flower, are similarly prolonged, the spur thus formed fitting into the perianth-spur and protected by it.

B L.S. of the flower. The reproductive organs, which are shielded on their upper side by the erect, lobed limb of the 2 united honey-leaves, are situated near the entrance to the tapering, nectar-secreting spur. Parts of the upper perianth-segment, which forms the enclosing spur and 3 of the 4 unspurred segments are shown.
Upper perianth-segment (limb): 18 × 10 mm
(spur): 16–17 mm
Unspurred perianth-segments: 17–18 × 10–20 mm
Limb of honey-leaves: 17–18 × 5–6 mm

C Some of the numerous stamens that are arranged spirally in 5 series round the superior ovary. The reproductive organs are shown here in an erect position though they are horizontal in nature. The anthers, some of which have reached maturity, are 2-celled and dehisce longitudinally. The scars on the receptacle show where perianth-segments have been removed.
Stamens: *c.* 9 mm

D Detail of a single stamen, showing the filament broadening considerably towards the base. The anther is attached at its base to the filament.
Filaments: 6–8 mm Anthers: 1.75 × 1.25 mm

E L.S. of the superior ovary terminated by a short, curved style. The pubescent ovary contains numerous ovules in the single loculus. Surrounding the base of the ovary are portions of the perianth-segments.
Ovary (at anthesis): 6 mm

F T.S. of the ovary, showing 3 of the numerous anatropous ovules attached to the marginal placenta.
Ovary: 2 mm in diameter.

G The fruit prior to dehiscence. When ripe, the follicle becomes erect and dehisces down its ventral suture to liberate the numerous black seeds.
Fruit: 20 × 5 mm

Fig. 5.1

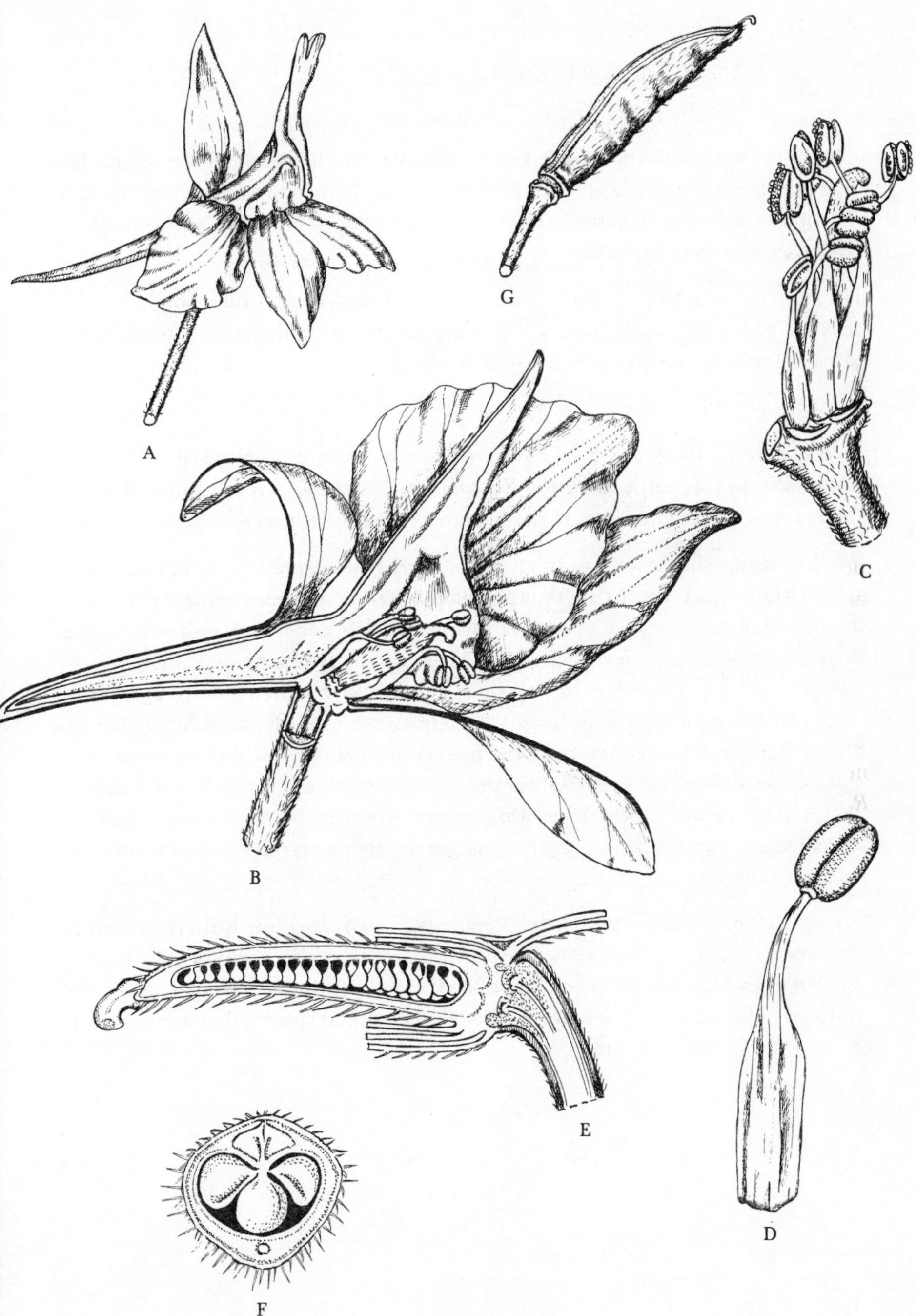

RANUNCULUS REPENS L.
Creeping Buttercup

Distribution. Native in a wide range of habitats in the whole of the British Isles and also occurring as a common weed in gardens. It is found throughout most of Europe and N. Asia to Japan, and has been introduced into N., Central and S. America, and New Zealand.

Vegetative characteristics. A perennial herb with long, stout roots, and stolons which root at the nodes. The hairy leaves are divided into 3 lobes, each of which is again divided into 3 toothed segments.

Floral formula. K5 C5 A∞ $\underline{G}\infty$

Flower and inflorescence. The cymose inflorescence bears glossy, yellow flowers from May until August or September. The pedicels reach from 60 to 100 mm in length and support the flowers in a more or less erect manner.

Pollination. The flowers of the Ranunculaceae are variously adapted to insect-pollination, e.g., *Clematis* (without nectar, visited by pollen-seeking short-tongued flies and bees), *Ranunculus* (nectar scarcely concealed, visited by hover-flies and small bees), *Nigella* (nectar in little closed cavities, visited chiefly by bees), *Aquilegia* and *Delphinium* (nectar in long spurs, visited chiefly by bumble-bees). In *Ranunculus repens*, insects are attracted by the glossy yellow petals and the nectar which is secreted from the pocket-like nectaries at the base of the petals. The pollen that is shed from the extrorse stamens falls on to the petals, and is easily picked up by the visiting insects. After the pollen has been shed, the stigmatic surface of the carpels enlarges in order to receive pollen from another flower.

Alternative flowers for study. *Ranunculus acris*, Meadow Buttercup, which flowers from May to July, sometimes until October and *R. bulbosus*, Bulbous Buttercup, which may begin to flower in March, but whose flowering period is normally May and June, are two of the more common species that are nearest to *R. repens* in floral structure.

Fig. 5.2 Ranunculaceae, *Ranunculus repens*

A L.S. of the flower with 2 of the 5 petals removed, exposing the numerous, free carpels surrounded by the numerous, spirally arranged stamens. The carpels and stamens are attached to a cone-shaped and hairy receptacle. The calyx is made up of 5 hairy sepals.
Sepals: 9 × 4 mm Petals: 16 × 12 mm

B Detail of petal base, showing the pocket-like nectary situated on the ventral surface.

C Stamen. C1: upper portion of young stamen, showing the 2-celled anther prior to dehiscence. C2: the whole stamen after dehiscence of anther.
Anthers: 2.25–2.5 mm Filaments: 5.5 mm

D L.S. of a carpel containing a single, anatropous ovule on a basal placenta. The stigma forms a beak-like structure which enlarges after the pollen has been shed from the stamens, favouring cross-pollination.

E T.S. of a carpel, showing the basal placentation of the ovule.

F The group of achenes after the other floral parts have been removed. The achenes eventually dry and separate from the receptacle for the purpose of fruit dispersal.
Achene (base to stigma): 2.35 mm Ovary: 1 mm
Ovule: 0.3 mm Group of achenes: 4 × 4 mm

G A carpel removed soon after fertilisation, showing the enlarged stigma. The carpel develops into a dry, indehiscent fruit known as an achene.

Fig. 5.2

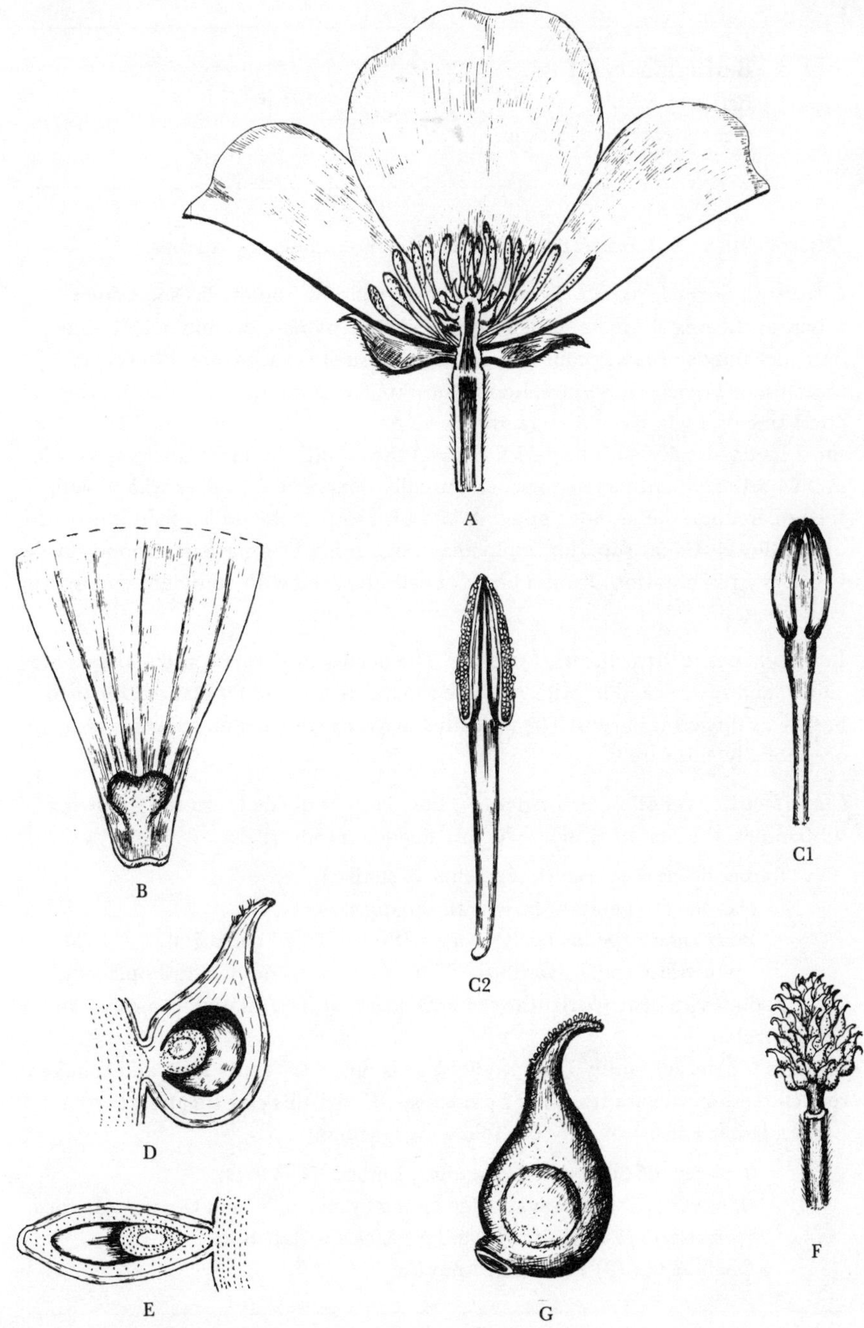

6 Berberidaceae Juss.

Barberry family

4 genera and 544 species

Distribution. N. temperate region, tropical mountains, S. America.

General characteristics. Perennial herbs or shrubs, rootstocks sometimes tuberous. Leaves alternate, usually spiny, simple, pinnate or pinnately 1–3 ternate, deciduous or less commonly persistent, usually exstipulate. Flowers in racemes, or fascicles or cymes, hermaphrodite, actinomorphic. Perianth typically consisting of 4 whorls of 3 segments, the 2 outer whorls often termed the calyx, the 2 inner (usually with nectaries at base) the corolla. Stamens distinct, usually in 2 whorls of 3; anthers introrse, and usually opening by 2 valves which, with the pollen upon them, move upwards and turn so that the pollen faces the centre of the flower. Ovary superior, unilocular, containing 1 or more anatropous ovules with basal placentation. Fruit a berry or capsule. Seed with abundant endosperm, embryo straight.

Economic and ornamental plants. The berries of *Berberis* and *Mahonia* are edible, and species and hybrids of these genera are commonly planted for their beauty as flowering shrubs. The perennial herbs *Epimedium* and *Vancouveria* are occasionally cultivated.

Classification. Following Engler (4), the family is divided into the following 2 subfamilies, the first of which is further divided into 3 tribes:

- Berberidoideae (perianth-segments all similar).
  - Nandineae (petals narrow with subapical nectary).
  - Berberideae (petals fairly broad with 2 subbasal nectaries).
  - Epimedieae (petals variously formed and nectaries variously placed).
- Podophylloideae (perianth-segments clearly differentiated into calyx and corolla).

In Willis (5) the subfamily Podophylloideae is raised to family rank and enlarged to include some genera from the Epimedieae. Under this classification the Berberidaceae consist of only the following 4 genera:

Berberis (450) N. and S. America, Eurasia, N. Africa.
Mahonia (70) Himalaya to Japan and Sumatra, N. and Central America.
Epimedium (21) S. Europe and N. Africa to Japan.
Vancouveria (3) Pacific N. America.

BERBERIS DARWINII Hook.

Distribution. A native of Chile, and named after Charles Darwin who discovered it while on the voyage of the 'Beagle'. Since its introduction into Britain in 1849 it has become a valuable and familiar evergreen, ornamental shrub and hedging plant.

Vegetative characteristics. An evergreen shrub of dense habit, with small, glossy, dark green leaves which are 3-spined at the apex, and arise in small groups from the axils of multiple spines.

Floral formula. K3+3 C3+3 A6 $\underline{G}1$

Flower and inflorescence. Pendulous racemes of 6–20 orange-yellow flowers protrude well beyond the evergreen, toothed leaves from April to June. It is not uncommon for a less prolific second flowering to occur in the summer and autumn although then the clusters of dark purple berries are the main attraction.

Pollination. The coloured petals and the nectar secreted from the glands near the base of the petals attract bees and wasps to the nodding flowers. When the petals begin to open out, the anther valves (see Fig. 6.E2) move up and re-orientate so that the pollen faces the centre of the flower. The stamens are sensitive at the base of their inner surface and when insects foraging for nectar touch this part the stamens spring towards the centre and dust the insect's head with pollen. The sudden movement causes the insect to fly away to another flower, where it transfers the pollen to the sticky margin of the stigma. If cross-pollination does not occur, self-pollination automatically takes place when the flowers fade, for then the anthers themselves come into contact with the stigma.

Alternative flowers for study. Any available member of the genus *Berberis* would be satisfactory, or any species of *Mahonia*, which is similar in floral structure, but has 3 whorls of sepals. Compare these with *Epimedium* which has differently formed nectaries and a capsule as fruit. Most species in these 3 genera are spring-flowering.

Fig. 6 Berberidaceae, *Berberis darwinii*

A Flower bud at bud burst, showing the 2 short bracts just below the free sepals.
Bud: 6×4.5 mm   Bract: 3mm

B The hermaphrodite, actinomorphic flower opened up to reveal 4 of the 6 stamens surrounding the superior ovary. The free perianth-segments are in 4 whorls, each whorl consisting of 3 segments. The 2 outer whorls form the calyx, and the 2 inner whorls with segments bearing nectar-secreting glands comprise the corolla. The segments forming the inner whorl of the calyx are larger than the rest.
Sepals (outer whorl): 5 mm
(inner whorl): 6 mm
Petals: 5 mm

C Base of a petal with the stamen removed to reveal the 2 nectar-secreting glands.

D A stamen with its adjacent petal. Note that the valves have dropped from the anthers, and that the filament is not attached to the petal.

E Stamen. E1: young stamen, showing the basifixed anthers prior to dehiscence. E2: mature stamen, showing method of dehiscence. The valves break away from the base to expose the pollen which adheres both to the connective (see D) and to the valves.
Stamen: 5 mm   Young anther: 1.25 mm
Valve (at dehiscence): 3 mm

F L.S. of ovary. The single loculus contains a few anatropous ovules attached to the basal placenta. The style is as long as the ovary and is capped by a well-developed stigma which has a sticky and papillose edge.
Style + ovary: 5 mm at flowering time

G T.S. of ovary cut near the base to show the parietal to basal placentation.
Ovary: 2 mm   Loculus: 1.2×1 mm   Ovule: 0.3 mm

H The fully developed fruit, a one or few-seeded berry with a dark purple epicarp and persistent style.
Fruit: 7.5×5 mm   Pedicel: 10 mm

Fig. 6

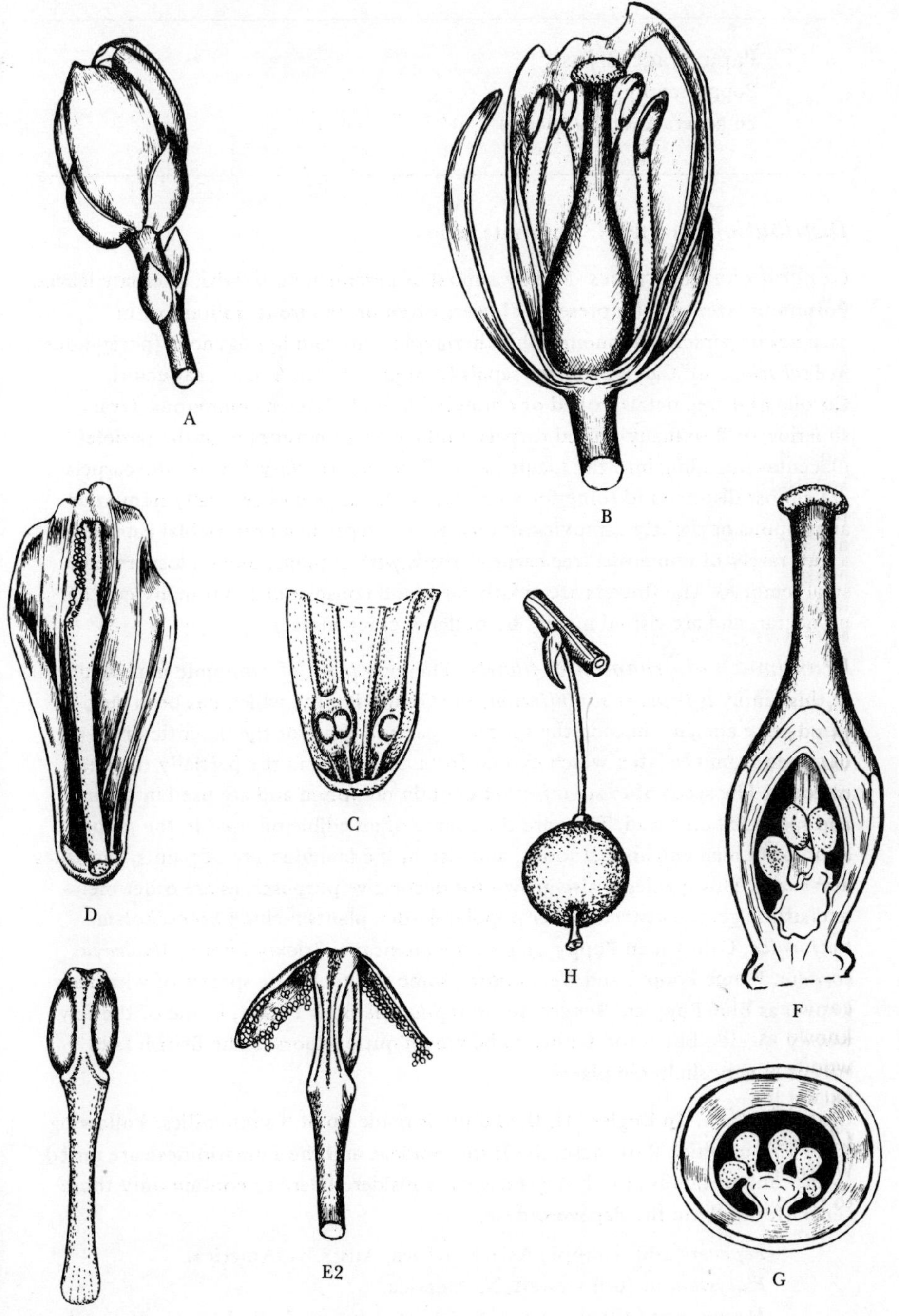

7 Papaveraceae Juss.

Poppy family

26 genera and 280 species

Distribution. Chiefly N. temperate region.

General characteristics. Mostly annual or perennial herbs with alternate leaves. Poisonous latex usually present. Flowers often protandrous, solitary or in racemes or panicles, actinomorphic, hermaphrodite and hypogynous (perigynous in *Eschscholzia*). Calyx of 2 free sepals (united in *Eschscholzia*), caducous. Corolla of 4 free petals, rolled or crumpled in bud. Stamens numerous. Ovary superior, of 2 to many united carpels, unilocular, sometimes with the parietal placentas intruding into the loculus (e.g., *Papaver*). In *Platystemon*, the carpels are almost distinct and sometimes separate in fruit. Ovules generally numerous, anatropous or slightly campylotropous. Fruit a septicidal or poricidal capsule or a nut, rarely of numerous free carpels. Seeds with copious, oily endosperm and small embryo. The flowers are mostly large and conspicuous, but many contain no nectar, and are visited mainly by pollen-seeking insects.

Economic and ornamental plants. The only plant of economic importance in this family is *Papaver somniferum*, the Opium Poppy, which has been cultivated since ancient times in the southern parts of Asia for the narcotic drug obtained from the latex which exudes from cuts made in the partially ripened capsules. The seeds of *P. somniferum* contain no opium and are used in baking and sprinkled on bread. They are the source of an edible oil used in the preparation of human and animal foods, and also in the manufacture of paints, varnishes and soaps. This species is also grown for decorative purposes, as are other members of the genus *Papaver.* Other popular garden plants include *Eschscholzia californica*, Californian Poppy, *Argemone mexicana*, Prickly Poppy, *Macleaya cordata*, Plume Poppy, and *Meconopsis*, some of the Asiatic species of which are known as Blue Poppies. *Dendromecon rigida*, the Bush Poppy, is one of the few woody species, but is too tender to be grown out-of-doors in the British Isles except in very sheltered places.

Classification. In Engler (4), the family is made up of 3 subfamilies. Following Willis (5), in which 2 of these, the Hypecooideae and the Fumarioideae are raised to the rank of family, the Papaveraceae is considered here to contain only those genera formerly in the Papaveroideae.

Papaver (100) Europe, Asia, S. Africa, Australia, America.
Platystemon (60) western N. America.
Meconopsis (42) Himalaya, W. China; 1 species in W. Europe, *M. cambrica.*

Glaucium (25) Europe, S.W. and central Asia.
Argemone (10) United States, Mexico, W. Indies.
Bocconia (10) Central and S. America, W. Indies.
Eschscholzia (10) Pacific N. America.
Dendromecon (3) California, Mexico.
Romneya (2) California, N.W. Mexico
Macleaya (2) E. Asia
Chelidonium (1) temperate and sub-arctic Eurasia, *C. majus.*
Sanguinaria (1) Atlantic N. America, *S. canadensis.*

PAPAVER RHOEAS L.
Common Poppy

Distribution. A weed, commonly found in waste places and cultivated ground in most of the British Isles and other parts of Europe except in the far north. It is also native in N. Africa and temperate Asia, and has been introduced into Australia, New Zealand and N. America.

Vegetative characteristics. A hairy annual or rarely biennial herb with pinnately cut or divided leaves.

Floral formula. $K2\ C2{+}2\ A\infty\ \underline{G}\infty$

Flower and inflorescence. The large, solitary, scarlet flowers are borne on long, hairy stems from June to August. As the flower opens, the crumpled petals expand causing the 2 hairy, protective sepals to fall away. After the flower has opened fully the petals are also shed, exposing the capsule.

Pollination. No nectar is secreted by the flower, although a large amount of pollen is shed by the numerous, bluish-black anthers. The abundance of pollen attracts bees, flies and beetles which can alight easily on the large, flattish, stigmatic disc in the centre of the flower. Cross-pollination is thus effected without difficulty. As the flower is homogamous, self-pollination can also take place, either in the course of insect visits or, because of the proximity of anthers and stigmas, without their assistance. However, pollen transferred to the stigmas of the same flower is ineffective since this species has been found to be self-incompatible.

Alternative flowers for study. Papaver dubium, Long-headed Poppy, and *P. argemone*, Long Prickly-headed Poppy, occur as arable weeds, and *P. somniferum*, Opium Poppy, is a well-known garden plant and readily obtainable. Comparison between *Papaver* and other genera such as *Meconopsis*, *Glaucium*, *Eschscholzia* and *Chelidonium* shows clearly the variation in fruit form within the family. *Macleaya* is an example of a genus where the inflorescence consists of a panicle of apetalous flowers. All the plants mentioned are summer-flowering.

Fig. 7 Papaveraceae, *Papaver rhoeas*

A The expanding flower bud, showing the 2 large, protective sepals clasping the unexpanded, crumpled petals. The sepals are caducous and fall away at an early stage.
Flower bud: 19 × 10 mm

B L.S. of the hermaphrodite, actinomorphic flower. The sepals have been shed. One of the inner petals and parts of the outer petals have been removed to expose the numerous stamens surrounding the ovary. The filaments are inserted on the receptacle at the base of the ovary.
Flowers: 7–10 cm in diameter Petals: 3–5 cm

C One of the free stamens which are arranged in several whorls on the receptacle. The anthers are 2-celled, extrorse and dehisce longitudinally.
Filaments: up to 10 mm Anthers: 2 mm

D L.S. of the superior ovary showing, on one side, the numerous ovules attached to a placenta. On the other side, the ovules have been removed exposing another placenta. There is no style, the well-developed sessile stigmas lying on top of the ovary and forming a stigmatic disc. (see F)

E T.S. of ovary, formed from an indefinite number of united carpels. The ovary is fundamentally unilocular, but is divided into several loculi by the growth of the parietal placentas towards the centre.
Ovary (at anthesis): 7 × 7 mm

F Part of the stigmatic disc, showing the radiating stigmas, usually 10–14 in number, ready to receive pollen.
Stigmatic disc: 8 mm in diameter

G The fruit, a capsule opening by pores. At maturity, the small flaps situated beneath the projecting stigmatic lobes bend back revealing a circle of oblong holes. As the capsule sways back and forth in the wind, the small seeds are gradually shaken out, a form of seed dispersal reminiscent of a censer. The vertical ribs visible on the outside of the capsule correspond to the placentas inside the fruit.
Fruit: *c.* 11 mm

Fig. 7

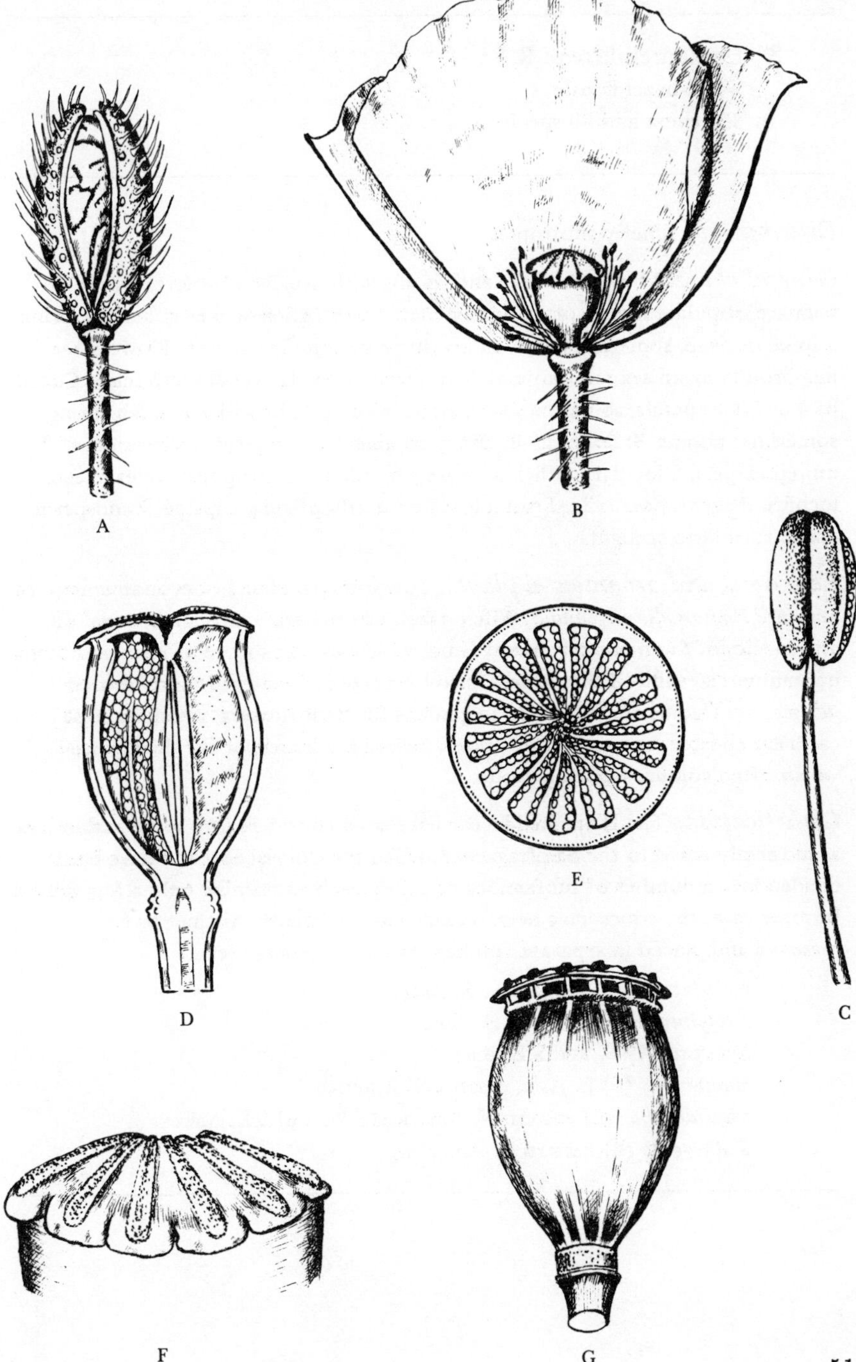

8 Hamamelidaceae R.Br.

Witch-hazel family

22 genera and 80 species

Distribution. Chiefly subtropics.

General characteristics. Trees and shrubs with usually alternate, simple or palmate, stipulate leaves, often with stellate hairs. Inflorescence racemose, often a spike or head, sometimes with an involucre of coloured bracts. Flowers hermaphrodite or unisexual. Calyx of 4 or 5 united sepals, usually imbricate. Corolla of 4 or 5 free petals, sometimes long and curled up in bud like a watch-spring, sometimes absent. Stamens 2–8. Ovary ranging from superior to inferior, of 2 united carpels, 2-locular, with 1 or more pendulous, anatropous ovules in each loculus. Placentation axile. Fruit a loculicidal or septicidal capsule. Endosperm present; embryo straight.

Economic and ornamental plants. The principal plants of economic importance are *Hamamelis virginiana*, Witch-hazel, whose bark yields a medicinal oil and species of *Liquidambar*, Sweet Gum, which are the source of resin and timber of commercial value. These genera, also *Corylopsis*, *Fothergilla* and *Parrotia persica*, are frequently cultivated in gardens for their flowers, which in most cases are conspicuous since they appear before the leaves, or for their foliage which often colours well in autumn.

Classification. The Hamamelidaceae have been considered by some authorities as indirectly allied to the Saxifragaceae and to the Corylaceae, and have been divided into a number of subfamilies or tribes, each containing only a few genera. In some cases the tribes have been regarded as sufficiently distinct to be removed and placed in separate families. The chief genera are:

Corylopsis (20) Himalaya, E. Asia.
Distylium (15) E. and S.E. Asia.
Sycopsis (7) E. and S.E. Asia.
Hamamelis (6) E. Asia, eastern N. America.
Liquidambar (6) eastern N. America, S.W. and S.E. Asia.
Fothergilla (4) eastern N. America

HAMAMELIS MOLLIS Oliv.

Chinese Witch-hazel

Distribution. Native of W. China and introduced into British gardens in 1879, where it succeeds best in sunny positions in the shrub border.

Vegetative characteristics. A deciduous shrub or small tree of spreading habit, with shortly stalked, stipulate, toothed leaves which are roundish to broadly obovate, shortly acuminate at the apex, obliquely cordate at base and densely stellate-hairy.

Floral formula. K(4) C4 A4 $\underline{G}(2)$

Flower and inflorescence. The hermaphrodite, actinomorphic, fragrant, golden-yellow flowers appear from December to March, before the leaves. Each inflorescence consists of a cluster of 3 or 4 sessile flowers on a short peduncle.

Pollination. Little is known about the pollination of *Hamamelis mollis*, but the conspicuous colour of the flowers and their fragrant scent suggest that this species is entomophilous. It is probable that insects would be readily attracted to these flowers since they appear at a time of the year when there are few other plants in bloom.

Alternative flowers for study. Hamamelis mollis is a commonly grown shrub chosen from a genus whose species and cultivars differ little in floral structure, flower-size and colour. Examination of other genera will reveal 5-merous flowers (e.g., *Corylopsis*) and apetalous flowers (e.g., *Fothergilla*). *Liquidambar* also has apetalous flowers, but in this genus the flowers are unisexual.

Fig. 8 Hamamelidaceae, *Hamamelis mollis*

A Two flower-clusters borne on a leafless twig. All the buds apart from the terminal bud have been removed for clarity.

B Anterior view of a single 4-merous, actinomorphic, hermaphrodite flower. The hairy, 4-lobed calyx has a purple inner surface and its short tube is adnate to the ovary. The 4 yellow, strap-like petals are alternate with the calyx-lobes. Prior to anthesis the petals are coiled within the bud.
Calyx: up to 8 mm in diameter Calyx-lobe: 4 × 3 mm
Petals: up to 18 × 1 mm

C The gynoecium and some of the adjacent floral parts. Numerous long hairs surround the superior ovary. Part of one of the petals is shown, and at its base a complete stamen and a staminode. On the opposite side of the gynoecium is the filament of another stamen.
Stamens: 1.5–3 mm (according to maturity)
Staminodes: 1.25–2 mm
Ovary: 1–2 mm (tips of stigmas 1 mm apart)

D The young carpels are only connate near the base and their styles are erect. In the course of development the ovaries become completely fused and the styles bend outwards to expose the stigmatic surface.

E Apical portion of stamen showing detail of the anther. The anther is 2-celled, each cell dehiscing by a flap-like valve.

F L.S. of the superior, 2-locular ovary formed from 2 united carpels. Within each loculus is a single, pendulous, anatropous ovule. Subtending the ovary are the lower portions of 2 of the petals, and below them, 2 of the hairy sepals.

G T.S. of ovary showing the single ovule in each loculus attached to the axile placenta. After fertilisation, the ovary develops into a 2-celled, woody capsule covered for a third to a half of its length by the calyx. The capsule dehisces downwards from the apex and expels the 2 shiny, black seeds a considerable distance.
T.S. of ovary (at anthesis): 0.8 × 0.6 mm

Fig. 8

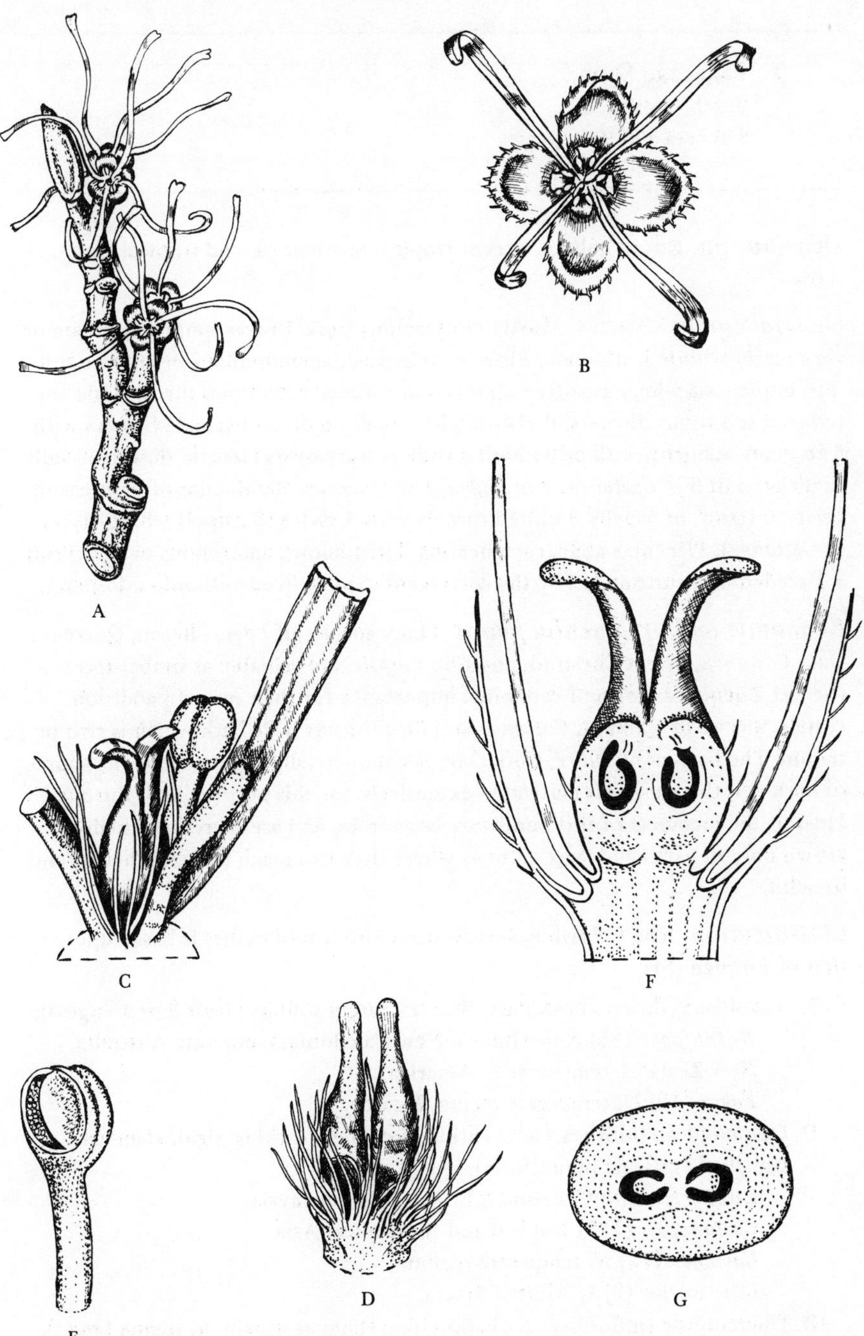

9 Fagaceae Dum.

Beech family

8 genera and 930 species

Distribution. Cosmopolitan, except tropical S. America, and tropical and S. Africa.

General characteristics. Mostly monoecious trees. Leaves simple, alternate or very rarely whorled, stipulate. Flowers unisexual, anemophilous or entomophilous, in dichasia which are often arranged in catkins; sometimes the dichasia are reduced to solitary flowers. Perianth with usually 6 divisions; male flowers with 5 to many stamens, with or without a rudimentary ovary; female flowers usually in dichasia of 3 in *Castanea*, 2 in *Fagus*, 1 in *Quercus*; staminodes often present. Ovary inferior, of usually 3 united carpels with 3 styles (6 carpels with 6 styles in *Castanea*). Placentas axile, each bearing 2 pendulous, anatropous ovules. Fruit a 1-seeded nut, surrounded by the accrescent cupule. Seed without endosperm.

Economic and ornamental plants. Many species of *Fagus*, Beech, *Quercus*, Oak, *Castanea*, Sweet Chestnut, and *Castanopsis*, are valuable as timber-trees: the last 2 genera are also of economic importance for their nuts. In addition, certain species of *Quercus*, *Castanea* and *Lithocarpus* have bark which is rich in tannin. The bark of *Q. suber*, Cork Oak, is commercially important as the source of cork and this species is cultivated extensively for this purpose in S. Europe. Most of the members of this family are large trees, and are therefore usually grown only in parks and large gardens where they can reach their full height and breadth.

Classification. The following classification into 3 subfamilies is based upon that of Forman (6):

I. Fagoideae (flowers in axillary clusters, rarely solitary; fruit 2 or 3-angled).
   - *Nothofagus* (35) New Guinea, New Caledonia, temperate Australia, New Zealand, temperate S. America.
   - *Fagus* (10) N. temperate region, Mexico.

II. Castanoideae (inflorescences catkin-like; male 'catkins' rigid, stamens usually 12; stigma punctiform).
   - *Lithocarpus* (300) E. and S.E. Asia, Indomalaysia.
   - *Castanopsis* (120) tropical and subtropical Asia.
   - *Castanea* (12) N. temperate region.
   - *Chrysolepis* (2) W. United States.

III. Quercoideae (inflorescences catkin-like; stamens usually 6; stigma broad).

Quercus (450) N. America to W. tropical S. America, temperate and subtropical Eurasia, N. Africa.
Trigonobalanus (2) S.E. Asia.

FAGUS SYLVATICA L.
Beech

Distribution. Native throughout most of Europe, including southern England and Wales; elsewhere planted and often naturalised. In S.E. England it is the characteristic dominant of chalk and soft limestone, often forming homogeneous woods.

Vegetative characteristics. A large, deciduous tree with a broad, dense crown, bearing ovate-elliptical leaves with sinuate, long ciliate margins.

Floral formula. Male: P(4–6) A8–12
Female: P(6) $\overline{G}(3)$

Flower and inflorescence. The flowers are unisexual and the inflorescences arise from the axils of the leaves in April and May. The numerous male or staminate flowers form a yellowish green dichasial cyme suspended like a tassel on a long peduncle. The inflorescences are discarded after the stamens have shed their pollen. The female or pistillate flowers are produced in pairs and are poised erect on a short peduncle. In the Fagaceae the female flowers typically form a 3-flowered dichasium, but in the genus *Fagus* the central flower is absent.

Pollination. The pendulous nature of the staminate inflorescences favours wind-pollination. The pollen shed by the anthers is carried to the styles which project from the involucre surrounding each pair of pistillate flowers.

Alternative flowers for study. No alternative is necessary owing to the common occurrence of this species. A number of garden forms, differing mainly in leaf-shape or colour, may be found. Comparison should be made with a genus from each of the other subfamilies, *Quercus* and *Castanea* being most readily available. In the anemophilous genus *Quercus*, the flowers are in separate inflorescences, the male in drooping catkins; the cupule only partially encloses the fruit (acorn), is not prickly and does not split into valves. In the insect-pollinated *Castanea*, the male flowers are in erect catkins, with the female in the lower part of the catkin; the prickly cupule completely encloses the fruit (chestnut) and splits into 2–4 valves. The native species of *Quercus* flower in April and May, while *Castanea sativa* flowers in June and July.

Fig. 9 Fagaceae, *Fagus sylvatica*

A A staminate inflorescence, shown upright in drawing, though pendulous in nature. There are about 15 flowers in each inflorescence.
Inflorescence: 10 × 7 mm

B A pistillate inflorescence, consisting of 2 flowers with protruding crimson styles. Each flower is surrounded by a hairy, usually 6-lobed perianth.
Inflorescence: 9 × 9 mm

C A single male flower, with a 4- to 6-lobed perianth surrounding 8–12 stamens.
Flower: 5.5 mm Perianth: 3.7 mm

D A male flower cut open to reveal the point of attachment of one of the stamens. The thread-like filament supports a basifixed, 2-celled anther which dehisces longitudinally.
Filaments: 9–11 mm Anthers: 1.2–1.5 × 0.4–0.7 mm

E L.S. of lower portion of a pair of female flowers. One of the ovaries has been opened showing 2 of the 3 loculi, with one of the 2 pendulous ovules in each loculus.
Ovary (at anthesis): 4 × 2.7 mm Loculus: 2 × 1 mm

F T.S. of the 2-flowered pistillate inflorescence. The 2 ovaries are surrounded by a hairy perianth. Each ovary is 3-locular, containing usually 2 ovules with axile placentation in each loculus.
T.S. of the pair of ovaries: 5 × 4.2 mm

G Portion of the cupule, exposing the 2 female flowers with inferior ovaries, each surmounted by 3 crimson styles.
Cupule (at anthesis): 6 mm in diameter Style: 4 mm

H The woody and slightly prickly, 4-valved cupule, cut through to show the 2 triangular fruits. One ovule in each ovary develops, at the expense of the others, into a seed with a hard pericarp. The whole fruit forms a single-seeded nut which is ripe in September or October. If the season is unfavourable, seeds may not be set, and only empty husks will be present.
Cupule: 30 mm wide when fully open Fruit: 15 × 9 mm

Fig. 9

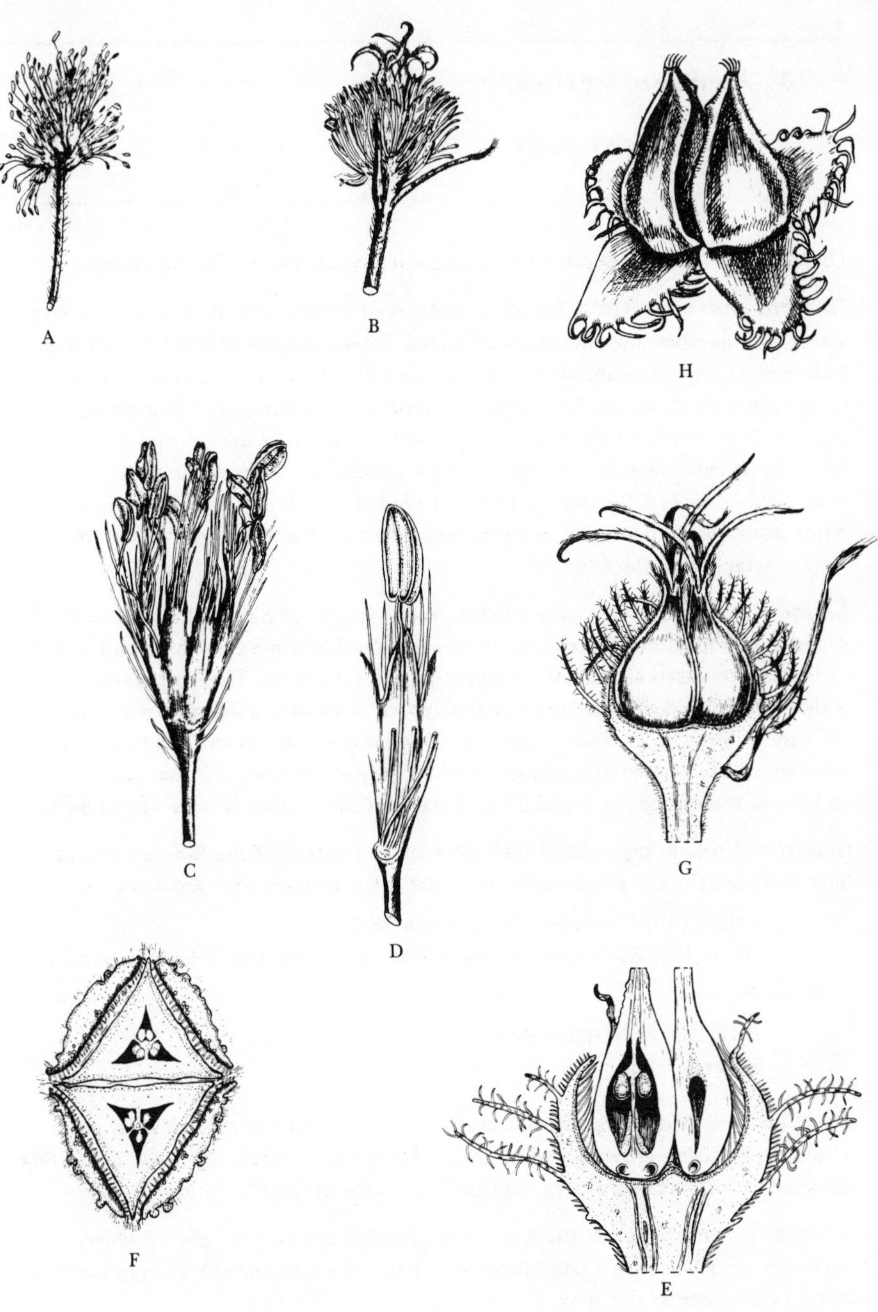

10 Betulaceae S.F. Gray

Birch family

2 genera and 95 species

Distribution. N. temperate region, tropical mountains, Andes, Argentina.

General characteristics. Monoecious trees or shrubs with alternate, undivided leaves and membranous, deciduous stipules. Flowers anemophilous, typically in 3-flowered cymules subtended by a bract and 2–4 bracteoles, and collectively forming terminal catkins. Male flower united to the bract, perianth minute; stamens 2–4. Female flower without perianth; ovary of 2 united carpels, 2-locular at base, each loculus containing 1 pendulous, anatropous ovule with axile placentation; 2 free styles. Fruit a 1-seeded nut. Seed without endosperm. After fertilisation, bract and bracteoles grow into a scale-like organ which may remain attached to the fruit.

Economic and ornamental plants. Many species of *Betula*, Birch, and *Alnus*, Alder, are of local importance for their wood, and in some cases bark and leaves have also been used as a source of dyes, oil and even food. They are much more widely valued as decorative trees and are often planted in gardens, parks and, in the case of birches, roadside verges also. Both birches and alders form graceful trees which are particularly attractive when they are bearing catkins. An additional reason for the popularity of many of the birches is their silvery bark.

Classification. Some authorities have a wider concept of the Betulaceae and have included the Corylaceae also. Here the 2 families are kept separate.

Betula (60) N. temperate and arctic regions.
Alnus (35) N. temperate region south to Assam and Indochina, Andes.

BETULA PENDULA Roth

Silver Birch

Distribution. Native throughout most of Europe, and common in most parts of the British Isles, growing mainly on the lighter soils, rarely on chalk, colonising heathland, and forming woods, especially in the Scottish Highlands.

Vegetative characteristics. A graceful, deciduous tree with silvery white, peeling bark, more or less pendulous branches and ovate-deltoid, sharply toothed leaves, acuminate at the apex.

Floral formula. Male: P(4) A2
Female: P0 G(2)

Flower and inflorescence. The unisexual flowers are borne during April and May in inflorescences known as catkins. The staminate inflorescences or male catkins are pendulous, while the pistillate inflorescences or female catkins are erect at first but also become pendulous in July and August when they are in fruit. Both male and female catkins have a central axis, with 3-flowered cymules arranged round it in a spiral manner. A perianth of 4 minute tepals subtends the male flower, but in the female flower the perianth is absent.

Pollination. The pendulous nature of the male catkins facilitates wind-dispersal of the abundant pollen on to the stigmatic surfaces of the female flowers.

Alternative flowers for study. Any other species of *Betula*, especially *B. pubescens*, Downy Birch, the other common native species, which also flowers in April and May, would be a suitable alternative. The genus *Alnus*, represented in the British Isles by *A. glutinosa*, differs from *Betula* in the following ways: the catkins appear before the leaves, the male flowers have 4 stamens, the female flowers are in pairs in the axil of the bract and the fruiting catkins are cone-like, each cone-scale being 5-lobed. The flowering period for *A. glutinosa* is from February to March.

Fig. 10 Betulaceae, *Betula pendula*

A A terminal shoot with a pendent, male catkin and a smaller, female catkin. Male catkins are formed during the autumn but remain immature until the following spring. They arise singly or in groups of 2 or 3 at the ends of the branches. The erect, female catkins are produced singly in April or May with the leaves on short lateral shoots below the male catkins.
Male catkin: 57 mm Female catkin: 12 mm

B Side view of a 3-flowered male dichasium or cymule attached to the main axis of the inflorescence. Adjacent dichasia have been removed for the sake of clarity. The group of 3 flowers is subtended by a bract and 2 bracteoles.
Axis of dichasium: 3 mm Bract: 2.25 mm

C A male dichasium of 3 flowers viewed from below. Each flower consists of 2 bifid stamens and a perianth which is often reduced from the typical 4 segments to 2 or even 1. The edges of the bracteoles can be seen behind the perianth-segments. The anthers are 2-celled and dehisce longitudinally.
Stamen: 1.5 mm Anther: 0.75 mm

D A 3-flowered female dichasium or cymule. Each of the female flowers consists of an ovary formed from 2 united carpels and is surmounted by 2 styles with entire stigmas. The group of 3 flowers is subtended by a bract and 2 bracteoles. Perianth-segments are absent. The bracteoles (not shown) are free from the bract until after fertilisation when they unite with it to form a 3-lobed scale (see G).
Bract: 0.6×0.7 mm Ovary: 0.2×0.25 mm Style: 0.6 mm

E T.S. of ovary after fertilisation. A portion of the developing wings can be seen. The ovary is 2-locular at the base but unilocular above and contains 2 pendulous, anatropous ovules with axile placentation.

F A single, indehiscent fruit detached from its scale. After fertilisation the ovary develops into a 1-seeded nut with broad, projecting wings.
Fruit: 6×7 mm

G The 3-lobed, woody scale resulting from the fusion of the bract and bracteoles. In autumn the catkins break up, allowing the winged nuts, 1–3 on the surface of each scale, to be dispersed by the wind.
Scale: 6×7 mm

Fig. 10

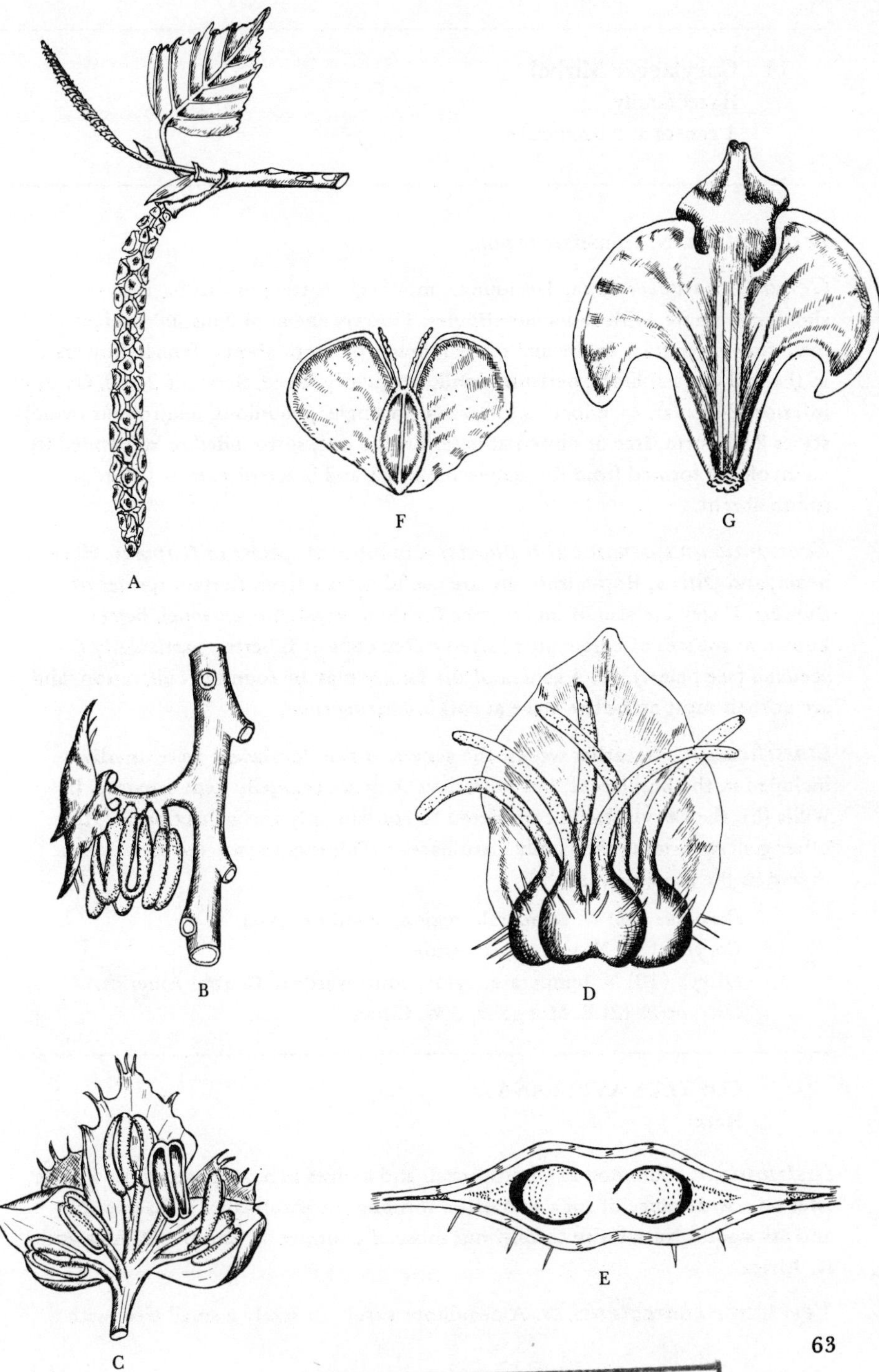

11 Corylaceae Mirbel

Hazel family

4 genera and 62 species

Distribution. N. temperate region.

General characteristics. Deciduous, monoecious trees or shrubs. Leaves simple, alternate, with caducous stipules. Flowers anemophilous, in catkins; male flowers solitary in the axil of each bract, perianth absent; female flowers 2 in the axil of each bract, perianth small, irregularly lobed. Stamens 2–20. Ovary inferior, 2-locular, each loculus containing a single, pendulous, anatropous ovule; styles 2, filiform, free or united at base. Fruit a nut, surrounded or subtended by an involucre formed from the accrescent bract and bracteoles; seed 1. Endosperm absent.

Economic and ornamental plants. A number of species of *Carpinus*, Hornbeam, and *Ostrya*, Hop-hornbeam, are useful timber-trees. Certain species of *Corylus*, Hazel, are also of importance for their wood, but are much better known as sources of edible nuts (often called cobs or filberts), particularly *C. avellana* (see below). All 4 genera of the family may be found in cultivation, and are at their most attractive stage at catkin-bearing time.

Classification. In earlier works, the genera of the Corylaceae were usually included in the Betulaceae, but nowadays they are generally kept separate. In Willis (5), the Corylaceae is considered to contain only the genus *Corylus*, the other genera forming the family Carpinaceae. This more restricted view is not shared in the present work.

Carpinus (35) N. temperate region, chiefly E. Asia.
Corylus (15) N. temperate region.
Ostrya (10) N. temperate region, southwards to Central America.
Ostryopsis (2) E. Mongolia, S.W. China.

CORYLUS AVELLANA L.

Hazel

Distribution. Common in woods, scrub and hedges in most parts of the British Isles on a wide range of soils, and often forming the shrub-layer of oakwoods and ashwoods. Native also throughout most of continental Europe, W. Asia and N. Africa.

Vegetative characteristics. A deciduous shrub, or rarely a small tree, with suborbicular, biserrate leaves, often shallowly lobed. The twigs are thickly clothed with reddish, glandular hairs.

Floral formula. Male: P0 A4
Female: Plobed $\overline{G}(2)$

Flower and inflorescence. The flowers are unisexual and appear before the leaves. The staminate inflorescences or male catkins are formed in the autumn but only become conspicuous during the months of January to April when the anthers develop a bright yellow colour. The male catkins are produced either singly or in groups of 2–4. The pistillate inflorescences are also formed during the autumn. These are sessile and resemble a bud, but become distinguishable from January onwards when the tufts of crimson styles begin to protrude.

Pollination. The pendulous nature of the male catkins and the production of copious pollen enables effective wind-pollination to take place. The catkins disintegrate after all the pollen has been shed.

Alternative flowers for study. C. avellana is sometimes represented in gardens by forms which have purple, yellow or laciniate leaves. Variation between species of *Corylus* occurs particularly with regard to dimensions of the fruits and form of the involucre surrounding them. In *C. americana* the involucre is twice as long as the fruit, and in *C. cornuta* and *C. maxima* it is also tubular in shape. The related tree *Carpinus betulus*, Hornbeam, differs from *Corylus* in the more numerous stamens of the male flower, more numerous female flowers which form drooping catkins and much smaller fruits. The latter are also subtended by a large, 3-lobed involucre, the middle lobe of which is considerably elongated. *C. betulus* flowers in April and May.

Fig. 11 Corylaceae, *Corylus avellana*

A The male catkin, and below it, the sessile, bud-like female inflorescence with its small tuft of crimson styles.
Male catkin: 70 mm Female inflorescence: 5 mm

B An enlarged view of the female inflorescence. The bud-like structure, crowned by a tuft of crimson styles, is made up of scales which enclose 3 or more bracts. In the axil of each bract are 2 flowers, each of which has 2 free styles.

C A single, male flower, consisting of 4 stamens, each of which is bifid below the anther. The anthers are just beginning to dehisce. Each male flower is subtended by a bract which has 2 bracteoles united with it.
Bract: 3 × 2.5 mm

D Dorsal view of an anther and part of the filament. The anther, which is terminated by a tuft of hairs, is 2-celled and dehisces longitudinally (see C).
Anther: 1.2 mm

E The bud-scales of the female inflorescence prised apart to reveal (in the illustration) 3 bracts subtending 6 flowers.

F L.S. of a pair of female flowers subtended by a bract. Prior to fertilisation only the crimson styles of the flowers are visible, protruding from the scales; afterwards the flowers develop into a leafy shoot bearing a terminal cluster of fruits.
Style: 4 mm

G Basal portion of a pair of female flowers attached to a bract. At the base of each flower is a very small involucre of connate bracteoles. After fertilisation the bud-scales wither away, and the bract and bracteoles fuse and enlarge, eventually forming a leafy involucre that surrounds the fruit (see I).
Bract: 1.5 mm wide Connate bracteoles: 0.25 mm wide

H T.S. of inferior ovary after fertilisation, showing the 2 loculi containing 1 ovule with axile placentation in each loculus.
T.S. of ovary: 0.8 × 0.45 mm

I The mature fruit, a large nut containing a single seed (the other fails to develop). Surrounding the nut is the leafy involucre formed by the fusion and subsequent enlargement of the bract and bracteoles.
Fruit: 15 × 18 mm

J L.S. of nut, showing the seed with its 2 well-developed cotyledons.

Fig. 11

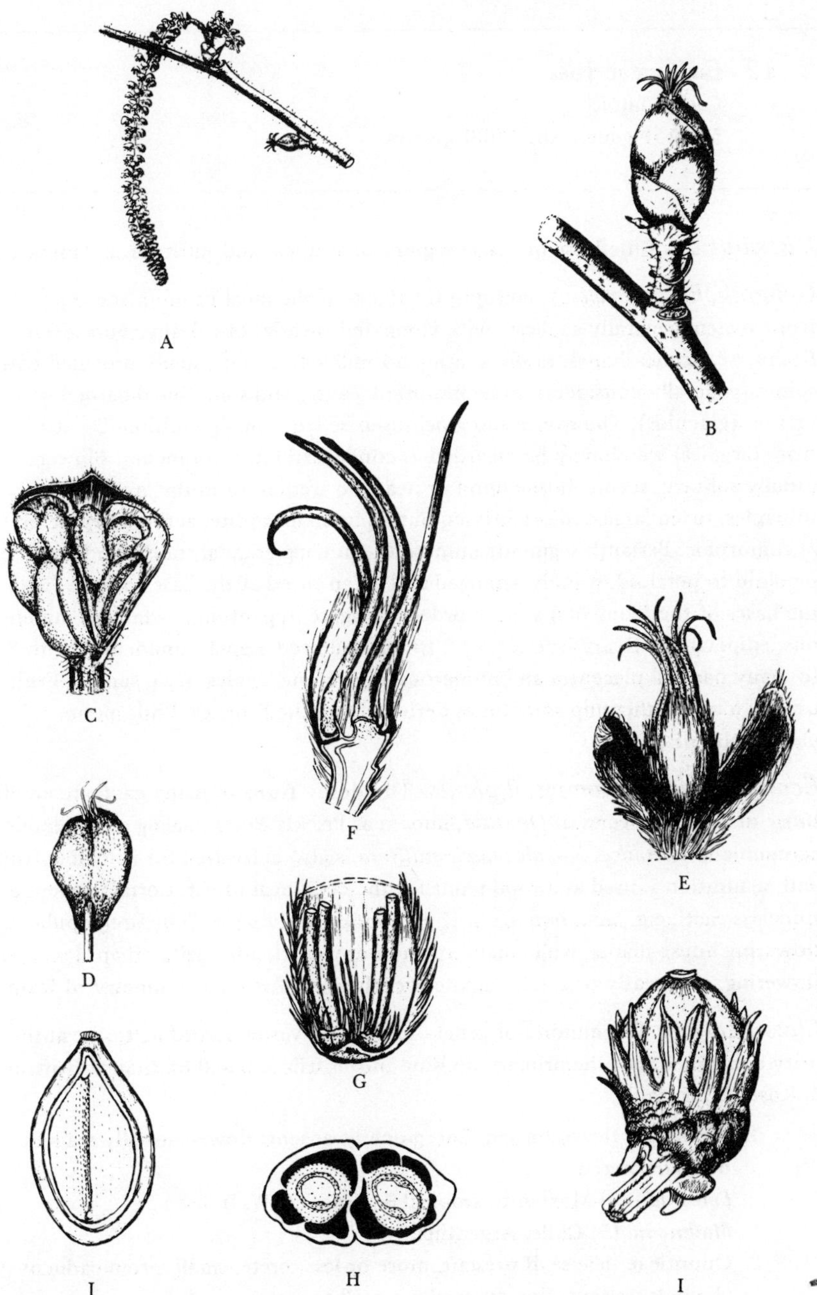

12 Cactaceae Juss.

Cactus family

50–150 genera and 2000 species

Distribution. Chiefly semi-desert regions of tropical and subtropical America.

General characteristics. Xerophytic plants of the most pronounced type. Root system generally shallow, with elongated, slender but fleshy roots. Stem fleshy, of various shapes, rarely bearing normal leaves, but usually provided with spines (generally considered to be modified leaves) and sometimed barbed bristles (glochids). The spines and glochids arise from small, cushion-like structures (areoles) which may be regarded as condensed lateral branches. Flowers usually solitary, sessile, borne upon or near the areoles, or in the 'axils' of tubercles, often large and brightly coloured, hermaphrodite, actinomorphic or zygomorphic. Perianth-segments numerous, showing gradual transition from sepaloid to petaloid, spirally arranged and often fused at the base to form (with the bases of the filaments) a more or less elongate hypanthium. Stamens numerous, epipetalous. Ovary inferior, of 2 to many united carpels, unilocular, with 2 to many parietal placentas and numerous anatropous ovules; style simple. Fruit usually a berry, the pulp sometimes derived from the funicles. Endosperm usually absent.

Economic and ornamental plants. The fleshy fruits of many cacti are edible, those of various species of *Opuntia*, known as Prickly Pears, having considerable economic importance. *Nopalea cochenillifera* is also cultivated for its edible fruit and in addition is used as a food-plant for the cochineal insect. Certain genera of spineless cacti, e.g., *Schlumbergera* (*Zygocactus*) and *Epiphyllum*, are popular as flowering house-plants, while many of the spiny cacti, although perhaps less free-flowering, are greatly prized by amateur enthusiasts for their symmetry of form.

Classification. The number of genera recognised varies according to the authority concerned, but the primary division into 3 tribes, based on that of Britton & Rose (7), is stable.

1. Pereskieae (leaves broad, flat; glochids absent; flowers usually stalked, often clustered).
   Pereskia (20) Mexico to tropical S. America, W. Indies.
   Maihuenia (5) Chile, Argentina.
2. Opuntieae (leaves, if present, more or less terete, small, often caducous; glochids present; flowers sessile, usually rotate).
   Opuntia (250) America.

3. Cacteae (Cereeae) (leaves absent or rudimentary; glochids absent; flowers sessile, generally with a tube).
   Mammillaria (200–300) S.W. United States to northern S. America, W. Indies.
   Echinocereus (75) S. United States, Mexico.
   Rhipsalis (60) W. Indies, Brazil, Argentina.
   Cereus (50) S. America, W. Indies.
   Cephalocereus (48) Florida to Brazil.
   Echinopsis (35) S. America
   Epiphyllum (21) Mexico to tropical S. America, W. Indies.
   Schlumbergera (2–5) Brazil

SCHLUMBERGERA × BUCKLEYI (T. Moore) D.R. Hunt
Christmas Cactus

Distribution. The genus *Schlumbergera*, which contains 2–5 species, is native in Brazil. The hybrid *S.* × *buckleyi* originates from a cross between *S. truncata* (*Epiphyllum truncatum*, *Zygocactus truncatus*) and *S. russelliana* (*Epiphyllum russellianum*) made by the horticulturist W. Buckley about 1850. It is a popular house plant, and is readily propagated by stem cuttings, but may be grafted on to *Pereskia aculeata* so as to show off the hanging branches to better advantage.

Vegetative characteristics. A much-branched, slow-growing perennial with a succulent stem composed of flattened segments, and areoles confined to the crenate margins. There is a tendency for the older, basal joints to become woody.

Floral formula. $P(\infty)\ A\infty\ \overline{G}(4)$

Flower and inflorescence. The showy, magenta to pink, hermaphrodite flowers are usually borne singly at the ends of the stems from December to February, though earlier or later flowering may sometimes take place.

Pollination. The bright pink colour of the flowers and their more or less oblique attachment to the terminal stem-joints suggest that birds may play a part in the pollination of this genus. Certainly insects are attracted to the flowers by the nectar which is secreted at the base of the tube formed by the perianth-segments. The style protrudes beyond the stamens and the flowers are protandrous, so that cross-pollination is favoured.

Alternative flowers for study. The parent species could be used as substitutes for *S.* × *buckleyi* or members of the closely related genus *Epiphyllum*, whose large flowers appear laterally on the stem during the summer months. Perhaps the best known species is *E. ackermannii* (sometimes placed in the genus *Nopalxochia*) which is a parent of a number of hybrids. Comparison should be made with such important genera as *Opuntia*, *Mammillaria*, etc., which differ considerably in habit and vegetative characteristics but rather less so in floral structure.

Fig. 12 Cactaceae, *Schlumbergera* × *buckleyi*

A A sessile, actinomorphic, hermaphrodite flower borne at the end of the segmented and flattened stem. The 12–15 perianth-segments are all coloured but vary in length, the outer ones (representing the sepals) being the shortest and the inner ones the longest. The single style is exserted beyond the numerous stamens.
Perianth-segments: 13–29 × 8–11 mm

B L.S. of a flower. The perianth-segments are united at the base into a tube and the stamens are also connate. These organs are borne on the hypanthium which encloses the inferior ovary. The long style arises from the apex of the ovary.
Style: 50 mm

C Detail of a group of stamens. The stamens are in 2 whorls, the outer whorl inserted on the perianth-tube and the inner whorl united at their bases to form a short tube round the style. The narrow filaments are exserted from the perianth, and the anthers are dorsifixed, 2-celled and dehisce longitudinally.
Filaments: *c.* 37 mm Anthers: 1.25 mm

D T.S. of the unilocular ovary formed from 4 united carpels, showing the numerous anatropous ovules attached to parietal placentas. The loculus is surrounded by well-developed fleshy tissue with several vascular bundles, and the ovary-wall is unevenly angled.
T.S. of ovary: 6 mm

E L.S. of the inferior ovary attached to a portion of the stem. Above the ovary are the bases of some of the floral parts borne on the hypanthium. At the centre is the style surrounded by the staminal tube formed by the inner whorl of stamens. The outer whorl is adnate to the base of the perianth-tube. On the margin of the hypanthium are 2 of the smallest perianth-segments shown in their entirety. *S.* × *buckleyi* does not appear to bear fruit, but its parent species produce berries, those of *S. russelliana* being strongly ribbed.
Ovary + hypanthium: 8 mm

F Detail of the 4-lobed stigma.
Stigma: 3.75 mm

Fig. 12

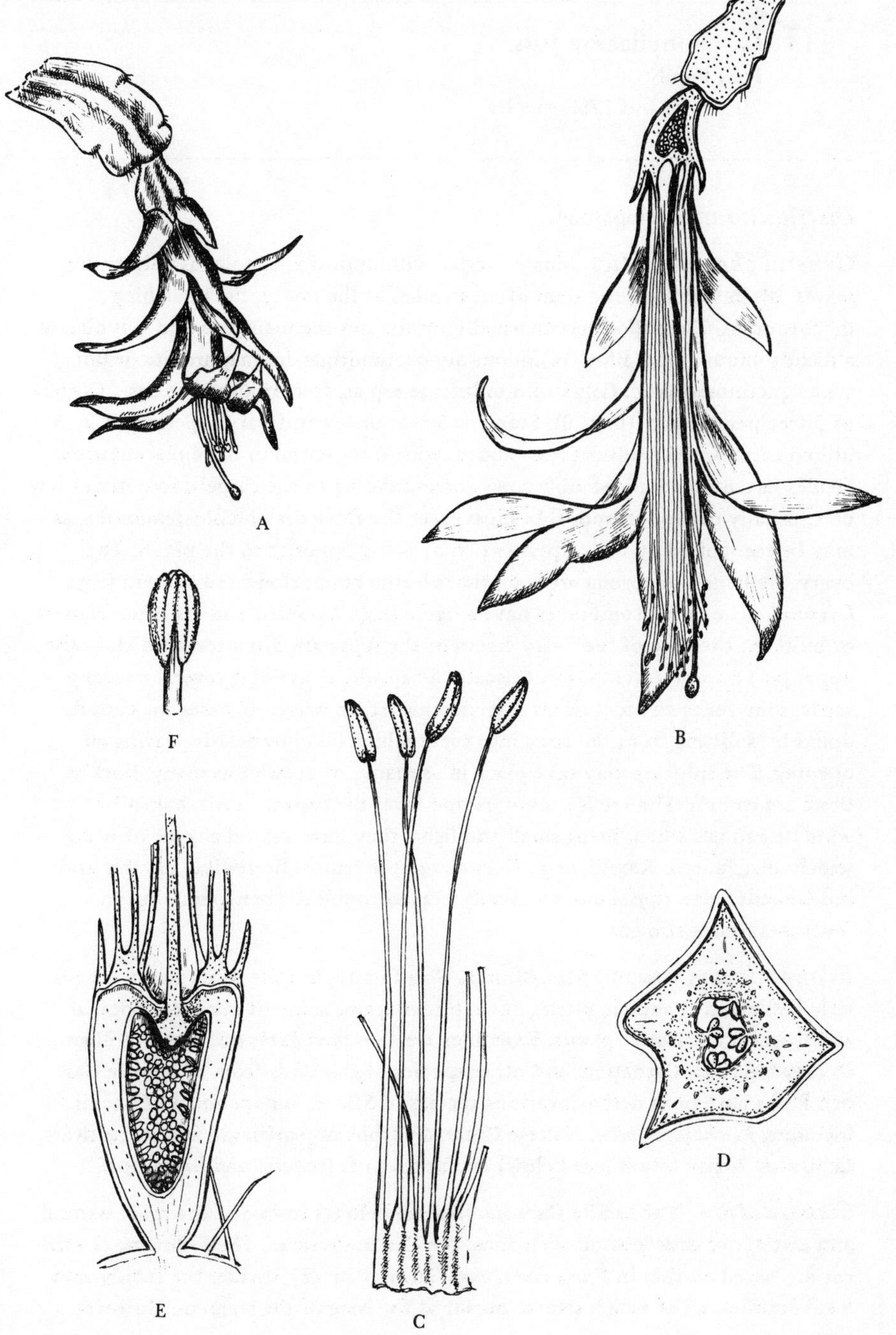

13 Caryophyllaceae Juss.

Pink family

70 genera and 1750 species

Distribution. Cosmopolitan.

General characteristics. Mostly herbs, with opposite, simple, usually entire leaves, often stipulate, the stem often swollen at the nodes, the branching dichotomous. The inflorescence usually terminates the main axis and is typically a dichotomous cyme. Flowers commonly protandrous, hermaphrodite or unisexual, actinomorphic. Calyx of 5 imbricate sepals, free or united at base. Corolla of 5 free petals (sometimes 0). Stamens 5 + 5, or fewer. Ovary superior, of 2–5 united carpels, unilocular at least above, with free-central or basal placentation. Ovules usually many, in double rows corresponding to the carpels, sometimes few or 1, usually campylotropous. In most cases the flower is obdiplostemonous, as may be recognised by the carpels (when 5) being opposite to the petals. The ovary, stamens and corolla are sometimes borne on an elongated column (e.g., *Lychnis*). The petals sometimes have a ligule (e.g., *Lychnis*) and are often clawed or bifid. At the base of the ovary traces of the septa are often seen, which in the upper part do not develop. Fruit usually a capsule, containing several or many seeds, sometimes an indehiscent 1-seeded nutlet. In nearly all cases the capsule opens by splitting from the apex into teeth which bend outwards, leaving an opening. The splitting may take place in as many, or in twice as many, lines as there are carpels. The seeds cannot escape from the capsule until shaken by wind or animals when, being small and light, they have a good chance of being widely distributed. Rarely, (e.g., *Cucubalus*) the fruit is berry-like, but dry and indehiscent when ripe. Embryo usually curved round the perisperm, but in a few cases nearly straight.

Economic and ornamental plants. This family, in spite of its size, is singularly lacking in economic plants, though it contains some of our most popular and fragrant decorative plants. Examples are *Dianthus barbatus*, Sweet William, *D. caryophyllus*, Carnation, and other species of *Dianthus* from which the Garden Pinks have been derived; various species of *Silene*, *Saponaria* and *Lychnis*, including *L. chalcedonica*, Maltese Cross, *Gypsophila paniculata* and its cultivars, *Cerastium tomentosum* (see below) and species of *Arenaria* and *Petrocoptis*.

Classification. The family Caryophyllaceae belongs to one of the more natural and distinctive orders of dicotyledons, the Centrospermae. The following classification, based on that in *Flora Europaea*, Tutin *et al.* (8), divides the family into 3 subfamilies, all of which secrete nectar at the base of the stamens. However, only long-tongued insects, mainly bees and Lepidoptera, can reach the nectar of the Silenoideae since the calyx-tube excludes short-tongued insects. In the other subfamilies, the sepals are not united into a tube, and the flowers open wide, so that the nectar is available to all insect visitors.

I. Alsinoideae (leaves opposite; stipules absent; petals usually well-developed; sepals free or joined only at base).
   Arenaria (250) N. temperate region.
   Stellaria (120) cosmopolitan.
   Cerastium (60) almost cosmopolitan.
   Sagina (20–30) mainly in N. temperate region.
   Scleranthus (10) Europe, Asia, Africa, Australia.

II. Paronychioideae (leaves opposite, alternate or verticillate; stipules present; petals often very small or absent; sepals free).
   Paronychia (50) cosmopolitan.
   Drymaria (50) mainly in Central and S. America and W. Indies.
   Spergularia (40) cosmopolitan.
   Herniaria (35) Europe, Mediterranean region to Afghanistan, S. Africa.
   Polycarpon (16) cosmopolitan.
   Corrigiola (10) cosmopolitan.
   Spergula (5) temperate regions.

III. Silenoideae (leaves opposite; stipules absent; epicalyx-scales sometimes present; sepals joined in a tubular or campanulate calyx; petals usually well-developed; stamens, petals and ovary often situated on a more or less elongated column).
   Silene (500) N. temperate region, especially Mediterranean.
   Dianthus (300) Europe, Asia and Africa, especially Mediterranean region.
   Gypsophila (125) temperate Eurasia, Egypt.
   Lychnis (12) temperate Eurasia.

CERASTIUM TOMENTOSUM L.

Snow-in-summer

Distribution. A commonly grown rock-garden and wall plant native in the Caucasus and S.E. Europe. It often escapes from cultivation, and has now become established in a number of places throughout the British Isles.

Vegetative characteristics. A perennial, mat-forming herb with densely white-tomentose stems and leaves, the latter linear-lanceolate, with slightly revolute margins.

Floral formula. K5 C5 A5+5 $\underline{G}$(5)

Flower and inflorescence. The inflorescence is a loose, dichasial cyme and the showy, white, actinomorphic, hermaphrodite flowers are borne on white-hairy stems from May to September.

Pollination. The flowers of *C. tomentosum* open wide, enabling even short-tongued insects, attracted by the white petals, to reach the nectar that is secreted by glands at the bases of the stamens. Cross-pollination normally takes place where the flowers are protandrous, but homogamous flowers have been observed, and it is likely that, in the absence of insect visits, automatic self-pollination may occur as in several other species of *Cerastium.*

Alternative flowers for study. C. fontanum ssp. *triviale* (*C. holosteoides*), the Common Mouse-ear Chickweed, which has a similar flowering period, is abundant throughout the British Isles and differs little in floral structure, but is perhaps less suitable than *C. tomentosum* owing to its smaller flowers. Comparison with a member of the Silenoideae will prove a useful exercise.

Fig. 13 Caryophyllaceae, *Cerastium tomentosum*

A The actinomorphic, hermaphrodite flower with 2 of the 5 free, bifid petals removed to expose the reproductive parts. The 5 free, tomentose sepals allow the petals to open wide, in contrast to those flowers in the family where the sepals are united into a calyx-tube. The superior ovary is surrounded by the free stamens which are in 2 rows of 5. Nectar is secreted at the bases of the stamens.
Corolla: 12–18 mm in diameter Sepals: *c.* 7 mm Petals: *c.* 27 mm

B A long and a short stamen attached to a teased-out portion of the receptacle. The longer stamen belongs to the outer row which stands opposite the petals, the shorter stamen to the inner row which is opposite to the sepals. The introrse anthers are 2-celled and dehisce longitudinally.
Anthers: 1.5 mm Filaments: 6–7 mm

C Sepals, petals and stamens have been cut back to reveal the superior ovary of 5 united carpels surmounted by the 5 free styles.
Ovary (at anthesis): 2×2 mm Styles: 4–5 mm

D Detail of upper portion of style, showing the papillose stigma.

E A single papilla cut from the stigmatic surface.

F T.S. of the ovary, showing the numerous ovules in the single loculus with free-central placentation.
Ovary (at anthesis): 2 mm in diameter

G L.S. of ovary. The ovules are arranged in double rows which correspond to the number of united carpels.

H The fruit is a capsule which dehisces apically by 10 small teeth (double the number of carpels). When the capsule is shaken by the wind or by animals, the numerous small seeds are dispersed. In the genus *Cerastium*, the capsule always exceeds the sepals in length.
Capsule: *c.* 10×4 mm Teeth: 1.25 mm

Fig. 13

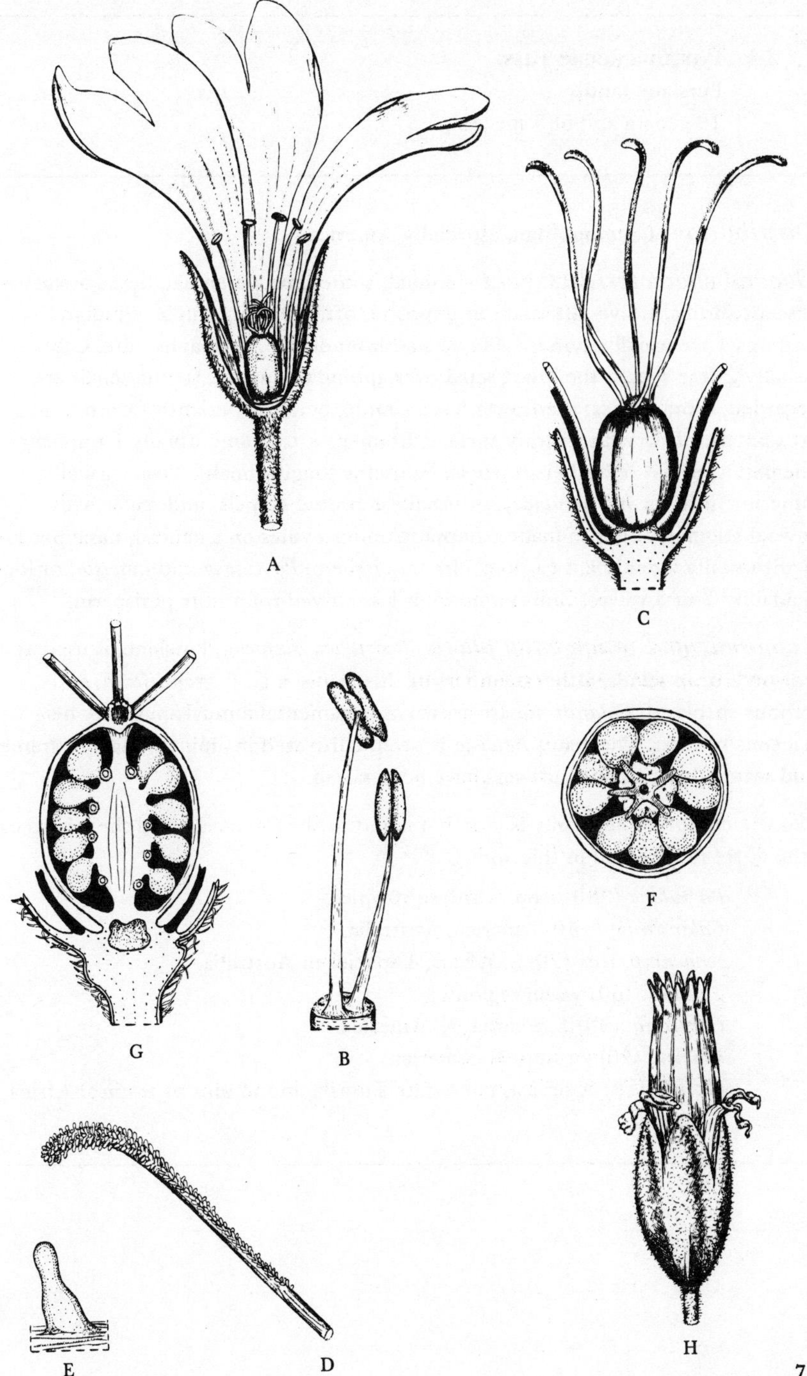

14 Portulacaceae Juss.

Purslane family

19 genera and 580 species

Distribution. Cosmopolitan, especially America.

General characteristics. Mostly annual, sometimes perennial, herbs or suffrutescent shrubs. Leaves alternate or opposite, often fleshy, simple, stipulate. Inflorescence usually cymose. Flowers actinomorphic, hermaphrodite. Calyx of usually 2 free sepals, the lower sepal overlapping the upper. (If the sepals are regarded as bracteoles, the flower has a simple, petaloid perianth.) Corolla of 5 free petals, often with a satiny surface. Stamens 4 to many, usually 5 opposite the petals, or 5 + 5; anthers introrse, dehiscing longitudinally. Ovary usually superior (inferior in *Portulaca*), of usually 3 united carpels, unilocular, with several stigmas, and 2 to many campylotropous ovules on a central, basal placenta. Fruit usually a dehiscent capsule, circumscissile in *Portulaca* and *Lewisia*, or loculicidal by 2 or 3 valves. Embryo more or less curved round the perisperm.

Economic and ornamental plants. Portulaca oleracea, Purslane, is used as a pot-herb or in salads. Other members of this genus, e.g., *P. grandiflora*, and various species of *Calandrinia* are grown as ornamental annuals in sunny herbaceous borders. The genus *Lewisia* is often cultivated in alpine houses or frames, and several hybrids and cultivars have been raised.

Classification. The family is closely related to the Cactaceae and the Aizoaceae (the latter not treated in this book).

Portulaca (200) tropics and subtropics.
Calandrinia (150) America, Australia.
Anacampseros (70) S. Africa, 1 species in Australia.
Talinum (50) warm regions.
Claytonia (35) E. Siberia, N. America.
Lewisia (20) western N. America.
Montia (15) America, temperate Eurasia, mountains of tropical Africa, Australia.

CALANDRINIA GRANDIFLORA Lindl.

Distribution. Native in Chile and introduced into Britain in the early nineteenth century. It is grown as a half-hardy annual suitable for rockeries and flower-borders that are well exposed to the sun.

Vegetative characteristics. A fleshy, glabrous, perennial herb, treated as an annual in gardens, and reaching a height of between 30 and 90 cm. Its leaves are alternate, ovate to rhomboid, acute at the apex and tapering gradually at the base into a winged petiole.

Floral formula. K2 C5 A∞ $\underline{G}$(3)

Flower and inflorescence. Pale purplish-pink, actinomorphic, hermaphrodite flowers are borne in lax racemes during July and August. The individual flowers are short-lived and only open in bright sunlight.

Pollination. In the genus *Calandrinia* the petals of the fading flowers become pulpy as moisture oozes out of the tissues and forms a thin layer on the surface. The fluid attracts flies which, in the course of their visits, effect cross-pollination.

Alternative flowers for study. Only a few of the many species of *Calandrinia* are commonly grown in gardens, but the related genera *Portulaca* and *Lewisia* would serve as suitable alternatives. *P. grandiflora*, Sun Plant, is readily obtainable in gardens during the summer months, but members of the genus *Lewisia* are best grown in an alpine house or cold frame. The floristic differences between the genera include a variation in the number of flower parts and the form of dehiscence of the fruit, the capsule being loculicidal in *Calandrinia* but circumscissile in *Portulaca* and *Lewisia*.

Fig. 14 Portulacaceae, *Calandrinia grandiflora*

A The flower-bud showing the 2 sepals, a particular feature of the family Portulacaceae, surrounding the emerging petals.
Bud: 25 mm

B The actinomorphic, hermaphrodite flower. The superior ovary is surrounded by numerous long stamens and 5 free, light purple or rose-red petals.
Corolla: 35–40 mm in diameter Petals: 20 × 15 mm
Sepals: 14–15 × 10 mm

C The superior ovary is surmounted by a short style with a 3-branched stigma and is surrounded by numerous introrse stamens attached to the well-developed receptacle.
Stamens: 8–10(–12) mm Ovary: 3.85–4.5 × 3 mm
Style: 4 mm

D Detail of stamen. D1: upper part, showing the introrse, 2-celled anther which dehisces longitudinally. D2: basal portion of filaments attached to the receptacle. Some of the filaments are hairy towards the base.
Anther: 2 mm

E L.S. of ovary on the well-developed receptacle. Numerous campylotropous ovules are borne on a central, basal placenta. The bases of the 2 sepals and 4 of the stamens are visible, but the petals have been removed for clarity.

F Detail of the 3-branched stigma at late bud stage.
Stigma: 2 × 2.5 mm

G T.S. of ovary showing the central placenta with ovules attached. The unilocular ovary has been formed from 3 united carpels.
Ovary: 2 mm in diameter

H Part of the fruit with one of the valves removed to reveal the numerous seeds. Surrounding the fruit are the remains of the 2 persistent sepals and the petals. The latter become mucilaginous as they fade, but eventually dry and form an irregular mass round the fruit.
Fruit: 4 × 2.5 mm

I Detail of part of the placenta with seeds attached.
Seeds: 0.2–0.3 mm in diameter

Fig. 14

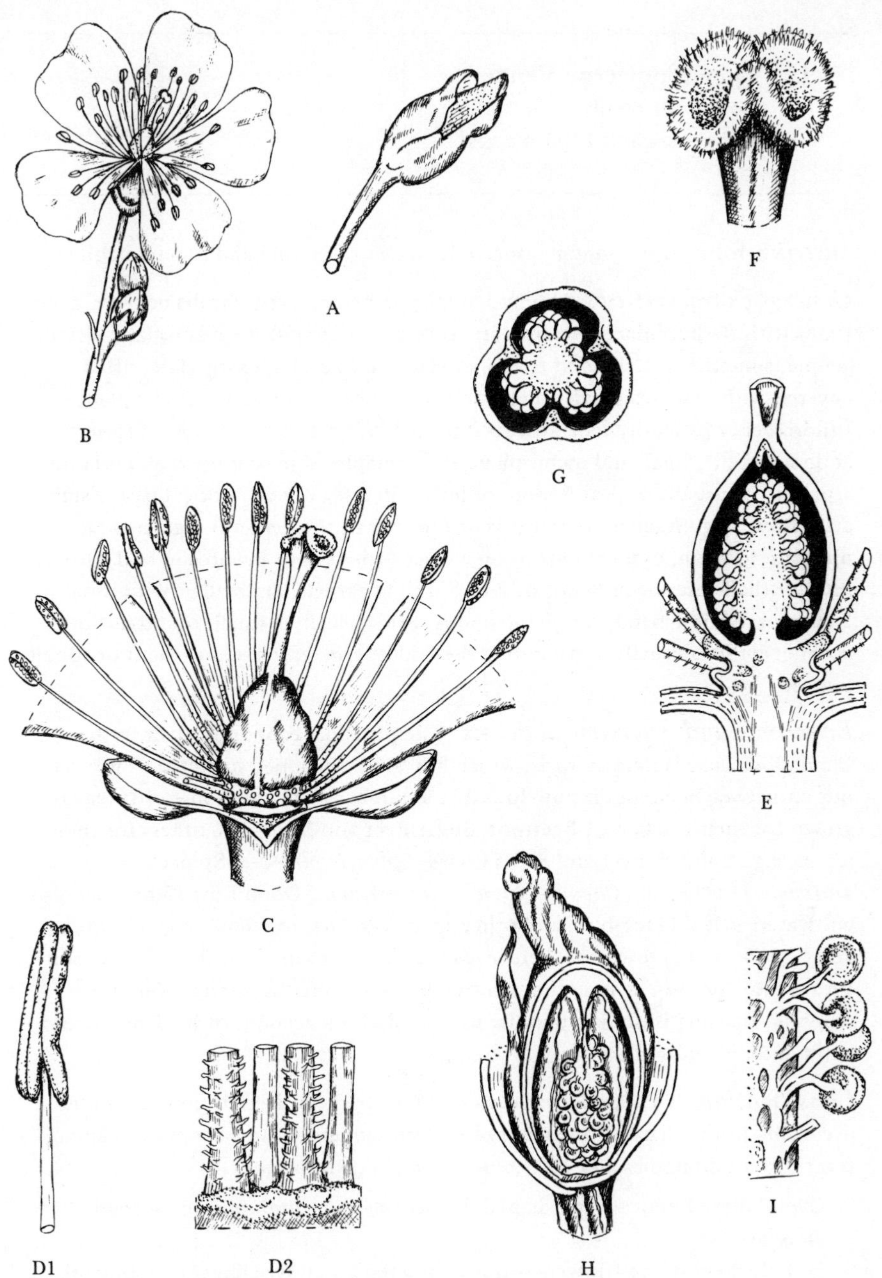

15 Chenopodiaceae Vent.

Goosefoot family

102 genera and 1400 species

Distribution. Cosmopolitan, especially in desert, coastal and ruderal habitats.

General characteristics. Mainly halophytic herbs, rarely shrubs or trees, sometimes with fleshy, jointed, leafless stems (e.g., *Salicornia*). Leaves usually alternate, simple, sometimes fleshy and terete, or reduced to scales, exstipulate, often covered with hairs which give a characteristic mealy appearance to the plants. Inflorescence primarily racemose, but partial inflorescences cymose. Flowers actinomorphic, small and inconspicuous, hermaphrodite or unisexual. Perianth sepaloid, of usually 2, 3 or 5 more or less united segments, imbricate, persistent after flowering. Stamens as many as or fewer than the perianth-segments and opposite to them, hypogynous or on a disc; anthers bent inwards in bud. Ovary inferior (half-inferior in *Beta*), of 2 or 3 united carpels, unilocular, with usually 2 stigmas; ovule 1 basal, campylotropous. Fruit usually a small, round nut or achene; embryo usually surrounding the endosperm, either simply bent or spirally twisted.

Economic and ornamental plants. The principal economic plant in the Chenopodiaceae is *Beta vulgaris*, which has been developed in cultivation from our native Sea Beet, and is now found in a number of forms. Some of these are grown for their roots, e.g., Beetroot, Sugar Beet and Mangolds, others for their leaves, e.g., Spinach Beet and Swiss Chard. *Spinacia oleracea*, Spinach, *Atriplex hortensis*, Orache, and *Chenopodium bonus-henricus*, Good King Henry, are also cultivated as leaf vegetables, and other species of *Chenopodium*, e.g., *C. album* and *C. quinoa* are grown for their seeds which are ground into flour. A crimson form of *A. hortensis* is sometimes grown as an ornamental annual. *Kochia scoparia*, Burning Bush, is a popular garden plant on account of its dense, ovoid habit and red autumn colour.

Classification. The family was divided by Ulbrich (in Ref. 9), into 2 groups according to the shape of the embryo and subsequently, on a variety of characters, into 8 subfamilies and 14 tribes. The main divisions are:

A. Cyclolobeae (embryo ring-shaped, horseshoe-like, conduplicate, or semicircular).

  I. Polycnemoideae (flowers solitary, bracteolate; leaves linear or subulate).
    Polycnemum (6–7) central and S. Europe, Mediterranean region to central Asia.

II. Betoideae (fruit opening by an operculum; stigma short, usually broad-lobed, papillose within).

Beta (6) Europe, Mediterranean region.

III. Chenopodioideae (flowers usually in glomerules, more rarely in spicate inflorescences; fruit usually remaining closed and surrounded at maturity by perianth or bracteoles).

Atriplex (200) temperate and subtropical regions.
Chenopodium (100–150) temperate regions.
Kochia (90) central Europe, temperate Asia, N. and S. Africa, Australia.
Spinacia (3) E. Mediterranean region to C. Asia.

IV. Corispermoideae (flowers spicate, ebracteolate; fruit naked at maturity).

Corispermum (60) N. temperate regions.

V. Salicornioideae (flowers in clavate, catkin-like inflorescences, ebracteolate; branches articulated).

Salicornia (35) temperate and tropical regions.

B. Spirolobeae (embryo spirally twisted).

VI. Sarcobatoideae (flowers unisexual, without bracteloes).

Sarcobatus (1–2) N. America

VII. Suaedoideae (flowers with small, scale-like bracteoles; leaves glabrous).

Suaeda (110) cosmopolitan

VIII. Salsoloideae (flowers with bracteoles usually as large or larger than perianth-segments; leaves usually covered with filiform hairs).

Salsola (150) cosmopolitan

SPINACIA OLERACEA L.
Spinach

Distribution. Spinach has been cultivated in gardens for so long that its original wild distribution is not known, but it was probably native in W. Asia. It grows well on a wide range of soils and occasionally escapes from cultivation.

Vegetative characteristics. An annual herb with large, succulent, edible leaves that form a basal rosette in the young plant. Later on, an erect, ribbed stem up to 1 m in height is produced on which the leaves are borne alternately. All the leaves are ovate to triangular hastate in shape, and the margins may be entire or dentate.

Floral formula. Male: P4 A4
Female: P0 $\underline{G}$1

Flower and inflorescence. The small, green, unisexual flowers are borne on separate plants from June onwards. The flowers on the male plant form a dense, spicate inflorescence (see Fig. 15.A), while those on the female plant appear in clusters in the axils of the leaves (see Fig. 15.B). The time of flowering depends very much on horticultural practice. The first sowing of *S. oleracea* can be made in March, with successional sowings during the spring and summer. Late summer sowings are made so that the crop can stand the winter. Overcrowding and lack of water cause the plants to run to seed, so manuring is useful both in retaining moisture and in encouraging leaf-growth.

Pollination. The inconspicuous colour of the flowers, the reduced or absent perianth, the well exposed though short stamens (see Fig. 15.C) and the protruding, filiform stigmas (see Figs. 15.B and D) suggest that the wind is the principal pollinating agent. Cross-pollination would occur as a matter of course since the plants are dioecious.

Alternative plants for study. S. oleracea is a readily obtainable and fast-growing annual which, besides having two kinds of flowers, has different types of fruit according to the cultivar concerned. Two groups of cultivars are recognised, the 'round-seeded' Summer Spinach, and the 'prickly-seeded' Winter Spinach. The species described must not be confused with other plants that have the vernacular name 'spinach', e.g., Mountain Spinach or Orache (*Atriplex hortensis*), Spinach Beet (a form of *Beta vulgaris*) and New Zealand Spinach (*Tetragonia tetragonioides*, a member of a different family, the Aizoaceae). Comparison may be made with other common, summer-flowering genera notably *Chenopodium*, Goosefoot, and *Suaeda*, Seablite, both of which produce hermaphrodite and female flowers, and *Beta* which has hermaphrodite flowers and a half-inferior ovary connate with the receptacle in fruit.

Fig. 15 Chenopodiaceae, *Spinacia oleracea*

A Groups of male flowers taken from the apex, middle and base of the flowering part of the stem. At the top the flowers form axillary clusters, while lower down they are in progressively longer, interrupted spikes in the axils of the leaves. The flowers on these spikes are arranged in clusters similar to those at the top of the stem.
Spikes: up to 55 mm

B A section of the stem of a female plant with a cluster of flowers in the axil of a leaf (petiole only shown). Five filiform stigmas protrude from each of the pistillate flowers.

C A sessile, male flower with several flower-buds at its base. The 4 stamens are situated opposite the 4 free perianth-segments. Two of the stamens have become erect in order to shed their pollen while the other 2 are still recurved. The dorsifixed anthers, which are 2-celled and dehisce longitudinally, are attached to comparatively short filaments for a wind-pollinated flower.
Male flower: 3 mm in diameter Perianth-segment: 2.5 mm
Filament: 4 mm Anther: 2×1 mm

D A sessile female flower with several flower-buds at its base. The female flower is without perianth-segments, but has 2 persistent bracteoles that become enlarged, connate and hardened in fruit. The 5 stigmas stand more or less erect between the 2 bracteoles.
Female flowers (excluding stigmas): $1.5–3 \times 1–2$ mm
Bracteole: 0.6 mm Stigma: 3.5 mm

E L.S. of the female flower, showing the ovary subtended by the 2 fleshy bracteoles. The ovary, which has a single loculus containing a solitary, campylotropous ovule, is terminated by a short style bearing 5 well-exserted stigmas.

F T.S. of the female flower. The ovule is enclosed within the ovary, which is itself protected by the 2 bracteoles. The placenta is not visible here as the ovule is basally attached.
Ovary: 0.9×0.8 mm

G One of the indehiscent fruits (the 'seed' of commerce) taken from a 'round-seeded' cultivar. The 2 persistent bracteoles become connate, enlarge, and harden round the fruit as it develops into an achene. The fruits are very variable in size.
Fruits: $2.5–5 \times 2.5–5$ mm

H The true seed removed from the protective case formed by the bracteoles. The seed is positioned vertically within the fruit with the embryo surrounding the endosperm.
Seed: 3×3 mm

Fig. 15

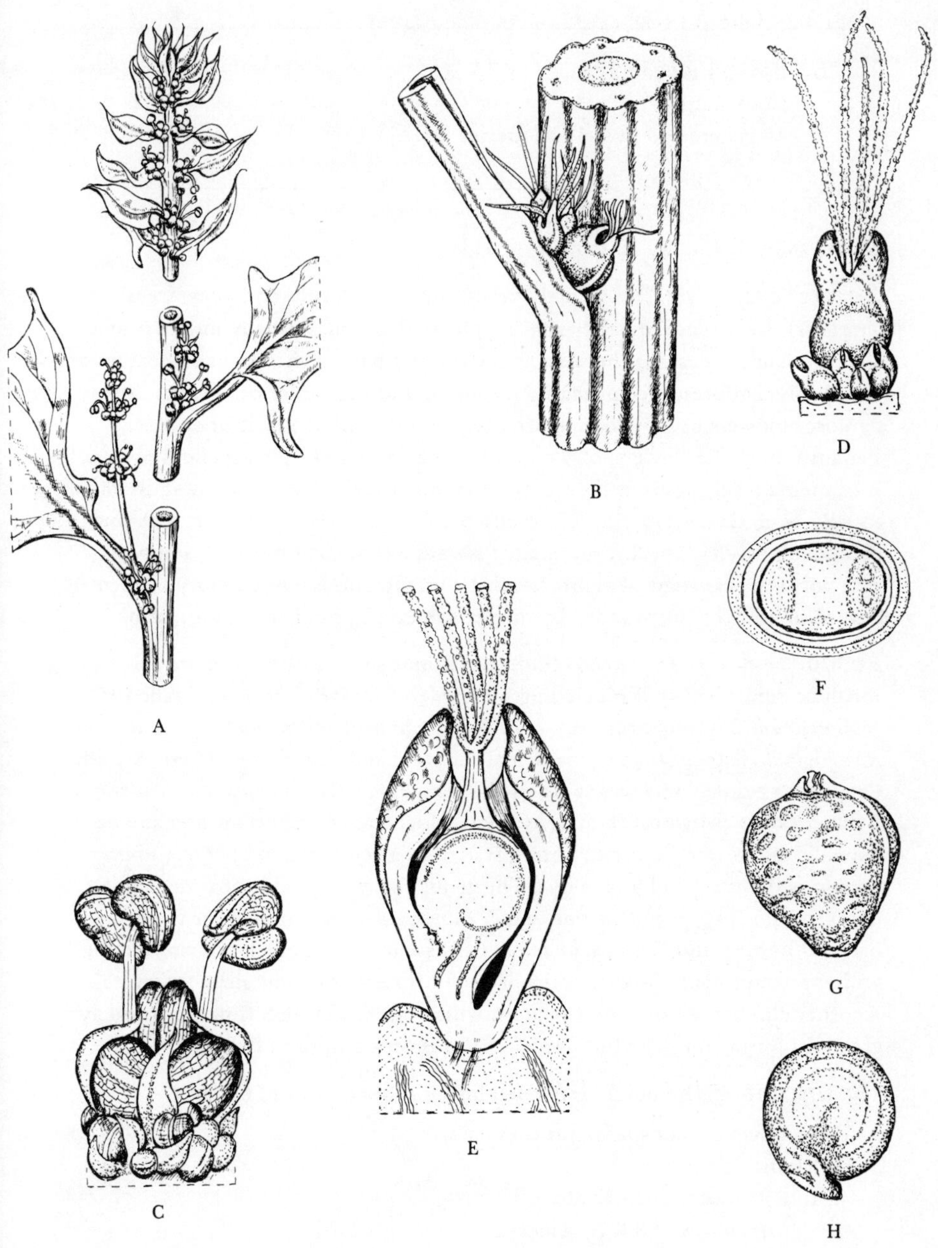

16 Polygonaceae Juss.

Dock family

40 genera and over 1000 species

Distribution. Chiefly N. temperate region.

General characteristics. Mostly herbs, but some shrubs and a few trees (e.g., *Triplaris*). Leaves usually alternate, simple, with stipules usually united into a sheath (ochrea) clasping the stem above the leaf-base, a characteristic feature of the family. Inflorescence primarily racemose, but partial inflorescences usually cymose. Flowers usually hermaphrodite, actinomorphic, cyclic or acyclic. Perianth, in cyclic flowers, of 6 segments in 2 whorls of 3; in acyclic flowers, of 5 segments resulting from the fusion of an inner and an outer segment. Stamens usually 6–9. Ovary superior, of usually 3 united carpels, unilocular, with one erect ovule; styles 2 or 3. Fruit almost always a triangular nut with smooth exterior, the persistent perianth forming a membranous wing. Embryo excentric, curved or straight, surrounded by mealy, sometimes ruminate, endosperm.

Economic and ornamental plants. Various species of *Rheum* are cultivated for their edible leaf-stalks including the garden Rhubarb, previously called *R. rhaponticum*, but more recently considered a hybrid and named *R.* × *cultorum. R. officinale* is the source of medicinal rhubarb, and *Coccoloba uvifera*, Seaside Grape, has edible fruits which are used for making jelly. *Fagopyrum esculentum*, Buckwheat, is grown for its starchy seeds which are an important human and animal food. *Rumex acetosa*, Sorrel, is used as a vegetable and *Polygonum tinctorium* is the source of a blue dye. The genus *Polygonum* also provides some ornamental plants, e.g., *P. affine* and *P. campanulatum*, suitable for the herbaceous border, and *P. baldschuanicum*, Russian Vine, a quick-growing climber with masses of white flowers which is useful for covering unsightly structures. Another climber, *Antigonon leptopus*, which has bright pink flowers, is widely grown in warm countries but needs glasshouse conditions in Britain.

Classification. The family is divided into 3 subfamilies and 6 tribes.

A. Flower cyclic; endosperm not ruminate.

- I. Rumicoideae
  1. Eriogoneae (ochrea absent).
     Eriogonum (200) N. America.
     Chorizanthe (50) western America.
  2. Rumiceae (ochrea present).
     Rumex (200) cosmopolitan.
     Rheum (50) temperate and subtropical Asia.

B. Flower acyclic (except for a few Coccoloboideae).
  II. Polygonoideae (endosperm not ruminate).
    3. Atraphaxideae (shrubs).
      Calligonum (80) S. Europe, N. Africa, W. Asia.
    4. Polygoneae (herbs)
      Polygonum (300) cosmopolitan.
      Fagopyrum (15) temperate Eurasia.
  III. Coccoloboideae (endosperm ruminate).
    5. Coccolobeae (flowers usually hermaphrodite).
      Coccoloba (150) tropical and subtropical America.
      Muehlenbeckia (15) Australasia, S. America.
    6. Triplarideae (plants usually dioecious).
      Triplaris (25) tropical S. America.

POLYGONUM PERSICARIA L.
Redshank, Willow-weed

Distribution. Native in the temperate regions of the northern hemisphere and common throughout the British Isles, where it is found on cultivated land, in waste places, by roadsides and near ponds and ditches.

Vegetative characteristics. An almost glabrous, annual herb, 25–80 cm in height, with a branched, reddish, erect or ascending stem swollen at the nodes. The alternate leaves have fringed, sheathing stipules (ochreae) at their base. The leaf-blades are lanceolate, often with a large, dark blotch above, and sometimes tomentose beneath.

Floral formula. P5 A6 $\underline{G}$(2–3)

Flower and inflorescence. The small, hermaphrodite, actinomorphic, pink flowers are grouped together to form a cylindrical, continuous or somewhat interrupted, spicate inflorescence, which is borne terminally or in the axils of the leaves from June to October.

Pollination. The flowers of *P. persicaria* are small and odourless, but they are rendered more conspicuous by being clustered into cylindrical inflorescences. Various kinds of insects have been observed visiting the flowers for their rather scanty nectar which is secreted at the bases of the stamens. Although cross-pollination is likely, self-pollination is equally possible owing to the proximity of stamens and stigmas.

Alternative flowers for study. All the species of *Polygonum* found wild in the British Isles have small flowers, which may make dissection somewhat difficult. *P. aviculare*, Knotgrass, is widespread and abundant in cultivated and waste places, but its flowers are without nectaries. *P. bistorta*, Bistort, is rather less common, but is interesting on account of its markedly protandrous flowers. Some of the cultivated species have larger flowers, e.g., *P. baldschuanicum* and *P. aubertii*, both vigorous climbers frequently planted in gardens. The closely related genus *Fagopyrum* includes *F. esculentum*, Buckwheat, an introduced crop plant, which sometimes occurs as a casual on waste ground and has heterostylous flowers. Comparison should be made with members of the genus *Rumex*, Dock, which are adapted to wind-pollination by having large numbers of pendulous, slender-stalked flowers with prominent, brush-like stigmas.

Fig. 16 Polygonaceae, *Polygonum persicaria*

A Two of the cylindrical, spicate inflorescences borne on stout stems. Each of the alternate leaves has at its base a tubular, sheathing stipule or ochrea, characteristic of the family.
Inflorescence: 15–24 × 4–8 mm

B Three of the hermaphrodite flowers from the base of the inflorescence at different stages of development. Below the flowers is the fringed ochrea and at one side of the swollen portion of the stem the base of a flower-stalk can be seen.
Mature flower: 3.75 × 2 mm

C The perianth, opened out to show how the stamens are attached. The 5 imbricate segments are connate at the base and are all petaloid. The 6 stamens are adnate to the basal portion of the perianth.
Perianth-segment: 2 × 2 mm Stamen: 1.5 mm

D Detail of a dorsifixed anther, with pollen-grains adhering to the anther-cells. The anthers are 2-celled and dehisce longitudinally.
Anther: 0.2 × 0.3 mm

E The sessile ovary bearing 2 styles each terminated by a capitate stigma. Some flowers have 3 styles and the ovary then develops into a 3-angled fruit (see G2).
Ovary (at early anthesis): 1 × 0.8 mm Style: 1 mm

F L.S. of an acyclic, hermaphrodite flower at late anthesis, as shown by the shrivelling styles and the relative size of the stamen that is behind the superior ovary. The ovary is composed of either 2 or 3 fused carpels (2 in the specimen illustrated, as indicated by the 2 styles), forming a single loculus containing a solitary, erect ovule on a basal placenta. The reproductive organs are protected by the persistent perianth.
Ovary: 1.5 × 1 mm Ovule: 0.75 × 0.25 mm

G T.S. of ovary at fruit formation. G1: a convex or lenticular ovary which would be terminated by 2 styles. G2: a 3-angled or trigonous ovary which would be terminated by 3 styles. The pericarp encloses a large seed with a well-developed endosperm. The embryo can be seen in the corner of each seed.

H The complete fruit, a black, glossy nut, bearing the remains of the 2 withered styles and surrounded by the persistent perianth.
Fruit: 3 × 2 mm

Fig. 16

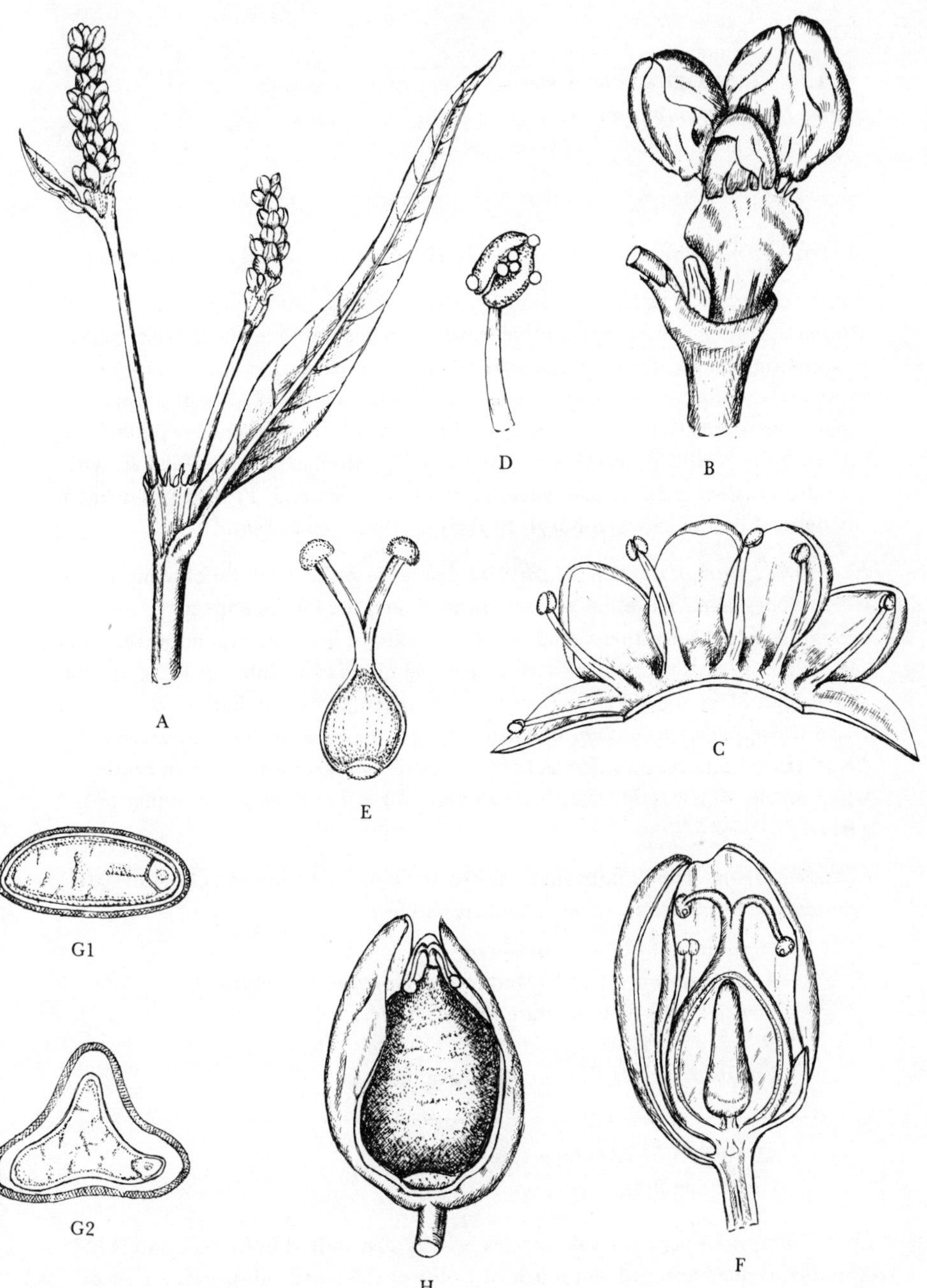

17 Plumbaginaceae Juss.

Sea Lavender family

19 genera and 775 species

Distribution. Cosmopolitan, especially along sea-shores and on salt steppes.

General characteristics. Perennial herbs or shrubs, sometimes scandent, with alternate, simple, entire, exstipulate leaves on whose surface occur water glands, or sometimes chalk glands (cf. *Saxifraga*). Inflorescence of various types, racemose and cymose. Flowers actinomorphic, hermaphrodite. Calyx of 5 united sepals, persistent. Corolla of 5 united petals, convolute. Stamens 5, epipetalous, opposite the corolla-lobes. Ovary superior, of 5 united carpels, unilocular, with 1 anatropous ovule on a basal placenta. Styles or stigmas 5. Fruit dry, 1-seeded, surrounded by the calyx. Embryo straight, in floury endosperm.

Economic and ornamental plants. Apart from the local use of some species of *Plumbago* and *Limonium* as food or medicine, the family appears to be of no economic value. Both these genera, however, are of horticultural importance. *P. capensis* and *P. rosea* are cultivated as semi-climbing subshrubs for the glasshouse and species of *Limonium*, Sea Lavender, are often planted in herbaceous borders. Dried flowering stems of *Limonium* (*Statice*) are used in floral decoration. *Ceratostigma plumbaginoides* and *C. willmottianum* are also grown in borders, while species of *Armeria*, Thrift, are suitable for either the border or the rock-garden.

Classification. The Plumbaginaceae are distinguished from the closely related Primulaceae by the solitary ovule and free styles.

Limonium (300) cosmopolitan.
Acantholimon (120) E. Mediterranean region to central Asia.
Armeria (80) N. temperate region, Andes.
Plumbago (12) warm regions.
Ceratostigma (8) E. tropical Africa, E. Asia.

ARMERIA MARITIMA (Mill.) Willd.

Thrift, Sea Pink

Distribution. A very variable species, widely distributed in Eurasia and N. America, though absent from much of E. Europe. Several subspecies are recognised. In the British Isles ssp. *maritima* is commonly found growing in coastal salt-marshes and pastures, on maritime rocks and cliffs and also on mountains inland. It is often cultivated in rock-gardens and borders for its attractive flowers and compact habit.

Vegetative characteristics. A tufted, cushion-like perennial with a basal rosette of linear, punctate, glabrous, single-veined leaves arising from a stout, woody rootstock.

Floral formula. K(5) C5 A5 $\underline{\text{G}}$(5)

Flower and inflorescence. An erect, usually shortly pubescent scape supports a terminal capitulum of fragrant, rose-pink or more rarely white, actinomorphic, hermaphrodite flowers from April to October.

Pollination. The conspicuous head of scented flowers is visited by various kinds of insects for pollen and nectar. The flowers are either homogamous or slightly protandrous. The latter condition obviously favours cross-pollination, but the homogamous flowers are also adapted for cross-pollination because of the dimorphism of their stigmas and pollen. In the case of those flowers with 'cob' stigmas, the stigmatic surface is covered with rounded bumps like a maize cob. Other flowers have papillate stigmas, the surface of these being composed of numerous, bluntly pointed papillae. Each type of flower has its pollen differently sculptured and is self-incompatible, requiring pollen from the other type of flower for effective pollination. Dimorphic stigmas and pollen occur not only in *A. maritima*, but in several other species of *Armeria* and also in certain species of the related genus *Limonium*.

Alternative flowers for study. *Armeria juniperifolia* and *A. pseudarmeria*, which are often cultivated, would be suitable alternatives. Comparison with other readily available genera, e.g., *Limonium*, *Plumbago* and *Ceratostigma*, will show differences in habit and type of inflorescence. In *Plumbago*, the persistent calyx, which aids seed dispersal, is glandular-hairy.

Fig. 17 Plumbaginaceae, *Armeria maritima*

A Upper portion of scape bearing the inflorescence, a capitulum made up of a close aggregate of 1- to 3-flowered, cymose spikelets (cincinni). Subtending the cincinni are scarious, scale-like bracts, the outermost ones forming an imbricate involucre. Several of the latter are prolonged downwards to form a tubular sheath surrounding the upper part of the scape.
Inflorescence: 15–20 mm in diameter Tubular sheath: 15–20 mm

B A single flower-bud at a late stage of development. The persistent calyx is funnel-shaped, and has 5 hairy ribs and a scarious, pleated limb. The calyx is inserted obliquely on the pedicel.
Ribs of calyx: 5–5.5 mm Calyx-teeth: up to 2 mm

C A single flower at anthesis with the 5 basally connate petals exposed beyond the funnel-shaped calyx. Below the flower are 3 scarious bracts, also 2 pedicels of aborted flowers.
Corolla: 8 × 4–5 mm

D L.S. of flower. The introrse stamens stand opposite the petals. The superior ovary has 5 styles (3 only are shown), which are hairy below and are situated opposite the calyx-teeth.
Stamens: 7–8 mm Styles: 5.5 mm Ovary: 1 × 0.75 mm

E Upper portion of a stamen with detail of anther. The anthers are versatile, 2-celled and dehisce longitudinally.
Anther: 2 × 0.5 mm

F The superior, unilocular ovary consisting of 5 united carpels. The basally connate styles are hairy towards the base.
Ovary: 1 × 0.75 mm

G L.S. of ovary. The single loculus contains a solitary ovule, pendulous from a basal funicle. Surrounding the ovary are the basally connate petals and hairy-ribbed calyx.

H The fruit at an early stage of development enclosed by the persistent calyx which, being light and membranous, will aid its dispersal when it is mature.
Fruit: 3 × 1 mm

Fig. 17

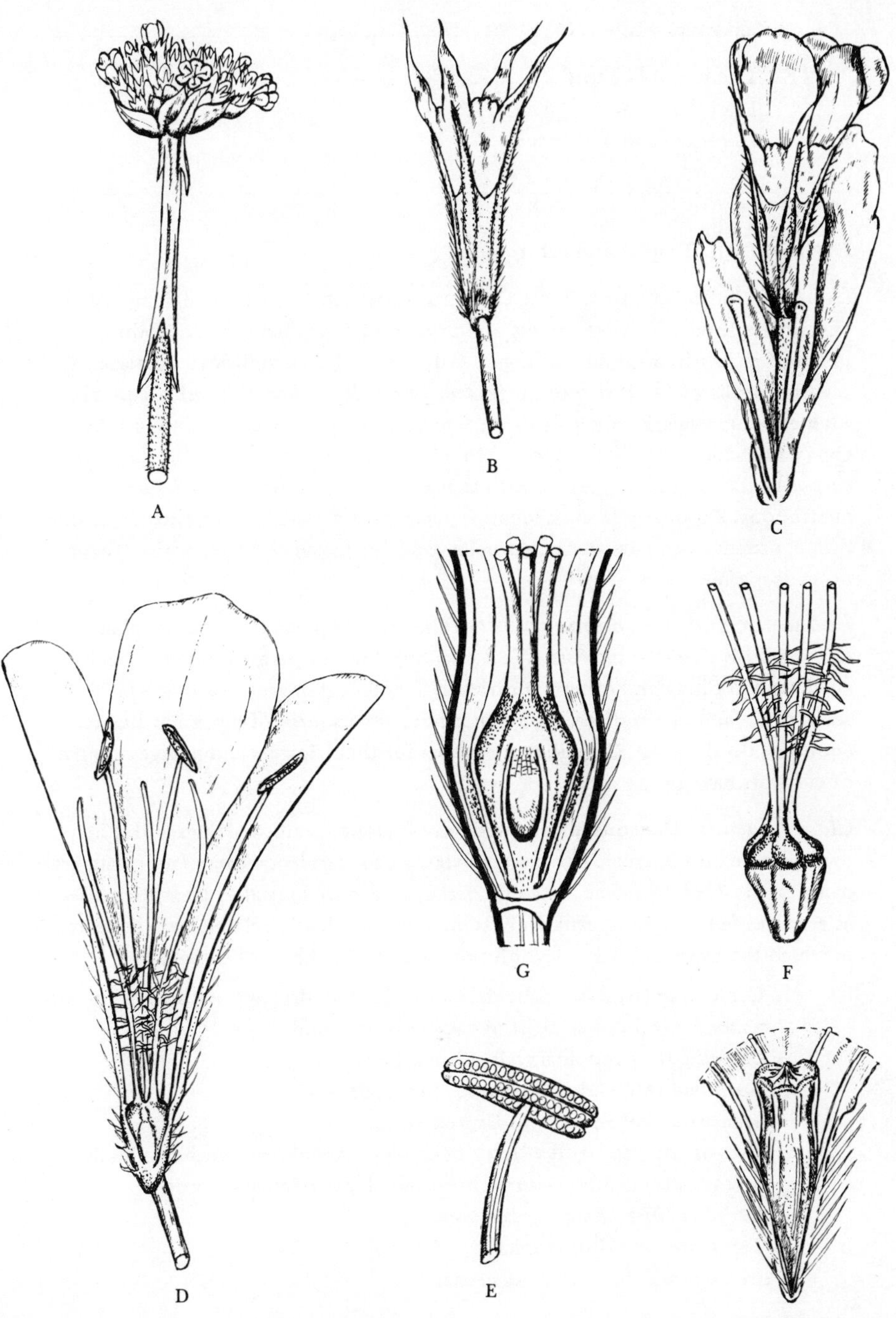

18 Theaceae D.Don

Tea family

18 genera and 550 species

Distribution. Tropics and subtropics.

General characteristics. Trees and shrubs with simple, alternate, leathery, exstipulate leaves. Flowers usually solitary, actinomorphic, hermaphrodite, perianth parts often spirally arranged. Calyx of 5–7 free, imbricate, persistent sepals. Corolla of (4–)5(–9 or more) free or basally connate, imbricate petals. Stamens numerous, rarely 5–10–15, free or in bundles or united into a tube. Ovary superior, rarely half-inferior (*Annesleya* and *Visnea*) or inferior (*Symplocarpon*) of 2–10 united carpels with as many free or united styles; ovules anatropous, 2 to many in each loculus; placentation axile. Fruit a capsule or dry drupe, usually with persistent columella. Embryo usually curved, with little or no endosperm.

Economic and ornamental plants. The most important economic plant in this family is *Camellia sinensis* (*Thea sinensis*), the Tea-plant, which has been cultivated in China and other countries in S. and E. Asia since early times. The seeds of *C. japonica* and certain other species are a source of oil in Far Eastern countries. In the West, Camellias are grown for their flowers and a large number of cultivars have been raised.

Classification. The concept of the family is along traditional lines and it is divided here into 2 tribes. However, in view of the tendency away from a superior ovary in the Ternstromieae, it would perhaps be more logical to treat the tribes as separate families, or alternatively to include the closely related Symplocaceae, in which the ovary is constantly inferior, as a third tribe of Theaceae.

1. Camellieae (fruit a loculicidal capsule, rarely drupaceous; anthers mostly versatile; embryo straight or radicle bent round).
   Camellia (82) Indomalaysia, China, Japan.
   Gordonia (40) China, Formosa, Indomalaysia.
   Stewartia (10) E. Asia, E. United States.
2. Ternstroemieae (fruit a berry or dry and indehiscent; anthers mostly non-versatile; embryo horseshoe-shaped, rarely almost straight).
   Eurya (130) E. Asia, Pacific area.
   Ternstroemia (100) tropics.
   Adinandra (80) E. and S.E. Asia.

CAMELLIA × WILLIAMSII W.W.Sm.

Distribution. *C.* × *williamsii* is a cross between *C. japonica*, a native of Japan and Korea, and *C. saluenensis*, which comes from western China. It was first produced in this country about 1930 by J.C. Williams at Caerhays Castle, Cornwall and by Col. Stephenson at Borde Hill, Sussex. The hybrid with its numerous cultivars are some of the most reliable of Camellias for planting in gardens on neutral and acid soils.

Vegetative characteristics. An evergreen shrub with glossy, dark green, leathery leaves which are elliptical or broadly elliptical and shallowly toothed.

Floral formula. K5 C5 A∞ $\underline{G}$3

Flower and inflorescence. The large, actinomorphic, solitary and erect flowers are almost sessile. The many cultivars vary considerably in the number (5 to many) and colour of their petals, which range through various shades of pink to pure white, and in flowering time, which occurs during the period November to May. In general, the flower-characters show the influence of *C. saluenensis*, while the vegetative features resemble *C. japonica*.

Pollination. Insects are attracted to the flowers by the conspicuous nature of the petals, and the numerous stamens. Nectar-secreting glands are absent.

Alternative flowers for study. *C. saluenensis*, which flowers in early spring, is nearest to *C.* × *williamsii* in floral structure, but failing this, any other species or single-flowered cultivar would be suitable. It should be noted that the flowers of *C.* × *williamsii* and *C. saluenensis* drop off when fading, while in the early to late spring-flowering *C. japonica* they remain on the stems.

Fig. 18 Theaceae, *Camellia* × *williamsii*

A Anterior view of the solitary, hermaphrodite flower, showing the 5 imbricate petals and the numerous stamens whose filaments are united towards the base into a fleshy cup.
Flowers: 7–9(–13) cm in diameter

B L.S. of flower. The relatively small sepals surround the basally connate petals. The partially connate stamens are adnate to the corolla. The superior ovary is surmounted by 3 connate styles which project well beyond the stamens. The floral parts tend to be spirally arranged.
Petal: 45 × 25 mm Stigma + style: *c.* 30 mm
Free portion of style: *c.* 9 mm

C Detail of a portion of the staminal cup showing the varying extent to which the filaments are united.
Anther + filament: *c.* 15 mm Free portion only: *c.* 12.5 mm

D Detail of upper portion of stamen. The anther is 2-celled and dehisces longitudinally.
Anther: 2 mm

E L.S. of the superior ovary at flowering time situated on the well-developed receptacle.
Ovary: 2 mm

F T.S. of ovary. Each of the 3 loculi contains 2 or more anatropous ovules on an axile placenta. The ovary in *C.* × *williamsii* is hairy like that of *C.* × *saluenensis*, in contrast to the ovary of *C. japonica* which is glabrous. *C.* × *williamsii* does not appear to fruit easily, and it has been observed that any fruit which has formed falls at an early stage. The parent species produce capsules which split into 3 valves to release the seeds.
Ovary: 2.75–3 mm in diameter

Fig. 18

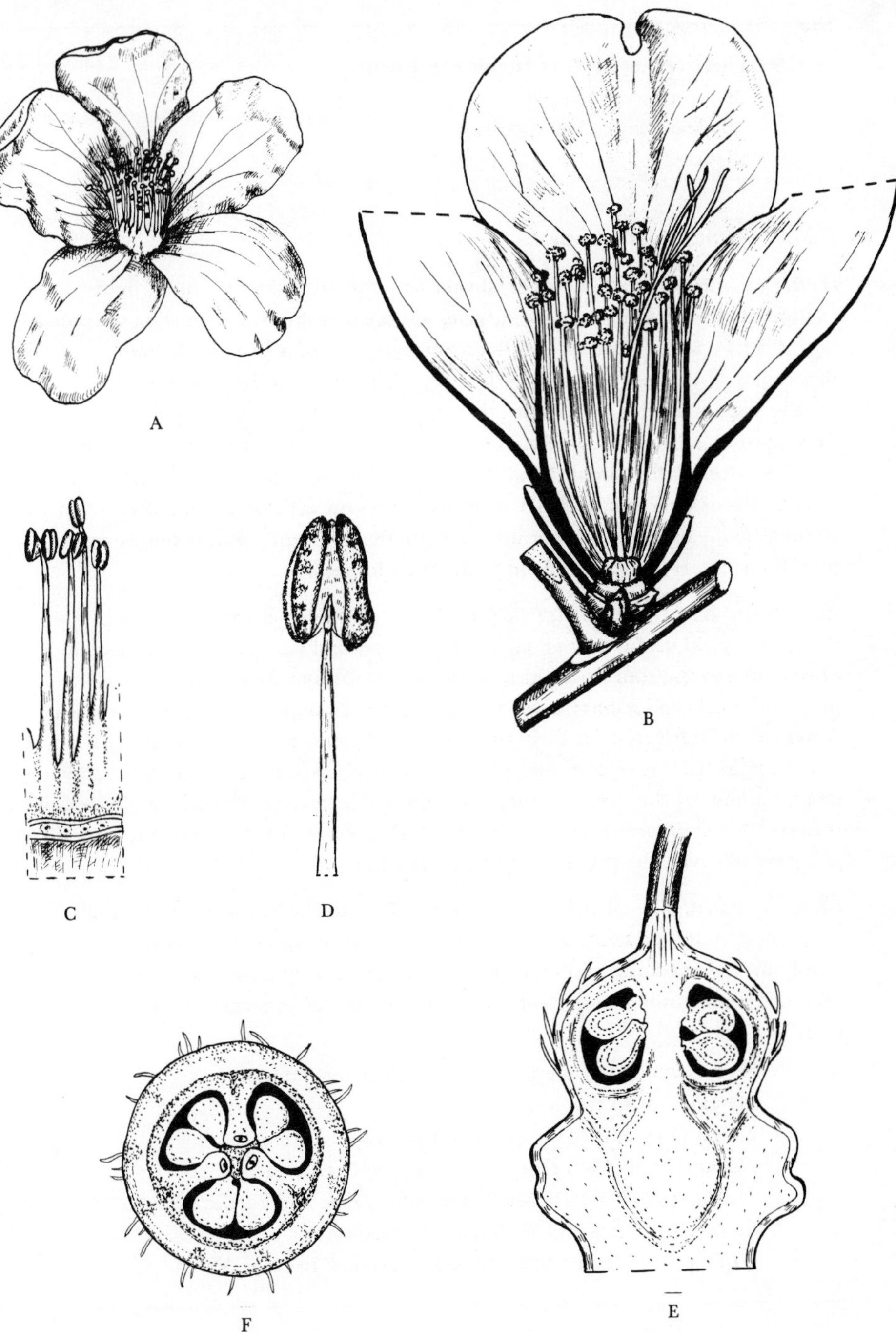

19 Guttiferae Juss. (Clusiaceae Lindl.)

St John's-wort family
40 genera and 1300 species

Distribution. Chiefly tropics.

General characteristics. Trees, shrubs or herbs with simple, entire, opposite, exstipulate leaves. Oil glands or passages always present, often showing as translucent dots upon the leaves. Inflorescence cymose, frequently umbellate. Bracteoles often close to calyx and hardly distinguishable from it. Flowers usually actinomorphic, hermaphrodite. Calyx of 2–10 free or united sepals. Corolla of 3–12 free petals. Stamens numerous, free or united, frequently in bundles, often partly staminodial. Ovary superior, of usually 3 or 5 united carpels, multi- or unilocular, with 1 to many ovules on axile or parietal placentas; styles as many as loculi, free or united, sometimes absent. Fruit often a capsule, sometimes a berry or drupe. Seed without endosperm.

Economic and ornamental plants. Many of the larger woody plants in the Guttiferae yield useful timber, e.g., *Calophyllum* and *Caraipa*. These genera and *Clusia* are also the source of resin. *Garcinia mangostana*, Mangosteen, a native of tropical Asia and *Mammea americana*, Mammee-apple, indigenous to tropical America, are cultivated for their fruit. The only genus of horticultural importance in the British Isles is *Hypericum*, some of the herbaceous species and probably a larger number of the shrubby ones being grown in gardens for their bright yellow flowers. *H. calycinum*, Rose-of-Sharon, which is more or less evergreen, is useful as a ground-cover, thriving even under large trees.

Classification. The Guttiferae are closely allied to the Theaceae, but are distinguished by the presence of oil glands. Some authorities have classified the family on various floral and vegetative characters into subfamilies and tribes, each containing only a comparatively small number of genera. The chief genera are:

Hypericum (400) temperate region, tropical mountains.
Garcinia (400) tropics.
Clusia (145) chiefly tropical and subtropical America.
Calophyllum (110) chiefly S.E. Asia to Pacific islands.
Mammea (50) chiefly Madagascar, S.E. Asia and Pacific islands, 1 in tropical America and W. Indies, *M. americana*.
Vismia (35) Mexico to tropical S. America, tropical Africa.

HYPERICUM PERFORATUM L.
Common St John's-wort

Distribution. Native in Europe, W. Asia, N. Africa, Madeira and the Azores, and common in grassland, hedgebanks and open woodland, especially on calcareous soils, in most parts of the British Isles.

Vegetative characteristics. An erect, glabrous perennial, with 2-ribbed stems, woody at the base. Its leaves are sessile, ovate to linear in shape and are abundantly furnished with translucent, glandular dots.

Floral formula. K(5) C5 A∞ $\underline{G}$(3)

Flower and inflorescence. The yellow, hermaphrodite, actinomorphic flowers are borne in branched cymes from June until September.

Pollination. The showy, homogamous flowers open to reveal 3 long styles and 3 bundles of stamens. Nectar is absent but many kinds of insects visit the flowers for the abundant pollen. The styles spread outwards between the groups of stamens, but at dehiscence the anthers do not touch the stigmas, though they are on the same level, so that pollination at this stage is dependent on insect visits. After anthesis, the petals and stamens are drawn inwards and the anthers, still covered with pollen, are brought into contact with the stigmas. In this way automatic self-pollination takes place if insect visits have failed. Recently, however, it has been found that apomixis occurs frequently in this species and indeed may be the usual form of reproduction.

Alternative flowers for study. Other species of *Hypericum* may show some variation from *H. perforatum*, mainly in the number of styles, bundles of stamens, and carpels. For example, the introduced species *H. calycinum*, Rose-of-Sharon, is entirely 5-merous, while the native *H. androsaemum*, Tutsan, has 3 styles, 5 bundles of stamens and an incompletely 3-celled ovary. The fruit of the latter is a red berry, turning purple-black when ripe, in contrast to the dry, dehiscent capsules of *H. perforatum* and *H. calycinum*.

Fig. 19 Guttiferae, *Hypericum perforatum*

A A flower-bud, showing the 5 short, subequal, entire, imbricate and basally connate sepals surrounding the corolla.
Sepals: 4 mm Flower-bud: 9 × 4.25 mm

B A fully opened flower, revealing the 5 persistent, yellow petals, the superior, 3-styled ovary and the stamens, which, in this species, are arranged in 3 bundles. The 3 free styles are alternate with the bundles of stamens.
Petals: 13 × 5.5–6 mm

C A fascicle of stamens detached from the flower. The long filaments are of unequal length and are connate near the base.
Free portion of stamen: up to 8 mm

D Detail of anther. The anthers are 2-celled and dehisce longitudinally, ripening at the same time as the stigmas.
Anther: 0.4 × 0.75 mm

E L.S. of ovary, with the basal parts of the surrounding organs. The numerous, anatropous ovules are attached to well-developed, axile placentas.
Style: 5 mm

F T.S. of the 3-locular ovary, showing the ovules attached to the axile placentas.
Ovary (at anthesis): 1.5 mm in diameter

G Detail of apex of style bearing a sessile stigma.
Stigma: 0.5 × 0.6 mm

H The fruit, a septicidal capsule, after dehiscence. The withering but persistent styles can be seen at the top of the 3 valves, and the persistent sepals at their base. The seeds are cylindrical, dark brown or black, and have a finely warted surface.
Fruit (prior to dehiscence): 8 × 7 mm

Fig. 19

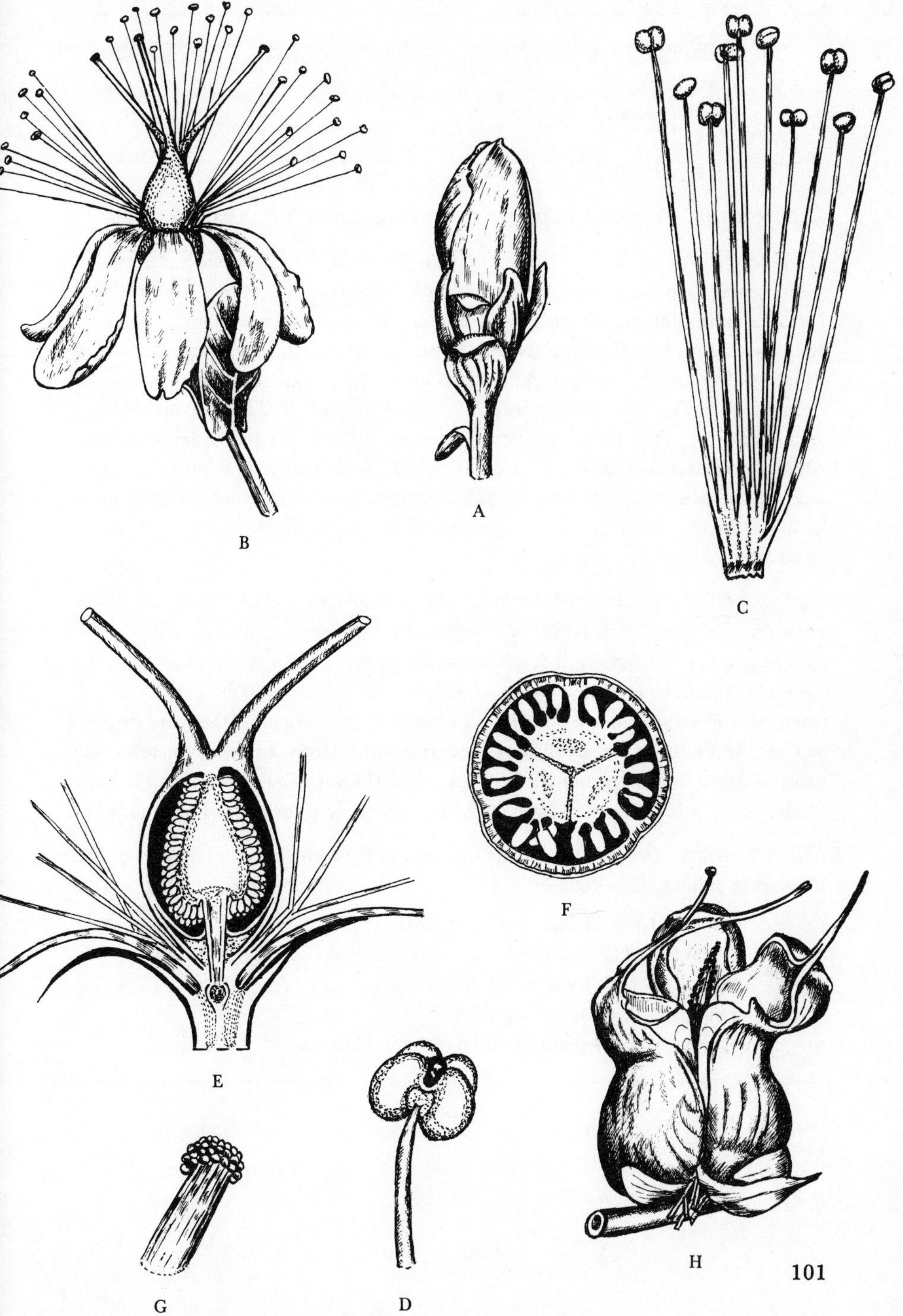

20 Tiliaceae Juss.

Lime family

50 genera and 550 species

Distribution. Tropical and temperate regions, chiefly S.E. Asia and Brazil.

General characteristics. Trees or shrubs, rarely herbs, often with stellate pubescence. Leaves usually alternate, simple, stipulate, sometimes asymmetric. Inflorescence cymose. Flowers usually hermaphrodite, actinomorphic, sometimes with epicalyx. Calyx of usually 5 free or united sepals, valvate. Corolla of usually 5 free petals, rarely absent, often glandular at base. Stamens usually numerous, free or united in groups, inserted at base of petals or on androphore; anthers 2-celled, dehiscing by apical pores or longitudinal slits. Ovary superior, of 2 to many carpels, as many loculi as carpels, with 1 to several more or less anatropous ovules in each loculus; placentation axile; style simple, with capitate or lobed stigma. Fruit usually a capsule or schizocarp; seed with straight embryo; endosperm scanty or copious.

Economic and ornamental plants. Several species of *Tilia*, Lime, are of economic importance in temperate regions for their timber, and 2 species of *Corchorus*, *C. capsularis* and *C. olitorius*, are widely grown in warm areas as the source of the fibre jute. A number of species and hybrids of *Tilia* are also of horticultural value and are planted for ornament in parks and large gardens, and as street trees, often on account of their fragrant flowers. In conservatories, the most widely grown member of the Tiliaceae is the shrub *Sparmannia africana*, House Lime, which bears clusters of white flowers with bright yellow stamens.

Classification. The most constant distinction from the closely related family Malvaceae is in the 2-celled anthers.

Grewia (150) Africa, Asia, Australia.
Triumfetta (150) tropics.
Corchorus (100) warm regions.
Tilia (50) N. temperate region.
Sparmannia (7) tropical and S. Africa, Madagascar.

TILIA PLATYPHYLLOS Scop.
Large-leaved Lime

Distribution. Native in central and S.E. Europe and W. Asia, and probably also in parts of the British Isles. This species is found in woods on limestone or base-rich soils and is often planted for ornament in parks and large gardens.

Vegetative characteristics. A deciduous tree with dark, fairly smooth bark and spreading branches, sometimes reaching a height of 30 m. Its leaves are broadly ovate, abruptly acuminate at the apex and obliquely cordate at the base, and have sharply serrate margins.

Floral formula. K5 C5 $A\infty$ $\underline{G}(5)$

Flower and inflorescence. The actinomorphic, hermaphrodite, yellowish white flowers appear in late June, forming pendulous, usually 3-flowered cymes. The axillary peduncles are adnate for about half their length to a pale, membranous bract (see Fig. 20.A).

Pollination. T. platyphyllos is pollinated mainly by various kinds of bees and by flies. These insects are attracted to the flowers by their strong scent and the copious nectar that collects in 2 small pits at the base of the sepals. The flowers are protandrous, favouring cross-pollination. Lime trees are considered by bee-keepers in many suburban areas to be the principal source of honey, and honey-bees may frequently be observed working the flowers. Some species of limes appear to contain in their flowers a substance which is toxic to bumble-bees.

Alternative flowers for study. Apart from *T. platyphyllos*, the limes most commonly found in Britain are the native *T. cordata*, Small-leaved Lime, and *T.* × *vulgaris* (*T.* × *europaea*), Common Lime, a hybrid between the other 2 species and almost certainly introduced into this country. These differ from *T. platyphyllos* in having cymes of 4–10 flowers, which begin to open in early July and in the fruits, which are not or only slightly ribbed. The familiar glass-house plant *Sparmannia africana*, which flowers in May, differs from *Tilia* in having the peduncles free, and 4-valved, dehiscent capsules as fruit.

Fig. 20 Tiliaceae, *Tilia platyphyllos*

A The inflorescence, a 3-flowered cyme. The peduncle is adnate in its basal half to a large, membranous bract. The bract is persistent, remaining while the fruits are formed, and acting as a wing to aid their dispersal.
Bract: 85–90 × 15–16 mm Peduncle: 30 mm
Pedicels: 9–20 mm

B An actinomorphic, hermaphrodite flower with some of the stamens and 3 of the perianth-segments removed to reveal the densely hairy, superior ovary. The 5 free sepals are valvate in bud and alternate with the 5 free petals. The numerous stamens are basally connate (see C3) and are longer than the perianth. The gynoecium consists of a superior ovary with a single style terminated by a 5-lobed stigma.
Flower: 15 mm in diameter Sepals: 6–7 × 3–3.5 mm
Petals: 7.5 × 3.5 mm Ovary: 3–3.5 mm
Style + stigma: 5.5 mm

C Detail of stamens. C1: ventral view of anther showing the 2 cells with longitudinal slits. C2: dorsal view of anther showing point of attachment of anther to filament. C3: basal portion of filaments showing fusion.
Stamens: 10 mm Anthers: 1–1.5 mm

D L.S. of ovary and receptacle with the bases of 4 stamens and 2 sepals (petals not shown). The shaded area on the sepals denotes the nectar-secreting glands. The hairy, superior ovary is cut open to expose 2 of the 10 ovules.

E Detail of stigma showing the 5 lobes (the same number as the fused carpels).
Stigma: 1.25 mm in diameter

F T.S. of ovary. The 5-locular ovary has axile placentation, with 2 ovules in each loculus.
Ovary (at anthesis): 3 mm in diameter

G L.S. of young fruit. One of the ovules is developing at the expense of the others. An abortive ovule can be seen squeezed to one side. The fruit forms an indehiscent nut with a single cell containing one seed. The pericarp is strongly 5-ribbed and woody. The whole infructescence is eventually shed with the membranous bract (see A).
Mature fruit: 7 × 5 mm Seed: 5 × 5.5 mm

Fig. 20

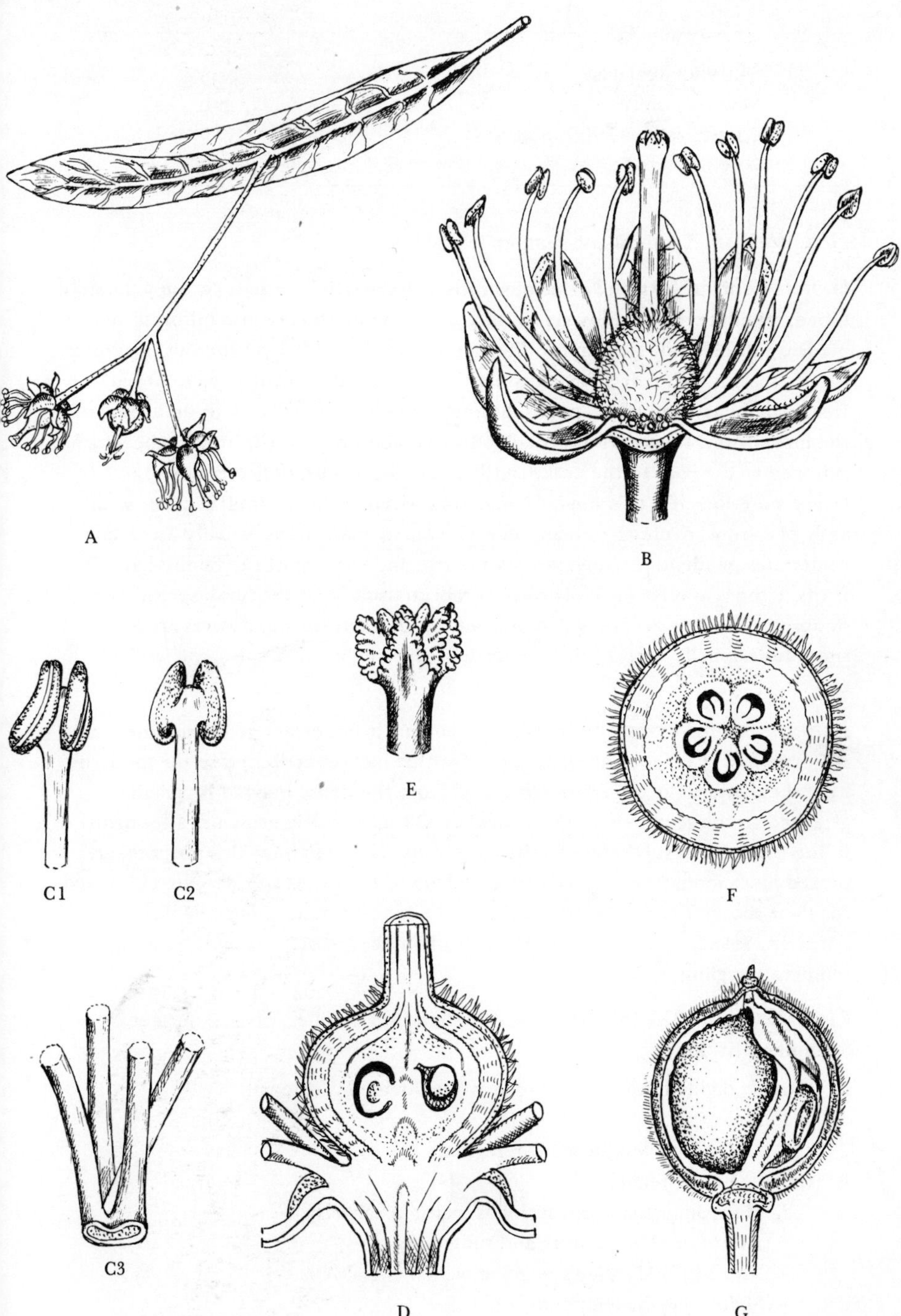

21 Malvaceae Juss.

Mallow family

75 genera and 1000 species

Distribution. Tropical and temperate regions.

General characteristics. Herbs, shrubs or trees with alternate, often palmately lobed, stipulate leaves. Flowers solitary or in compound cymose inflorescences made up of cincinni, hermaphrodite, actinomorphic, usually 5-merous. Epicalyx often present. Calyx of 5 free or united sepals, valvate. Corolla convolute, of 5 free, usually asymmetrical, petals. Stamens usually many owing to branching of the inner whorl (outer whorl usually absent), all united below into a tube which is joined at the base to the petals; anthers monothecous, pollen grains spiny. Ovary superior, of 1 to many united carpels (frequently 5), multilocular, with axile placentas; ovules 1 to many in each carpel, anatropous, usually ascending, sometimes pendent. *Malvaviscus* has a berry, and the rest of the family have dry fruits, either capsules or schizocarps. Embryo usually curved; endosperm scanty or absent. Flowers generally protandrous. In *Plagianthus* the flowers are sometimes functionally unisexual, the female with reduced petals, 1 carpel and 1 ovule.

Economic and ornamental plants. The most important genus commercially is *Gossypium*, Cotton, several species of which have contributed to the modern hybrid cultivars. Cotton is manufactured from the dense mass of long hairs attached to the seeds. *Hibiscus esculentus*, Okra, is widely grown as a foodstuff in the tropics and subtropics for its fruit or 'pods', which, for this purpose, are picked in an immature stage. Other members of the genus *Hibiscus* are cultivated for their showy flowers, while species of the hardier genera *Malva*, Mallow, *Lavatera*, *Sidalcea*, *Althaea* and *Alcea*, Hollyhock, are often found in gardens in temperate regions.

Classification. The following classification into 4 tribes is based on that by Schumann (in Ref. 9):

A. (Carpels divided into vertical rows of one-ovuled portions).

1. Malopeae.
   Malope (4) Mediterranean region.

B. (Carpels in one plane).

2. Malveae (schizocarp; styles as many as carpels).
   Abutilon (100) tropics and subtropics.
   Alcea (60) Mediterranean region to central Asia.
   Malva (40) N. temperate region.

Lavatera (25) Canary Islands, Mediterranean region to E. Asia, Australia, California.
Althaea (12) W. Europe to N.E. Siberia.
Anoda (10) tropical America.

3. Ureneae (schizocarp; styles twice as many as carpels).
Pavonia (200) tropics and subtropics.
Urena (6) tropics and subtropics.
4. Hibisceae (capsule).
Hibiscus (300) tropics and subtropics.
Gossypium (20) tropics and subtropics.

MALVA SYLVESTRIS L.
Common Mallow

Distribution. Native throughout most of Europe including the British Isles, and commonly found growing on waste places and roadsides.

Vegetative characteristics. A perennial herb with hairy, long-stalked leaves, which have 5–7 shallowly crenate lobes and are cordate at the base.

Floral formula. K(5) C5 A(∞) $\underline{G}(\infty)$

Flower and inflorescence. The flowers arise in clusters in the axils of the leaves. They are large, hermaphrodite and actinomorphic, with a tendency to asymmetry, and make an attractive show from June to September with their rose-purple, emarginate petals. Outside the calyx of each flower is an epicalyx consisting of 3 free segments.

Pollination. Many kinds of insects are attracted by the showy petals and by the nectar which is secreted at their base. The flowers are protandrous and in the first stage of anthesis the anthers form a pyramid at the centre of the flower above the immature stigmas. After the stamens have shed their pollen they bend back, exposing the developing stigmas which spread out and in their turn become the dominant feature at the centre of the flower.

Alternative flowers for study. M. sylvestris is easily obtainable in most localities, but *M. neglecta*, the Dwarf Mallow, which is normally self-pollinated, may be found in similar habitats, also from June to September. The genera *Althaea*, *Alcea* and *Lavatera* are often represented in gardens and could also be used, but comparison with genera in other tribes would be valuable, e.g., *Malope trifida*, a summer-flowering annual and *Hibiscus syriacus*, a hardy shrub which flowers from July to September. At other times of the year the flowers of *Hibiscus rosa-sinensis* and other species grown in glasshouses may be available.

Fig. 21 Malvaceae, *Malva sylvestris*

A The actinomorphic, hermaphrodite flower, with one of the 5 petals removed to reveal the withering anthers and, above them, the protruding stigmas. The 5 petals are twisted in bud and their bases, which are adnate to the staminal tube, bear fine hairs which protect the nectar-secreting pits. The 5 hairy sepals, which form the persistent calyx, are united in their lower half. Below the calyx can be seen part of the epicalyx, which consists of 3 free segments, one-third shorter than the calyx.
Flower: *c.* 32 mm in diameter Petal: 17 mm
Calyx-lobe: 3 × 3 mm

B L.S. of centre of flower, showing the staminal tube surrounding the connate styles which arise from the centre of the gynoecium. At the top, the developing stigmas are taking the place of the withering anthers.
Staminal tube: 8–9 mm Ovary (at anthesis): 1.25 mm

C T.S. of the superior ovary, showing the numerous loculi. Each loculus contains an anatropous ovule with axile placentation.
Ovary (at anthesis): 2 mm in diameter

D Detail of a portion of the central column formed by the connate styles terminating in the free stigmas, which are equal in number to the carpels.

E The fruit, a schizocarp which separates into many indehiscent one-seeded nutlets with reticulate markings (see G). Surrounding the nutlets is the persistent calyx.
Fruit: 7 mm in diameter

F The free, upper portion of 2 of the numerous stamens whose filaments are united below into a staminal tube, a condition described as monadelphous. The one-celled, reniform anther dehisces by a slit across the top. During dehiscence, the anthers turn outwards shedding their characteristic large, round, spiny pollen grains.
Free portion of stamens: 1–2 mm
Anthers: 0.5 × 0.5–0.75 mm

G A single, wedge-shaped nutlet showing the reticulate markings on the back. When the nutlets are ripe, they break away from each other and from the persistent central column formed by the apex of the receptacle.
Nutlet: 3 × 3 mm

Fig. 21

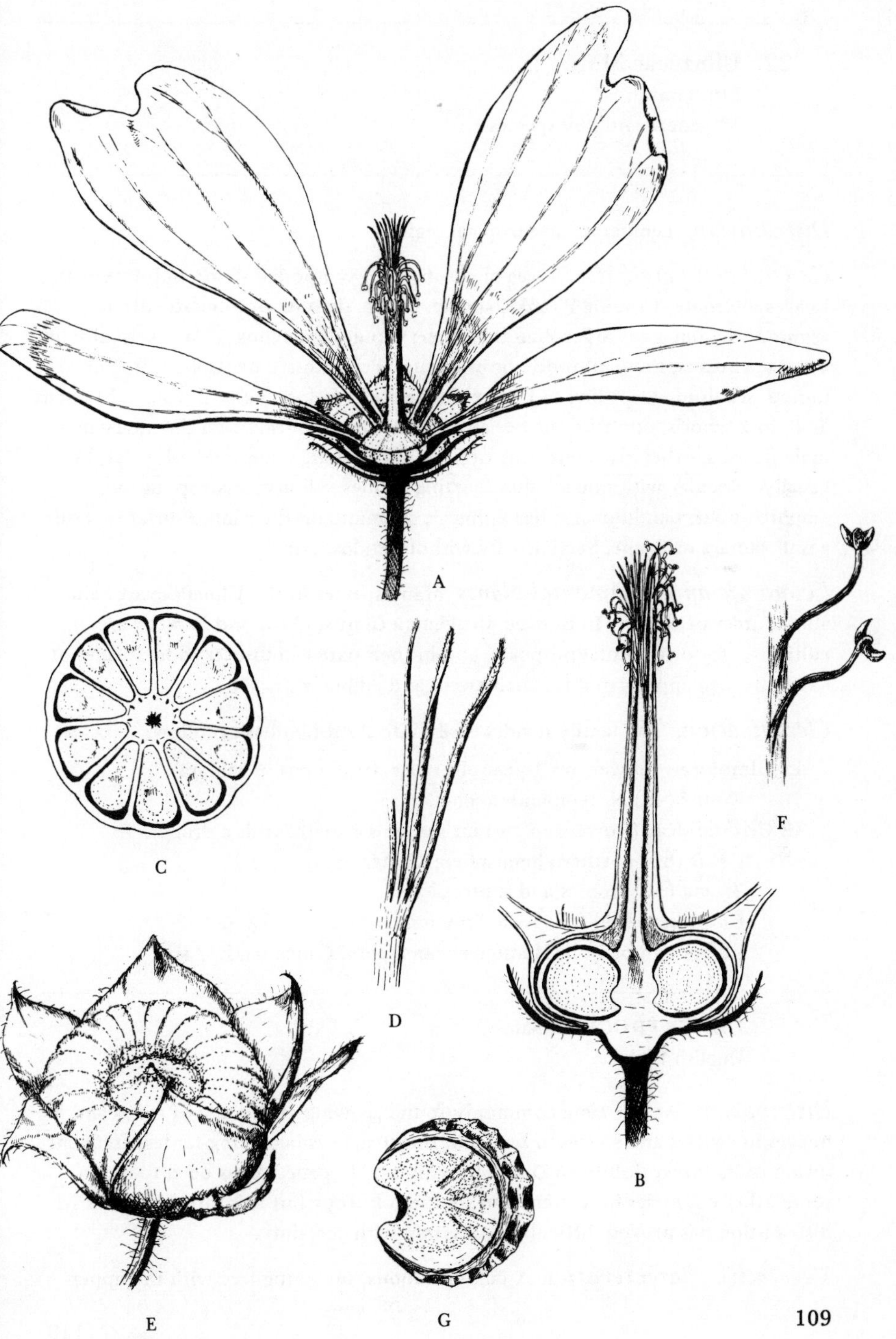

22 Ulmaceae Mirb.

Elm family

15 genera and 200 species

Distribution. Temperate and tropical regions.

General characteristics. Trees or shrubs with sympodial stems and watery sap. Leaves alternate, forming 2 ranks, simple, entire, dentate or lobulate, often asymmetric, one side larger than the other; stipules caducous. Flowers usually in cymose clusters, hermaphrodite or unisexual. Perianth-segments 4–8, free or united, sepaloid, according to Engler theoretically belonging to 2 whorls. Stamens 4–8, in 2 whorls, opposite the perianth-segments. Gynoecium rudimentary in male flowers, otherwise consisting of 2 united carpels, sometimes 2-locular, but usually 1-locular with one loculus aborting; ovules solitary, anatropous or amphitropous, pendulous; styles 2, linear, stigmatic on their inner surfaces. Fruit a nut, samara or drupe. Seed usually without endosperm.

Economic and ornamental plants. Many species of the Ulmaceae are valuable sources of timber. In Europe, the genera *Ulmus*, *Celtis* and *Zelkova* are cultivated for ornamental purposes, and in their native countries some species of *Celtis* are also appreciated for their sweet and edible fruit.

Classification. The family divides easily into 2 subfamilies as follows:

I. Ulmoideae (flowers on 1-year-old twigs, fruit a nut or samara).
   Ulmus (45) N. temperate region.

II. Celtidioideae (flowers on current season's growth, fruit a drupe).
   Celtis (80) northern hemisphere, S. Africa.
   Trema (30) tropics and subtropics.
   Gironniera (15) S.E. Asia, Polynesia.
   Zelkova (6 or 7) E. Mediterranean region, Caucasus, E. Asia.

ULMUS PROCERA Salisb.

English Elm

Distribution. At one time commonly found growing at the side of roads and in hedges in central and southern England, this species is becoming far less frequent owing to its susceptibility to Dutch Elm disease. *U. procera* has been found in some other countries in western and southern Europe but it is rare and its native distribution has proved difficult to ascertain with certainty.

Vegetative characteristics. A tall, deciduous, suckering tree with the upper branches spreading and forming a crown of rounded lobes. The suborbicular to ovate, toothed leaves are scabrid above, acute at the apex, and asymmetrical at the rounded or subcuneate base.

Floral formula. P4–5 A4–5 $\underline{G}$(2)

Flower and inflorescence. The small bundles or fascicles of flowers arise from the one-year-old twigs, emerging during the months of February and March before the leaves appear. The fruits develop during May and June.

Pollination. The flowers of the Elm family are all wind-pollinated, hence the usual attractions for insects, nectaries and a conspicuous, brightly coloured perianth, are missing. The relative position of the male and female parts of the flower promotes cross-pollination. The anthers are situated at the ends of long filaments, so that at dehiscence the pollen is less likely to fall on the styles of the same flower.

Alternative flowers for study. In the absence of *Ulmus procera*, *U. glabra*, Wych Elm, or *U. carpinifolia*, Small-leaved Elm, will suffice for the purpose of floral examination. These species, too, flower in February and March. *U. glabra* is scattered throughout the British Isles, but is commoner in the W. and N., while *U. carpinifolia* is found in many parts of S. and E. England but is less frequent outside this area.

Fig. 22 Ulmaceae, *Ulmus procera*

A A single, hermaphrodite flower separated from the inflorescence illustrated in Fig. 22.B. The campanulate, short-lobed perianth encloses the superior ovary and the surrounding stamens with their conspicuous red anthers. Unisexual flowers occur in other genera within the family, this condition being brought about by the abortion of stamens or carpels.

B A complete fascicle of flowers surrounded by the bud scales. The flowers themselves are borne in the axils of the inner scales.
Inflorescence: 8–10 × 10–16 mm

C L.S. of the carpel exposing the single, anatropous ovule suspended from the apical portion of the loculus. The 2 styles are divergent and bear stigmatic papillae on their inner surfaces. The carpel is 1.3 mm wide at anthesis.

D T.S. of ovary. A section has been taken at the point where the ovule is attached to the placenta, well above the centre of the ovary. The point of sectioning accounts for the reduced width of the ovule.

E The whole fruit forms a flattened, winged, single-seeded, dry structure known as a samara. Attached to the base of the samara are the persistent perianth-segments. The apex of the samara is slightly notched.
Fruit: 13–18 × 11–15 mm

F Detail of the upper portion of a stamen showing the 2-celled anther attached to an erect filament. The anthers open to liberate their pollen by splitting lengthwise. The stamens are of the same number as the perianth-segments and are situated opposite them.

Fig. 22

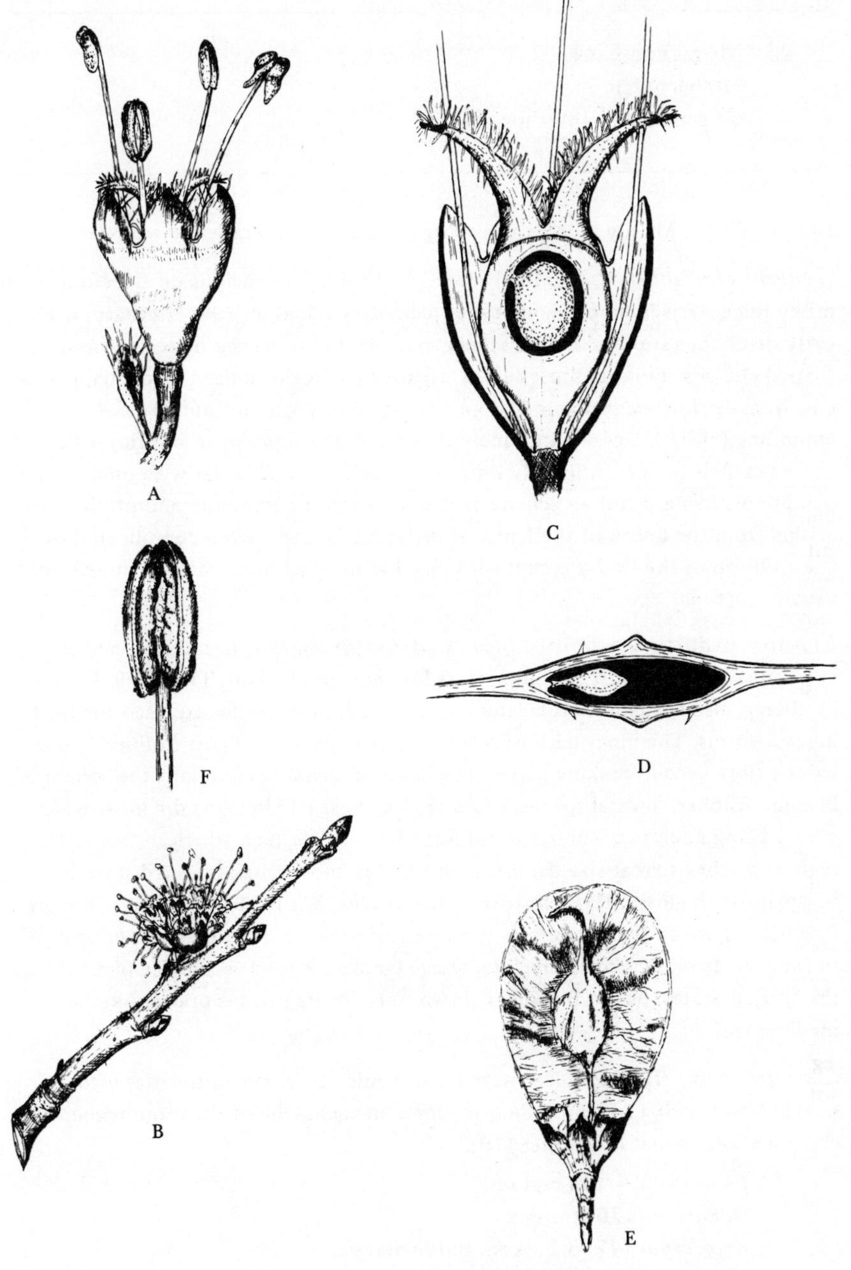

23 Moraceae Link

Mulberry family

53 genera and 1400 species

Distribution. Mainly tropics and subtropics, a few in temperate regions.

General characteristics. Mostly trees or shrubs, monoecious or dioecious, with milky juice. Leaves simple, sometimes palmately lobed, usually alternate, with early deciduous stipules. Inflorescence cymose, the flowers grouped in variously shaped clusters. Flowers unisexual. Perianth of 4 free or united segments, persistent. In male flowers, stamens 4, opposite perianth-segments, anthers not exploding (cf. Urticaceae). In female flowers, ovary superior to inferior, of 2 united carpels (of which usually only one develops), unilocular with one, usually pendulous, ovule. Fruit an achene or drupe-like, but commonly a multiple fruit arising from the union of the fruits of different flowers, often complicated by the addition of the fleshy receptacles. Seed with or without endosperm; embryo usually curved.

Economic and ornamental plants. *Artocarpus altilis*, Breadfruit, and *A. heterophyllus*, Jackfruit, are important for their edible fruit. The genus *Morus*, Mulberry, also has edible fruit, and the leaves of some species are used for feeding silkworms. The inner bark of *Broussonetia papyrifera*, Paper Mulberry, provides a fibre used in making paper. The latex of *Castilloa elastica* is the source of Panama Rubber. Several species of *Ficus*, Fig, bear edible fruit, the most widely grown being *F. carica.* Rubber is obtained from *F. elastica*, which in tropical regions reaches a great size. In this country it is known mainly as a decorative pot-plant with glossy, leathery leaves. *Morus alba*, White Mulberry, and *M. nigra* (see below) are cultivated in large gardens and parks as ornamental trees as well as for their fruit. *Maclura pomifera*, Osage Orange, is used as a hedge plant in the United States, but in Britain is grown only for its curious orange-like but inedible fruit.

Classification. The Moraceae have been divided by some authorities into several tribes with a small number of genera in each. One of the more recent classifications is that by Corner (10).

Ficus (800) warm regions.
Dorstenia (170) tropics.
Artocarpus (47) S.E. Asia, Indomalaysia.
Maclura (12) warm regions.
Morus (10) N. and S. America, tropical Africa, S. and E. Asia.
Broussonetia (7 or 8) E. Asia, Polynesia.

MORUS NIGRA L.
Black Mulberry

Distribution. Cultivated for so long that the original limits of its distribution have been obliterated. It is generally considered to have been native in W. Asia, and subsequently introduced into Europe where it has become naturalised in the southern areas. It is known to have been cultivated in Britain since the sixteenth century, and has been grown in parks and gardens mainly in the south of the country.

Vegetative characteristics. A deciduous, monoecious or dioecious tree up to 9 m in height with a dense, spreading head of branches and a short trunk. The leaves are alternate, glossy green above, broadly ovate, deeply cordate at the base and with dentate or lobed margins.

Floral formula. Male: P(4) A4 Female: P(4) $\underline{G}$(2)

Flower and inflorescence. The actinomorphic, unisexual flowers are borne in short, drooping inflorescences that arise during May from the axils of the leaves. The staminate flowers form catkins which fall away soon after they have shed their pollen, while the pistillate flowers are in spikes that later develop into fleshy, blackberry-like infructescences.

Pollination. When the anthers of the male inflorescence dehisce, their pollen is carried by the wind to the prominent paired stigmas of the female flowers. In the British Isles, the Black Mulberry is usually dioecious, female trees being more common than male. The female trees are parthenocarpic and are therefore able to produce fruit without being pollinated. The pollination of *Ficus carica*, Fig, is unusual and is worth a brief mention. There are 2 forms of this species, the edible fig and the wild fig (caprifig), both of which produce 3 generations of flowers per year inside a fleshy receptacle. The edible fig produces only fertile female flowers, but the wild fig has different kinds of flowers on each occasion. In late spring, male flowers develop near the apical opening of the receptacle and sterile female flowers (gall-flowers) elsewhere. In summer, fertile female flowers are produced, and in autumn only sterile female flowers. A small wasp, *Blastophaga psenes*, lays its eggs in early spring inside the gall-flowers. When the insects become mature, the winged females emerge from the caprifig, taking pollen from the male flowers as they go. On entering a caprifig of the summer generation or an edible fig, the pollen is transferred to the fertile female flowers and pollination takes place. Like the female Mulberry, some cultivars of the edible fig are parthenocarpic, these varieties being grown particularly where *B. psenes* is not found. A fuller account may be found in *The Pollination of Flowers* by Proctor and Yeo (1973) and an even more detailed one (in German) in Hegi's *Illustrierte Flora von Mittel-Europa.*

Alternative flowers for study. **M. nigra** is the most commonly cultivated species in the British Isles, though *M. alba*, White Mulberry, may occasionally be found. The family Moraceae contains many diverse and interesting genera, including the large and important genus *Ficus* (see above), but flowering specimens for study may be difficult to obtain.

Fig. 23 Moraceae, *Morus nigra*

A Two male inflorescences in the form of pendent, green catkins which arise from the leaf-axils. The flowers in the older inflorescence have begun to shed their pollen.
Male inflorescences: 20–30 × 10 mm

B A female inflorescence, drawn erect for clarity. Numerous flowers are set closely together forming a dense, green spike. Each of the flowers has 2 prominent, white stigmas. A well-developed bract arises on one side of the hairy peduncle. On the other, a section through the petiole indicates where a leaf has been removed.
Bract: 10 mm Peduncle: 6–7 mm
Female inflorescence: 8 × 6.5 mm

C Dorsal view of a single male flower cut in half at anthesis. The flower consists of 4 free stamens with versatile anthers arranged opposite the 4 connate perianth-segments. The filaments prior to anthesis are incurved, but do not subsequently explode as in the closely related family Urticaceae.

D L.S. of male flower showing the filaments of 2 of the 4 stamens opposite the perianth-segments.
Flower: 4 mm in diameter Perianth-segment: 4 mm

E Upper portion of stamen. The prominent filament bears a versatile, 2-celled anther that dehisces longitudinally.
Filament: 4 mm Anther: 1.5 × 1.75 mm

F A single female flower detached from the inflorescence. Two of the 4 perianth-segments can be seen surrounding the superior ovary which bears 2 spreading, white stigmas.

G L.S. of the female flower. The ovary consists of one carpel containing a single, anatropous, pendulous ovule. The ovary was originally 2-carpelled, but one carpel has been lost by abortion.
Ovary: 2 × 2.5 mm Styles: 2–4 mm

H The infructescence, an aggregate of drupes forming a syncarp. Both carpels and perianth-segments contribute to the formation of the fruits which ripen in late July. As they ripen, the fruits become orange-scarlet in colour, but they turn black just before falling.
Infructescence: 17 × 13 mm

Fig. 23

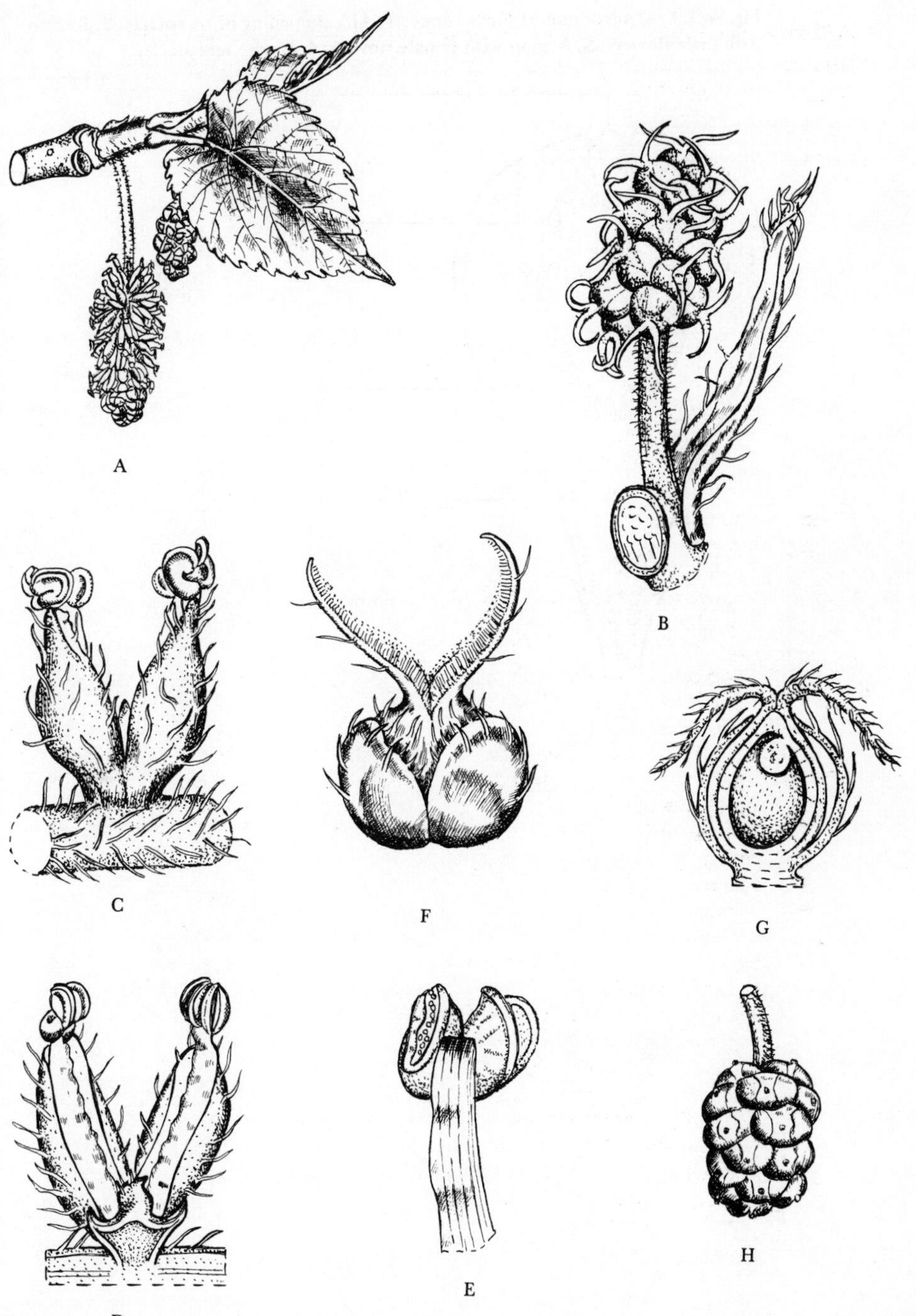

Fig. A. L.S. of syconium of *Ficus carica*: 1, Apical opening of receptacle; 2, Region with male flowers; 3, Region with female flowers; 4, Fleshy receptacle.

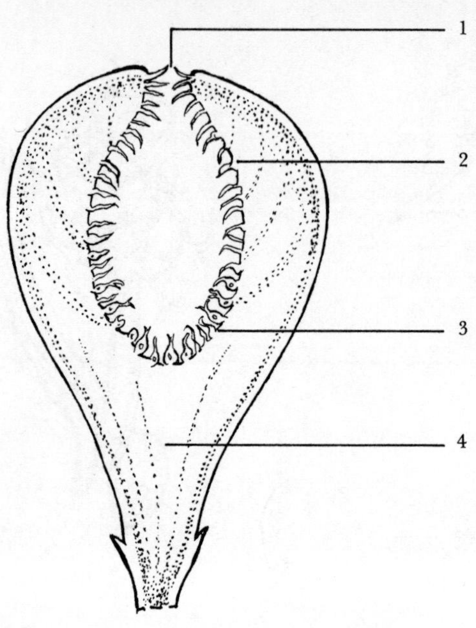

24 Urticaceae Juss.

Nettle family
45 genera and 1060 species

Distribution. Tropical and temperate regions.

General characteristics. Mostly herbs or undershrubs without latex. Leaves alternate or opposite, stipulate. Inflorescence cymose, often condensed. Flowers usually unisexual and actinomorphic. Perianth of 4 or 5 free or united segments, sepaloid. Stamens 4 or 5, straight (in the Conocephaleae) or bent down inwards in bud and exploding when ripe. Ovary superior, unilocular, with 1 erect, basal, orthotropous ovule (sometimes apical in the Conocephaleae) and 1 style. Fruit an achene. Seed usually with rich, oily endosperm; embryo small and straight.

Economic and ornamental plants. The family is of economic value as a source of fibre, the most important species in this respect being *Boehmeria nivea*, which is cultivated throughout most of the tropics and some of the cooler regions, and *Maoutia puya* which is confined to the tropical Himalaya and Burma. A few members of the Urticaceae are grown for their decorative foliage, chiefly species and cultivars of the genera *Pilea* and *Pellionia*, while *Soleirolia* (*Helxine*) *soleirolii*, native in the islands of the W. Mediterranean, is often cultivated for its habit of forming a dense, moss-like ground-cover.

Classification. The Urticaceae is closely related to the Ulmaceae, which differs in the form of the inflorescence, aestivation of stamens, position of ovule and type of fruit and to the Moraceae, which in addition to having a pendent ovule and a curved embryo, contains latex. The family is divided here into 6 tribes:

A. With stinging hairs. Perianth of female 4 to 5-merous.

1. Urticeae (Urereae).
   Urtica (50) mostly N. temperate, a few in tropical and S. temperate regions.
   Urera (35) Hawaii, warm parts of America, tropical and S. Africa, Madagascar.
   Laportea (23) tropics and subtropics, temperate E. Asia, eastern N. America, S. Africa, Madagascar.

B. No stinging hairs.

2. Procrideae (perianth of female 3-merous, stigma paintbrush-like).
   Pilea (400) tropics.
   Elatostema (200) Old World tropics.
   Pellionia (50) tropical and E. Asia, Polynesia.

3. Boehmerieae (male flower with usually 4 or 5 stamens; no involucre).
   Boehmeria (100) tropics and N. subtropics
   Maoutia (15) Indomalaysia, Polynesia.
4. Parietarieae (perianth present; bracts often united in involucre).
   Parietaria (30) temperate and tropical regions.
5. Forskohleeae (male flower reduced to 1 stamen)
   Forskohlea (6) Canary Islands, S.E. Spain, Africa, Arabia, W. India.
6. Conocephaleae (filament in male flower straight).
   Cecropia (100) tropical America.
   Poikilospermum (20) E. Himalaya to S. China and Malaysia.

URTICA DIOICA L.
Common Nettle

Distribution. A common plant throughout all the temperate regions of the world. In the British Isles it grows in a wide range of habitats, including hedge-banks, grassy places and near buildings from sea-level to over 800 m (2700 ft) on Ben Lawers in Scotland.

Vegetative characteristics. A hispid perennial with much-branched, yellow roots, creeping stems which root at the nodes and opposite, ovate, serrate leaves.

Floral formula. Male: P4 A4
Female: P(4) $\underline{G}$1

Flower and inflorescence. The plants are normally dioecious and numerous, small, unisexual flowers are borne in a spike-like inflorescence of clustered cymes from June until August.

Pollination. Nettle flowers are wind-pollinated and in the female flowers the ovary bears a tuft of stigmatic hairs for the reception of the pollen. While the male flower is in bud (see Fig. 24.B) the stamens are bent so far inwards that the anthers lie at the base of the flower. The tension increases until the resistance offered by the perianth is overcome, whereupon the filaments suddenly straighten, the anthers dehisce simultaneously and a small cloud of pollen is liberated to be carried by the wind on to the receptive stigma of a mature, female flower.

Alternative flowers for study. An alternative to *Urtica dioica* is scarcely needed, but a useful comparison may be made with the less common and rather local *U. urens*, Small Nettle, which is an annual and monoecious. The only other native member of the Urticaceae is *Parietaria diffusa*, Pellitory-of-the-wall, which also has unisexual flowers. All three have approximately the same flowering period.

Fig. 24 Urticaceae, *Urtica dioica*

A A portion of the stem with 3 complete inflorescences. The inflorescence, which consists of a spike of clustered cymes, is axillary and arises from a suppressed lateral shoot.
Inflorescence: up to 10 cm

B The actinomorphic, male flower in bud. There is no division of the perianth into calyx and corolla. The 4 green perianth-segments are free and equal in size.
Male flower (in bud): 1 mm in diameter

C The male flower at anthesis. The perianth-segments have been pushed open by the stamens; the filaments are now fully extended and the anthers have discharged their pollen. In the centre is the rudimentary ovary.
Male flower (at anthesis): 3 mm in diameter

D The upper portion of a stamen, before (D1) and after (D2) dehiscence. The anthers are 2-celled and dehisce longitudinally.
Filament: 2.25 mm Anther: 0.75 mm

E The female flower. The 4 unequal perianth-segments are united at the base; the 2 larger segments, which envelop the gynoecium, have been removed so as to expose the superior, unilocular ovary, which contains a single, orthotropous ovule. The ovary is surmounted by a large, brush-like stigma for catching the wind-borne pollen grains.
Ovary (at anthesis): 0.6 × 0.4 mm Stigma: 0.4 mm

F T.S. of the mature ovary showing the single, centrally-placed ovule.

G1 The 4 hairy perianth-segments, the 2 larger of which enclose the fruit.

G2 One of the larger segments has been removed to reveal the achene with its persistent, withered stigma.
Fruit: 1.2 × 1 mm

Fig. 24

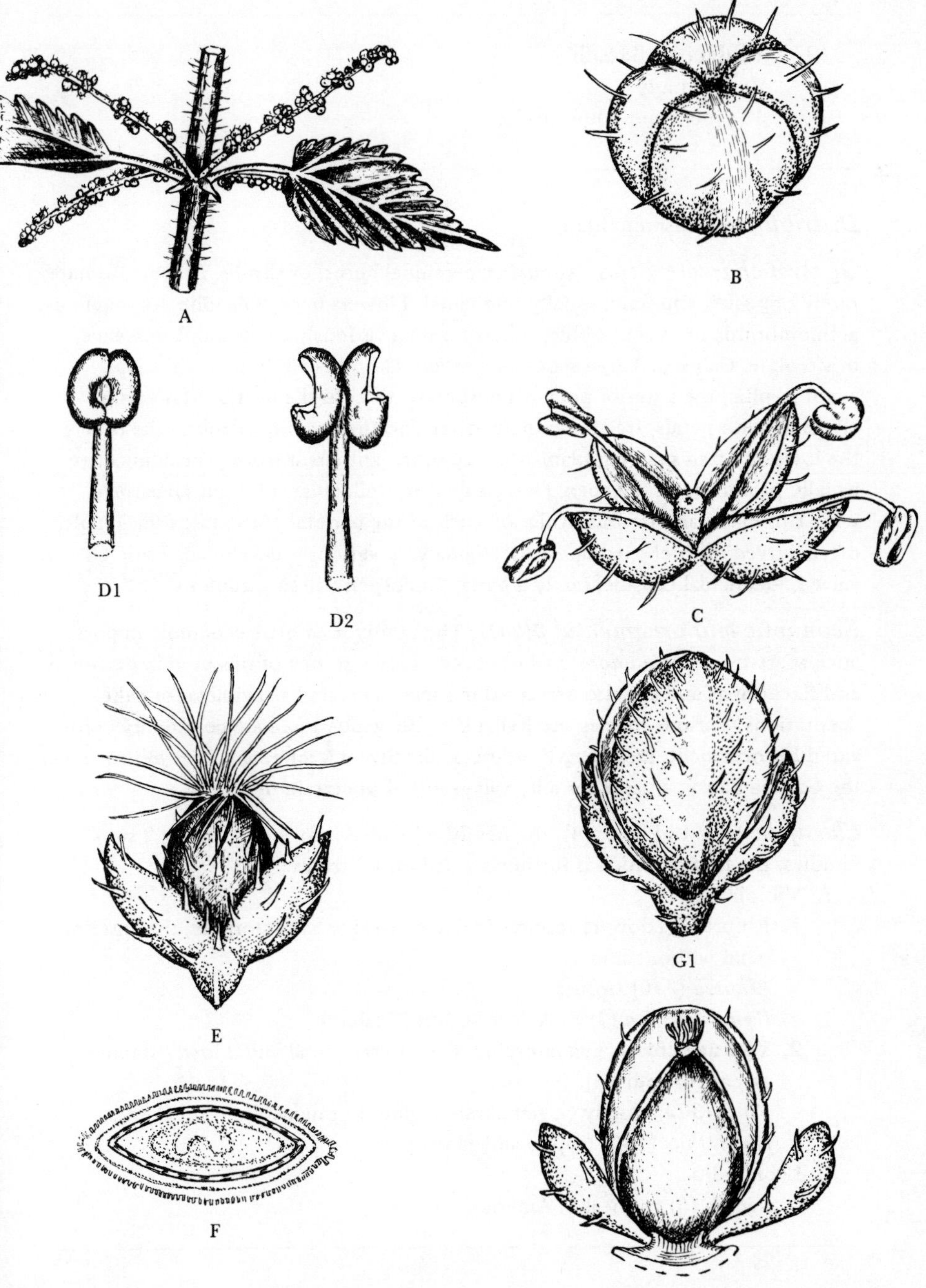

25 Violaceae Batsch.

Violet family

22 genera and 1000 species

Distribution. Cosmopolitan.

General characteristics. Annual or perennial herbs, or shrubs. Leaves alternate, rarely opposite, stipulate, usually undivided. Flowers hermaphrodite, hypogynous, actinomorphic or zygomorphic, solitary or in a variously formed inflorescence, bracteolate. Calyx of 5 free sepals, persistent. Corolla of 5 free petals, usually zygomorphic, the anterior petal often spurred to hold the nectar. Stamens 5, alternate with petals, free or connate, often forming a cylinder round the ovary, the lower 2 often spurred; filaments very short, anthers introrse, the connective usually prolonged at the apex. Ovary superior, unilocular, of 3 united carpels, with 1 to many anatropous ovules on each of the parietal placentas; style simple, often curved or thickened upwards; stigma very variously developed. Fruit a 3-valved, loculicidal capsule, rarely a berry. Endosperm fleshy, copious.

Economic and ornamental plants. The family is of little economic importance apart from *Viola odorata* whose flowers are a source of oils used in perfumes and flavourings and are also preserved in sugar as crystallised violets for cake decorations. Various species and hybrids of the genus *Viola* are commonly cultivated for ornament, including *V. odorata*, the Sweet Violet, and *V.* × *wittrockiana*, the Garden Pansy, a name for a hybrid group of uncertain origin.

Classification. In Engler (4), the family is divided into the following 2 subfamilies, the first of which is further divided into 2 tribes:

I. Violoideae.

1. Rinoreeae (flowers actinomorphic or weakly zygomorphic, the anterior petal without spur).
   Rinorea (340) tropics.
   Hymenanthera (7) E. Australia, New Zealand.
2. Violeae (flowers zygomorphic, the anterior petal and 2 lower stamens saccate or spurred).
   Viola (500) chiefly N. temperate region, but many from the Andes.
   Hybanthus (150) tropics and subtropics.

II. Leonioideae.

Leonia (6) tropical S. America.

VIOLA RIVINIANA Rchb.

Common Dog-violet

Distribution. A common plant throughout most of Europe including the British Isles, also found in the Atlas Mountains of Morocco and in Madeira. It grows on all types of soils, providing they are not too wet, and may be found in pastures, heaths, hedgebanks, woods and on mountain rocks.

Vegetative characteristics. A perennial herb with alternate, cordate, shallowly toothed leaves with long petioles.

Floral formula. K5 C5 A5 $\underline{G}(3)$

Flower and inflorescence. The solitary, blue-violet, zygomorphic flowers are usually found from April until June. Occasionally the plant produces a few flowers during the months of August to October.

Pollination. Cross-pollination always takes place in the normal flowers of *V. riviniana*, even though the plant is self-fertile. Insects, mainly bees, are attracted to the flower by the colour, scent and the presence of nectar, which is secreted by the spurs of the anterior anthers and then stored in the spur of the front petal. The stigmatic surface is so designed that it only allows pollen from another flower to be deposited on it, not that from the same flower. The bee alights on the flower and directs its proboscis towards the nectar situated just below the stigma which lies along a groove in the anterior petal. As the proboscis passes down the groove, it becomes coated with pollen from the anthers; on being withdrawn, the proboscis presses back a small valve on the stigmatic surface, preventing this pollen from being deposited on the receptive area. On entering a second flower, the bee deposits the pollen on the upper surface of the valve (the receptive area of the stigma) and in this way cross-pollination is accomplished. Self-pollination takes place in apetalous flowers which are fertile, but small and inconspicuous. These flowers, which never open, but remain in a bud-like state, are termed cleistogamous. They are produced in summer after the normal flowers have ripened seed.

Alternative flowers for study. Any other *Viola* species would be suitable but some of these may begin to flower later than *V. riviniana.*

Fig. 25 Violaceae, *Viola riviniana*

A L.S. of flower. The flower is supported on a curved pedicel which has 2 bracteoles attached. Two of the 5 petals have been removed to reveal 2 of the 5 stamens surrounding the ovary. The 2 anterior stamens have long, nectar-secreting spurs which protrude into the hollow spur of the front petal. The anthers are introrse and are attached to a short filament. The corolla is surrounded by 5 sepals which are prolonged below their point of insertion.
Inflorescence: 12 mm from bracteole to flower. Bracteole: 5.9 mm
Calyx: Long spurred sepal 8 mm Short spurred sepal 7 mm
Corolla: Spur base to tip 18 mm Spur base to receptacle 5.6 mm
Anterior petal: 14 × 5.2 mm Lateral petals: 14 × 6 mm

B Detail of a section of the anterior petal showing the hairs which line the groove in which the stigma lies.
Hairs: *c.* 0.75 mm

C Dorsal view of the 2 anterior anthers showing the spurs. Above the pollen-cells of each anther is a triangular structure, orange in colour, which is an apical extension of the connective. Behind the anthers can be seen part of the stigma and style.
Spurred stamen: Extension of connective: 3.2 mm
Spur: 4.5 mm

D Ventral view of a single anther showing the dehiscence of the 2 pollen sacs; the elongated connective can be seen above.

E T.S. of the single-chambered ovary with the 3 parietal placentas. The ovary is formed by the fusion of 3 carpels.
Ovary (at anthesis): 3 mm in diameter.

F L.S. of part of the ovary showing the large ovules attached to a placenta. The place of insertion of one of the petals can be seen at the base of the ovary.
Ovary (at anthesis): 2–2.45 mm

G Part of the ovary cut away to expose the numerous ovules and also details of the style and stigma. The style is thickened above, and has a hooked tip with a rather wide stigmatic opening.
Stigma + style: 2.1 mm Stigmatic beak: 0.3 mm
Ovules: 0.2–0.3 mm in diameter.

H The mature fruit (capsule) prior to dehiscence, surrounded by the 5 sepals, which are prolonged into appendages below their point of attachment.
Fruit: 8 × 4.75 mm

I Dehiscence of the mature fruit into 3 boat-shaped valves, the sides of which contract on drying and eject the hard, smooth seeds one by one. In some species, e.g., *V. odorata*, the seeds are provided with an elaiosome which is attractive to ants and aids dispersal.
Valve: *c.* 6.5 mm Seed: 2 × 1 mm

Fig. 25

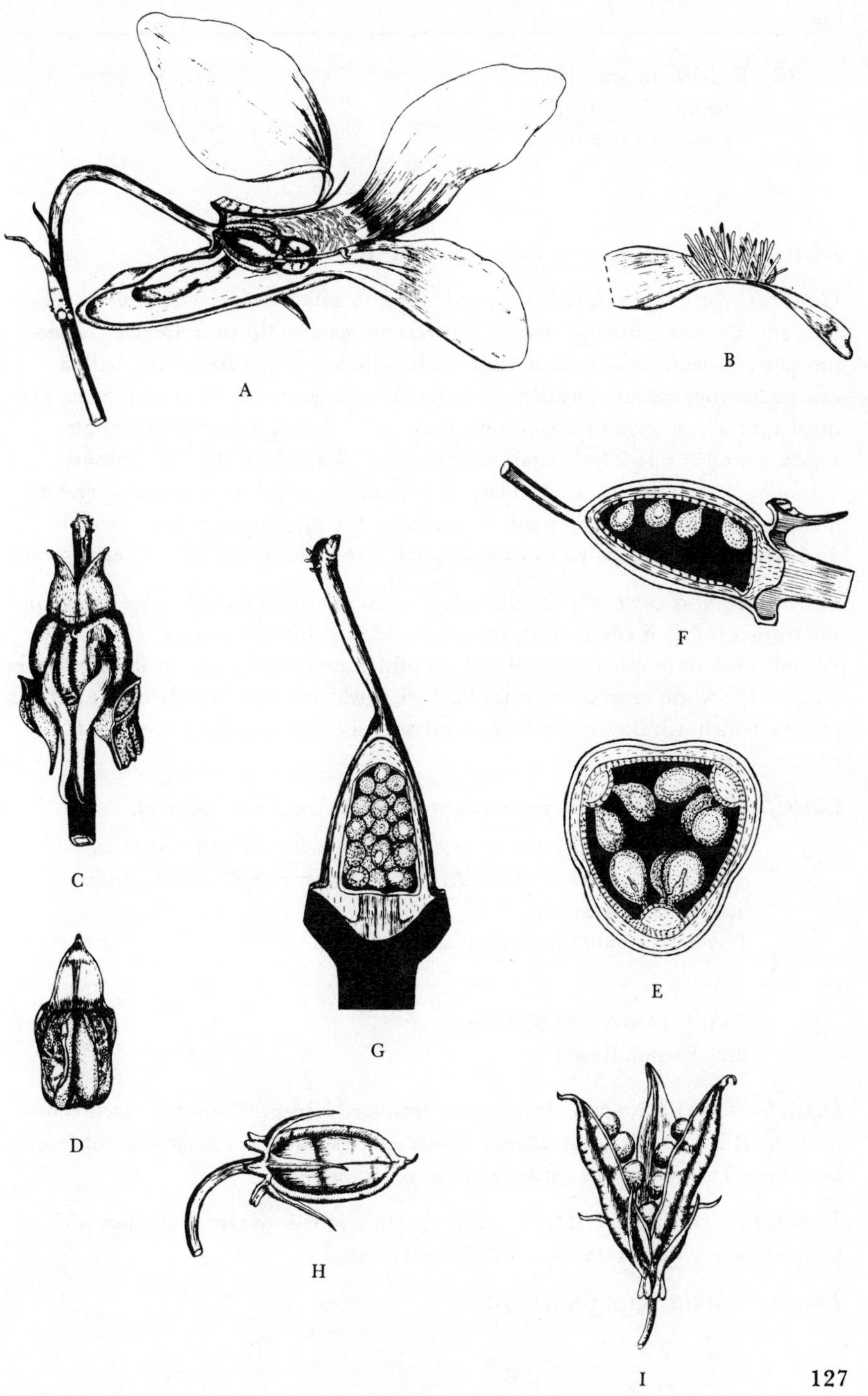

26 Passifloraceae Juss.

Passion-flower family

12 genera and 650 species

Distribution. Tropical and warm temperate regions.

General characteristics. Shrubs and herbs, mostly climbers with axillary tendrils and alternate, stipulate leaves. Flowers hermaphrodite or unisexual, actinomorphic. Receptacle of various shapes, often hollowed and frequently with a central androgynophore; usually terminated by outgrowths of petaloid or staminodial appearance forming the corona. Calyx of 3–5 free or united, imbricate sepals. Corolla of 3–5 free petals, rarely absent. Stamens 3–5(–10), anthers versatile. Ovary superior, unilocular, of 3–5 united carpels containing several or many anatropous ovules on parietal placentas. Style 1, simple or branched, or 3–5 separate styles. Fruit a capsule or berry. Seed with fleshy aril and endosperm.

Economic and ornamental plants. Several species of *Passiflora* are grown in the tropics for their edible fruit, the most widely cultivated being *P. edulis*, the Passion Fruit or Purple Granadilla. Many others are grown outdoors in the warmer parts of the world or in glasshouses for their attractive flowers. One of the hardiest species, which will flower and fruit in suitably sheltered British gardens, is *P. caerulea* (see below).

Classification. The family is a small one and not usually subdivided.

Passiflora (500) chiefly America, but a few in Asia and Australia.
Adenia (92) tropical and S. Africa, Madagascar, S.W. Arabia, Indomalaysia, N. Australia.
Tryphostemma (35) tropical and S. Africa.

PASSIFLORA CAERULEA L.

Blue Passion-flower

Distribution. Native in S. Brazil and introduced into Britain in the seventeenth century. It is a popular, but slightly tender species which can be grown outdoors in sheltered aspects of the garden or in a cool glasshouse.

Vegetative characteristics. A vigorous, more or less evergreen climber with palmate, 5- or 7-lobed leaves and axillary tendrils.

Floral formula. K(5) C5 A5 $\underline{G}$(3)

Flower and inflorescence. The showy, hermaphrodite, actinomorphic flowers arise from the axils of the leaves and bloom during the summer and early autumn. The inflorescence is a modified dichasial cyme in which the central flower is represented by a tendril, one of the lateral outgrowths develops into a flower, and the opposite lateral outgrowth remains undeveloped. The 3 bracts on the flowering peduncle persist for a while into the fruiting stage (see Fig. 26.G). The prominent corona and androgynophore are distinctive features of the flower.

Pollination. The corona has the appearance of differently coloured concentric rings which act as nectar-guides to bees (and in America to humming-birds also) and lead them to the nectar secreted by a ring within the operculum (see Fig. 26.B). In the first stage of anthesis the insect probing for nectar receives pollen on its back from the downward curving anthers. Later, the styles bend down even further than the anthers and the now receptive stigmas are able to collect pollen from other insect visitors.

Alternative flowers for study. The genus *Passiflora* contains a wide range of species for observation and comparison. For example, *P. antioquiensis* has a more elongated androgynophore than many but a relatively inconspicuous corona, while *P. edulis* has a shorter central column but a corona which extends almost to the tips of the petals; *P. racemosa* bears its red flowers in terminal racemes of 8–13, in contrast to the solitary flowers borne by most other species. The flowering period is generally the summer and early autumn.

Fig. 26 Passifloraceae, *Passiflora caerulea*

A Anterior view of flower. The outer whorl of basally connate, white or pinkish perianth-segments is made up of 5 sepals, each with a small horn near the tip. The inner whorl, alternate with the sepals, consists of the 5 petals. Radiating from points nearer the centre are the narrow thread-like structures or filaments, collectively called the corona. The longer blue-tipped filaments, which form the outer ring of the corona, are termed the radii, while the shorter ones are known as the pali. Arising from the centre of the flower is a column bearing both male and female reproductive organs, the androgynophore.
Flowers: 7–10 cm in diameter Sepals and petals: *c.* 34 × 12 mm
Radii: 14 mm

B L.S. of the central part of the flower. At the base are the bracts, and above these the hollow cup-like structure termed the hypanthium which bears the perianth-segments, corona and the operculum, a delicate membrane which acts as a cover over the nectar-secreting ring. At the base of the central column and sheltered by the operculum, is a rim, probably staminodial in nature, called the limen. Higher up the column are the 5 stamens and above them the ovary surmounted by 3 styles with capitate stigmas (see diagram).
Androgynophore: *c.* 13 mm Filament: 9 mm
Anther: 13 mm Style + stigma: 15 mm

C Detail of the upper portion of a stamen, showing the versatile anther attached to the filament. The anther is 2-celled and dehisces longitudinally.

D T.S. of the superior, unilocular ovary. The numerous ovules are attached to 3 parietal placentas.
Ovary (at anthesis): 4 × 2 mm

E T.S. of a portion of a mature fruit, showing a placenta with mature seeds attached to it by well-developed funicles. The seeds are covered by a fleshy, bright red aril.
Seeds (including aril): 8–9 × 4 mm

F Detail of upper portion of a style and its capitate stigma.

G The whole fruit, a many-seeded berry, which when ripe has an orange-coloured skin. The berry is edible but less palatable than fruits of *P. edulis*. Attached to the fruit-stalk are the 3 persistent bracts.
Fruits: 40 × 25–30 mm Bract: 15 × 15 mm

Fig. 26

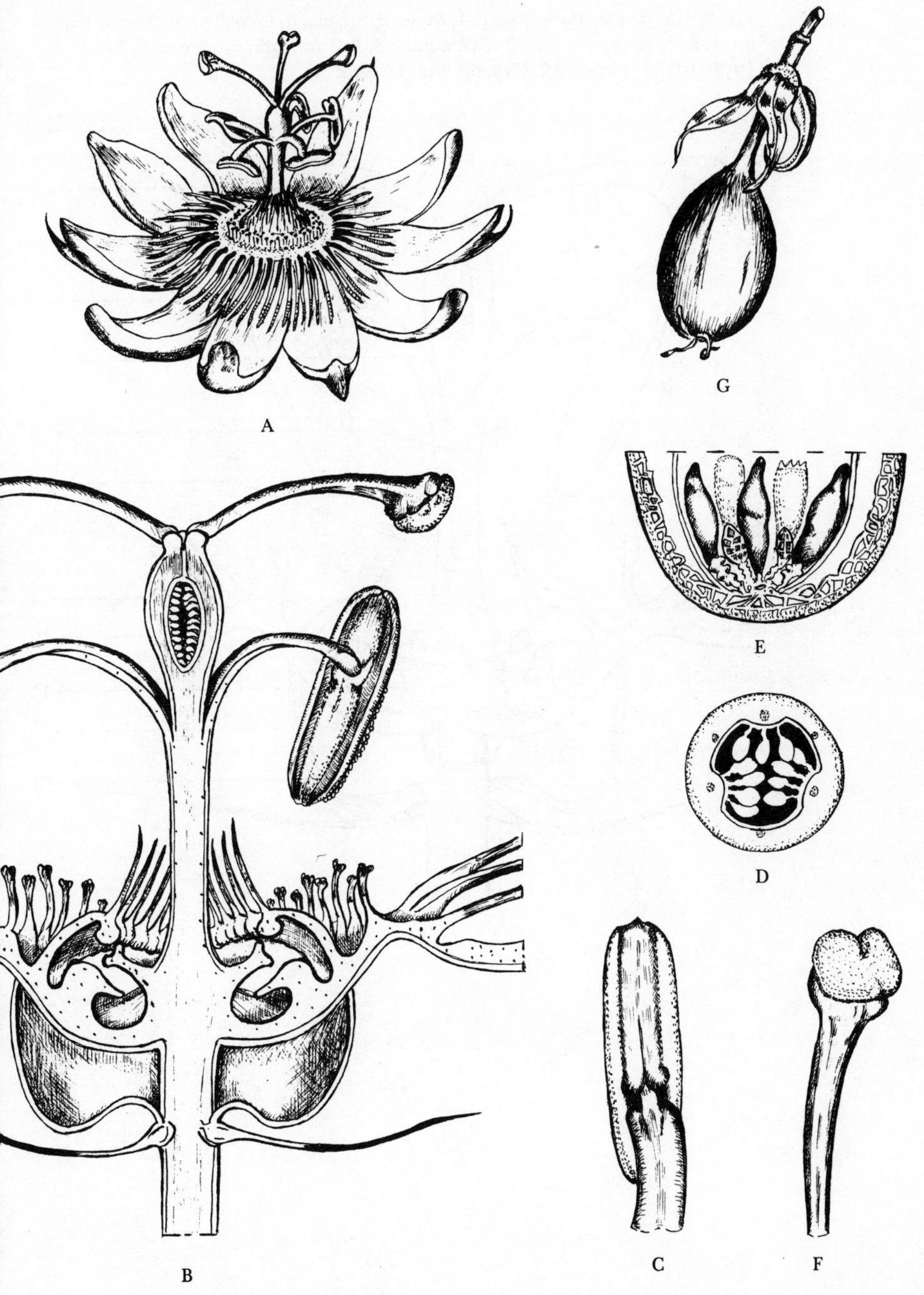

Fig. B. L.S. of *Passiflora* flower: 1, Style; 2, Stigma; 3, Ovary; 4, Anther; 5, Filament; 6, Androgynophore; 7, Operculum; 8, Pali and radii of corona; 9, Petal; 10, Sepal; 11, Limen; 12, Hypanthium; 13, Bracts.

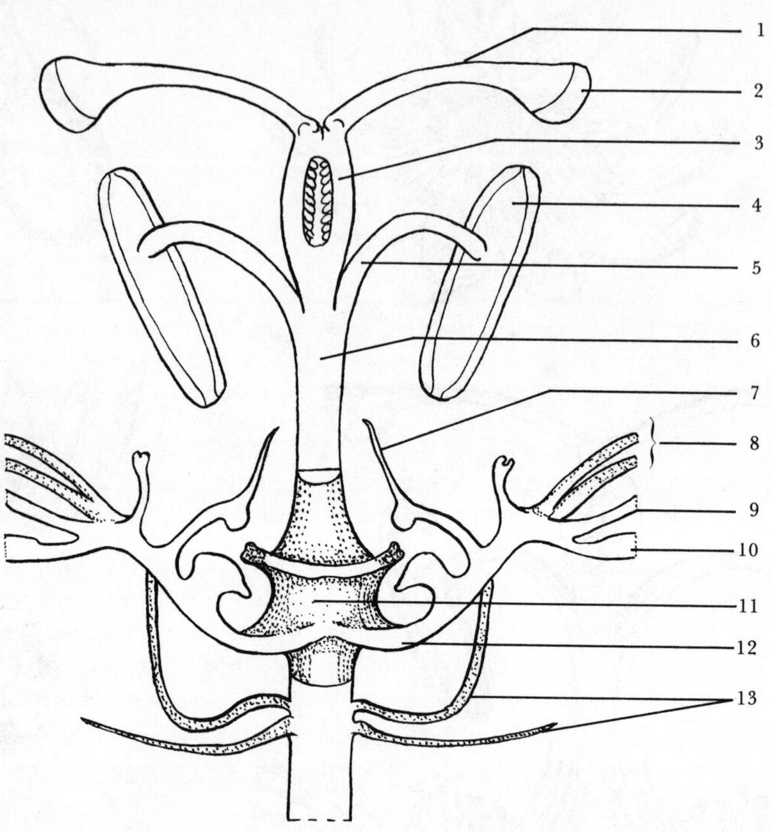

27 Cistaceae Juss.

Rock-rose family

8 genera and 200 species

Distribution. Mainly N. temperate region, a few in S. America.

General characteristics. Shrubs and herbs with opposite, rarely alternate, simple leaves, often inrolled, sometimes stipulate, usually with glandular hairs. Flowers solitary or in cymes, hermaphrodite, actinomorphic. Calyx of 5 free sepals, the 2 outer usually smaller than the inner. Corolla of 3–5 (0 in cleistogamous flowers) free petals, usually crumpled in bud, convolute, the convolutions of calyx and corolla in opposite directions. Stamens usually numerous on a hypogynous disc; anthers introrse. Ovary superior, of 3–5(–10) united carpels, unilocular, with parietal, often projecting placentas, each bearing 2 to many orthotropous or anatropous ovules. Fruit a capsule, usually loculicidal. Endosperm abundant; embryo curved.

Economic and ornamental plants. Several species of *Cistus*, including *C. ladaniferus*, contain the fragrant resinous substance ladanum which is used commercially in perfumery. The genus *Cistus* is also of horticultural value and a number of species and hybrids are suitable for planting as evergreen flowering shrubs in the milder parts of the country. The hardier genus *Helianthemum* also contains several species and cultivars which are useful for the rock-garden or sunny borders.

Classification. The Cistaceae are distinguished from related families by the usually opposite leaves, the reversely convolute perianth series, the caducous petals, the numerous stamens, and the seed with copious endosperm.

Helianthemum (100) W. Europe to central Asia, N. Africa.
Crocanthemum (30) America, W. Indies.
Cistus (20) Canary Islands, Mediterranean region to Transcaucasus.
Lechea (20) N. and Central America, W. Indies.
Halimium (14) Mediterranean region.

HELIANTHEMUM NUMMULARIUM (L.) Mill.
(H. CHAMAECISTUS Mill.)
Common Rock-rose

Distribution. Native in Europe and W. Asia and locally common in grassland and scrub in Britain, though very rare in Ireland.

Vegetative characteristics. A procumbent or ascending dwarf shrub with opposite, short-stalked leaves which are oblong or lanceolate to ovate or orbicular, subglabrous to pubescent above and white-tomentose beneath. The stipules are lanceolate and longer than the petioles.

Floral formula. K5 C5 A∞ $\underline{G}$(3)

Flower and inflorescence. The actinomorphic, hermaphrodite flowers are in 1- to 12-flowered, unilateral cymes and appear from June until September. The flowers are usually golden-yellow, but rarely they may be orange, pale yellow, cream, pink or white.

Pollination. Various types of insects are attracted to the flowers by their bright colour and abundance of pollen. In the absence of cross-pollination, self-pollination readily takes place because of the close proximity of some of the anthers to the stigma, which becomes receptive at the same time as the anthers dehisce.

Alternative flowers for study. Any of the garden rock-roses, which are hybrids derived from *H. nummularium* and allied species, would be suitable alternatives. Our other native species, *H. apenninum* and *H. canum*, are rare in Britain and should not be collected. The closely related genus *Cistus* is frequently grown in gardens, and differs principally in having larger flowers, and 5- or 10-valved capsules.

Fig. 27 Cistaceae, *Helianthemum nummularium*

A A flower-bud, with 5 hairy sepals protecting the corolla. The 2 outer sepals are shorter than the 3 inner, which remain until the fruit is formed.
Bud: 7 × 3.5 mm

B The inflorescence, a unilateral cyme, with one of the hermaphrodite flowers fully open, exposing the 5 golden-yellow petals and numerous stamens. The petals are quickly caducous and fall within a day or two.
Corolla: 3 cm in diameter Petals: 11–13 × 13 mm
Inner sepals: 7 × 4 mm

C L.S. of a flower showing the arrangement of the essential organs. The numerous stamens are borne on a well-developed disc below the ovary and develop from the top downwards. The older stamens are therefore found just beneath the ovary. The filiform style is either sigmoid (S-shaped) or straight, thickening out towards the apex and terminating in a large, capitate stigma.
Stamens: up to 5 mm Ovary: 1.25 mm
Style + stigma: 3 mm

D Detail of anther. D1 (ventral view): the 2 cells dehiscing longitudinally. D2 (dorsal view): the point of attachment of the anther to the filament.
Anther: 0.75 × 0.5 mm

E L.S. of the superior ovary and the hypogynous disc, showing how the sepals, petals and stamens are attached.

F T.S. of ovary. Placentation is parietal and 2 or more orthotropous ovules are attached by stout funicles to each of the placentas which project into the single loculus.
Ovary: 1.1 mm in diameter

G The whole fruit prior to dehiscence, a 3-valved capsule which dehisces loculicidally. The withering, filiform style is still attached to the apex of the capsule. At the base are the persistent sepals.
Fruit: 6 × 5 mm

H The mature fruit, cut open to expose the inner surface of one of the valves. The placenta bears small seeds that owe their variable shape to compression.

Fig. 27

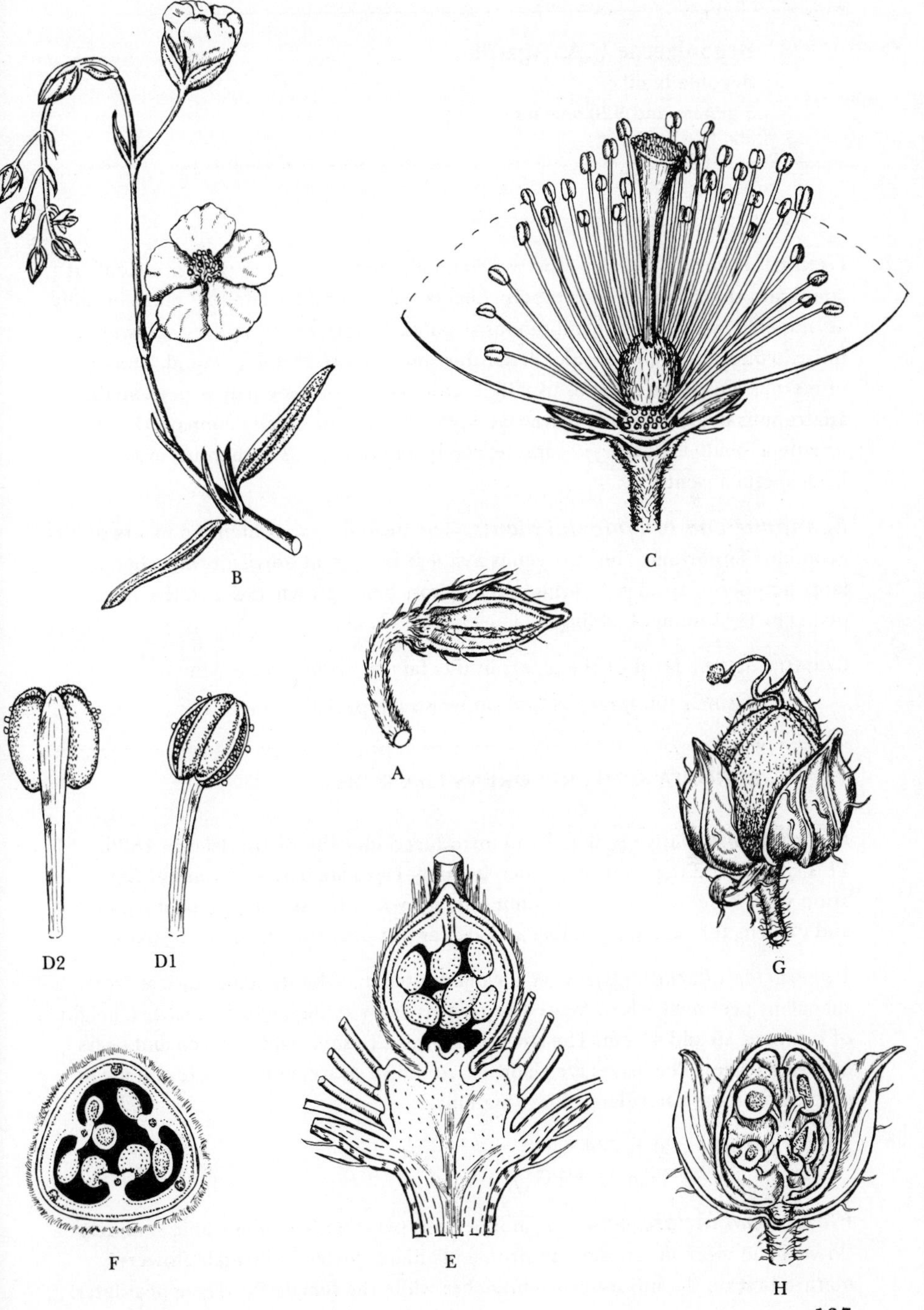

28 Begoniaceae C.A. Agardh

Begonia family

5 genera and 920 species

Distribution. Tropics and subtropics.

General characteristics. Mostly perennial, monoecious herbs, with succulent stems and thick rhizomes or tubers. Leaves radical or alternate in 2 ranks, usually asymmetric, with large, membranous stipules. Inflorescence axillary. Flowers unisexual, often zygomorphic. Perianth-segments 2 to many, petaloid. Stamens numerous. Ovary inferior, of usually 3 united carpels, 3-locular, with numerous anatropous ovules on axile placentas; styles distinct or basally connate. Fruit usually a loculicidal, winged capsule, rarely a berry. Seeds numerous, minute. Endosperm absent.

Economic and ornamental plants. The Begoniaceae contain no plants of economic importance, but the genus *Begonia* is of great horticultural value, a large number of species, hybrids and cultivars being grown as decorative house plants or for summer bedding.

Classification. Most of the plants in this family belong to the genus

Begonia (900) tropics and subtropics, especially America.

BEGONIA SEMPERFLORENS Link & Otto

Distribution. Native in Brazil and introduced into the British Isles in 1829. This species and the cultivars, which in many cases are derived from hybridisation with *B. schmidtiana*, are commonly grown in the sunnier parts of parks and gardens for summer bedding and also as pot-plants or in window-boxes.

Vegetative characteristics. A fibrous-rooted, half-hardy, more or less erect, succulent perennial, often treated in gardens as an annual, which reaches a height of between 15 and 45 cm. The stem is glabrous, fleshy, reddish green and bears opposite leaves. The leaves are glabrous, roundish ovate, more or less oblique at the base and have serrulate, ciliate margins.

Floral formula. Male: P4 A∞

Female: P5 $\overline{\text{G}}$(3)

Flower and inflorescence. A monoecious plant with zygomorphic staminate flowers and more or less actinomorphic pistillate flowers. The male flowers mature first on the inflorescence-branches while the female flowers appear later.

The showy flowers are borne in axillary, dichasial cymes from June to October. They are rose-red or white in the typical form, though in some cultivars they are scarlet or orange-scarlet. A number of cultivars have double flowers.

Pollination. The brightly coloured perianth and the spreading habit of the inflorescence suggest that the flowers are insect-pollinated. The earlier development of the staminate flowers helps to achieve cross-pollination. The pollinating insects seek out the pollen from the numerous stamens at the centre of the male flowers. The female flowers have yellow, branched styles, resembling superficially the numerous stamens of the male flowers. This resemblance may possibly induce pollen-gathering insects to visit the pistillate flowers also, causing pollen to be transferred to the stigmas.

Alternative flowers for study. Horticulturally, begonias are divided into 3 groups according to their rootstock. The rhizomatous-rooted group, which includes the *B. rex-cultorum* hybrids, is cultivated primarily for its brightly coloured foliage and the flowers are usually small and inconspicuous. The fibrous-rooted and tuberous-rooted groups, on the other hand, are grown mainly for their flowers and any of the numerous species or single-flowered cultivars would be suitable alternatives to *B. semperflorens.* It should be noted that, although begonias are monoecious, it may not always be possible to find both staminate and pistillate flowers at the same time on a particular plant.

Fig. 28 Begoniaceae, *Begonia semperflorens*

A The upper portion of a flowering shoot, showing 2 dichasial cymes, each consisting of 3 flowers and arising from a leaf-axil. The central staminate flower of the lower cluster is fully open while the 2 lateral flowers are still in bud. The male flower is zygomorphic, with 2 broad and 2 narrow perianth-segments and a central cluster of yellow stamens.
Broad perianth-segments: 20–27 × 15–25 mm
Narrow perianth-segments: 16–20 × 6–8 mm
Pedicels: 3–6 cm

B Posterior view of the more or less actinomorphic pistillate flower, which has 5 perianth-segments and a 3-winged, inferior ovary subtended by a single bract.
Perianth-segments of female flower: 10–15 × 11–15 mm

C The central cluster of stamens. The filaments are basally connate and the basifixed anthers are 2-celled and dehisce longitudinally.
Stamen: 5 mm

D A single stamen at dehiscence showing the longitudinal slit and pollen grains. The connective is extended to the top of the stamen.
Slit: 2.5 mm

E T.S. of the 3-winged, 3-locular ovary with numerous, anatropous ovules attached to the often forked, axile placenta within each loculus.

F L.S. of the inferior ovary, showing 2 of the 3 wings extending from the ovary-wall. The numerous ovules are attached to the well-developed placentas. Above the ovary are papillose stigmas and the remains of the perianth-segments. After fertilisation the ovary develops into a capsule which dehisces loculicidally to liberate the numerous, very small seeds.
Ovary: *c.* 15 mm wide extending to 19 mm across wing-tips
Loculus: 10 mm

G Detail of the basally connate, yellow styles with their well-branched, papillose stigmas which, from a distance, resemble a cluster of stamens.
Style + stigma: 5–6 × 8 mm

Fig. 28

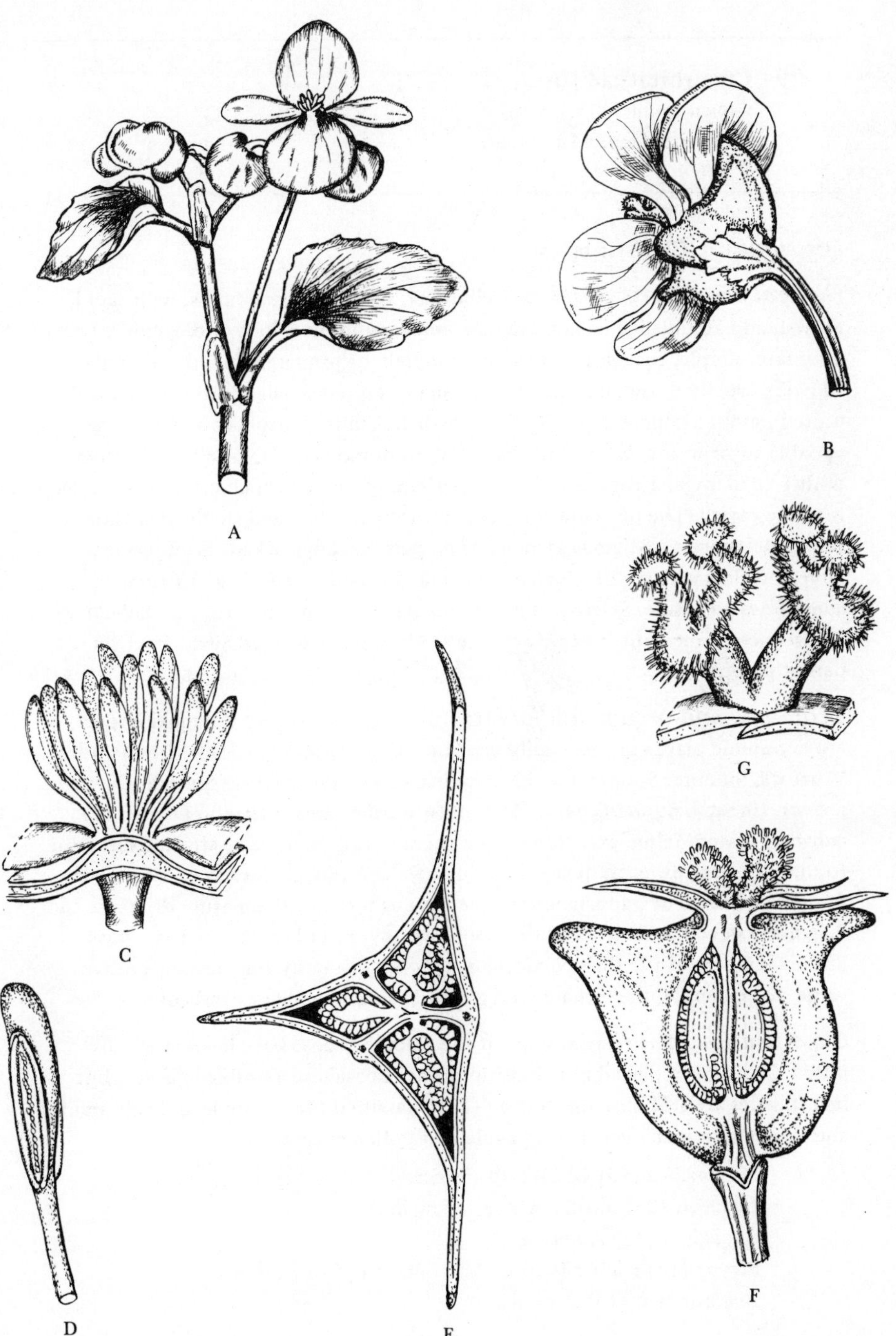

29 Cucurbitaceae Juss.

Gourd family

110 genera and 640 species

Distribution. Chiefly tropics.

General characteristics. Mainly climbing, tendril-bearing herbs, with rapid growth and abundance of sap. Plants commonly monoecious or dioecious. Leaves alternate, simple, roundish, entire or palmately or pinnately lobed, exstipulate. Flowers usually unisexual. Calyx of 5 usually united sepals. Corolla of 5 usually united petals. Stamens typically 5, more or less united into a column but very variable in structure. Ovary inferior, of 3–5 united carpels, usually unilocular, with 1 to many anatropous ovules on parietal placentas which often project deep into the cavity. The placental situation is often complicated by the intrusion of the carpel margins. Stigmas as many as carpels, usually forked. Fruit usually large and fleshy, berry-like, when firm-walled called a pepo (e.g., *Cucumis*); sometimes a capsule. Seeds without endosperm. In some genera, e.g., *Ecballium*, dehiscence of the fruit is explosive, the seeds being projected away from the parent plant.

Economic and ornamental plants. In temperate regions the most important economic plants in this family are *Cucurbita pepo*, a species which includes Marrows, Summer Squashes and Pumpkins, *C. maxima*, Winter Squash, *Cucumis sativus*, Cucumber, and *C. melo*, Melon. In warmer areas *Citrullus lanatus* (*C. vulgaris*) Water Melon, and *Momordica charantia*, Balsam Pear, are cultivated for food. Plants for other purposes include *Luffa cylindrica*, Loofah or Vegetable Sponge, which is of importance for the fibrous mesh of the mature dry fruit and *Lagenaria siceraria*, Bottle Gourd, whose woody outer layer is used as a flask. Some members of the Cucurbitaceae are grown primarily for their decorative value, for example, the 'ornamental gourds' of various shapes and colours.

Classification. The relationships of the Cucurbitaceae have been much disputed. It has been placed near Passifloraceae, Loasaceae and Begoniaceae, but its affinities are still obscure. Jeffrey (11), classified the family into 2 subfamilies and 9 tribes, in many cases using ovule and pollen characters.

Momordica (45) Old World tropics.
Cucumis (25) mostly Africa, a few in Asia.
Cucurbita (15) America.
Sicyos (15) Pacific Islands, Australia, tropical America.
Echinocystis (15) America.

Lagenaria (6) chiefly tropical Africa.
Bryonia (4) Europe, Asia, N. Africa, Canary Islands.

CUCURBITA PEPO L.
Vegetable Marrow

Distribution. The Marrow probably originated in America, but it has long been cultivated elsewhere for its large, fleshy fruits. It can be grown on a wide range of soils and is sometimes self-sown.

Vegetative characteristics. A trailing, tendril-bearing annual with an angular, bristly-hairy stem. The large, alternate, palmately-veined leaves are broadly ovate in outline, cordate at the base and variously lobed.

Floral formula. Male: K(5) C(5) A3 Female: K(5) C(5) $\overline{G}3$

Flower and inflorescence. The plant is monoecious and the large, yellow, unisexual flowers arise singly from the axils of the leaves during the summer months. The pistillate flower is somewhat smaller than the staminate flower and the elongated, inferior ovary is clearly visible below the perianth (see Fig. 29.B).

Pollination. Bees and other insects are attracted by the bright yellow corolla and by the nectar that is secreted from a disc at the base of the 3-branched style. Night-flying insects are also attracted by the abundance of pollen produced by the staminate flowers between 10 p.m. and 3 a.m. An insect foraging for nectar and visiting a male flower will be dusted with pollen which will be transferred to the broad stigmas of the 3-branched style when it visits a female flower. Cross-pollination is promoted by the unisexuality of the flowers.

Alternative flowers for study. The Marrow is usually readily available, but in its absence any one of the other forms of *Cucurbita pepo* would be suitable. Comparison between genera will show marked differences in fruits but less so in floral structure. The staminate flowers of *Cucumis sativus* have 3 free stamens and there may be more than one flower in the axil of a leaf. The only British member of the Cucurbitaceae is *Bryonia dioica*, White Bryony, which is a locally common tendril-climber in hedgerows and scrub where it flowers from May to September. The plant is dioecious, with a cymose inflorescence of small, green flowers. The fruits are small, red, poisonous berries.

Fig. 29 Cucurbitaceae, *Cucurbita pepo*

A The male (staminate) actinomorphic flower is borne on a long peduncle arising from the axil of a leaf. At the top of the peduncle is a cup-shaped receptacle which bears the narrowly lobed calyx and the broadly lobed corolla.
Receptacle: 12 × 15 mm Calyx-lobe: 15 × 2 mm
Corolla-tube: 45 mm Corolla-lobe: 45 × 24 mm

B The female (pistillate) flower is also actinomorphic and solitary, but is borne on a short peduncle that gradually broadens into the ribbed, inferior ovary. The ovary is terminated by the 5-lobed calyx and the 5-lobed corolla (the latter shown closed). A branched, spirally coiled tendril arises from the stem near to the peduncle.
Ovary (at anthesis): 40 × 13 mm
Receptacle + perianth: 40 mm

C The lower part of a staminate flower showing the 3 stamens that form an arch over the nectar-secreting disc. The 2-celled anthers are coherent into a central column supported by the free, broad-based filaments.
Young anther: 11 mm Mature anther: 14 mm
Anther column: 6 mm in diameter Filament: 11 mm
Width of filament at base: 11 mm

D L.S. of the lower part of a female flower showing the reproductive organs. The inferior ovary is hairy when young and contains numerous ovules. The ovary is surmounted by a 3-branched style, each branch bearing a prominent 2-lobed stigma. Surrounding the base of the style is the nectar-secreting disc.
Ovary: 40 × 12 mm

E T.S. of the young fruit, a pepo, showing the unilocular ovary composed of 3 fused carpels and containing numerous ovules borne on the 3 parietal placentas. In the mature fruit the large, flat seeds are enclosed in a soft, pulpy endocarp, which in turn is surrounded by the fleshy mesocarp. The epicarp consists of a firm rind which acts as a protective skin.
Young fruit: 90 mm in diameter

F L.S. of a portion of a placenta, showing the numerous anatropous ovules at different stages of development embedded in the soft, pulpy tissue.

G A germinating seed, showing the small peg which is situated at the base of the hypocotyl. The peg holds one side of the testa (seed-coat) while the other side is split off by the expanding plumule.
Seeds: 13–14 × 9 mm

Fig. 29

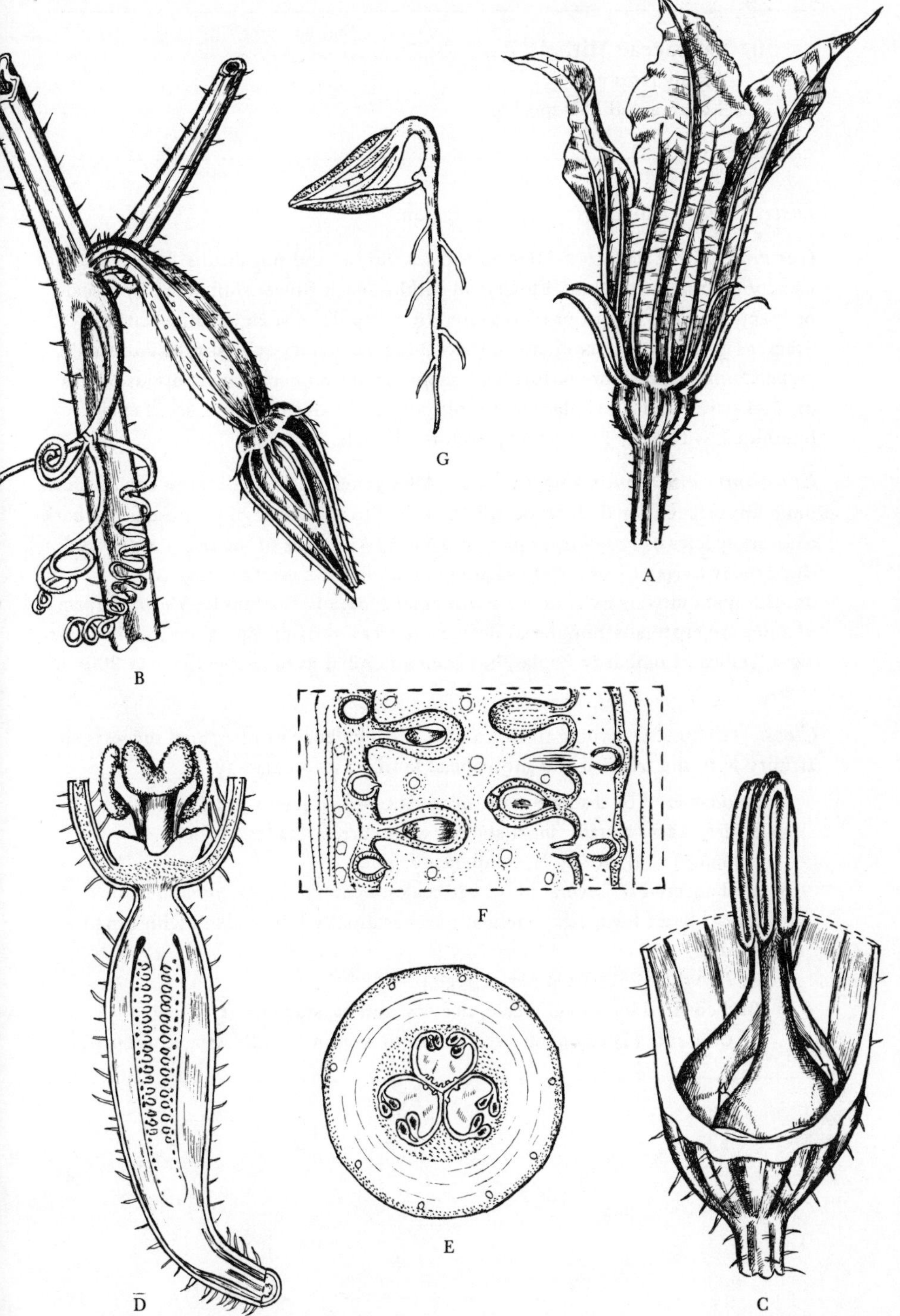

30 Salicaceae Mirbel

Willow family

3 genera and 530 species

Distribution. Chiefly N. temperate region.

General characteristics. Dioecious trees, shrubs and sub-shrubs. Leaves simple, usually alternate, stipulate. Flowers in catkins, each flower subtended by a scale or bract. Perianth absent or represented by a cupular disc or small nectary. Stamens 2–30, free or connate, anthers 2-locular. Ovary superior, of 2 united carpels, unilocular, more or less flask-shaped, with numerous anatropous ovules on 2–4 parietal or basal placentas. Fruit a 2- to 4-valved capsule; seeds exalbuminous, with silky hairs arising from the funicle.

Economic and ornamental plants. All 3 genera in the Salicaceae are of economic importance for their wood which is used for a variety of purposes. The bark of some species of *Populus*, Poplar, and *Salix*, Willow, is of medicinal value and the slender twigs of some Willows are in demand for basket-making. Many hybrids and cultivars exist of the 2 principal genera in the family. Weeping forms of *Salix* are especially popular as decorative trees in damp situations and *Populus nigra* 'Italica', Lombardy Poplar, has been a familiar avenue tree for over 200 years.

Classification. According to Holm (12) the incidence of rust fungi suggests an affinity with the tropical and subtropical family Flacourtiaceae.

Bud covered by a single scale, nectaries 1–4, stamens usually 2, bract entire, leaves usually short-stalked with narrow blades:
Salix (500) chiefly N. temperate region.

Bud covered by several imbricate scales, nectary absent, stamens 4 to many, bract toothed or laciniate, leaves usually long-stalked with broad blades:
Populus (35) N. temperate region.

Bud covered by a single scale, nectary absent, stamens 5:
Chosenia (1) temperate and subarctic region of N.E. Asia, *C. arbutifolia.*

SALIX CAPREA L.
Goat Willow

Distribution. Common in woods, hedges, scrubland and the margins of ponds, lakes and rivers throughout the British Isles and most of Europe. Native also in temperate Asia eastwards to Turkestan.

Vegetative characteristics. A deciduous shrub or small tree with alternate, ovate to obovate leaves, dark green above and grey-tomentose beneath.

Floral formula. Male: P 0 A 2
Female: P 0 $\underline{G}$ (2)

Flower and inflorescence. The unisexual asepalous and apetalous flowers are solitary in the axil of a scale or bract. They are arranged spirally round a central axis and collectively form the type of inflorescence known as a catkin. At the base of the catkins, which reach maturity in March and April, are small bracts (see Fig. 30.G).

Pollination. The genus *Salix* is pollinated by insects in contrast to the related genus *Populus* which is wind-pollinated. Various kinds of insects, especially early bees, are attracted to the catkins at a time of the year when there are relatively few sources of nectar available. All the catkins are rendered conspicuous by appearing before the leaves and the male catkins particularly so by the presence of the bright yellow pollen at the tips of the stamens. Both male and female flowers produce abundant nectar from the well-developed nectar-glands at their base. The dioecism of the trees ensures cross-pollination within species and also leads to frequent hybridisation between different species.

Alternative flowers for study. Any species of *Salix* that is readily to hand would be suitable since the structure of the flowers is fundamentally similar. The main differences between the species occur in time of flowering, dimensions of the catkins, colour of the scales and the number of stamens which may vary from 2 to 5 (rarely to 12). Comparison with *Populus* would be valuable, since this genus exhibits anemophilous characters, including pendulous male catkins with more numerous stamens and the absence of nectaries. Most species flower in March and April.

Fig. 30 Salicaceae, *Salix caprea*

A A single female or pistillate flower. The stalked gynoecium is composed of 2 fused carpels and is subtended by an entire, hairy bract, at the base of which is the prominent nectary.
Bract: 2–3 mm Ovary: 4 mm

B L.S. of the superior, unilocular ovary, which contains numerous anatropous ovules.

C T.S. of ovary, showing the 2 parietal placentas with ovules attached.
T.S. of ovary (at anthesis): 1.25 × 0.75 mm

D Detail of style, branched at the apex to form 2 prominent, stigmatic surfaces.
Branch of style: 0.5 mm

E A single male or staminate flower. The 2 free stamens are subtended by an entire, hairy bract, at the base of which is a prominent nectary. (In some species the stamens are connate at the base.)
Bract: 2 mm
Stamens in young flower: 4–5 mm, in mature flower: 10 mm

F A stamen prior to dehiscence. The anther dehisces longitudinally and sheds abundant pollen.
Anther: 1.25 mm

G A male catkin. The hairy appearance of the catkin is due to the closely-packed, hairy bracts subtending the individual flowers.
Male catkins (at maturity): 19–20 mm
Female catkins (not illustrated): 25 mm

H Detail of the solitary nectary situated at the base of both male and female flowers.
Nectary: 1 mm

I The fruit at dehiscence, a capsule which opens by 2 valves to liberate the numerous seeds.
Fruit: 6 mm

J The seed, bearing long hairs that arise from the funicle and aid wind-dispersal. The seeds are quickly dehydrated owing to their thin testa and early germination is essential.
Seed: 1.5 mm Hairs: 7 mm

Fig. 30

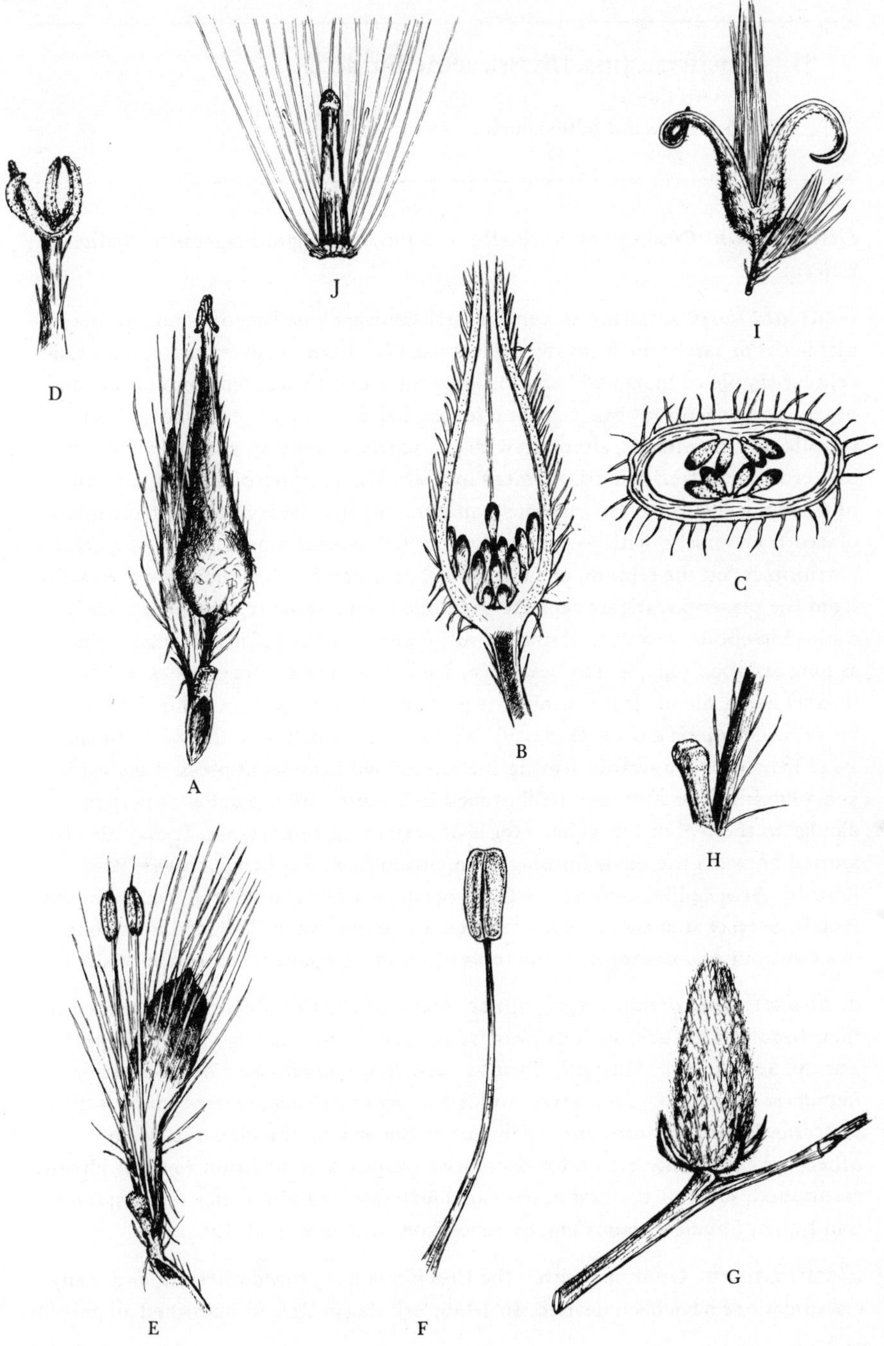

31 Cruciferae Juss. (Brassicaceae Burnett)

Mustard family

375 genera and 3200 species

Distribution. Cosmopolitan, chiefly N. temperate region, especially Mediterranean.

General characteristics. A very natural family, consisting of annual to perennial herbs or rarely small shrubs. Leaves usually alternate, exstipulate, with unicellular simple or branched hairs. Inflorescence usually a raceme or corymb and nearly always without bracts or bracteoles. Calyx of 4 free sepals in 2 whorls. Corolla of 4 free petals, alternate with the sepals, usually spreading in the form of a cross and often clawed. Stamens in 2 whorls, an outer of 2 short, an inner of 4 long, stamens (tetradynamous); anthers introrse. Ovary of 2 united carpels placed transversely, with parietal placentas, but 2-locular on account of a central partition called the replum, or false septum, formed by the union of outgrowths from the placentas; stigma capitate to bilobed, on a short style; ovules usually campylotropous. Fruit a pod-like capsule, called a siliqua if it is at least 3 times as long as broad (e.g., *Arabis caucasica*, see below, and *Cheiranthus cheiri*, Wallflower) and a silicula if it is shorter (e.g., *Capsella bursa-pastoris*, Shepherd's Purse, and *Lunaria annua*, Honesty). When the fruit dehisces, the valves break away from below upwards, leaving the replum with the seeds pressed against it and adhering. The fruit may be flattened in 2 ways, either parallel or perpendicular to the replum; this character is of systematic importance. It may also be jointed between the seeds forming a lomentum (e.g., *Raphanus sativus*, Wild Radish). Achene-like, one-seeded fruits occur in a few genera (e.g., *Isatis tinctoria*, Woad). Seed characters are also of systematic importance. The seed is exalbuminous (without endosperm) and the testa often mucilaginous; embryo curved.

Economic and ornamental plants. Many of the Cruciferae are cultivated for their food value. These include *Brassica oleracea* vars., Cabbage, etc., *B. rapa*, Turnip, *Sinapis alba*, Mustard, *Rorippa nasturtium-aquaticum*, Water-cress, *Lepidium sativum*, Garden Cress, and *Armoracia rusticana*, Horse-radish. *Isatis tinctoria*, Woad, was formerly cultivated as the source of a blue dye. Many others are commonly grown for decorative purposes. In addition to those already mentioned, some of the best known are *Matthiola* and *Malcolmia*, Stocks, *Iberis*, Candytuft, *Lobularia maritima*, Sweet Alison, *Aubrieta*, and *Alyssum*.

Classification. Grouping within the Cruciferae has proved difficult, and many classifications have been devised. In 1936, Schulz (in Ref. 9) published an entirely new scheme based on a wide variety of characters, dividing the family into 19 tribes. Another, more recent classification is that by Janchen (13).

Draba (300) N. temperate and arctic regions, mountains of Central and S. America.
Cardamine (160) cosmopolitan.
Lepidium (150) cosmopolitan.
Alyssum (150) Mediterranean region to Siberia.
Arabis (120) temperate Eurasia, Mediterranean region, mountains of tropical Africa, N. America.
Erysimum (100) Mediterranean region, Europe, Asia.
Sisymbrium (90) temperate Eurasia, Mediterranean region, S. Africa, N. America, Andes.
Heliophila (75) S. Africa.
Rorippa (70) almost cosmopolitan.

ARABIS CAUCASICA Schlecht.
Garden Arabis

Distribution. A native of mountains of the Mediterranean region and Near East to Iran. It was introduced into the British Isles as an ornamental plant, and is often grown in rock gardens and on walls. It sometimes escapes from cultivation and becomes naturalised.

Vegetative characteristics. A perennial mat-forming herb, with greyish green or whitish leaves. The basal leaves have 2–5 obtuse teeth along each margin, and the cauline leaves are auriculate to sagittate at the base.

Floral formula. K2+2 C4 A4+2 $\underline{G}(2)$

Flower and inflorescence. Several hermaphrodite flowers with white petals are borne in more or less erect racemes. Each flower is about 14 mm in diameter and is attached to a pedicel up to 11 mm in length. The main flowering period is from March until June, but there may be a second show of flowers later in the year.

Pollination. As in most members of the Cruciferae, nectar secreted into the bases of the pouched sepals attracts insects and promotes cross-pollination. However, self-pollination of the flowers is also a frequent occurrence.

Alternative flowers for study. Arabis alpina, Alpine Rock-cress (a smaller-flowered species), *Sinapis arvensis*, Charlock, and *Cheiranthus cheiri*, Wallflower, are suitable alternatives which flower in the late spring and early summer. Various forms of fruits are mentioned under General characteristics of the family, but interesting variations in the flower are exhibited by some genera. In *Iberis*, the flowers are zygomorphic, the 2 outer petals being much larger than the 2 inner (cf. Umbelliferae). The petals are often absent in *Coronopus didymus*, Lesser Swine-cress, *Lepidium ruderale*, Narrow-leaved Pepperwort, and *L. densiflorum* and occasionally absent in *Capsella bursa-pastoris.*

Fig. 31 Cruciferae, *Arabis caucasica*

A L.S. of the flower showing 2 of the 4 petals and sepals. The free petals are clawed and the inner sepals are pouched or saccate at the base. Surrounding the ovary are the tetradynamous stamens, typical of the family. Both of the short stamens and 2 of the long ones are illustrated.
Sepals: 5 mm Petals: 12 mm Petal claw: 7 mm
Long stamens: 8 mm Short stamens: 6 mm

B Anterior view of the flower showing the 4 free petals which alternate with the 4 free sepals.

C Petals and sepals removed to expose 2 of the long stamens and the superior ovary. The anthers are dorsifixed and introrse, and consist of 2 cells which dehisce longitudinally. At the base of the filaments and joined to the receptacle are nectar-secreting glands. The very short style has a knob-like stigma which matures at the same time as the stamens.
Gynoecium (at flowering time): 9 mm Stigma: 0.35 x 0.85 mm

D L.S. of the ovary showing the 2 carpels with numerous ovules.

E and F T.S. of the ovary, cut in 2 different ways to show the parietal placentation of the ovules.
Ovary: 0.5 mm in diameter

G The fruit (siliqua) formed from the fertilised carpels. When the siliqua dehisces, the 2 valves open from below upwards, exposing the slightly winged seeds arranged along each of the placentas.
Fruit: 25 x 2 mm

Fig. 31

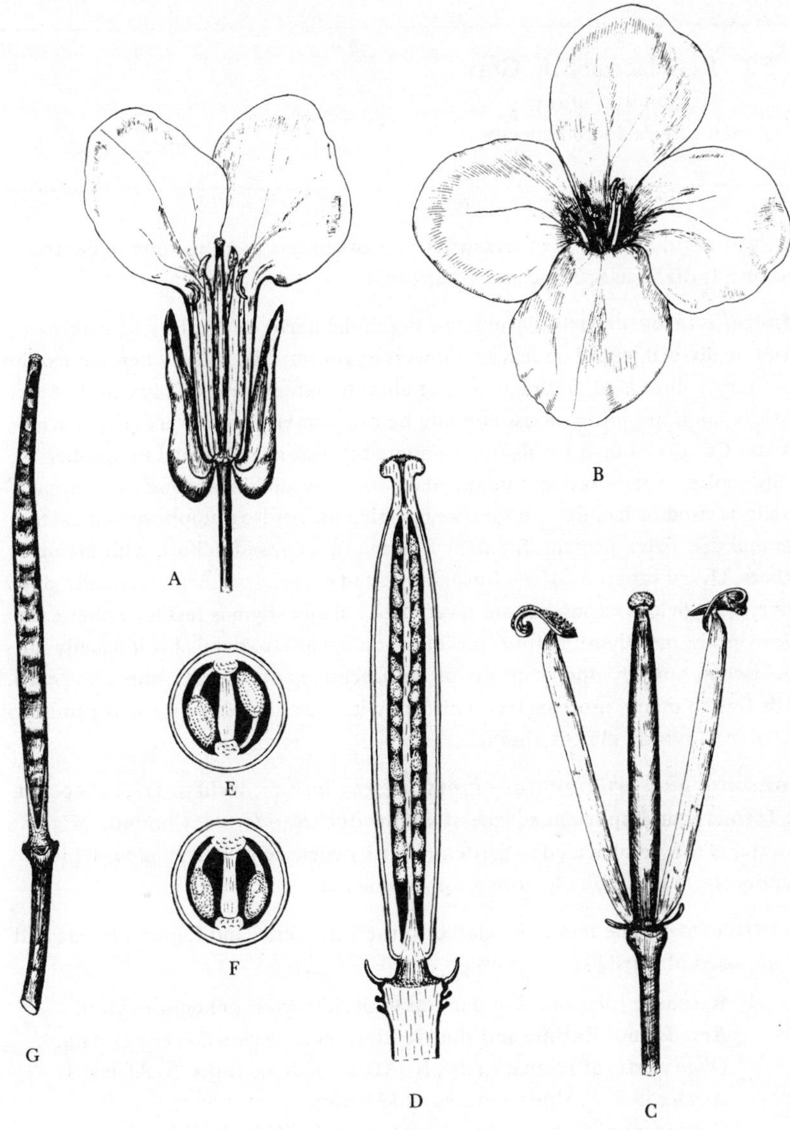

32 Resedaceae S.F. Gray

Mignonette family

6 genera and 81 species

Distribution. Chiefly Mediterranean, but extending from Europe to central Asia and India, also S. Africa and California.

General characteristics. Annual or perennial herbs or shrubs with alternate, entire or divided, stipulate leaves. Flowers zygomorphic, usually hermaphrodite, more rarely unisexual, in bracteate but ebracteolate racemes. Calyx of 2–8 sepals, sometimes more or less connate below, sometimes unequal, slightly imbricate. Corolla of 0–8 petals (rarely connate), not always equal in number to the sepals, mostly broadly unguiculate with a scale-like appendage; lamina usually incised or laciniate, sometimes persistent. Androgynophore and extrastaminal disc often present. Stamens 3 to many, exposed in bud, with introrse anthers. Ovary superior, of 2–7 usually united carpels, unilocular, usually open above, with ovules exposed from a very early stage; stigmas sessile; ovules campylotropous, usually numerous; placentation usually parietal. Fruit usually an indehiscent, apically open capsule, often baccate or of separate, spreading carpels. Seeds few to many, more or less reniform with carunculoid outgrowth; embryo curved; endosperm almost absent.

Economic and ornamental plants. Reseda luteola, Weld or Dyer's Rocket, was formerly an important source of yellow dye. *R. odorata*, Common Mignonette, is often cultivated in gardens for its sweet scent, and *R. alba*, White Mignonette, is occasionally grown for ornament.

Classification. The family is related to the Cruciferae and Capparidaceae and is composed of 3 tribes:

1. Resedeae (placentation parietal with numerous pendent ovules).
   Reseda (60) Europe and the Mediterranean region to central Asia.
   Oligomeris (80) Canary Isls., N. Africa to N.W. India, S. Africa; 1 species in S.W. United States and Mexico.
   Ochradenus (5) N.E. Africa and Socotra to N.W. India.
   Randonia (3) N. Africa, Somalia, Arabia.
2. Cayluseeae (placentation basal with 10–18 erect ovules).
   Caylusea (3) Cape Verde Islands, N. and E. Africa to India.
3. Astrocarpeae (1, rarely 2, pendent ovules in centre of abaxial wall of carpel).
   Sesamoides (*Astrocarpus*) (2) S.W. Europe.

RESEDA LUTEA L.
Wild Mignonette

Distribution. A plant of waste places, disturbed ground and arable land, especially on chalk or limestone; native in central and S. Europe northwards to Britain, Denmark and Sweden, also found in N. Africa and W. Asia. It has been introduced into N. America.

Vegetative characteristics. A glabrous, bushy annual to perennial herb with branched stems. The leaves are pinnately divided into 1–3 pairs of narrow segments with undulate margins.

Floral formula. K6 C6 A12–20 $\underline{G}(3)$

Flower and inflorescence. The hermaphrodite, greenish-yellow flowers are borne in spike-like racemes from June to August. The reproductive organs are situated on a zygomorphic disc.

Pollination. The conspicuous stamens and the nectar secreted by the well-developed posticous portion of the hypogynous disc attract sawflies, wasps and bees. When the flower first opens, the stamens are curved over the gynoecium, but during dehiscence they bend back towards the disc, ensuring that insect visitors are dusted with pollen. The stigmatic surface is papillose and well adapted for receiving pollen from insects which alight on the conveniently placed ovary. Self-pollination can occur, owing to the proximity of stigmas and anthers, but this has been found to have little or no result.

Alternative flowers for study. Reseda is the only genus normally available in Britain; the biennial *R. luteola* which flowers simultaneously with *R. lutea* or *R. odorata*, whose flowering period extends from May until September, are suitable alternatives.

Fig. 32 Resedaceae, *Reseda lutea*

A The inflorescence, a spike-like raceme. At the apex are flowers still in an early stage while at the base fruits are developing.

B An individual flower at early anthesis, with some of the floral parts removed to reveal the point of attachment of the ovary and the stamens. At the base of the flower are 2 of the 6 linear, persistent sepals and, above them, 3 of the 6 petals. In the centre is the disc, with 2 of the stamens attached to the more developed portion. Two more stamens are shown in the anterior region, the others having been removed for clarity. The styles and stigmas are not yet fully developed.

C Petal shapes. C1: One of the 2 trifid upper petals. C2 and C3: 2 forms of the 2 bifid (or sometimes trifid) lateral petals. C4 and C5: 2 forms of the 2 linear lower petals.
Upper petal: 3 × 1.5 mm
Lateral petals C2: 3 × 1 mm C3: 2.75 × 1.25 mm
Lower petals C4: 2.25 × 0.5–0.75 mm C5: 2 × 0.5 mm

D One of the 12–20 stamens with 2-celled, introrse anthers which dehisce longitudinally.
Stamens: 2–2.75 mm Anthers: 0.75–1 mm

E L.S. of the superior ovary, showing the numerous campylotropous ovules which develop into black, kidney-shaped seeds.
Ovary (at early anthesis): 1.25–2 × 0.75 mm
(at later stage): 8.5 × 4 mm

F T.S. of ovary after fertilisation. The ovary is unilocular, and is composed of 3 fused carpels. The ovules are on parietal placentas.
T.S. of ovary: 3.5 × 3.5 mm

G Detail of the apex of the 3-horned gynoecium during the flowering stage. Each horn represents a separate style and papillose stigma.
Stigma + style: 0.75 mm

H The mature fruit, a 3-horned capsule, with the remains of the other floral parts persisting at its base.
Fruit: 12–18 mm Seeds: 1.5–2 mm

Fig. 32

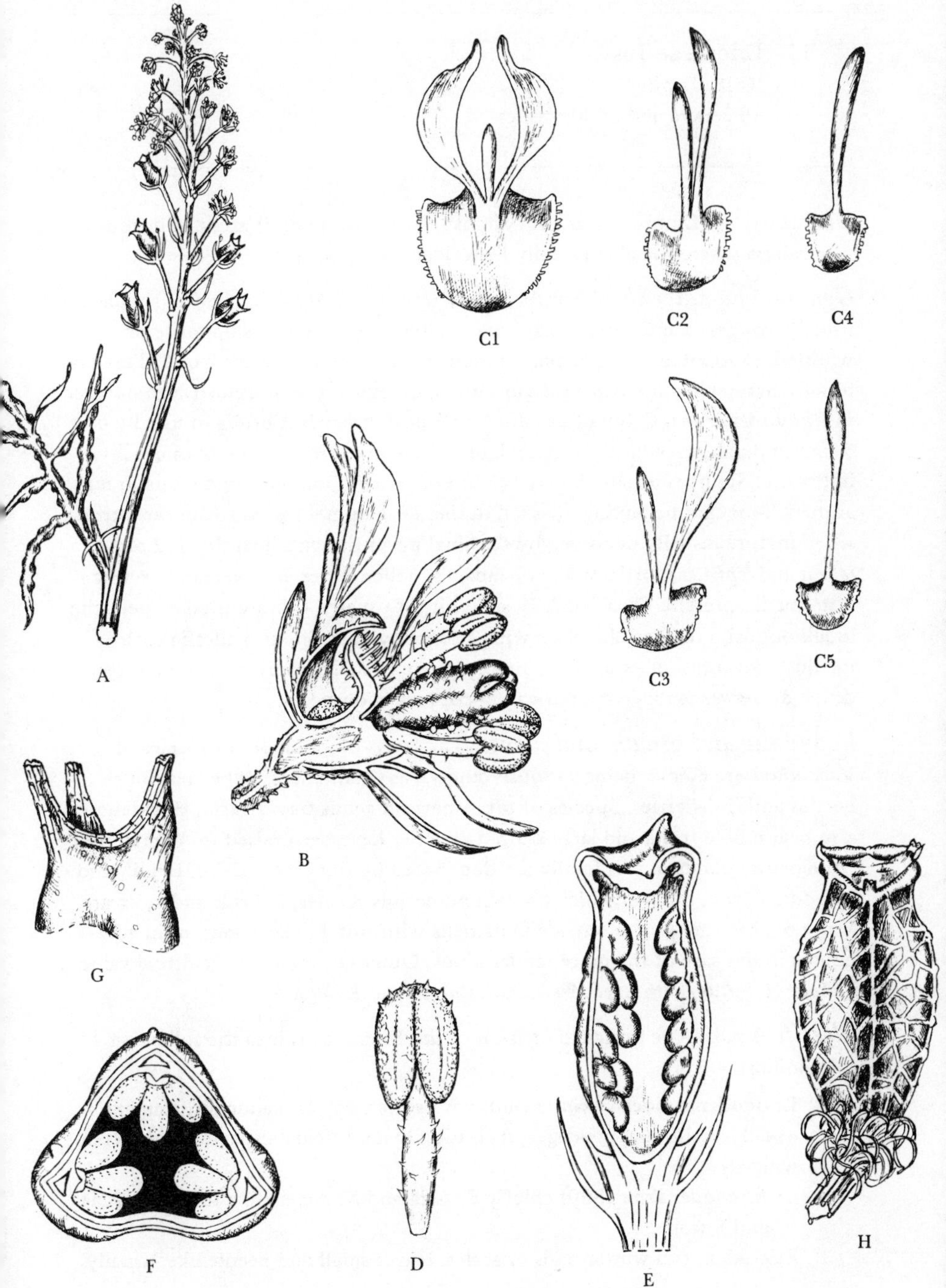

33 Ericaceae Juss.

Heath family

80 genera and 2500 species

Distribution. Cosmopolitan, but almost absent from desert areas, and from Australasia where the allied family Epacridaceae replaces it.

General characteristics. Shrubs, rarely trees (e.g., *Arbutus*). Leaves simple, usually evergreen and coriaceous, usually alternate, sometimes opposite or whorled, exstipulate. Inflorescence bracteate. Bracteoles usually 2 or 3. Flowers usually hermaphrodite (*Epigaea* dioecious), usually actinomorphic (zygomorphic in *Rhododendron*). Calyx of usually 4 or 5 united sepals. Corolla of usually 4 or 5 united petals, urceolate, campanulate or hypocrateriform, the lobes usually imbricate. Stamens usually 8 or 10 (5 in *Loiseleuria*), obdiplostemonous, free; anthers 2-locular, becoming inverted during development so as to appear introrse when mature, usually dehiscing by terminal pores (longitudinal slits in *Loiseleuria* and *Epigaea*), often with appendages; pollen in tetrads. Nectar-secreting disc usually present. Ovary usually superior, of usually 4 or 5 united carpels, the loculi opposite the corolla-lobes, with 1 to many anatropous ovules in each loculus; placentation axile. Style simple, stigma usually capitate. Fruit a capsule, drupe or berry. Embryo cylindrical, in copious endosperm.

Economic and ornamental plants. The fruits of a number of species of *Vaccinium* are edible, being variously known as Cranberries, Bilberries, Blueberries and Cowberries. Species of the American genus *Gaylussacia*, Huckleberry, also bear edible fruits and large-fruited varieties have been raised for commerce. Ornamental plants in this family are dominated by the genera *Erica*, Heath, and *Rhododendron*, both of which contain numerous species, hybrids and cultivars. Many of these are grown outside in gardens with suitably acid soils, or in glasshouses in the case of the more tender kinds. Other genera of horticultural value include *Arbutus*, *Pernettya*, *Pieris*, *Gaultheria* and *Kalmia*.

Classification. The principal division of the Ericaceae is into the following 4 subfamilies:

I. Rhododendroideae (winter buds with scales, corolla caducous, stamens usually without appendages, fruit usually a septicidal capsule, seeds often winged).

Rhododendron (800) chiefly E. Asia and N. America, a few in W. Asia and Europe.

II. Ericoideae (no winter buds or scales, leaves small and needle-like, usually whorled, corolla usually persistent, stamens usually appendaged, fruit usually a loculicidal capsule or nut, seeds not winged).

Erica (500) Europe, Atlantic islands, Africa, W. Asia.

III. Vaccinioideae (winter buds with scales, leaves usually broad and flat, inflorescence usually leafless, corolla caducous, stamens usually appendaged, ovary often inferior, fruit a capsule, drupe or berry, seeds not winged).

Vaccinium (300) N. temperate region, tropical mountains, Andes, S. Africa, Madagascar.

Gaultheria (200) circumpacific.

Arctostaphylos (70) chiefly America, 2 species also in Eurasia.

IV. Epigaeoideae (plants dioecious, leaves cordate, stamens without appendages, anthers with longitudinal slits, stigma greatly expanded, 5-lobed, ovary densely pubescent, placentas double).

Epigaea (3) Japan, W. Asia, E. United States.

ERICA HERBACEA L. (E. CARNEA L.)
Winter Heath

Distribution. Native in coniferous woods and on stony slopes in the Alps and S. central Europe northwards to E. central Germany and eastwards to E. Austria, extending locally southwards to central Italy and Macedonia. It was introduced into Britain by the Earl of Coventry in 1763, and since this date numerous cultivars have been raised. These are often grown in gardens for their free-flowering habit and for their tolerance of a wide range of environmental conditions, including calcareous soils and atmospheric pollution.

Vegetative characteristics. A dwarf, evergreen shrub with procumbent stems and ascending flowering branches. The young stems are almost glabrous and have ridges running downwards from the base of the leaves. The linear leaves, in whorls of 4, have revolute margins.

Floral formula. K4 C(4) A8 $\underline{G}(4)$

Flower and inflorescence. The hermaphrodite, actinomorphic flowers are borne in a leafy, one-sided raceme at the end of the previous year's shoots (see Fig. 33.A). The species and its cultivars offer a wide range of colours and flowering-times. The flowers in nature are pink or rarely white, but cultivars range from pure white through numerous shades of pink to a deep rosy red. The main flowering time is from January to March, but some early flowering cultivars may bloom in November and late ones in April.

Pollination. Various kinds of bees and butterflies are attracted by the brightly coloured flowers and by the nectar secreted by a disc surrounding the base of the ovary. The visiting insect first comes into contact with the capitate stigma which receives any pollen brought from another flower. When the stamens are disturbed, fresh pollen falls from the apical pores of the anthers on to the insect's body. Self-pollination is avoided by the relative positions of the stigma and anthers.

Alternative flowers for study. In the absence of *E. herbacea* or its cultivars (some of which may not set seed), a number of species are available as alternatives, including the native *E. cinerea*, Bell Heather, that grows on dry moorland, and *E. tetralix*, Cross-leaved Heath, that prefers wet moorland and bogs. Both species flower from July to September. The closely related *Calluna vulgaris*, Heather or Ling, differs from *Erica* in having a petaloid calyx larger than the corolla. The fruit of *Calluna* is a few-seeded, septicidal capsule, whereas in *Erica* it is a many-seeded, loculicidal capsule. Many of the Ericaceae are calcifuge and therefore difficult to obtain in certain parts of the country but, if possible, comparison should be made with members of the other subfamilies which differ in various ways.

Fig. 33 Ericaceae, *Erica herbacea*

A An inflorescence at the end of an ascending, leafy shoot. The one-sided raceme is made up of several actinomorphic, hermaphrodite flowers with urceolate corollas.
Inflorescence: 15 mm

B A single, pendulous flower, borne on a short, glabrous pedicel furnished with 2 bracteoles about half way along. The 4 free sepals are about half as long as the 4-lobed corolla. In this species, both stamens and stigma are exserted. The scars on the ridged stem and the 2 cut stalks indicate where adjacent flowers have been removed for clarity.
Pedicel: 3 mm Sepal: 3.25 x 1 mm
Corolla-tube: 5 x 2 mm Corolla-lobe: 0.5 x 1 mm

C A flower shown in an erect position, with calyx and corolla cut lengthwise to reveal the stamens with their long, flattened filaments surrounding the gynoecium. The stamens are adnate to the edge of the nectar-secreting disc.
Filament: 4 mm Anther: 1.5 mm Style: 5.25 mm

D The upper portion of a stamen, showing the anther which is 2-celled and dehisces by pores to release the pollen grains that are grouped into tetrads.

E The base of the flower with stamens removed to reveal the superior ovary. The wall of the ovary appears lobed, the degree of configuration depending on the flower concerned and its stage of development. There is pronounced lobing in the flower illustrated. Below the ovary is the nectar-secreting disc and at its apex the lower portion of the single style can be seen.

F Detail of the capitate stigma that terminates the simple style. Its diameter is only slightly greater than that of the style.
Stigma: 0.2 mm in diameter

G L.S. of the ovary and the basal portion of the calyx and corolla (the stamens have been removed). The well-developed nectar-secreting disc is clearly visible below the ovary. The many, anatropous ovules are attached to axile placentas.
Ovary: 0.7 x 0.8 mm Disc: 0.3 x 0.8 mm

H T.S. of ovary formed from 4 fused carpels. Each carpel has a single loculus containing several ovules on axile placentas. In the wild the ovary develops into a loculicidal capsule enveloped by the persistent perianth, but fruit is not always produced by the cultivars usually grown in gardens.
Ovary: 0.8 mm in diameter

Fig. 33

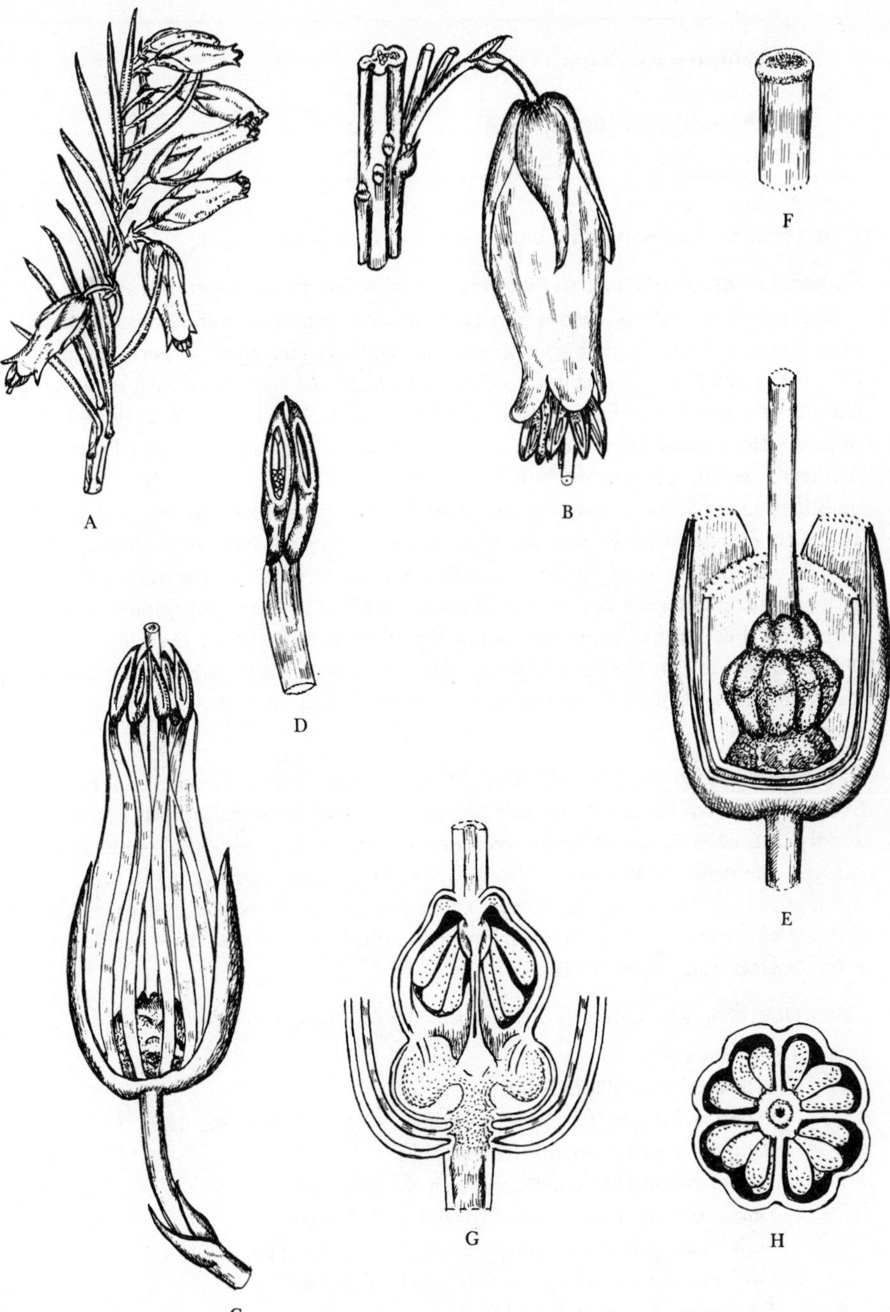

34 Primulaceae Vent.

Primrose family
20 genera and 1000 species

Distribution. Cosmopolitan, but especially N. temperate region.

General characteristics. Herbaceous plants, perennial or sometimes annual, with rhizomes or tubers. Leaves mostly opposite or whorled, sometimes alternate, exstipulate. Flowers often borne on scapes, usually actinomorphic, hermaphrodite, often heterostyled and usually 5-merous. Calyx of 5 united sepals, persistent. Corolla of 5 usually united petals, or absent (*Glaux*). Stamens 5, epipetalous and opposite the corolla-lobes, occasionally 5 staminodes opposite the calyx-lobes. Anthers 2-celled, introrse, dehiscing longitudinally. The presence of the staminodes here (as in Myrsinaceae) explains the antipetalous position of the stamens as due to abortion of the originally outer whorl. Ovary typically of 5 united carpels, superior or half-inferior (*Samolus*), unilocular with free-central placentation, style and stigma simple, ovules usually numerous, arranged spirally or in whorls on the placenta, semi-anatropous. Fruit a usually 5-valved capsule, dehiscing by teeth at the tip or a pyxis, dehiscing transversely, the top coming off like a lid (*Anagallis*). Seeds few or many. Embryo small, in fleshy or hard endosperm.

Economic and ornamental plants. The family is of no economic importance but is well known for the decorative plants it contains. In addition to the species and hybrids of *Primula*, which are commonly cultivated for rock and woodland gardens, the genera *Androsace*, *Dodecatheon*, *Lysimachia* and *Cyclamen* are often grown. *C. persicum*, the florists' Cyclamen, is a popular house-plant, as are *Primula obconica* and *P. malacoides*, but it should be noted that these species of *Primula* often cause dermatitis.

Classification. The family is divided into the following 4 tribes:

I. (Ovary superior).

1. Primuleae (corolla imbricate).
   Primula (500) chiefly in hilly regions of the northern hemisphere.
   Androsace (100) northern hemisphere.
   Dodecatheon (50) chiefly western N. America.
   Dionysia (41) mountains of central and W. Asia.
   Soldanella (11) mountains of central and S. Europe.
2. Cyclamineae (corolla convolute, plant tuberous).
   Cyclamen (15) central and S. Europe, N. Africa, W. Asia.

3. Lysimachieae (corolla convolute, plant not tuberous).
   Lysimachia (200) cosmopolitan, especially E. Asia and N. America.
   Anagallis (30) mainly W. Europe and Africa.

II. (Ovary half-inferior).

4. Samoleae (corolla imbricate).
   Samolus (10–15) cosmopolitan, especially southern hemisphere.

PRIMULA VULGARIS Huds.
Primrose

Distribution. A common plant found growing in open woodlands, hedgerows and meadows throughout the British Isles. Its distribution extends from W. Europe to the Balkans and S. Russia, Asia Minor and N. Africa.

Vegetative characteristics. A perennial herb with a short, stout rhizome and a rosette of obovate, irregularly-toothed leaves.

Floral formula. K(5) C(5) A5 $\underline{\text{G}}$(5)

Flower and inflorescence. The yellow, actinomorphic, hermaphrodite flowers, borne singly on a pedicel, appear from January to May. They are of two types, differing in the relative positions of the style and stamens. 'Pin-eyed' flowers have a long style reaching the mouth of the corolla-tube, with the stamens attached about half-way up the tube, while in 'thrum-eyed' flowers the positions are reversed. This variation in the arrangement of these organs is known as heterostyly and is connected with the avoidance of self-fertilisation (see also Lythraceae).

Pollination. Cross-pollination is ensured by the presence of the two different types of flower, each type pollinating the other. The main pollinating agents appear to be long-tongued bees and long-tongued flies, although many other kinds of insects have been observed visiting the flowers.

Alternative flowers for study. Primula veris, Cowslip, which flowers in April and May, or other species of *Primula* would be suitable. A hybrid between *P. vulgaris* and *P. veris* occurs throughout the British Isles but is considerably less common than the parent species. However, the same cross, with probable admixture of other species, has resulted in the well-known Polyanthus of gardens, which is readily available in flower from March to May.

Fig. 34 Primulaceae, *Primula vulgaris*

A L.S. of a pin-eyed flower. The hairy calyx composed of 5 sepals surrounds the corolla-tube. The superior ovary is situated at the base of the tube, and is surmounted by the long style. Above the tube, the corolla broadens out into 5 showy lobes.
Calyx-tube: 16 mm Calyx-lobe: 6 mm
Corolla: 30 mm in diameter
Stigma + style: 10.5 mm Ovary: 2 × 2 mm
Anther: 2 mm

B L.S. of corolla-tube of a thrum-eyed flower, showing the shorter style and the stamens at the mouth of the tube.
Stigma + style: 6 mm Ovary: 2 × 2 mm Anther: 2.5 mm

C Section of petal, cut down the centre to show how the stamen is attached. The 2-celled anther is introrse and dehisces longitudinally.

D T.S. of ovary with free-central placentation. The vascular bundles supplying the ovary are arranged centrally in the placenta.
Ovary: 2 mm in diameter

E L.S. of ovary, showing how the ovules are attached to the placenta and the position of the ovary in relation to the corolla and the hairy calyx.

F The capitate stigma and apical region of style.
Stigma: 0.6 × 1 mm

G The calyx has been partly cut away to show the fruit, a capsule, which even at maturity only just reaches the base of the calyx-lobes. As the capsule ripens the pedicel bends towards the ground and at dehiscence the 5 valves, each of which is bifid at the apex, split open allowing the seeds to fall to the ground around the plant. The seeds of *P. vulgaris*, unlike most other species of *Primula*, contain an oil in their outer layer and in a specially developed oil-body (elaiosome) which is attractive to ants. These insects quickly carry away the seeds, but after robbing them of their oil, leave them alone to germinate when conditions are suitable. In this way the seeds are dispersed over a wider area.
Mature capsule: 11 × 10 mm

Fig. 34

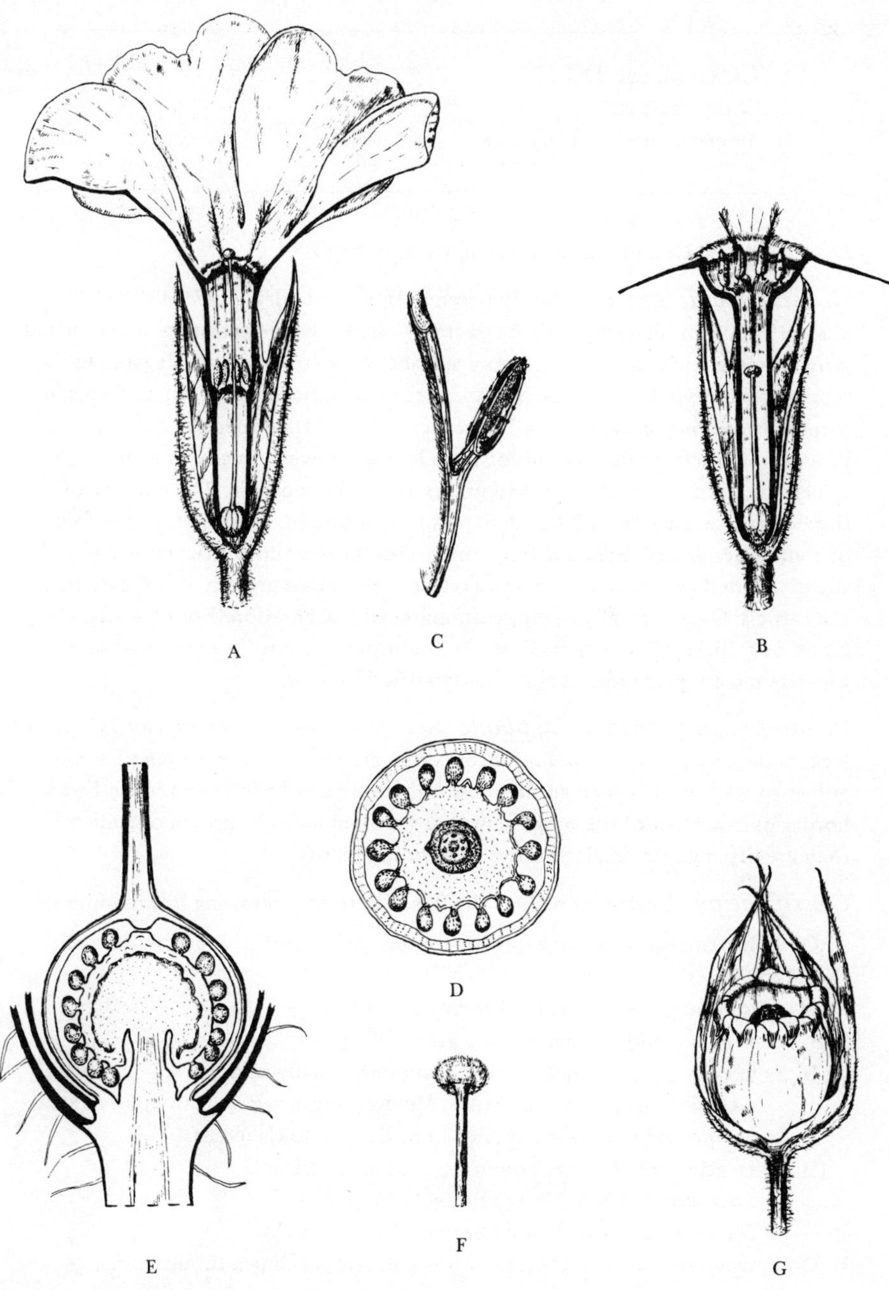

35 Crassulaceae DC.

Stonecrop family

35 genera and 1500 species

Distribution. Cosmopolitan, but chiefly S. African.

General characteristics. Mainly perennials, living in dry, especially rocky places. They exhibit xerophytic characters – fleshy leaves and stem, often tufted growth, close packing of leaves, waxy surface, sunk stomata, etc. Vegetative reproduction frequent, usually by rhizomes or offsets; some species of *Crassula* form adventitious buds in the inflorescence, while certain species of *Kalanchoe* produce plantlets on the margins of their leaves. Flowers usually in cymes (cincinni), hermaphrodite or rarely unisexual, actinomorphic. The number of floral parts can vary from 3 to 30. Sepals free or united, persistent. Petals free or sometimes united. Stamens frequently obdiplostemonous. Carpels often slightly united at the base. Nectar-secreting scales often present at the base of the carpels. Ovules usually many, with marginal placentation. Fruit usually a group of follicles with very small seeds. Endosperm almost or entirely absent. Flowers mostly protandrous and chiefly visited by flies.

Economic and ornamental plants. Several species of *Sempervivum*, Houseleek, and many species of *Sedum*, Stonecrop, are cultivated as garden plants, either on walls or in a rock-garden. The larger species of Stonecrop make fine border plants. Most of the other genera are too tender to be grown outside, but they are often grown in glasshouses or as house plants.

Classification. Engler (4) divides the family into the following 6 subfamilies:

A. Obdiplostemonous, rarely haplostemonous, petals free or if connate usually only at base:

I. Sedoideae (flowers usually 5-merous).

Sedum (600) N. temperate region.

II. Sempervivoideae (flowers rarely 5-merous, usually more).

Aeonium (40) Atlantic islands, Mediterranean region, Ethiopia, Arabia.

Sempervivum (25) mountains of S. Europe to Caucasus.

III. Echeverioideae (flowers 5-merous, genera American).

Echeveria (150) S. United States to Argentina.

Dudleya (40) S.W. United States, N.W. Mexico.

B. Obdiplostemonous, petals more or less connate forming a tubular or campanulate corolla:

IV. Cotyledonoideae (flowers 5-merous).

Adromischus (50) S. Africa.

Cotyledon (40) S. Africa.

V. Kalanchoideae (flowers 4-merous).

Kalanchoe (200) tropical and S. Africa to China and Java.

C. Haplostemonous, petals free or only basally connate.

VI. Crassuloideae.

Crassula (300) cosmopolitan, especially S. Africa.

KALANCHOE TUBIFLORA (Harv.) Hamet
(BRYOPHYLLUM TUBIFLORUM Harv.)

Distribution. A native of Madagascar and a popular pot-plant.

Vegetative characteristics. A succulent, perennial herb with more or less cylindrical leaves (mostly in whorls of 3) which frequently produce plantlets from the adventitious buds at their tips.

Floral formula. K(4) C(4) A4+4 $\underline{G}$4

Flower and inflorescence. The actinomorphic, hermaphrodite red or reddish flowers may be produced at any time during the year, and are borne in a more or less pendulous manner in terminal cymes.

Pollination. Little is known about the pollination of the genus *Kalanchoe*, though it has been suggested that birds may be the pollinating agents. The flowers of *K. tubiflora* produce a large quantity of nectar secreted by the scales situated at the base of the carpels. British members of the Crassulaceae are occasionally self-pollinated, but more frequently pollination is carried out by small flies or hymenopterous insects.

Alternative flowers for study. The flowers of the genus *Kalanchoe* are comparatively large and in the absence of *K. tubiflora* the equally common *K. daigremontiana*, which produces plantlets all round the edge of its lanceolate leaves, could be used. It should be noted that if another genus is chosen, there may be a considerable difference in the number of floral parts and the degree of attachment of the sepals and petals. One of our 6 native Stonecrops, *Sedum acre*, which flowers in June and July, is found throughout the British Isles and is an example of a member of the Crassulaceae with free petals. The introduced genus *Sempervivum* and the native genera *Crassula* and *Umbilicus* are each represented by one species in Britain.

Fig. 35 Crassulaceae, *Kalanchoe tubiflora*

A L.S. of flower. The 4-lobed calyx and corolla, a particular feature of *Kalanchoe*, have been cut to reveal the carpels which are joined at the base and have long, linear styles. The superior ovary is surrounded by the still immature stamens. The sepals, connate only towards the base, surround the petals which are united for most of their length into an urn-shaped tube before opening out into 4 spreading lobes.

B The gamopetalous corolla, cut down one side and flattened out to show the introrse stamens attached to the corolla-tube. The stamens are in 2 whorls of 4, one whorl, opposite the petals (antipetalous) and the other opposite the sepals (antisepalous). Each of the anthers is 2-celled and dehisces longitudinally.
Calyx lobe + tube: 14 mm
Corolla (in bud): 13 mm
(immature flowers): 19–20 mm
(mature flowers): 25–35 mm

C Two of the stamens at an advanced stage, showing elongation of the filaments.
Anther (in bud): 2 mm Filament (in bud): 3 mm
Mature anther + filament: 20 mm

D L.S. of one of the 4 carpels showing the numerous ovules attached to a well-developed placenta situated on the ventral surface. After fertilisation, the carpel develops into a dehiscent follicle. This fruit splits down the ventral suture to liberate the light seeds which are carried away by the wind.
Stigma + style + ovary: 20–21 mm Ovary: 5.25 mm

E T.S. of complete ovary showing the 4 carpels containing numerous ovules with marginal placentation. At anthesis the carpels show some degree of fusion but later they separate entirely.

F Close-up of the stigma situated at the apex of a linear style.
Stigma: 0.75 mm across

G One of the 4 carpels with the nectar-secreting scale at its base. Each of these scales forms an integral part of the carpel and reaches a length of 0.75 mm.

Fig. 35

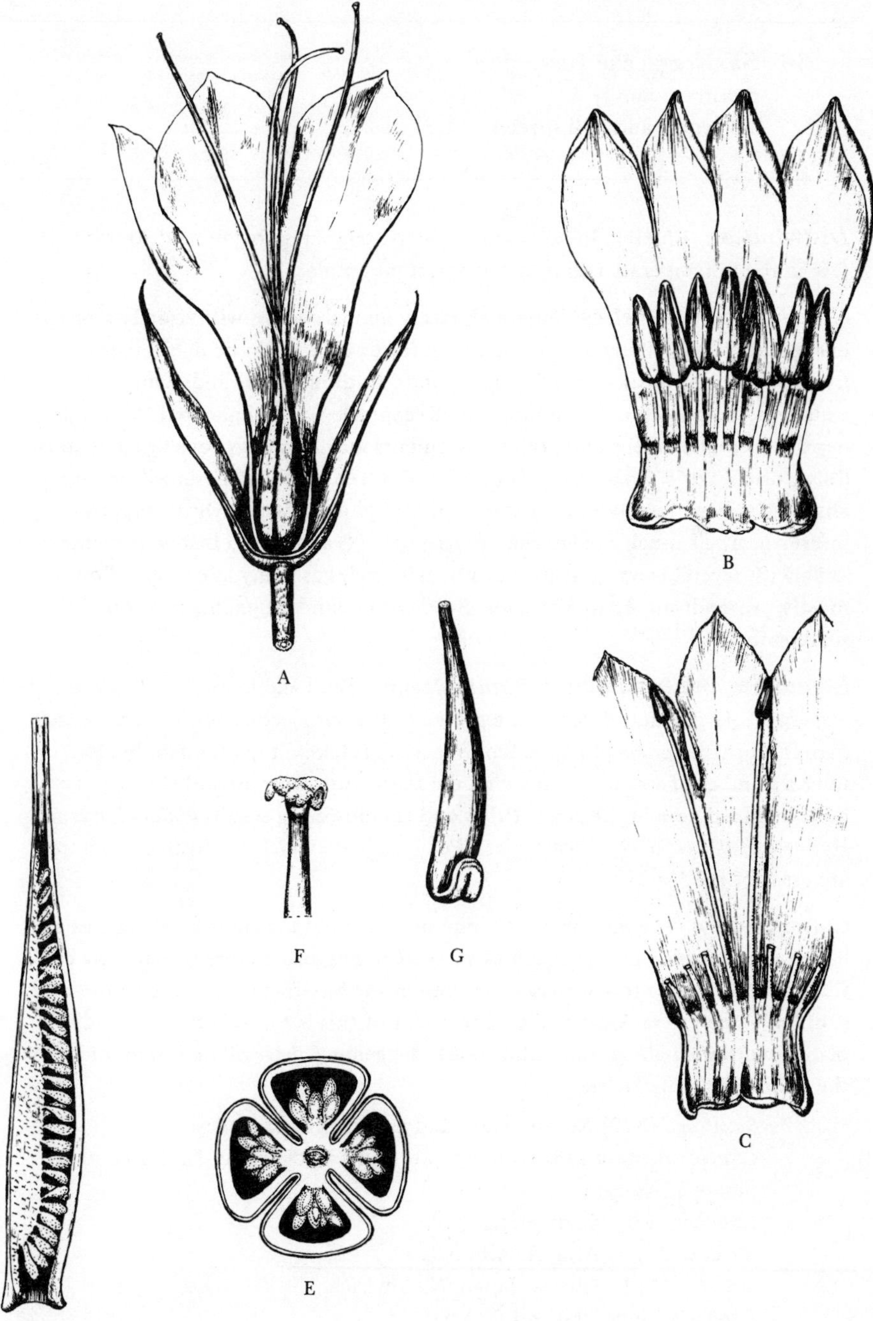

36 Saxifragaceae Juss.

Saxifrage family

30 genera and 580 species

Distribution. Chiefly alpine plants in the N. temperate region and Arctic, but a few in the S. temperate region and tropical mountains.

General characteristics. Perennial, rarely annual, herbs, with usually alternate, exstipulate leaves. Many alpine and arctic forms of xerophytic habit, many hygrophilous. Inflorescence of various kinds, both racemose and cymose. Flowers usually hermaphrodite, actinomorphic. Receptacle flat or hollowed to various depths, so that stamens and perianth-segments may be perigynous or epigynous. Calyx usually of 5 sepals. Corolla usually of 5 free petals, sometimes connate or absent. Stamens usually in 2 whorls of 5, obdiplostemonous. Ovary superior to inferior, carpels usually fewer than petals, often 2 and joined below, placentation axile with several rows of anatropous ovules, styles as many as carpels. Flowers mostly protandrous. Fruit a capsule. Seed with abundant endosperm round a small embryo.

Economic and ornamental plants. Many species and hybrids of *Saxifraga*, *Heuchera*, *Astilbe* and *Bergenia* are grown in the herbaceous border or rock garden. If a broader concept of the Saxifragaceae is taken, then the family also contains a number of woody plants, e.g., the bush fruits Currant and Gooseberry, both included here in the genus *Ribes* and the flowering shrubs *Philadelphus*, *Hydrangea*, *Deutzia* and *Escallonia*, all of which are widely cultivated in temperate areas of the world.

Classification. The Saxifragaceae may be treated in the traditional way, as a large family divided into as many as 17 subfamilies, but a more recent view confines the family to those genera contained in the Saxifragoideae, raising the other subfamilies to the rank of family. One effect of this is to exclude the woody plants mentioned above and in this book the genus *Ribes* will be found under the family Grossulariaceae.

Saxifraga (370) N. temperate and arctic regions, Andes.
Chrysosplenium (55) N. temperate and arctic regions, N. Africa, temperate S. America.
Heuchera (50) N. America.
Astilbe (25) E. Asia, N. America.
Mitella (15) E. Siberia, Japan, N. America.
Bergenia (6) central and E. Asia.

SAXIFRAGA GRANULATA L.
Meadow Saxifrage

Distribution. Native in many parts of the British Isles but considered to have been introduced in some western counties and in Ireland. It is also found in most parts of Europe.

Vegetative characteristics. A perennial herb with a basal rosette of reniform, crenately lobed leaves with long petioles.

Floral formula. K5 C5 A5+5 $\overline{\mathrm{G}}(2)$

Flower and inflorescence. The white flowers are borne on long stalks which are pilose below and glandular above, the whole inflorescence forming a loose terminal cyme. The flowers appear from April until June while the fruits mature from June until August. Though the form of the inflorescence varies considerably within the family, the flowers are almost always actinomorphic. One of the few exceptions is *Saxifraga stolonifera* (*S. sarmentosa*), a fairly common house plant which has zygomorphic flowers.

Pollination. Many kinds of small insects are attracted by the nectar that is secreted by the outer wall of the ovary, as well as by the loose but conspicuous inflorescence of white flowers.

Alternative flowers for study. Many members of the genus *Saxifraga* are rare in the British Isles or at least local, so some caution is required when picking flowers for study. *S. granulata* itself has a widespread but rather local distribution, which is unfortunate considering the convenient size of the flower. However, several garden Saxifrages are readily available and suitable as alternatives and wherever possible these should be chosen in the interests of conservation.

Fig. 36 Saxifragaceae, *Saxifraga granulata*

A Detail of one of the actinomorphic, hermaphrodite flowers. One of the 5 imbricate, glandular sepals has been removed, also 2 of the 5 petals, to reveal the reproductive organs. The ovary is ½–¾ inferior and its lower part is adnate to the receptacle. The 2 styles are erect in the early stages of floral development but spread out later (see D).
Receptacle (at anthesis): 2.5 mm in diameter Sepal: 4 × 1.7 mm
Petal: 14 × 5 mm

B Two of the 10 stamens at different stages of development. The 2-celled anthers are dorsifixed and dehisce longitudinally, ripening before the stigmas. The stamens are obdiplostemonous.
Filament: 4 mm Anther: 1 mm

C T.S. of the ovary composed of 2 carpels showing the numerous ovules attached to the swollen, axile placentas.
Ovary: 1 × 1.5 mm at anthesis, enlarging to 5.75 × 6 mm after fertilisation.

D L.S. of ovary showing the perigynous position of the sepals which are attached opposite the middle of the ovary. The styles, bearing glandular stigmas, spread out during development and persist throughout the life of the flower.
Style + ovary (including receptacle) at anthesis: 4 mm
Style + stigma after petal fall: 6 mm

E Close-up of the glandular-hairy, ventral surface of the stigma.
Stigmatic surface: 1 × 0.75 mm

F The fruit, a capsule, after seed dispersal. It is surrounded by the persistent calyx and the filaments of the stamens which are attached opposite the middle of the ovary. On drying, the capsule splits open at right angles to the main axis in order to disperse the seeds.
Fruit: *c.* 7 mm

Fig. 36

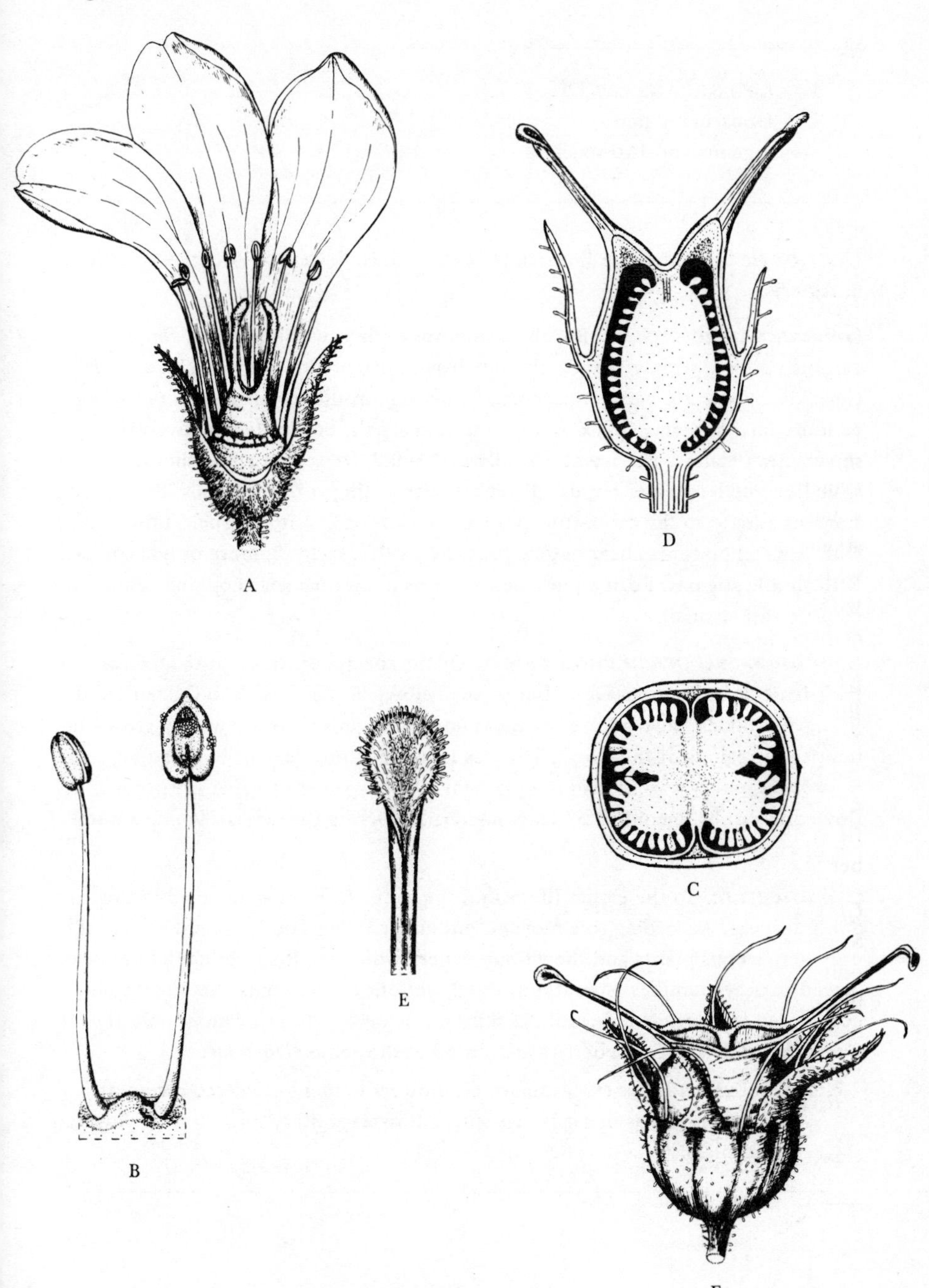

37 Grossulariaceae DC.

Gooseberry family

1 genus and 150 species

Distribution. Temperate Eurasia, N.W. Africa, N. & Central America, Pacific S. America.

General characteristics. Shrubs, sometimes spiny, with alternate, simple, variously lobed, stipulate or exstipulate leaves, often with resinous glands; sometimes dioecious. Flowers actinomorphic, hermaphrodite or unisexual, racemose or sometimes solitary. Calyx of 4 or 5 united sepals, imbricate or subvalvate, sometimes petaloid, persistent. Corolla of 5 small, free, obovate or subulate, scale-like petals. Stamens 5, usually short, alternating with the petals. Petals and stamens adnate to the calyx-tube. Ovary inferior, of 2 united carpels, unilocular, with parietal placentas bearing few to many ovules; styles 2, more or less connate, with simple stigmas. Fruit a juicy berry. Seeds numerous with copious endosperm; embryo rather small.

Economic and ornamental plants. Of the species of *Ribes* cultivated for their fruit, *R. uva-crispa*, Gooseberry (see below), *R. rubrum*, Red Currant, and *R. nigrum*, Black Currant, are the most important and these are widely grown in temperate regions of the world. The less familiar White Currant is a seedling from *R. rubrum* which is lacking in red pigment. Other species of *Ribes* are popular as flowering shrubs, including *R. sanguineum*, Flowering Currant, of which a number of cultivars are now available.

Classification. In the earlier literature, the genus *Ribes* was included in the Saxifragaceae. According to a more recent concept this family is held to contain only herbaceous plants and the woody genera which are thus excluded have been placed in other families. Sometimes the 2 sections of the genus *Ribes* are each considered to merit generic rank. If this view is taken, *Ribes* includes only the Currants, while the Gooseberries are placed in the genus *Grossularia.*

Ribes Subgenus Ribes (twigs unarmed, flowers in many-flowered racemes).
Subgenus Grossularia (twigs spiny, flowers solitary or in 2- to 4-flowered racemes).

RIBES UVA-CRISPA L.
Gooseberry

Distribution. Native in southern, central and western Europe and widespread in the British Isles, though not native in Ireland. It is commonly cultivated for its fruit and sometimes escapes from gardens and becomes naturalised in hedgerows and woodlands.

Vegetative characteristics. A deciduous, much-branched, spiny shrub, with small 3- to 5-lobed leaves which are toothed in the upper part, and broadly cuneate to subcordate at the base.

Floral formula. K(5) C5 A5 $\overline{\mathrm{G}}$(2)

Flower and inflorescence. Axillary clusters of 1–3 hermaphrodite flowers with small, white petals appear from March to May.

Pollination. Hymenopterous insects are attracted to the flowers by the nectar secreted from the epigynous disc at the base of the bell-shaped receptacle. The flowers are protandrous and in the first (male) stage of anthesis the stigmas are still immature, so that cross-pollination occurs during insect visits. Later the stigmas become receptive and stand at the same level as the anthers. It is then possible for some self-pollination to take place.

Alternative flowers for study. In the absence of *R. uva-crispa*, the Red and Black Currants, *R. rubrum* and *R. nigrum*, both of which flower in April and May, are suitable substitutes. These differ from *R. uva-crispa* in being homogamous.

Fig. 37 Grossulariaceae, *Ribes uva-crispa*

A A single, actinomorphic, hermaphrodite flower with 2 bracteoles attached to the pedicel. Adnate to the campanulate receptacle is the hispid calyx whose purplish tinged lobes are considerably larger than the small, white petals.
Calyx-lobe: 3.5 × 2.5 mm Petal: 2 × 1.5 mm
Receptacle: 4 × 4 mm

B The upper portion of the receptacle. The petals are inserted on the hispid edge of the receptacle. The stamens are alternate with them and stand opposite the sepals. The anthers are introrse, 2-celled and dehisce longitudinally.
Filament: 2.5 mm Anther: 0.5 mm

C L.S. of flower. The broadly campanulate receptacle is formed by extension and fusion with the calyx. Below the receptacle is the inferior ovary. The 2 styles, connate below, arise from the centre of the nectar-secreting disc.

D T.S. of the unilocular ovary, showing the ovules attached to the 2 parietal placentas.
T.S. of ovary: 2.25 × 1.7 mm

E L.S. of ovary with part of the disc and connate styles. The numerous anatropous ovules are shown attached to one of the placentas.
Ovary (at anthesis): 2 mm

F Detail of upper (free) portion of the styles showing the entire stigmas.

G The mature fruit, a hispid or smooth berry with numerous seeds and a persistent calyx. Many cultivated forms have been produced, and their fruit varies considerably in regard to size, colour and hairiness.
Fruit: 22 × 18 mm

Fig. 37

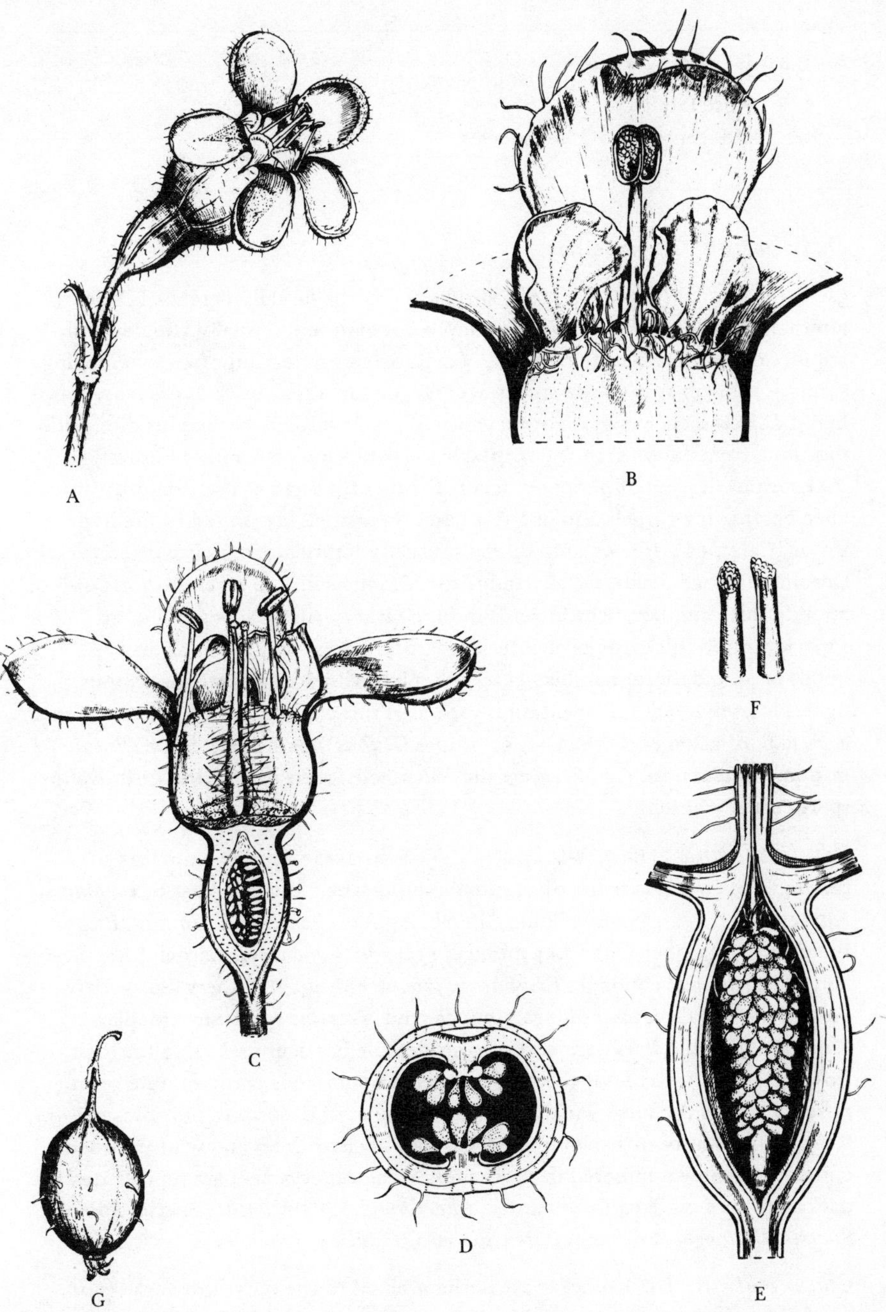

38 Rosaceae Juss.

Rose family

100 genera and 3000 species

Distribution. Cosmopolitan.

General characteristics. Trees, shrubs and herbs, usually perennial. Leaves alternate (except in *Rhodotypos*), simple or compound, usually stipulate, the stipules often adnate to the petiole. Vegetative reproduction often by creeping stems – runners as in *Fragaria*, Strawberry, or suckers as in *Rubus idaeus*, Raspberry. Inflorescence cymose or racemose. Flowers usually hermaphrodite, actinomorphic, terminal on stems. Receptacle (hypanthium) generally hollowed so that various degrees of perigyny occur. Frequently there is a central protruberance bearing the carpels and in a few cases the carpels are united to the hypanthium. Calyx of 5 free or united sepals, usually imbricate; epicalyx often present. Corolla of 5 free petals, usually imbricate. Stamens in 1 to several whorls of 5 or an indefinite number, bent inwards in bud. Ovary usually apocarpous and superior, rarely syncarpous or inferior; carpels 1 to several times as many as petals or an indefinite number, 1- to 5-locular, with 1 to several anatropous ovules in each loculus. Placentation various. Fruit various, dry or fleshy, often an aggregate of achenes (*Potentilla*) or drupes (*Rubus*), or a single drupe (*Prunus*) or pome (*Malus*). In, e.g., *Fragaria* the receptacle forms part of the fruit. Endosperm usually absent.

Economic and ornamental plants. The Rosaceae contain a number of genera which provide fruits of economic importance, the principal being *Malus*, Apple, *Pyrus*, Pear, *Prunus*, Plum, Cherry, Apricot, Peach and Almond, *Rubus*, Blackberry, Raspberry and Loganberry, *Fragaria* × *ananassa*, Garden Strawberry, and in subtropical countries *Eriobotrya japonica*, Loquat. Others, less widely grown, include *Cydonia oblonga*, Quince, and *Mespilus germanica*, Medlar. *Prunus*, *Malus* and *Pyrus* are also of considerable horticultural value for their flowers, and in the case of the first two genera numerous cultivars have been raised. Perhaps the most widely cultivated genus for decorative purposes is *Rosa*, Rose, which has been grown in gardens since ancient times and whose named cultivars are now numbered in thousands. Many other genera are represented in parks and gardens, e.g., *Cotoneaster*, *Chaenomeles*, *Pyracantha*, *Kerria*, *Spiraea*, *Sorbus*, *Crataegus*, *Sorbaria*, *Potentilla* and *Geum*.

Classification. The Rosaceae are perhaps allied to the Calycanthaceae (not treated here) and to the Myrtaceae. Some authorities regard the following 4 subfamilies as being sufficiently distinct to be treated as separate families.

I. Spiraeoideae (carpels usually 2–5, free or connate at base, not sunk in hypanthium; fruit of 1–5 follicles).

Spiraea (100) N. temperate region.
Neillia (13) E. Asia.
Sorbaria (10) central and E. Asia, N. America.
Physocarpus (10) N.E. Asia, N. America.
Exochorda (5) central Asia, China.

II. Rosoideae (carpels usually many, sometimes enclosed in the hypanthium in fruit).

Potentilla (500) nearly cosmopolitan, chiefly N. temperate region and arctic.
Rosa (250) N. temperate region, tropical mountains.
Rubus (250) cosmopolitan.
Alchemilla (250) N. temperate region, tropical mountains.
Acaena (100) chiefly southern hemisphere.
Fragaria (15) Eurasia, N. America, Chile.
Filipendula (10) N. temperate region.

III.Pyroideae (Pomoideae) (carpels 2–5, united to inner wall of hypanthium; fruit a pome).

Crataegus (200) N. temperate region.
Sorbus (100) N. temperate region.
Cotoneaster (50) N. temperate region.
Malus (35) N. temperate region.
Pyrus (30) temperate Eurasia.

IV.Prunoideae (carpel usually 1, free of the hypanthium; fruit a drupe).

Prunus (430) cosmopolitan.

FILIPENDULA ULMARIA (L.) Maxim. (SPIRAEA ULMARIA L.)
Meadow-sweet

Distribution. Native in most parts of Europe except the extreme south and in temperate Asia. It is commonly found throughout the British Isles on river banks, in damp hedgerows, wet woods and meadows, fens, marshes, swamps and on wet rock ledges.

Vegetative characteristics. A perennial, rhizomatous herb with leafy stems reaching between 50 and 200 cm in height. Leaves usually tomentose beneath and consisting of 2–5 pairs of large, toothed leaflets with smaller leaflets between them.

Floral formula. K5 C5 A∞ $\underline{G}$6–10

Flower and inflorescence. Numerous hermaphrodite, actinomorphic, creamy white flowers are borne in cymose panicles from June until September.

Pollination. Numerous types of insects have been observed visiting the homogamous flowers. Nectar is absent, but the insects are attracted by the large, showy inflorescence, the strong scent, and the abundance of pollen. As the flowers are crowded together pollination among themselves can occur as easily as self-pollination, which may itself take place with or without the aid of insect-visitors.

Alternative flowers for study. *Filipendula ulmaria* is a very common plant and an alternative is hardly necessary. It was at one time included in the genus *Spiraea* and therefore a member of the subfamily Spiraeoideae. The transfer to the genus *Filipendula* now places the plant in the subfamily Rosoideae. *Spiraea* differs from *Filipendula* in having 5 carpels, several ovules per carpel and a fruit consisting of a group of follicles. *Filipendula*, on the other hand, has from 6 to 12 carpels, only 2 ovules in each carpel and a fruit composed of a head of achenes. The only other species of *Filipendula* native in Britain is *F. vulgaris*, Dropwort, which is widespread but local in dry, sunny turf on chalk and limestone and flowers between May and August.

Fig. 38.1 Rosaceae, *Filipendula ulmaria*

A The actinomorphic, hermaphrodite flower with some of the parts removed to reveal the gynoecium situated on the flattened hypanthium. The sepals are reflexed and pubescent and the petals are clawed. The carpels, which are more or less erect at anthesis, are surrounded by numerous stamens with filaments that narrow towards the base.
Flower: 8 mm in diameter Sepals: 1.5 x 1 mm
Petals: 3–4 x 2.5 mm Stamens: 5 mm

B Detail of an anther, which is dorsifixed, 2-celled and dehisces longitudinally.
Anther: 0.6 x 0.5 mm

C A single carpel, showing the short style, well-developed stigma and the point of attachment just above the base of the ovary.
Ovary: 1.25 x 0.75 mm Style + stigma: 1 mm

D L.S. of ovary, the single loculus containing 2 pendulous, anatropous ovules.

E T.S. of a whorl of carpels, showing ovules in 5 of the 6 ovaries. The number of carpels may be as many as 10.
Whorl of carpels: 1.2 mm in diameter

F The whorl of carpels at fruiting stage when they become spirally twisted and form a head of achenes, each achene containing a single seed.
Head of achenes: 3 x 3 mm

Fig. 38.1

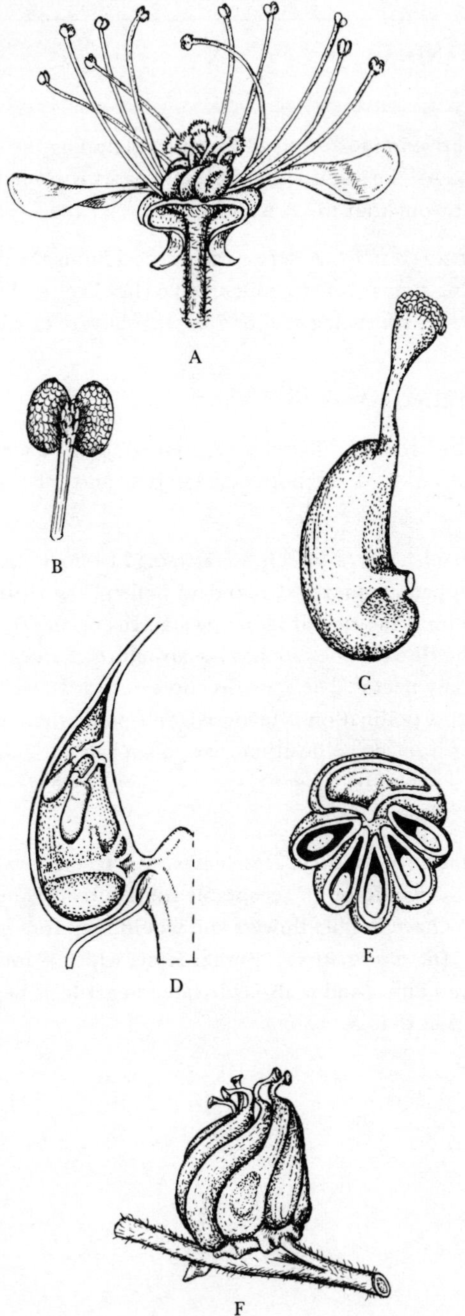

ROSA CANINA L.

Dog Rose

Distribution. Native in most parts of Europe (excluding the extreme north), N. Africa and S.W. Asia; naturalised in N. America. It is common in hedges, woods and scrub throughout the British Isles.

Vegetative characteristics. A very variable, deciduous shrub from 1 to 3 m in height, with arching stems armed with curved or hooked prickles. The leaves are stipulate and pinnate, consisting of 5 or 7 ovate, obovate or elliptical, toothed leaflets.

Floral formula. $K5\ C5\ A\infty\ \underline{G}\infty$

Flower and inflorescence. The showy, scented, pink or white flowers, either solitary or in groups of 2–4, are borne during June and July on relatively long pedicels.

Pollination. Several types of insects are attracted to the homogamous flowers by their fragrance, petal colour and abundant pollen. The visiting insects usually land on the disc within the ring of radiating stamens or on the exposed stigmas at the centre of the flower. The disc has the structure of a nectary but does not appear to secrete any nectar. The stamens curve outwards, well away from the stigmas, so that cross-pollination is favoured. Self-pollination, however, may occur if the flower is in such a position that pollen can fall on to a stigma, or if an insect, in the course of its foraging, causes pollen to be transferred to the stigma of the same flower.

Alternative flowers for study. Rosa canina is a readily available shrub but any other single rose would be an acceptable substitute. It should be noted that many rose cultivars have double flowers and would therefore not be typical. *R. pimpinellifolia*, (*R. spinosissima*), Burnet Rose, which is found growing wild on sandy heaths and dunes and is also cultivated in gardens, begins flowering about a month earlier than *R. canina.*

Fig. 38.2 Rosaceae, *Rosa canina*

A L.S. of the actinomorphic, hermaphrodite flower with some of the sepals and petals removed to reveal the numerous stamens that are adnate to the margin of the prominent disc. A large number of styles protrude in a bunch from the orifice in the centre of the disc. The styles arise from the apex of free carpels enclosed within the urceolate hypanthium (see D). After fertilisation, the hypanthium ripens into a fleshy, red pseudocarp, 'hip', which is attractive to birds. The reflexed sepals, which are sometimes narrowly lobed, are deciduous after anthesis.
Sepals: 25 x 6.5–7 mm Petals: 35 x 30 mm
Stamens: up to 10 mm

B The prominent disc with a central orifice through which protrude numerous styles with simple stigmas. Round the edge of the disc are the numerous stamens, only the bases of which are shown.
Disc: 6–6.5 mm in diameter

C Detail of the basifixed anther, which is 2-celled and dehisces longitudinally.
Anther: 2 x 2 mm

D L.S. of the hypanthium showing some of the numerous carpels attached to its inner surface. Each carpel contains a single, pendulous ovule. The styles bunch together to pass through the narrow orifice in the disc, but then spread out to expose the maximum stigmatic area to insect-visitors.
Hypanthium: *c.* 10 mm Disc: 2 mm

E A single carpel attached to part of the inner wall of the hypanthium. The carpel bears several long hairs and terminates in a long style with a well-developed stigma. The length of the style varies according to the position of the carpel within the hypanthium. After fertilisation the ovules develop into hairy achenes enclosed within the fleshy pseudocarp, 'hip', which has developed from the hypanthium.
Carpels: 8–11 mm Ovary: 1 mm in diameter

Fig. 38.2

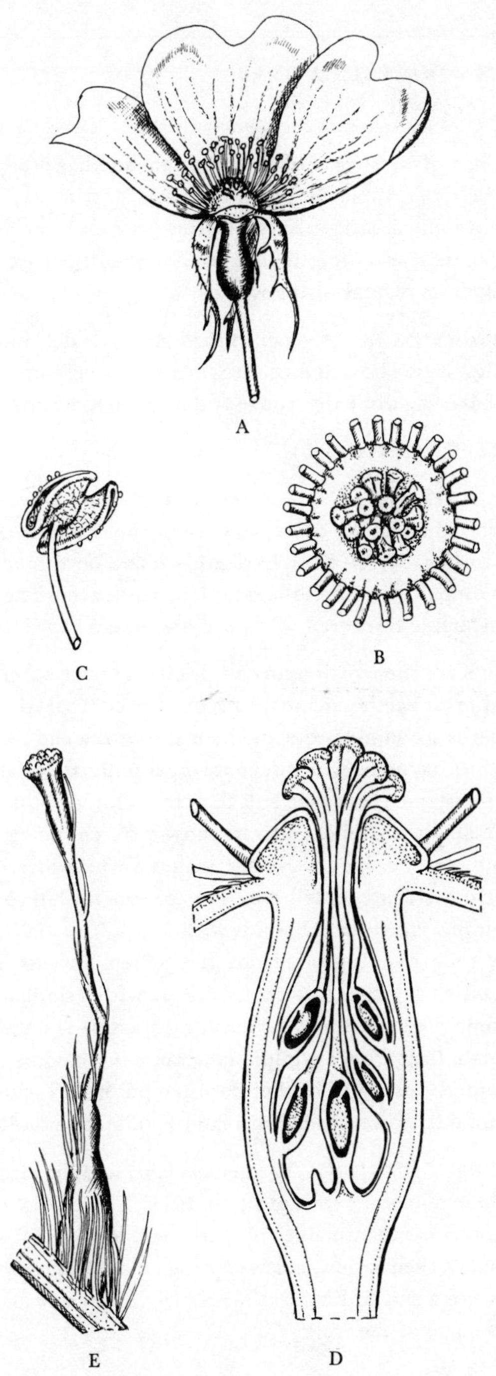

MALUS × DOMESTICA Borkh.
Cultivated Apple

Distribution. *M.* × *domestica* is of hybrid origin and has probably been derived from *M. sylvestris*, Crab Apple, *M. dasyphylla* and *M. praecox*, all native in Europe, and from some Asiatic species. It often escapes from cultivation, sometimes becoming naturalised. More than a thousand cultivars are grown for their fruits in the temperate regions of the world.

Vegetative characteristics. A small to medium-sized deciduous tree with tomentose young twigs. The dull green leaves are ovate-elliptical, serrate, usually rounded at the base, slightly hairy above and more densely so beneath.

Floral formula. K5 C5 A∞ $\overline{G}$(5)

Flower and inflorescence. The scented white to pink flowers are borne in umbels on fruiting spurs on the older wood or at the end of shoots formed in the spring prior to fruiting. The pedicel, hypanthium and outer surface of the calyx are more or less tomentose. The flowering time varies according to the particular cultivar, but most apple trees are in flower between late April and the end of May.

Pollination. Bees are the most important pollinators but other insects, including flies, midges and small beetles are attracted by the scent, petal-colour, pollen and by the nectar that is secreted on the inner surface of the cup-like hypanthium. The flowers are protogynous, which favours cross-pollination, although in those cultivars where the stamens are long and therefore closer to the stigmas, self-pollination can take place. Apple cultivars show different degrees of self-compatibility and better crops are usually achieved when different cultivars cross-pollinate. Domestic apples fall into 2 major genetic categories, diploids with 2 sets of chromosomes and triploids with 3 sets. The diploid apples have viable pollen but the triploids produce inferior pollen. Successful pollination is promoted by planting together different cultivars with a similar flowering period. Two different diploid cultivars may be used, each acting as a pollinator for the other, but to obtain fruit from a triploid cultivar it is recommended that at least 2 diploids be planted close by. The diploids then pollinate each other and the triploid also. Triploids, however, do not readily pollinate diploid cultivars.

Alternative flowers for study. The genus *Malus* and the closely related genus *Pyrus*, Pear, include numerous species and hybrids, but as *M.* × *domestica* is represented by many easily available cultivars there is no real need for an alternative. Comparison between various cultivars of *M.* × *domestica* will reveal numerous variations in flowering and fruiting times, relative flower size and especially the size, colour, and shape of the fruit.

Fig. 38.3 Rosaceae, *Malus* × *domestica*

A L.S. of the actinomorphic, hermaphrodite flower at early anthesis. The 5 sepals and the 5 free, clawed petals are adnate to the rim of the cup-like hypanthium. The 5 styles are connate below and gradually increase in length until at maturity, when the stigmas are receptive, they are as long as or longer than the stamens. The inferior ovary is embedded in the fleshy hypanthium.
Corolla: 40–42 mm in diameter Petals: 18–25 × 10–14 mm
Sepals: 8 × 2.75 mm Style + stigma: 3 mm

B Two stamens, prior to dehiscence, attached to a portion of the hypanthium. The stamens are usually whorled and are adnate to the hypanthium at a point just below the petals. The anthers are versatile, 2-celled and dehisce longitudinally.
Stamens: 5–8 mm Anthers: 2.25 × 1.75 mm

C T.S. of ovary and hypanthium at anthesis. The ovary consists of 5 united carpels, each carpel containing 2 ovules attached to an axile placenta.
Hypanthium: 3 mm in diameter

D L.S. of the mature, fleshy fruit, a pome, showing the persistent calyx within the apical hollow. The edible part is formed by the development of the ovary wall and the hypanthium. The 'core' consists of the cartilaginous walls of the united carpels which enclose the seeds and protect them from desiccation. The carpels are often 1-seeded owing to abortion of the other seed in the loculus.
Fruit: 55 × 55 mm

Fig. 38.3

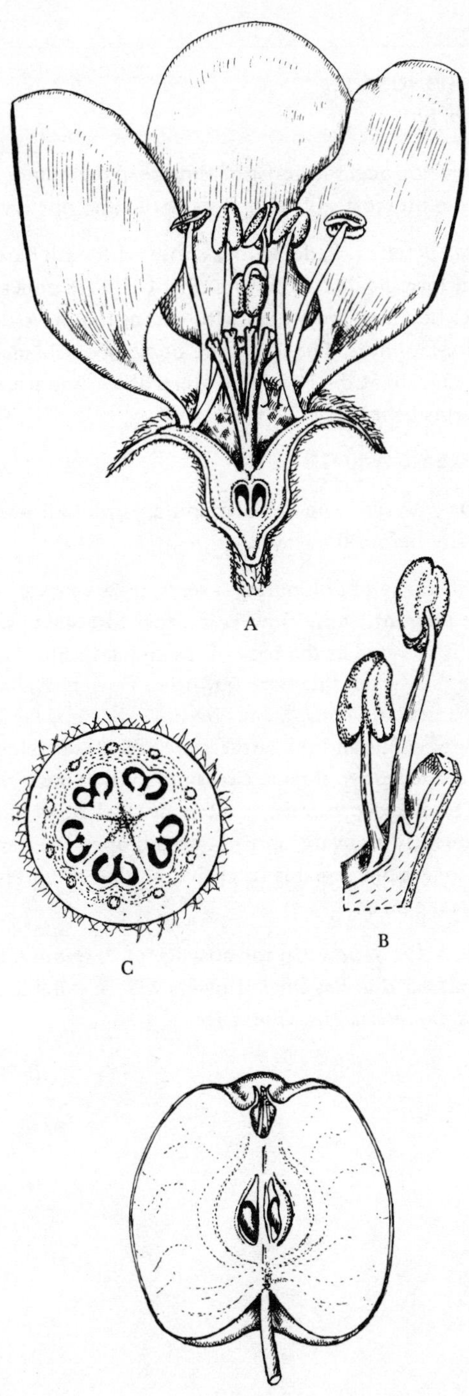

PRUNUS SPINOSA L.
Blackthorn, Sloe

Distribution. Native throughout most of Europe and commonly found growing in the British Isles on most soils in hedges, scrub and open woodland.

Vegetative characteristics. A deciduous shrub with dark bark, 1–4 m high, often suckering and forming dense thickets. The twigs are pubescent when young and the older branches are armed with thorns that have developed from numerous short, lateral shoots. The leaves are obovate or oblanceolate, finely crenate or serrate, cuneate at the base, dull green above and usually more or less pubescent on the veins beneath.

Floral formula. K5 C5 A20 $\underline{G}$1

Flower and inflorescence. The axillary, solitary, white flowers appear from March to May, usually before the leaves.

Pollination. A wide range of pollinating insects, including various kinds of bees, are attracted to the fragrant, white flowers in order to obtain pollen and seek out the nectar that is secreted at the base of the hypanthium. The flowers begin to open early in the year when there are few other flowers available for insects to visit and they are particularly conspicuous because they appear before the leaves. The flowers are protogynous and the anthers are still closed when the buds expand and reveal the receptive stigma. Cross-pollination must therefore be effected by insects that have previously visited an older flower in the male stage. The stamens subsequently elongate, spread out and open their anthers. The style also becomes longer but remains receptive, so that should insect-visits fail, self-pollination can take place.

Alternative flowers for study. In the absence of *P. spinosa*, any other *Prunus* species or cultivar that has single flowers will be suitable including the edible Plum, *Prunus domestica* ssp. *domestica.*

Fig. 38.4 Rosaceae, *Prunus spinosa*

A L.S. of the actinomorphic, hermaphrodite flower. The spreading petals, reflexed sepals and numerous stamens are borne on the rim of the hypanthium. In the centre of the hypanthium is the ovary surmounted by a long style.
Corolla: 22 mm in diameter Sepals: 3 x 2 mm
Petals: 8.5–11 x 5.5–6.5 mm Hypanthium: 3 x 4 mm
Gynoecium: 11 mm Stamens: 6–9.5 mm

B Detail of the dorsifixed anther which is 2-celled and dehisces longitudinally.
Anther: 0.6 x 0.5 mm

C L.S. of hypanthium and ovary. The unilocular ovary contains 2 pendulous ovules (only 1 shown). Surrounding the ovary is the cup-like hypanthium which secretes droplets of nectar. The hypanthium is discarded when the fruit develops.

D Detail of the capitate stigma with pollen grains.
Stigma: 0.7 mm in diameter

E T.S. of ovary at late anthesis. Two pendulous ovules are situated off centre within the single loculus. After fertilisation one of the ovules develops at the expense of the other so that a single-seeded fruit is produced.
Ovary: 0.9 mm in diameter Loculus: 0.6 mm in diameter

F L.S. of the mature fruit, a bluish black drupe with a whitish bloom. The fruit is composed of 3 distinct layers – the outer skin (epicarp), a fleshy, green middle layer (mesocarp) and a stony inner layer (endocarp) which encloses the seed. When the seed germinates the endocarp splits along a longitudinal groove.
Fruit: 13 x 11–12 mm

Fig. 38.4

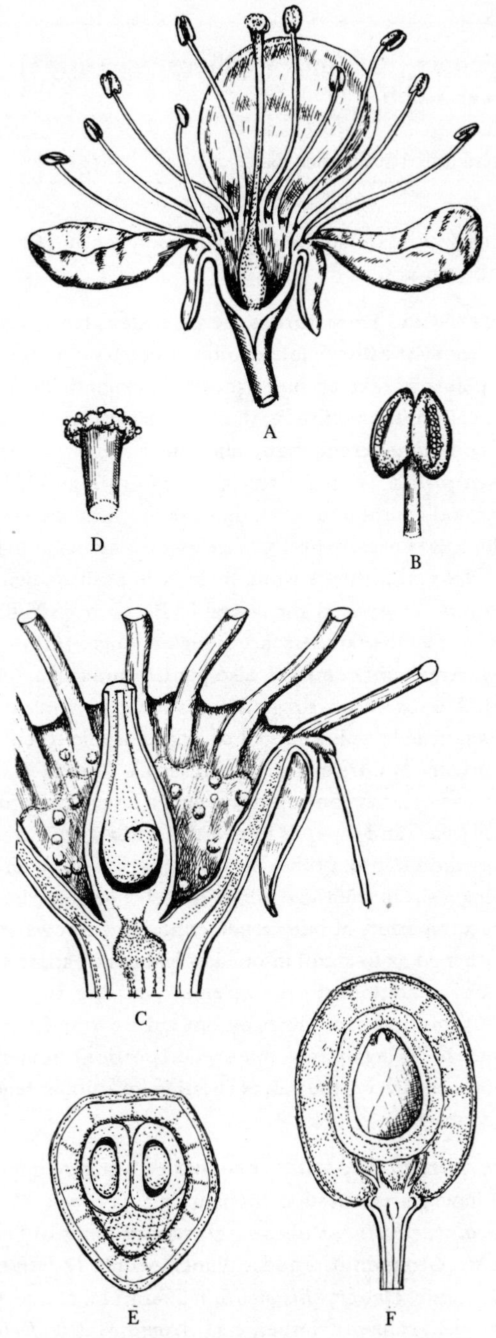

39 Leguminosae Juss. (incl. Papilionaceae Giseke) (Fabaceae Rchb.)

Pea family
600 genera and 12000 species

Distribution. Cosmopolitan (subfamilies Mimosoideae and Caesalpinioideae are mostly found in the tropics).

General characteristics. Trees, shrubs, herbs, water plants, xerophytes, climbers, etc. The roots of most plants exhibit tubercles containing bacteria which enable the plants to take up much more atmospheric nitrogen. Since they enrich the soil instead of impoverishing it, many species are valuable as crops on poor soils. Stems commonly erect, many climbing by leaf- or stem-tendrils or by twining round other plants. Thorns often present (e.g., *Acacia*). Leaves usually alternate, stipulate and compound, sometimes reduced to scales (e.g., *Carmichaelia*). The leaves usually perform sleep-movements at night, but in some (e.g., *Mimosa* and *Neptunia*) they assume their sleep position when touched. In *Desmodium motorium* (*D. gyrans*) the lateral leaflets move steadily round in elliptical orbits as long as the temperature is high enough. Inflorescence usually racemose, simple raceme very common, also panicle and spike, heads of sessile flowers common in Mimosoideae, single flowers sometimes occur. Flowers actinomorphic (then usually polygamous) or zygomorphic (then usually hermaphrodite). Calyx of usually 5 more or less united sepals. Corolla of usually 5 free or united petals; in many cases zygomorphic, having a large standard petal (vexillum), 2 wing petals (alae) and 2 petals joined to form a keel (carina). Stamens 10 to many, free or united into a tube, in the latter case one stamen often remains free, leaving a slit in the tube; where a keel is present the stamens are enclosed in it. Ovary superior, of one carpel, containing 2 rows of ovules (alternate with one another so as to stand in one vertical rank), anatropous or amphitropous, ascending or pendulous. Fruit extremely variable, typically a legume opening by both sutures, but sometimes by one suture or quite indehiscent, dry or fleshy, sometimes breaking up into one-seeded portions (lomentum). Pods frequently opening explosively, the valves twisting up spirally (e.g., *Ulex*, *Cytisus*). Endosperm usually absent.

Economic and ornamental plants. Economically a very important family with many genera having seeds used as foodstuffs, e.g., *Pisum*, *Cajanus*, *Cicer*, *Lotus*, *Vicia*, *Phaseolus*, *Glycine*, *Dolichos*, all various kinds of Peas and Beans, *Lens*, Lentil, *Arachis*, Groundnut. Fodder plants include *Trifolium* spp., Clover, *Medicago sativa*, Lucerne, *Onobrychis viciifolia*, Sainfoin. Many tropical and subtropical species yield valuable timber, e.g., *Acacia*, Wattle, *Dalbergia*, Rosewood, *Robinia pseudoacacia*, Locust-tree, *Gleditsia*, Honey Locust, *Pericopsis* (*Afrormosia*), *Pterocarpus*, Padauk. *Crotalaria*, Sunn Hemp, and others are sources of fibre. *Acacia*, *Genista*, *Haematoxylum*, *Indigofera*, etc., yield dyes. Gums and resins are obtained from *Acacia*, *Copaifera*, *Hymenaea*, etc. Oil is expressed from the seeds of *Arachis* and *Voandzeia*, Groundnuts. The roots of *Glycyrrhiza* spp. are the source of Liquorice. Many of the Leguminosae are used for decorative purposes. In temperate regions the most commonly cultivated include *Lathyrus odoratus*, Sweet Pea, *Lupinus* spp., Lupins, *Genista* and *Cytisus*, Broom, *Spartium junceum*, Spanish Broom, *Colutea*, Bladder Senna, *Cercis*, Judas-tree, *Laburnum*, and *Wisteria*.

Classification. The Leguminosae is divided here into 3 subfamilies, but some authorities treat these as separate families. The subfamilies are further divided into a large number of tribes (not indicated here), about half of these being in the Papilionoideae.

I. Mimosoideae (Flowers actinomorphic. Sepals valvate, rarely imbricate. Petals valvate in bud. Stamens 4–10 or numerous. Leaves often bipinnate).
  - *Acacia* (750–800) tropics and subtropics.
  - *Mimosa* (450–500) mainly tropical and subtropical America.
  - *Inga* (200) tropical and subtropical America, W. Indies.
  - *Pithecellobium* (200) tropics.
  - *Albizia* (100–150) warm regions of Old World.
  - *Calliandra* (100) Madagascar, warm regions of Asia, America.

II. Caesalpinioideae (Flowers usually more or less zygomorphic. Sepals imbricate, rarely valvate. Petals imbricate-ascending in bud. Stamens 10 or fewer, rarely numerous. Leaves usually pinnate, occasionally bipinnate, rarely simple).
  - *Cassia* (500–600) tropical and warm temperate regions.
  - *Bauhinia* (300) warm regions.
  - *Caesalpinia* (100) tropics and subtropics.
  - *Gleditsia* (11) tropics and subtropics.
  - *Cercis* (7) N. temperate region.

III. Papilionoideae (Flowers usually zygomorphic. Sepals imbricate. Petals imbricate-descending in bud. Stamens generally 10. Leaves pinnate, digitate, trifoliolate or simple).
  - *Astragalus* (2000) cosmopolitan, except Australia.
  - *Indigofera* (700) warm regions.
  - *Crotalaria* (600) tropics and subtropics.
  - *Desmodium* (450) tropics and subtropics.
  - *Dalbergia* (300) tropics and subtropics, S. Africa.

Oxytropis (300) N. temperate region
Trifolium (300) temperate and subtropical regions.
Phaseolus (200–240) tropics and subtropics.
Lupinus (200) America, Mediterranean region.
Hedysarum (150) N. temperate region.
Vicia (150) N. temperate region, S. America.
Lathyrus (130) N. temperate region, mountains of tropical Africa and S. America.
Lotus (100) temperate Europe, Asia and Africa.
Medicago (100) temperate Eurasia, Mediterranean region, S. Africa.
Genista (75) Europe, N. Africa, W. Asia.
Cytisus (25–30) Atlantic islands, Europe, Mediterranean region.

ACACIA ARMATA R. Br. (Mimosoideae)
Kangaroo Thorn

Distribution. Native in Australia, but often grown elsewhere as a glasshouse shrub. In Britain it can only be cultivated outside in the mild S.W. region, preferably against a wall.

Vegetative characteristics. A spreading, evergreen shrub, reaching a height of 3 m in its native habitat. Its leaves are represented by phyllodes which are semi-ovate, the straight edge being close to the stem and the curved outer edge more or less undulate; the nerve of the phyllode is excentric and ends in a sharp point. Stipules, situated at the base of the phyllodes, are reduced to slender spines.

Floral formula. $K(5)\ C5\ A\infty\ \underline{G}1$

Flower and inflorescence. The flowers appear in April. Each inflorescence arises on a peduncle from the axil of a phyllode and consists of a strongly-scented, ball-shaped cluster of predominantly staminate flowers together with a few hermaphrodite ones. Each individual flower is actinomorphic with valvate sepals and petals.

Pollination. Various kinds of insects are attracted by the numerous yellow stamens and the scent of the flowers. In the course of its visit the insect's body becomes coated with pollen from the prominent stamens and this is readily transferred from one inflorescence to another. Small birds may also act as pollinating agents by dispersing the pollen while hunting out the insect visitors.

Alternative flowers for study. In the absence of *A. armata*, *A. dealbata*, the 'Mimosa' of the florist, is readily obtainable but differs from *A. armata* in that the heads of pale yellow flowers form a panicle and the leaves are of the normal kind and bipinnate. The individual flowers in this subfamily are generally 5-merous, though 3-, 4- or 6-merous flowers are not uncommon. The well-known

Mimosa pudica, Sensitive Plant, has 4-merous, purplish flowers. The bipinnate leaves droop when touched and the pinnae close together.

Fig. 39.1 Leguminosae, *Acacia armata*

A Apical portion of a flowering stem, showing the phyllodes with the central nerve ending in a short point and the persistent stipules which are reduced to short, slender spines. Arising from the axils of the phyllodes are the peduncles bearing the strongly scented, globular inflorescences.
Peduncles: *c.* 14 mm Inflorescences: 6–8 mm in diameter

B The arrangement of some of the flowers in a capitulum. All the flowers illustrated are staminate apart from 1 hermaphrodite flower which is identified by the single style protruding beyond the stamens.

C A single flower, showing the 5-lobed calyx surrounding the more or less free petals and the basal portion of 5 of the numerous stamens.
Calyx: 1 mm Corolla: 2 mm

D L.S. of a single hermaphrodite flower. The numerous free stamens are inserted round the base of the superior ovary, which is terminated by a long style. The ovary is unilocular with marginal placentation. The ovules form a row along the placenta and each is attached by a funicle which becomes silvery in colour and ends in a cup-shaped aril. The fertilised ovary develops into a hairy, straight or slightly curved legume, 35–50 mm long.
Stamens: 3–3.75 mm Ovary (at anthesis): 0.9 × 0.25–0.3 mm
Style: 3.75–4 mm

E Stamen. E1: upper portion of young stamen prior to dehiscence. E2: upper portion of mature stamen. The anthers are 2-celled and dehisce longitudinally.
Anther: 0.125 mm

Fig. 39.1

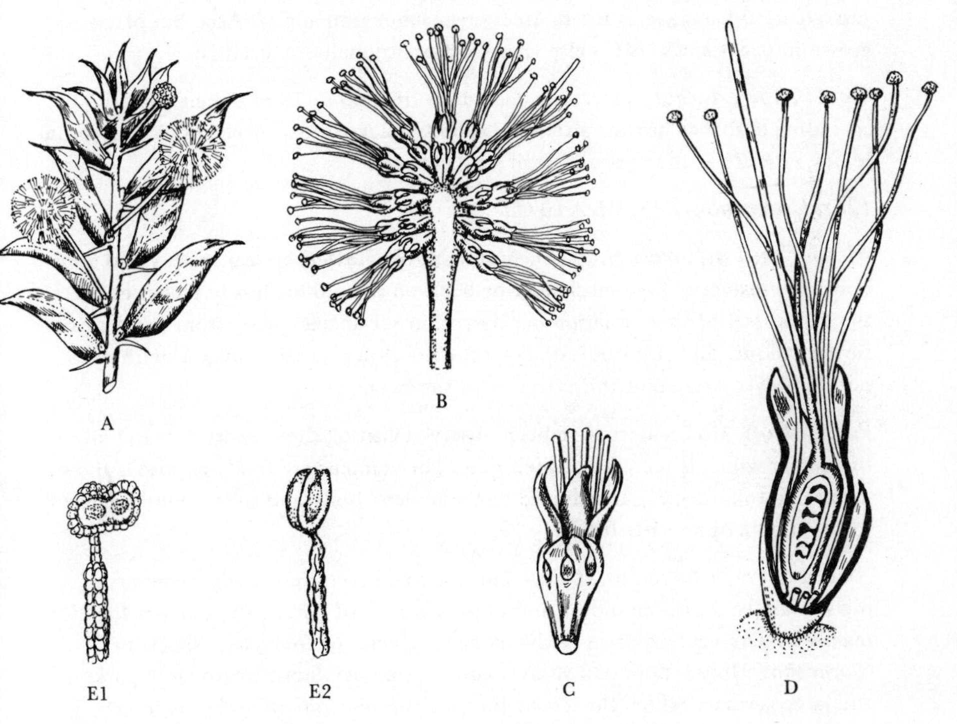

CERCIS SILIQUASTRUM L. (Caesalpinioideae)
Judas-tree

Distribution. Native in the E. Mediterranean region and W. Asia, but often grown in parks and gardens elsewhere and occasionally naturalised.

Vegetative characteristics. A deciduous tree, up to 10 m in height, with low, spreading branches bearing alternate, suborbicular, obtuse or emarginate, glabrous leaves, cordate at the base and with long petioles.

Floral formula. K(5) C5 A10 $\underline{G}$1

Flower and inflorescence. The purplish-rose flowers appear in May and are borne in clusters of 3–6 on slender pedicels on the old leafless branches or on the trunk itself. Little variation occurs within the species apart from the depth of flower-colour. The cultivars 'Alba' with white flowers and 'Bodnant' with deep purple flowers represent the extremes of the range.

Pollination. Honey-bees have been observed visiting the flowers in search of the nectar which is secreted at their base. The stamens are freely exposed, allowing pollen to be readily transferred to the insect's body and subsequently carried to the stigma of another flower.

Alternative flowers for study. There are relatively few hardy members of the subfamily Caesalpinioideae and in the absence of *Cercis siliquastrum* the most readily available alternatives are likely to be species of *Gleditsia*, especially *G. triacanthos*, Honey Locust, a spiny-stemmed tree occasionally grown in parks and large gardens, which flowers in July. *Caesalpinia gilliesii* and *C. japonica*, which are less hardy, may also be found in flower in July in gardens in the milder parts of the British Isles. Some of the N. American species of *Cassia* can also be cultivated outside in this country, but other species of *Cassia* and *Caesalpinia* require glasshouse conditions.

Fig. 39.2 Leguminosae, *Cercis siliquastrum*

A A single flower with a zygomorphic corolla of 5 petals and a campanulate calyx with 5 equal teeth.

B L.S. of flower. Both petals and sepals are imbricate. The petals are clawed and vary considerably in width. Five of the 10 free stamens are exposed to reveal their point of insertion. The superior ovary with marginal placentation is terminated by a curved style.
Pedicels: 15–17 mm Calyx: 8 mm Petals: 10–11 × 4.5–7 mm
Ovary + style (at anthesis): 16 mm

C Stamen. C1: ventral view of apex of stamen showing longitudinal dehiscence of anther with pollen grains. C2: dorsal view, showing dorsifixed attachment of anther to filament.
Stamens: 13–14 mm Anthers: 1.25 mm

D T.S. of ovary with 2 ovules visible. The ovules are arranged in 2 rows and alternate with each other so as to form 1 vertical rank.
T.S. of ovary: 1 × 0.4 mm

E The fruit, a legume which is much compressed and has a narrow wing on its ventral suture. In autumn the fruits are more or less crimson in colour but they turn brown during the winter and persist for a long time on the tree. Each fruit contains up to 12 black seeds.
Fruits: 50–65 × 10–12 mm
Seeds: 5 × 3–3.5 mm

Fig. 39.2

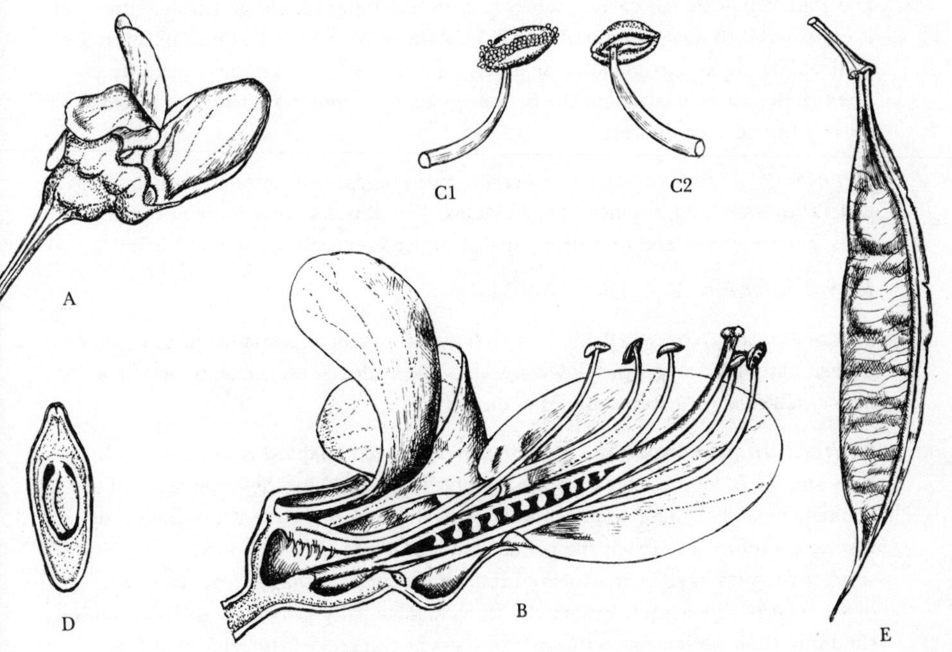

VICIA FABA L. (Papilionoideae)

Broad Bean

Distribution. One of the most ancient of all Old World cultivated vegetables and grown as food for cattle, horses and human beings since prehistoric times. It is considered to have been native in S.W. Asia or N. Africa, but no undoubtedly wild plants are known from these areas. Alternatively, it may have been developed under cultivation from the S. European *V. narbonensis* which it resembles closely in many characters.

Vegetative characteristics. An erect, entirely glabrous annual with square, usually unbranched, slightly winged stems. The stipulate leaves are without tendrils and are composed of 2 or 3 pairs of bluish-green, fleshy, ovate leaflets.

Floral formula. K(5) C5 A(10) $\underline{G}$1

Flower and inflorescence. Two to 6 zygomorphic, hermaphrodite, white or occasionally lilac or purple flowers with a black blotch on the wing petal are borne in axillary clusters from May until July.

Pollination. The Papilionoideae are generally well adapted to pollination by bees and *Vicia faba* is no exception. Nectar is secreted by the inner sides of the stamens near their base and accumulates in the staminal tube. Only long-tongued insects therefore can reach the nectar in a legitimate way, although some with shorter tongues have been observed to bite through the base of the flower in order to steal the nectar. Before the flowers have fully grown the anthers dehisce, shedding their pollen on to the tuft of hairs at the end of the style (see Fig. 39.3.G). When an insect alights on an open flower, it presses down the wings and the keel petals, which are connected at the base, and exposes the rigid style. The pollen adhering to the apical hairs is then transferred to the underside of the insect and the rubbing of the insect's body causes the stigma to become sticky, and thus receptive to pollen from another flower.

Alternative flowers for study. The Broad Bean is represented by many cultivars which have only minor floristic differences and it is so widely grown that an alternative during its flowering period is scarcely necessary. During mild winters and particularly in the early spring *Ulex europaeus*, Gorse, may be found in flower. In the autumn *Vicia sativa*, Common Vetch, and species of *Ononis*, Restharrow, may still be flowering.

Fig. 39.3 Leguminosae, *Vicia faba*

A L.S. of the zygomorphic, 5-merous flower, showing the staminal tube enclosed within the 2 connate petals which form the keel. One of the black-blotched wing petals can be seen, also a portion of the standard, the largest petal.
Calyx-tube: 7–8 mm Calyx-lobe: 8 mm

B The petals. Top: the standard (vexillum) 24–30 × 12.5–15 mm. Centre: one of the wing petals (ala) – claw 10–12 mm, limb 10–11 × 5–6 mm. Bottom: the 2 joined petals forming the keel (carina) – claw 10–11 mm, limb *c.* 7 × 5.5 mm.

C The apical region of the staminal tube, showing the free, curved portions of the 10 stamens. The upturned end of the style can be seen projecting from the staminal tube.
Free part of stamen: *c.* 9 mm

D Detail of free part of stamen. The protandrous stamens have 2-celled anthers which dehisce by longitudinal slits.
Anther: 3.5 × 1.5 mm

E L.S. of the superior ovary terminated by the brush-like style. The ovary is unilocular, consisting of one carpel with marginal placentation. The ovules are arranged alternately to form a single row and are attached by well-developed funicles to the placenta situated along the ventral suture.
Ovary (at anthesis): *c.* 14 mm

F T.S. of ovary, showing one of the ovules joined by its funicle to the single placenta.
T.S. of ovary (in drawing): 7 × 5 mm Ovule: 3.25 × 1.5 mm

G Detail of the apex of the style, showing the tuft of hairs and stigmatic surface.

H Fruit at an early stage of development. The fruit is a pod (legume) which at maturity splits down the ventral and dorsal sutures. The seeds are large and have a leathery testa. The length of the mature legume is variable and depends on the cultivar concerned. It normally reaches 10–20 cm.
Fruit (in drawing): 4.5 cm

Fig. 39.3

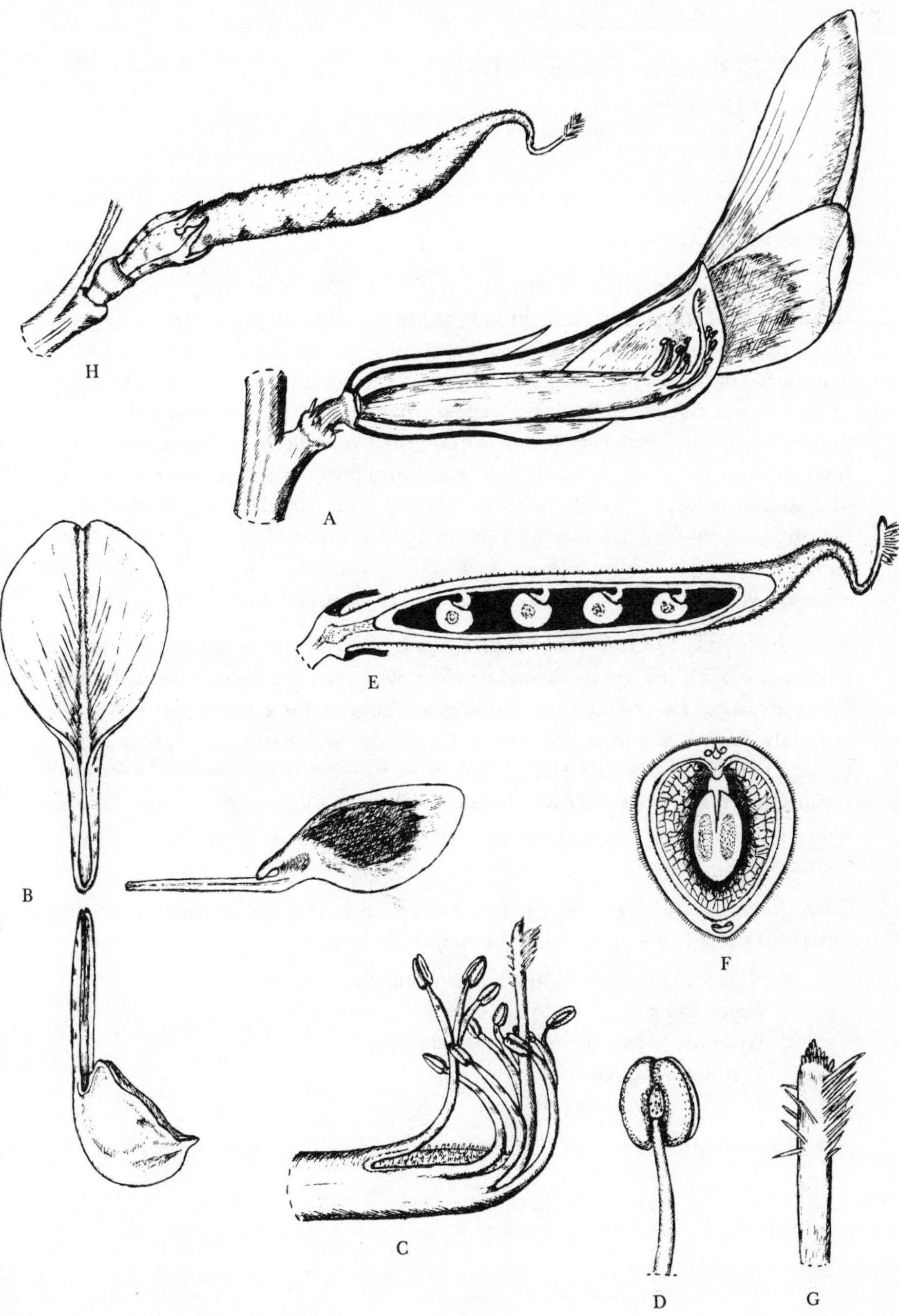

40 Lythraceae Jaume St.-Hil.

Loosestrife family

25 genera and 550 species

Distribution. Cosmopolitan.

General characteristics. Herbs, shrubs or trees. Leaves usually opposite, entire, simple, exstipulate or with only minute stipules. Inflorescence a raceme, panicle or cyme. Flowers hermaphrodite, actinomorphic or zygomorphic. Epicalyx of combined stipules (cf. *Potentilla* in Rosaceae) frequently present. Calyx valvate, of 4–8 sepals. Corolla of 4–8 free petals, crumpled in bud, sometimes absent. Stamens inserted (often very low down) on the hypanthium or 'calyx-tube', typically twice as many as sepals, but sometimes fewer or more. Ovary superior, of 2–6 united carpels, usually with as many loculi as carpels and with axile placentation, rarely unilocular with parietal placentas; ovules usually numerous and anatropous; style simple; stigma usually capitate. Heterostyly occurs in some genera, e.g., *Lythrum* (see below). Fruit usually a capsule. Endosperm absent.

Economic and ornamental plants. The Lythraceae are of little economic importance but a few are medicinal or yield dyes, e.g., *Lawsonia inermis*, Henna. Some species of *Lagerstroemia* provide good timber. This genus is also known horticulturally on account of *L. indica*, Crape Myrtle, a tree or shrub with pink to deep red (or occasionally white) flowers which is widely cultivated in warm regions, but requires a sunny wall in cooler climates. *Cuphea ignea*, Cigar-flower, with its tubular, bright scarlet flowers, has become a popular flowering house plant.

Classification. In Engler (4) the Lythraceae are divided into 2 tribes, according to whether or not the septa reach the top of the ovary:

1. Nesaeeae (ovary completely partitioned).
   Lagerstroemia (53) tropical Asia to N. Australia.
2. Lythreae (ovary incompletely partitioned).
   Cuphea (250) America.
   Lythrum (35) cosmopolitan.

LYTHRUM SALICARIA L.
Purple Loosestrife

Distribution. Native throughout Europe, except the extreme north, and in W. and N. Asia, N. Africa and N. America. It is locally abundant in England, Wales and Ireland, but less frequent in Scotland. It grows in reed-swamps, ditches, fens, marshes, by riversides, at the margins of lakes and in other damp places.

Vegetative characteristics. A perennial, more or less pubescent herb with erect stems 50–150 cm in height. Its leaves are mostly opposite or in whorls of 3, sessile, ovate to lanceolate-oblong, acute at the apex and truncate and semi-amplexicaul at the base.

Floral formula. K6 C6 A6+6 $\underline{G}$(2)

Flower and inflorescence. The inflorescence consists of a dense spike of numerous, small, whorl-like cymes that arise from the axils of small bracts. The conspicuous, purple flowers appear from June until August and are trimorphic (see below).

Pollination. Bees and hover-flies are the chief pollinators of *L. salicaria*, being attracted to the purple flowers by the nectar secreted at the base of the hypanthium or 'calyx-tube'. The flowers are trimorphic and have stamens and styles of 3 different lengths, the kind of flower produced being determined genetically. They occur on average in approximately equal numbers, although within any one population there may be considerable variation in the proportions. Flowers with a long style have short and medium length stamens (see Fig. 40.B1) and large pollen grains; those with a short style have long and medium length stamens (see Fig. 40.B2) and small pollen grains; the third kind, with a medium length style, have long and short stamens (see Fig. 40.B3) and medium sized pollen grains. The difference in length between styles and stamens is of great importance in 'legitimate' (successful) cross-pollination of the flowers, since there is a much higher seed-set when a style receives pollen from stamens of the same length as itself. The following table summarises the situation, legitimate pollinations being indicated by arrows:

Long-styled flower	Medium-styled flower	Short-styled flower
(LONG) style ←-------	LONG stamens	LONG stamens (- - - → (LONG) style)
MEDIUM stamens ----→	(MEDIUM) style ←-----	MEDIUM stamens
SHORT stamens (- - - → (SHORT) style)	SHORT stamens ------→	(SHORT) style

Alternative flowers for study. *L. salicaria* occurs so frequently that an alternative is hardly necessary. The other genera mentioned in the section on Classification are in cultivation and may be used for comparison. *Lagerstroemia*, usually represented by *L. indica*, has large, showy petals and numerous stamens, and *Cuphea*, best known by *C. ignea*, has zygomorphic flowers and variation in the number of petals. In *C. ignea* the petals are absent and it is the bright red calyx tipped with black and white which has made this species so popular as a house plant.

Fig. 40 Lythraceae, *Lythrum salicaria*

A An actinomorphic, hermaphrodite flower arising from the axil of a bract. The ribbed hypanthium or 'calyx-tube' has 6 valvate sepals attached to its margin. Alternating with these are the 6 longer and narrower portions of the epicalyx. Opposite the epicalyx-segments (i.e., alternate with the sepals) are the 6 free petals also inserted on the rim of the hypanthium. The petals are crumpled in bud. The measurements of the following are the same in all 3 forms of the flower:
Hypanthium: 5.5 mm Sepals: 2 mm Petals: 10–12 mm

B L.S. of each of the 3 forms of the flower (B1, B2 and B3), illustrating heterostyly. The pubescent, hollow hypanthium encloses the superior ovary. The 12 stamens are attached to the inner surface of the hypanthium, near its base, and are in 2 whorls of 6, the outer whorl alternating with the petals.
B1 (long-styled flower)
Short stamen: 4 mm Medium stamen: 7 mm
Ovary: 2.5 mm Style + stigma: 9 mm
B2 (short-styled flower)
Medium stamen: 7 mm Long stamen: 12 mm
Ovary: 2.5 mm Style + stigma: 1.75 mm
B3 (medium-styled flower)
Short stamen: 5 mm Long stamen: 10–11 mm
Ovary: 2 mm Style + stigma: 6 mm

C Detail of the introrse anthers. C1 (ventral view): the anther is 2-celled and dehisces longitudinally. C2 (dorsal view): the anther is attached in a dorsifixed manner to the filament.
Anther: 1.25 × 1 mm (the same in all 3 forms of the flower)

D L.S. of the lower part of the flower. The ovary is terminated by a single style with a capitate stigma. Numerous anatropous ovules are attached to the prominent placentas.
Ovary: 2.5 × 1 mm

E T.S. of ovary formed from 2 fused carpels. The 2 loculi contain numerous ovules attached to axile placentas.
T.S. of ovary: 0.8 × 1.1 mm

F The fruit, a 2-valved capsule which dehisces septicidally. The hypanthium and the remains of the stamens persist round the fruit after the petals have withered away.
Fruits: 3–3.5 × 2 mm

Fig. 40

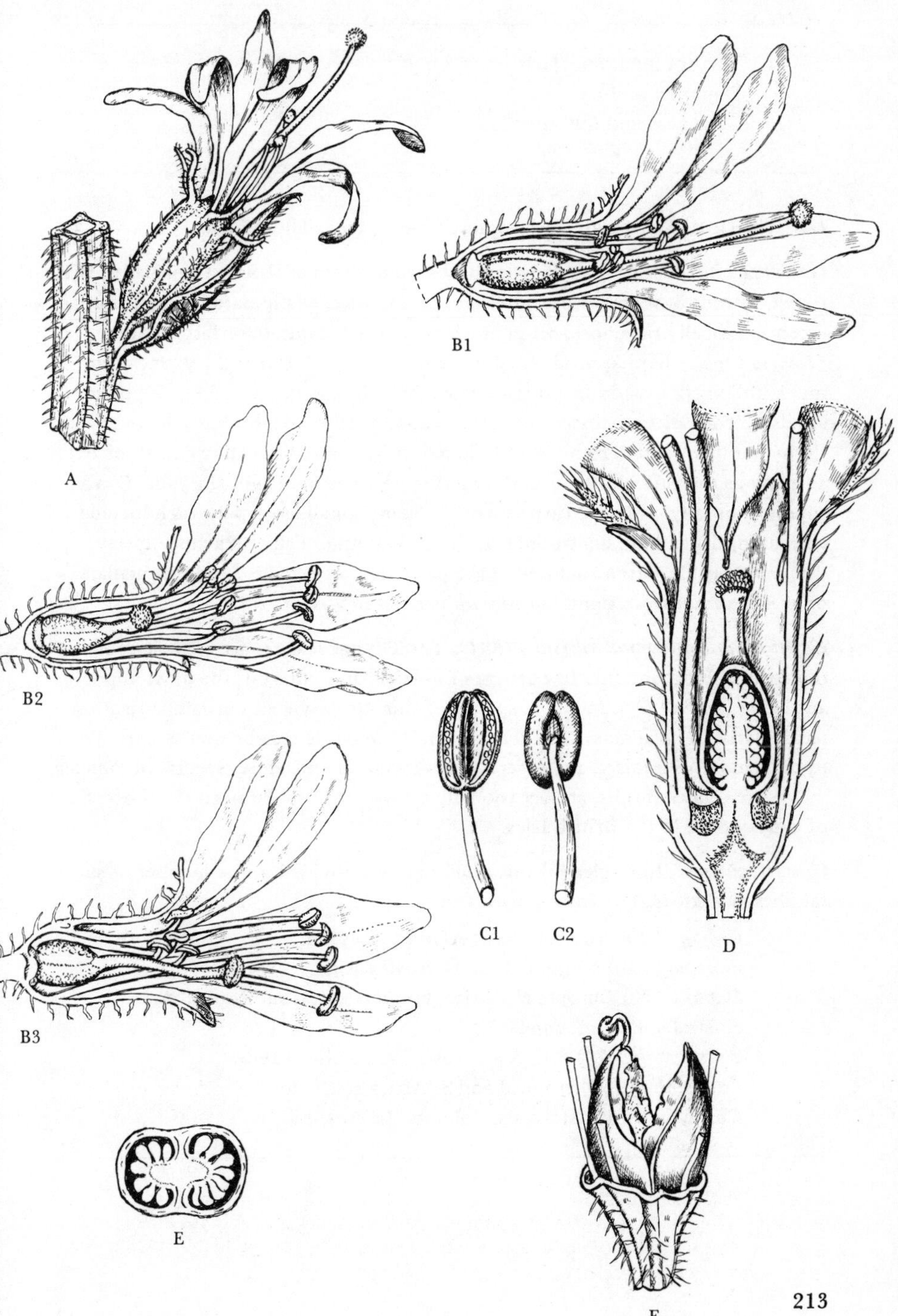

41 Thymelaeaceae Juss.

Daphne family

50 genera and 500 species

Distribution. Temperate and tropical regions, especially Africa.

General characteristics. Mostly shrubs, some trees or lianes, a few herbs, with tough, fibrous bast and entire, alternate or opposite, exstipulate leaves. Inflorescence basically racemose, often much condensed, capitate or fasciculate. Flowers usually hermaphrodite, actinomorphic, 4- or 5-merous. Hypanthium much hollowed, usually forming a deep tube. Calyx petaloid, of 4 or 5 united sepals. Corolla often reduced to scales or absent. Disc membranous or somewhat fleshy, annular, cupular or variously lobed or laciniate. Stamens as many or twice as many as the calyx-lobes, usually inserted near the mouth of the tube. Ovary superior, usually of 1 or 2 carpels, with as many loculi as carpels, each loculus containing 1 pendent, anatropous ovule; style simple. Fruit an achene, berry, drupe or capsule, often enclosed in the persistent receptacle. Seed carunculate or arillate; embryo straight, endosperm present or absent.

Economic and ornamental plants. The Thymelaeaceae are of little economic value, but contain a few genera of horticultural interest, the most important being *Daphne*. A number of species of this shrubby genus are in cultivation, particularly *D. mezereum*, *D. cneorum* and *D. laureola* (see below). Some hybrids have been raised and several cultivars are in existence. Species of *Pimelea*, which are tender shrubs, are occasionally grown, but usually need the protection of a glasshouse in the British Isles.

Classification. In Engler (4) this small family is divided into a number of subfamilies and tribes. The chief genera are:

Gnidia (100) tropical and S. Africa to Ceylon.
Pimelea (80) Philippine Islands, Australasia.
Daphne (70) Europe, N. Africa, temperate and subtropical Asia, Australia, Pacific islands.
Wikstroemia (70) S.E. Asia, Australia, Pacific islands.
Lasiosiphon (50) tropical and S. Africa to Ceylon.
Gonystylus (30) Malaysia, Solomon Islands, Fiji.

DAPHNE LAUREOLA L.
Spurge Laurel

Distribution. Native in W. and S. Europe, W. Asia, N. Africa, and the Azores. It is found in woods mainly on calcareous soils in England and Wales, also in Scotland where it is considered to have been introduced.

Vegetative characteristics. A poisonous, evergreen shrub, 40–100 cm high, with sparingly branched stems bearing alternate, shortly stalked, leathery leaves concentrated near their tips. The glossy, dark green leaves are obovate-lanceolate in shape, and acute or subacute at the apex.

Floral formula. K(4) C0 A4+4 $\underline{G}$1

Flower and inflorescence. The yellowish-green, actinomorphic, hermaphrodite flowers are borne in short, axillary racemes on the previous year's growth, appearing between February and April.

Pollination. The homogamous flowers are adapted to pollination by bees and Lepidoptera. These long-tongued insects are attracted by the scent and by the nectar secreted by an annular disc at the base of the ovary. The perianth and hypanthium form a narrow tube, and short-tongued insects are unable to reach the nectar because of the stamens attached to the upper part of the tube, and the bulging ovary in the lower part. When a long-tongued insect visits the flower, its head and body become dusted with pollen as it probes for nectar. The pollen is subsequently carried away and transferred to the stigma of another flower.

Alternative flowers for study. The other species of *Daphne* found in Britain is the deciduous *D. mezereum*, but this is rare in the wild and should never be picked. It is, however, common in cultivation. It has pinkish-purple, fragrant flowers from February to April, borne on leafless stems. The only other genus likely to be available is *Pimelea* which is sometimes represented in cool glasshouses by the spring-flowering *P. ferruginea. Pimelea* has 2 well-exserted stamens and a long style in contrast to *Daphne*, which has 8 stamens, usually not exserted and a short or even absent style.

Fig. 41 Thymelaeaceae, *Daphne laureola*

A An axillary raceme, showing 2 of the 5–10 hermaphrodite, actinomorphic flowers at anthesis, subtended by several green bracts.
Bracts: 9–10 × 3 mm

B The upper portion of the flower opened out to reveal the stamens which are in 2 whorls of 4. The flower is without petals but has 4 imbricate, petaloid sepals which terminate the tubular hypanthium. Both sepals and hypanthium are the same colour and give the appearance of a single structure.

C L.S. of the flower. The superior ovary is situated within the lower part of the hypanthium and is terminated by a short style with a prominent, capitate stigma. The 2 whorls of stamens occupy the upper part of the hypanthium. The nectar-secreting disc can be seen at the base of the ovary.
Hypanthium: 6–7 × 2.5 mm Sepals: 2.5–4 × 2.5–3 mm

D Detail of the free portion of a stamen. The anthers are introrse, 2-celled and dehisce longitudinally. A few pollen-grains can be seen adhering to the anther-cells.
Anther: 1.25–1.3 mm

E L.S. of the superior ovary enclosed within the hypanthium. The ovary is unilocular and contains a single, pendent, anatropous ovule. The small circles at the base of the ovary show where the nectar-secreting disc has been cut through.
Ovary (at anthesis): 2.2 × 0.8 mm Style: 0.4 mm
Stigma: 0.35 mm

F T.S. of the ovary as it develops into a fruit. The centre is occupied by a well-developed 'stone', which consists of the woody endocarp enclosing the large seed. Surrounding this 'stone' is the fleshy mesocarp and the thin, outer skin or epicarp.
Ovary: 1.15 mm in diameter

G The fruit at maturity, an ovoid, black, poisonous drupe. The short stalks are the pedicels of discarded fruits. Most of the bracts have fallen off, leaving one at the base of the peduncle.
Fruits: 8 × 5–6 mm

Fig. 41

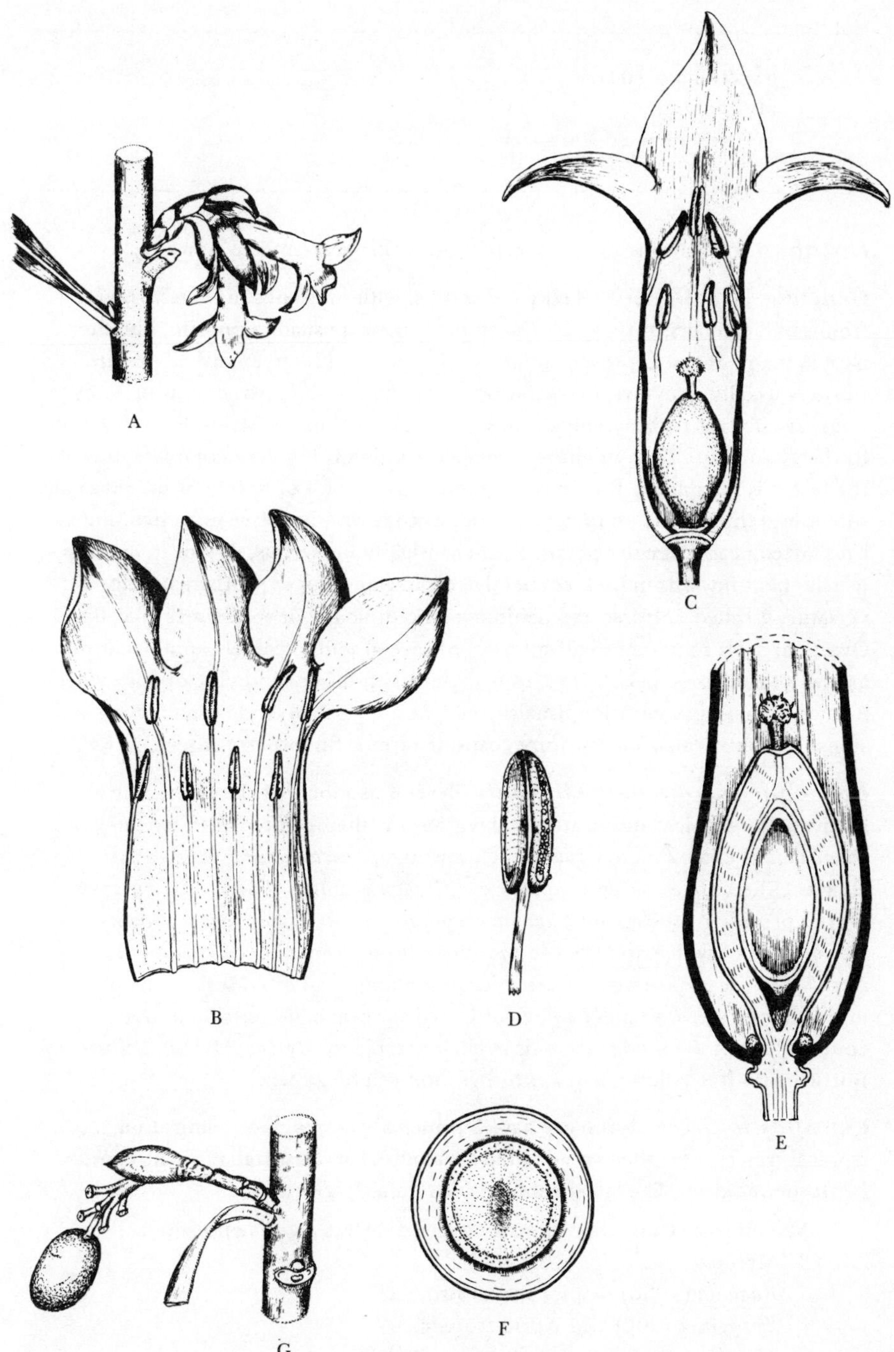

42 Myrtaceae Juss.

Myrtle family

100 genera and 3400 species

Distribution. Warm regions, especially Australia and tropical America.

General characteristics. Trees and shrubs, with oil-glands in leaves, varying from small creepers to the giant *Eucalyptus*. Leaves usually opposite, but alternate in many of the Leptospermoideae, exstipulate, evergreen, usually entire. Flowers usually in cymes, hermaphrodite, actinomorphic, perigynous or epigynous, usually with 2 bracteoles at base. Receptacle more or less hollow and united to the gynoecium. In the Leptospermeae the union is not very complete, but in the rest it is so, and the flower is epigynous. Calyx of 4 or 5, free or united sepals, sometimes thrown off unopened as a lid. Corolla of 4 or 5 free or united, imbricate, often nearly circular petals. Stamens usually numerous, free or in bundles, usually bent inwards in bud; connective often gland-tipped; anthers usually versatile, 2-celled, introrse and dehiscing longitudinally or sometimes apically. Ovary inferior, sometimes half-inferior, of several united carpels, unilocular or as many loculi as carpels, with 2 to many anatropous or campylotropous ovules in each loculus; placentation usually axile, rarely parietal; style usually long and stigma simple. Fruit a berry, drupe, capsule or nut. Endosperm absent.

Economic and ornamental plants. Several members of the Myrtaceae are cultivated in tropical and subtropical regions for their edible fruits, e.g., *Feijoa sellowiana*, Feijoa, *Psidium guajava*, Guava, and *P. cattleianum*, Strawberry Guava. Others are a source of spices, e.g., *Pimenta dioica*, whose dried, unripe berries provide Allspice, and *Eugenia caryophyllus* (*Syzygium aromaticum*) the flower-buds of which are dried to produce Cloves. A number of species of *Eucalyptus* are important sources of timber, tannin and oils. Most of the plants in this family are too tender to be cultivated for ornament outside in this country, but in the milder areas or in sheltered places *Myrtus*, Myrtle, *Callistemon*, Bottle-brush (see below), and *Leptospermum* can be grown.

Classification. The Myrtaceae have 2 principal centres of concentration, tropical America for the berry-fruited Myrtoideae and Australia for the capsular Leptospermoideae. The latter subfamily contains 2 tribes.

I. Myrtoideae (fruit a berry, rarely a drupe; leaves always opposite).
   1. Myrteae.
      Eugenia (1000) tropics and subtropics.
      Syzygium (500) Old World tropics.
      Myrcia (500) tropical S. America, W. Indies.

Psidium (140) tropical America, W. Indies.
Myrtus (100) tropics and subtropics, especially America.
Calyptranthes (100) tropical America, W. Indies.
Pimenta (18) tropical America, W. Indies.

II. Leptospermoideae (fruit dry; leaves opposite or alternate).

2. Leptospermeae (ovary multilocular, at least when young; fruit a loculicidal capsule).
Eucalyptus (500) Australia.
Melaleuca (100) Australia, Pacific.
Baeckea (65) Australia, New Caledonia.
Metrosideros (60) S. Africa, E. Malaysia, Australasia, Polynesia.
Tristania (50) Malaysia, Queensland, New Caledonia, Fiji.
Leptospermum (50) Malaysia, Australasia.
Callistemon (25) Australia, New Caledonia.

3. Chamaelaucieae (ovary unilocular; fruit a 1-seeded nut; endemic in Australia).
Calycothrix (40)
Verticordia (40)
Darwinia (35)
Chamaelaucium (12)

CALLISTEMON CITRINUS (Curt.) Stapf

Bottle-brush

Distribution. Native on the coasts of E. Australia and introduced into Britain in 1788. It is popular as an ornamental greenhouse shrub, but is only suitable for outdoor cultivation in this country in the milder areas or in very sheltered gardens.

Vegetative characteristics. An evergreen shrub with alternate, rather narrow leaves tapering at both ends. A peculiarity of *Callistemon* (and *Melaleuca*) is that the point of growth does not cease with flower production, but continues to grow on, producing more leaves and flowers on the new extension to the stem.

Floral formula. K(5) C5 A∞ $\overline{G}$(3)

Flower and inflorescence. The inflorescence, which appears in June, consists of a spike of sessile flowers borne in the axils of deciduous floral bracts. The sepals and petals are small and it is the numerous, long, red stamens which form the most conspicuous feature. The tufts of stamens make up a cylindrical inflorescence reminiscent of a 'bottle-brush' – hence the popular name for the genus *Callistemon.*

Pollination. In common with *Eucalyptus* (Myrtaceae), *Banksia* (Proteaceae) and other Australian plants with brush-like inflorescences, *Callistemon* is pollinated by birds. Also, the colour red is much more frequently found in bird-pollinated plants than in those pollinated by insects, and this is present to a striking degree in the stamens of *C. citrinus.*

Alternative flowers for study. The related genus *Melaleuca*, which is a suitable alternative, is distinguished from *Callistemon* in having the stamens in 5 bundles, situated opposite the petals. If possible, comparison should be made with a species of the important genus *Eucalyptus*, also with the only European member of the family, *Myrtus communis*, which is in a different subfamily, and usually flowers in July and August.

Fig. 42 Myrtaceae, *Callistemon citrinus*

A The complete inflorescence, with the stem above it continuing to grow and bearing newly-formed leaves.
Inflorescence: *c.* 10 cm

B The complete infructescence, a dense mass of fruits which have developed on the older part of the stem.
Infructescence: *c.* 11 cm

C An individual, sessile, actinomorphic, hermaphrodite flower. The calyx consists of a campanulate tube with 5 free lobes. The 5 free imbricate petals, of which 3 are shown, are deciduous. The numerous stamens with long, red filaments surround and mask the single, simple style.
Stamens: 25–30 mm

D L.S. of lower part of flower, showing the campanulate calyx-tube adnate at its base to the inferior ovary. The filaments are attached to the rim of the calyx-tube. The single style arises centrally from the ovary, which contains numerous ovules on the well-developed, axile placenta.
Calyx: 3.25 mm Ovary: 2 mm

E T.S. of ovary with axile placentas. The ovary is formed by the fusion of 3 carpels, each with a single loculus containing numerous ovules.
Ovary: 2.75 mm in diameter

F Portion of style terminated by the simple stigma.
Stigma: 0.5 mm in diameter

G Upper portion of stamen, showing the versatile, 2-celled anther dehiscing longitudinally.
Anther: 1 × 0.4 mm

H One of the fruits, a woody, loculicidal capsule enclosed within the calyx-tube.
Fruit: 5 × 7 mm

Fig. 42

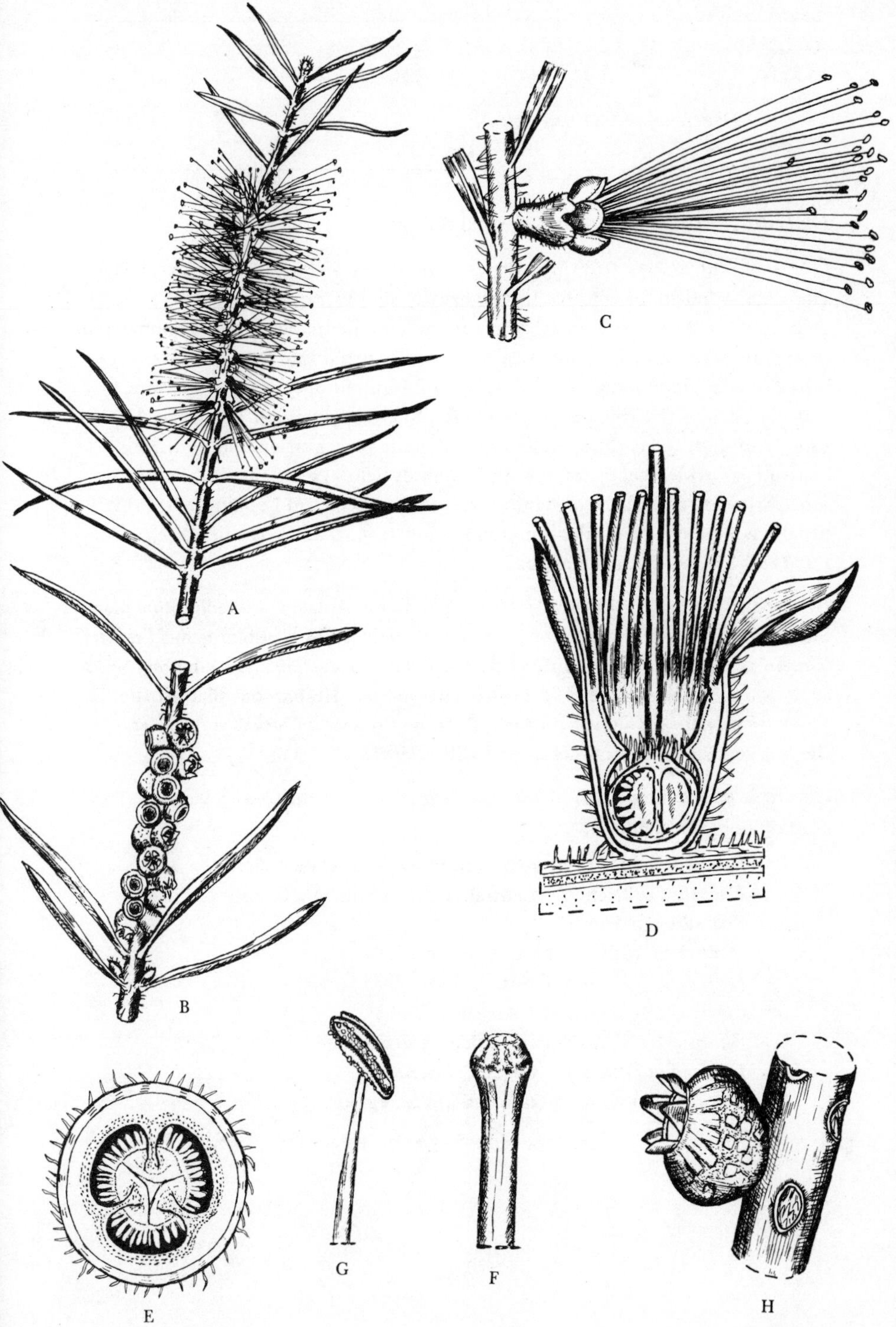

43 Onagraceae Juss.

Willow-herb family

21 genera and 640 species

Distribution. Temperate and tropical regions.

General characteristics. Mostly perennial herbs, a few shrubs or trees, with alternate, opposite or whorled leaves, usually simple, rarely stipulate. Flowers solitary in axils or in spikes, racemes, or panicles, hermaphrodite, actinomorphic or zygomorphic, usually 4-merous. Axis usually prolonged beyond the ovary into a hypanthium or 'calyx-tube'. Calyx of 4 united or free sepals, valvate. Corolla usually of 4 free petals, convolute. Stamens usually 4 + 4. Ovary inferior, usually of 4 united carpels, 4-locular, with axile placentation and numerous anatropous ovules; septa commonly imperfect below; style simple, stigmas 1 or more. Flowers often very protandrous, suited to bees and Lepidoptera. Fruit usually a loculicidal capsule, sometimes a nut (e.g., *Gaura*) or a berry (e.g., *Fuchsia*). Endosperm little or absent.

Economic and ornamental plants. The family is lacking in economic plants but it contains several genera of horticultural value. The shrubby genus *Fuchsia* is mainly represented by hybrids deriving from *F. magellanica* (see below), *F. coccinea* and various species from Central America. Herbaceous plants often grown in gardens include *Oenothera*, Evening Primrose, *Clarkia* and *Godetia* – the last genus now sometimes united with *Clarkia*.

Classification. Engler (4) divides the family into 9 tribes, but several of these contain only one genus.

Epilobium (215) N. and S. temperate and arctic regions.
Fuchsia (100) chiefly Central and S. America, a few species in New Zealand and Tahiti.
Oenothera (80) America, W. Indies.
Ludwigia (75) cosmopolitan, especially tropical America.
Clarkia (36) western N. America, Chile.
Gaura (18) N. America, Mexico, Argentina.
Lopezia (17) Mexico, Central America.
Circaea (12) N. temperate and arctic regions.

FUCHSIA MAGELLANICA Lam.

Distribution. Native in S. Chile and Argentina, *F. magellanica* is widely grown in gardens for its colourful, pendent flowers and is used as a hedge-plant in the milder districts of the British Isles. In some places it has escaped from cultivation, and has become naturalised locally in S.W. England and W. Ireland.

Vegetative characteristics. A medium-sized shrub, with toothed, red-veined ovate-oblong, acuminate leaves which are arranged in pairs or in whorls of 3.

Floral formula. K(4) C4 A4+4 $\overline{\text{G}}$(4)

Flower and inflorescence. Scarlet and purple, usually solitary flowers are produced from June to September on flexible pedicels which arise from the leaf-axils (see Fig. 43.G).

Pollination. In their native country the pendulous, protandrous flowers are usually ornithophilous. The perianth forms a tube, at the base of which is a nectar-secreting disc. The red colour of the calyx attracts birds and the presence of nectar insects also. When birds seek out the nectar, they sometimes pierce the 'calyx-tube'. The flexible pedicel reduces the chances of excessive damage, though at the same time it may hinder the birds in their attempts to obtain the nectar. Pollen is readily transferred from the long-exserted stamens to the bird's feathers, and is subsequently carried away to another flower.

Alternative flowers for study. Any other species of *Fuchsia* will act as an acceptable alternative, also the *Fuchsia* hybrids which are popular as pot-plants. A comparison should be made with *Oenothera*, Evening Primrose, or *Epilobium*, Willow-herb, both of which have capsular fruits and 4-lobed stigmas. *Circaea*, Enchanter's Nightshade, is an example of a genus with 2-merous flowers. It has a 1- or 2-seeded, nut-like fruit covered with barbed bristles. *Chamaenerion angustifolium*, Rosebay Willow-herb, shows a tendency to zygomorphic flowers. All the plants mentioned flower during the summer and early autumn.

Fig. 43 Onagraceae, *Fuchsia magellanica*

A A pendulous, actinomorphic, hermaphrodite flower, with long, red, petaloid sepals terminating the narrowly cylindrical 'calyx-tube' or hypanthium. The 8 stamens are exserted well beyond the 4 sepals and the 4 relatively small purple petals. The style protrudes even beyond the stamens.
Hypanthium: 10 mm Sepal: 20 mm Petal: 10 mm
Stamens: 15–24 mm Style + stigma: 30 mm

B L.S. of the centre of the flower, showing the hypanthium situated above the inferior ovary. At the base of the hypanthium is the nectar-secreting disc which surrounds the long style that arises from the top of the ovary. The free stamens and petals are attached to the rim of the hypanthium.

C Stamen. C1: upper portion of stamen, showing the anther prior to dehiscence. C2: the anther at dehiscence, exposing the pollen grains.
Anthers: 2–2.25 × 1 mm

D L.S. of the inferior ovary, showing the numerous ovules attached to the axile placenta. Surmounting the ovary is the style, surrounded by the nectar-secreting disc at the base of the hypanthium.

E Detail of the 4-lobed stigma.
Stigma: 1.5 × 1 mm

F T.S. of the ovary formed from 4 united carpels. The ovary is 4-locular and has axile placentation.
Ovary (at anthesis): 2–3 mm in diameter

G The fruit, a 4-celled, black berry containing numerous seeds.
Pedicel: up to 50 mm Fruit: 13 × 5 mm

Fig. 43

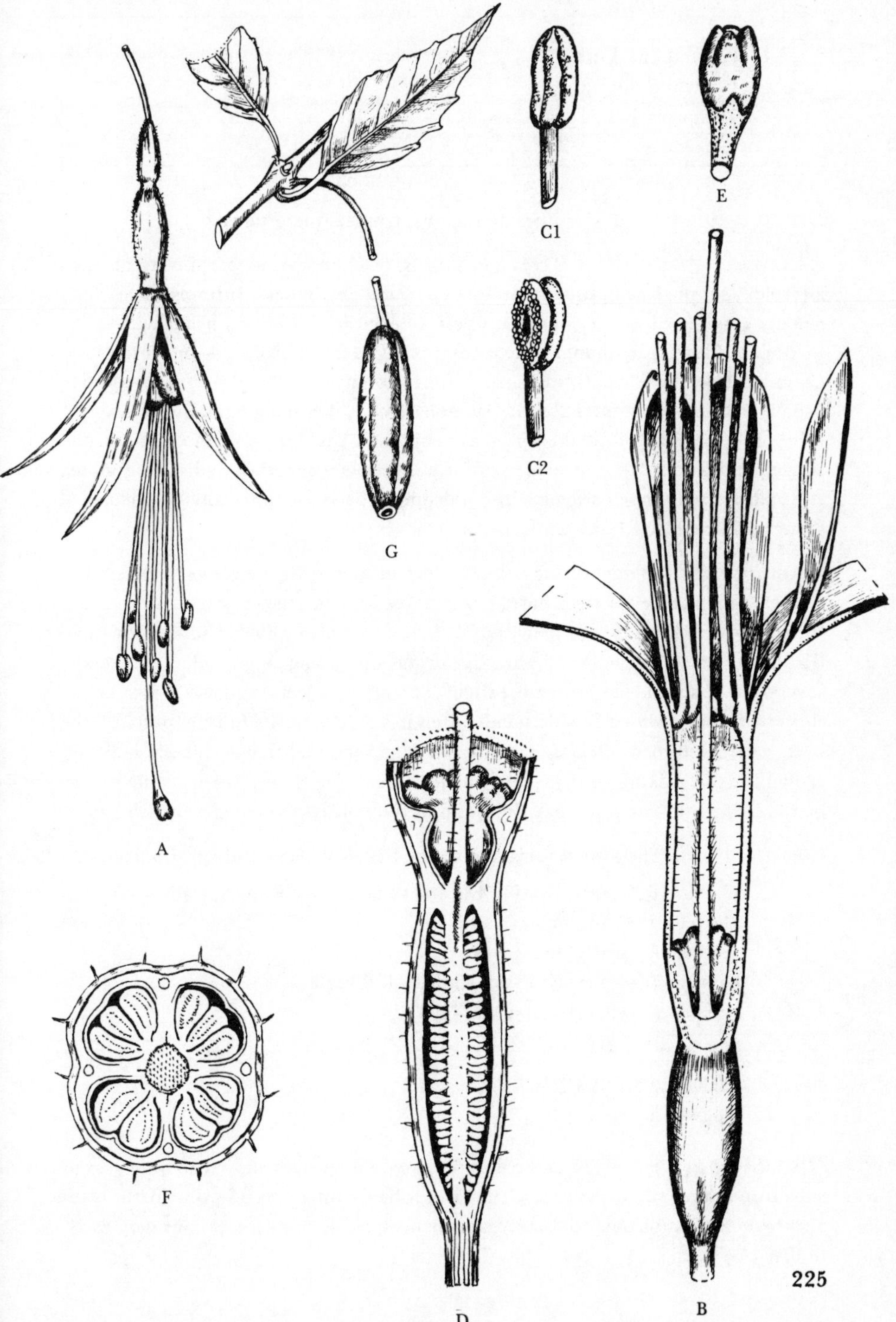

44 Cornaceae Dum.

Dogwood family
12 genera and 110 species

Distribution. N. and S. temperate regions, tropical mountains.

General characteristics. Trees and shrubs, rarely herbs, with opposite or alternate, simple leaves, usually petiolate, entire, exstipulate. Inflorescence usually condensed to corymbs or umbels, or even (e.g., *Cornus*) heads with involucre. Flowers hermaphrodite or unisexual, actinomorphic, 4- or 5-merous. Calyx of 4 or 5 free sepals, small, sometimes absent. Corolla of 4 or 5 free petals, usually valvate, sometimes absent. Stamens 4 or 5, dehiscing laterally. Ovary inferior, of usually 2 united carpels, as many loculi as carpels, each containing a single pendent, anatropous ovule with usually axile placentation; disc epigynous; style simple with lobed stigma. Fruit a drupe or berry, with usually a 2-locular stone or 2 separate stones. Endosperm present.

Economic and ornamental plants. Several species of the genus *Cornus* are of local importance for their hard, close-grained wood used in tool-making and for their edible fruit. Some species of *Griselinia* are also valuable sources of timber. *Cornus mas*, Cornelian Cherry, *C. florida*, Flowering Dogwood, *C. kousa* and several other species are grown for their decorative flowers, and the evergreen shrub *Aucuba japonica* is widely cultivated in gardens for its foliage, particularly in its variegated form. *Davidia involucrata*, Handkerchief-tree, is occasionally grown for the striking white bracts which subtend the flower-heads, while species of the evergreen *Griselinia* may be found in the milder parts of the British Isles.

Classification. The family is related to the Caprifoliaceae and Escalloniaceae.

Cornus (62) central and S. Europe to Caucasus, E. Asia, California.
Mastixia (25) Indomalaysia.
Melanophylla (8) Madagascar.
Griselinia (6) New Zealand, Chile, S.E. Brazil.
Aucuba (3 or 4) Himalaya to Japan.

CORNUS SANGUINEA L.

Dogwood

Distribution. Widespread and common in woods, hedges and scrub on calcareous soils throughout the British Isles, though probably introduced in the north. Native also from S. Scandinavia to Spain and Portugal, Sicily and Greece, but very rare in S.W. Asia.

Vegetative characteristics. A deciduous shrub bearing opposite, ovate leaves, with the main veins curving round towards the apex.

Floral formula. K4 C4 A4 $\overline{\text{G}}$(2)

Flower and inflorescence. The dull white flowers are borne in flat-topped, corymbose cymes in June and July.

Pollination. Dogwood flowers are homogamous and secrete nectar from a disc which forms a ring round the style. Various kinds of insects are attracted by the conspicuous inflorescence, the heavy odour of the flowers and the nectar which lies freely exposed on the nectar-secreting disc. Self- as well as cross-pollination may take place as the insects walk over the surface of the flowers. Occasionally pollination may occur between neighbouring flowers of the same inflorescence owing to the spreading nature of the stamens.

Alternative flowers for study. In the absence of *C. sanguinea*, any other species of *Cornus* is acceptable. *C. mas* bears yellow flowers in an axillary umbel subtended by 4 small, yellowish-green bracts and the flowers appear in February before the leaves. In some species, the involucral bracts are very prominent, e.g., in *C. canadensis*, *C. florida* and *C. kousa*, which show conspicuously large, white, petaloid bracts during the summer months. Comparison should be made with the evergreen shrub *Aucuba japonica*, which flowers in March and April, and with the summer-flowering rhizomatous herb *Cornus suecica*, Dwarf Cornel, which exhibits annual flowering stems.

Fig. 44 Cornaceae, *Cornus sanguinea*

A A single, actinomorphic, hermaphrodite flower. The 4 free petals are attached above the ovary and hide the minute sepals. The spreading stamens alternate with the petals and arise from just below the nectar-secreting disc that surrounds the single style.
Flower: 9 mm in diameter Petal: 4 mm Stamen: 4 mm

B A flower with petals and stamens removed to expose 2 of the 4 minute sepals that are attached above the pubescent ovary. The sepals protrude to just above the nectar-secreting disc.
Ovary: 1.7 mm Style: 3 mm

C Upper portion of stamen. The dorsifixed anther is 2-celled and dehisces laterally, as is usual in this family.
Anther: 2 × 0.5 mm

D L.S. of the inferior ovary showing the 2 loculi, each containing a pendulous, anatropous ovule. Above the ovary, the lower portion of the style can be seen with a part of the nectar-secreting disc visible on each side.
Ovary: 1.7 × 1.6 mm

E T.S. of ovary formed from 2 united carpels. Each carpel has a single loculus containing one ovule with axile placentation. The point of attachment of the ovule can be seen in D.
Ovary: 1.6 mm in diameter Ovule (at anthesis): 0.7 × 0.4 mm

F Detail of the 4-lobed stigma and its receptive surface.
Stigma: 1 mm in diameter

G Part of the infructescence consisting of black drupes.
Peduncle: 30 mm Pedicel: *c.* 4 mm Fruit: 5 × 6 mm

Fig. 44

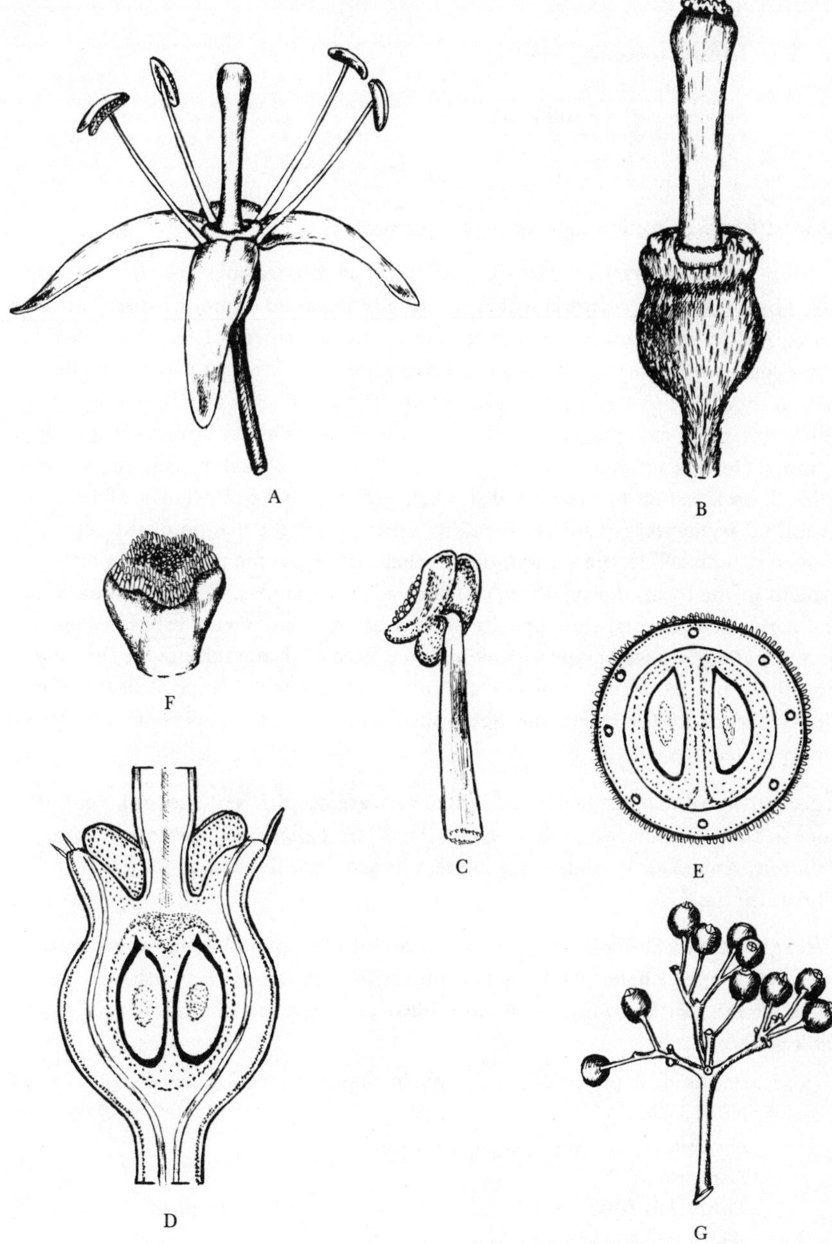

45 Loranthaceae Juss.

Mistletoe family

36 genera and 1300 species

Distribution. Tropical and temperate regions.

General characteristics. Mostly small, semi-parasitic shrubs attached to their hosts by suckers or haustoria which are usually regarded as modified adventitious roots. A few, e.g., *Nuytsia*, root in the earth. An outgrowth often of considerable size occurs where the parasitic root joins the host and the root often branches within the tissue of the host as in the genus *Viscum*. Leaves usually evergreen, leathery, entire, exstipulate, sometimes reduced to scales. Inflorescence usually cymose, flowers actinomorphic, hermaphrodite or unisexual, usually in groups of 3 (or 2, by abortion of the central flower). Perianth in 1 or 2 whorls, of 4–6 free or united segments, sepaloid or petaloid, arising from the margin of the cup-shaped receptacle. In the Loranthoideae there is, below the perianth, an outgrowth in the form of a small rim or fringe – the calyculus. Stamens as many as the perianth-segments, and opposite and adnate to them. Ovary inferior, unilocular, of 3–5 united carpels, sunk in and united with the receptacle, the ovules not differentiated from the placenta. Fruit a 1- to 3-seeded berry or drupe, the fleshy part really the receptacle. Seeds surrounded by a layer of viscin, a very sticky substance.

Economic and ornamental plants. There are no plants of economic importance in the Loranthaceae, but *Viscum album* (see below) and species of *Phoradendron*, American Mistletoe, are in great demand for decorative purposes at Christmas time.

Classification. The following classification into 2 subfamilies and 6 tribes is based on that of Engler (4), but some authorities restrict the Loranthaceae to those genera contained in the Loranthoideae and raise the Viscoideae to the rank of family.

I. Loranthoideae (calyculus or 2 bracteoles present; pollen usually trilobate).

1. Nuytsieae.

   Nuytsia (1) W. Australia, *N. floribunda.*

2. Lorantheae.

   Loranthus (600) mainly Old World tropics and subtropics, a few in temperate Eurasia and Australasia.

   Struthanthus (75)
Phthirusa (60)
Psittacanthus (50) } tropical America.

II. Viscoideae (calyculus absent; pollen spherical).

3. Eremolepideae.
   Eremolepis (7) tropical America, W. Indies.
4. Phoradendreae.
   Phoradendron (190) America, W. Indies.
   Dendrophthora (55) tropical America, W. Indies.
5. Arceuthobieae.
   Arceuthobium (15) N. America, W. Indies, Mediterranean region, S. Asia.
6. Visceae.
   Viscum (60–70) mostly warm regions of the Old World, a few in temperate Eurasia.

VISCUM ALBUM L.
Mistletoe

Distribution. Common in the southern part of Britain, but absent from Scotland and Ireland. Native throughout most of Europe and N. Africa eastwards to central Asia and Japan.

Vegetative characteristics. A woody, evergreen plant, semi-parasitic on the branches of a great variety of deciduous trees. It is repeatedly branched in a dichotomous manner, each branch bearing 2 green, leathery leaves, narrowly obovate in shape.

Floral formula. Male: P4 A4
Female: P4 $\overline{\mathrm{G}}$(4)

Flower and inflorescence. The plants are dioecious and the unisexual, greenish flowers, which appear from February to April, are borne in sessile groups of 3 in the forks of the branching stems.

Pollination. Bees and flies have been observed visiting the flowers in search of nectar, presumably attracted by the characteristic odour of orange which is said to occur in both male and female flowers. The pollen grains, which are produced from the cellular tissue of the stamens adnate to the inner surface of the perianth of the male flower, are covered with short spines, causing them to adhere to each other and to the bodies of visiting insects. The pollen masses are subsequently transferred to the capitate stigmas of female flowers as the insects continue their search for nectar.

Alternative flowers for study. Viscum album is the only native representative of the family within the British Isles and an alternative is not readily available.

Fig. 45 Loranthaceae, *Viscum album*

A A male inflorescence situated at the apex of the stem and subtended by 2 lateral shoot-buds. Two of the 3 actinomorphic, 4-merous flowers are shown. The 4 perianth-segments, to which the stamens are joined, spread out at anthesis. Rarely, the perianth may consist of only 3 segments, owing to the fusion of 2 adjacent parts.
Inflorescence: 8 mm T.S. of bud: 7 × 5 mm

B L.S. of a male flower at anthesis, showing a facial view of one of the 4 stamens which are adnate to the inner surface of the perianth-segments. The staminal area is paler than the rest of the flower and is composed of numerous loculi that yield a large quantity of sulphur yellow pollen.
Terminal male flower: 5 × 5 mm Perianth-segment: 3 × 2.5 mm
Lateral male flower: 4 × 3.5 mm Perianth-segment: 2 × 2 mm

C The female inflorescence, made up of 3 actinomorphic, 4-merous flowers, smaller than the male and embedded in the surrounding tissue.
Bud + inflorescence: 8 mm T.S. of bud: 5 × 4.75 mm

D Receptacular tissue partly cut away to reveal the embedded portion of the female flower. The 4 perianth-segments, unlike those of the male flower, have a tendency to point inwards.
Female flower (excluding perianth-segments): 2.25 × 1.5 mm
Perianth-segment: 1 × 1 mm

E L.S. of the female flower, showing 2 of the 4 perianth-segments. The inferior ovary, which is surmounted by a sessile, capitate stigma, is embedded in and adnate to the receptacle. The ovules arise from a central placental area on the base of the ovary, but they are not differentiated from the placenta. Nectar is secreted from a poorly differentiated glandular ring situated between the stigma and the base of the perianth.
Stigma: 0.75 mm

F L.S. of fruit, a berry-like pseudocarp formed from both receptacle and 4-carpelled ovary. A sticky substance known as viscin forms between the fleshy receptacle and the membranous pericarp which encloses the 2 or 3 seeds. The viscid layer surrounding the pericarp helps in the dispersal of the seeds by causing them to cling to the beaks of birds, which eventually rub them off on the branches of the host tree.
Fruit: 8 × 7 mm

G A portion of the stem from a female plant during December, showing the 3 mature fruits formed from an inflorescence. A newly formed inflorescence can be seen between the 2 leathery leaves. These leaves represent the termination of a year's growth from one of the 2 lateral buds that subtended the previous inflorescence. Only part of the growth from the other lateral bud is shown.
Base of fruit to apex of bud: 30 mm Leaf: 50 × 10 mm

Fig. 45

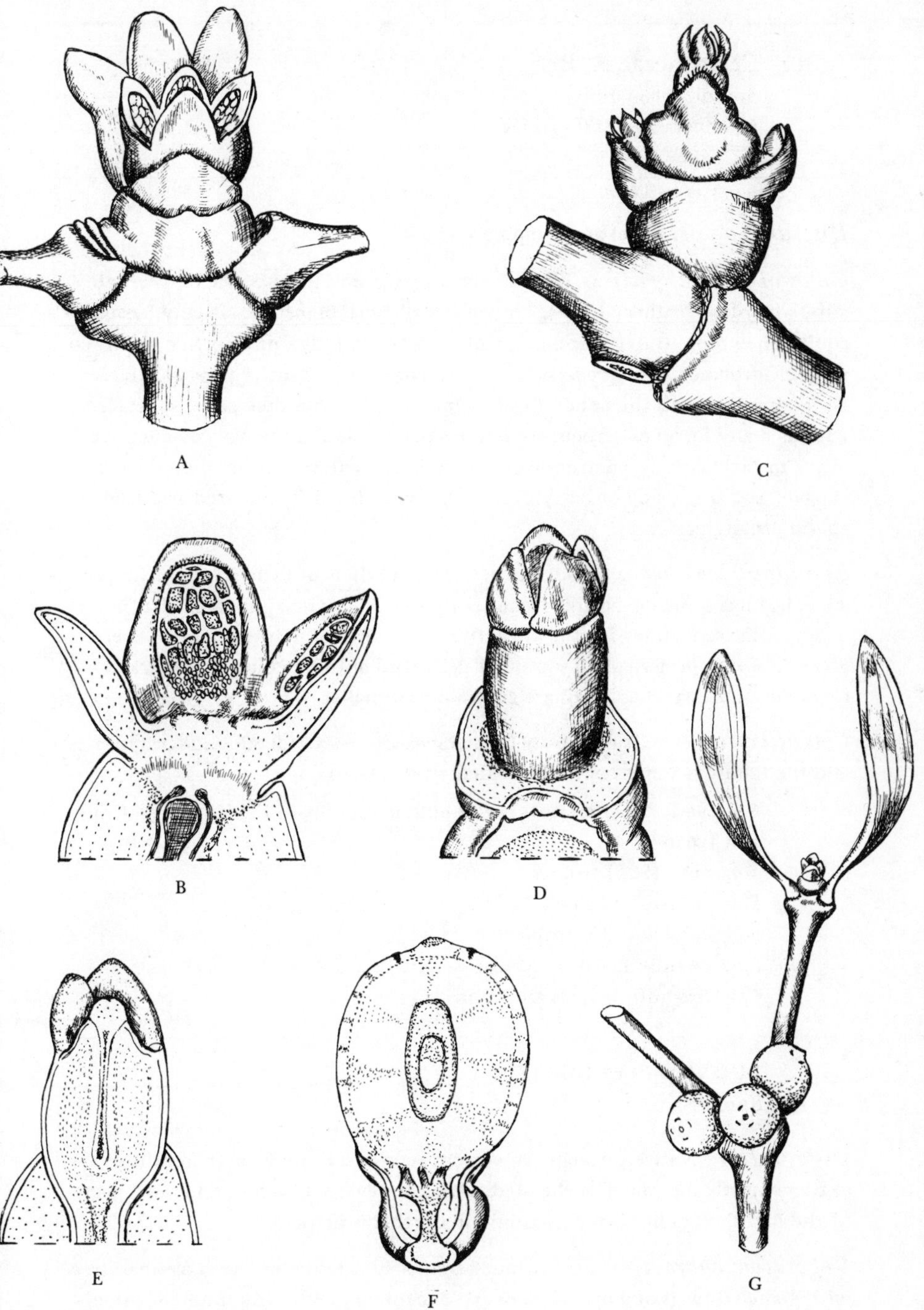

46 Celastraceae R. Br.

Spindle-tree family

55 genera and 850 species

Distribution. Tropical and temperate regions.

General characteristics. Trees or shrubs with simple, opposite or alternate, stipulate, often leathery leaves. Flowers small, actinomorphic, usually hermaphrodite, in cymose (rarely racemose) inflorescences. Calyx of 3–5 free or united sepals. Corolla of 3–5 free petals. Well-developed disc usually present. Stamens 3–5, alternate with the petals. Ovary superior, of 2–5 united carpels, usually with as many loculi as carpels, sometimes partly sunk in the disc, ovules generally 2 in each loculus, anatropous, with axile placentation. Fruit a loculicidal capsule, samara, drupe or berry. Seed often with brightly coloured aril. Endosperm usually present.

Economic and ornamental plants. The family is of little economic importance, but has some decorative value. The genus *Euonymus*, in particular, is often cultivated, either for the attractive fruits or for the foliage, a rich evergreen in some species, and a source of delightful autumn colour in others. *Celastrus* and *Maytenus*, also, are grown occasionally.

Classification. The fruits of the Celastraceae vary considerably in structure and the family is sometimes sub-divided on this basis.

Euonymus (176) almost cosmopolitan, mainly in the Himalaya, China and Japan.
Maytenus (225) tropics.
Salacia (200) tropics.
Hippocratea (120) tropics.
Cassine (40) S. Africa, Madagascar, tropical Asia to the Pacific.
Celastrus (30) tropics and subtropics.

EUONYMUS EUROPAEUS L.

Spindle-tree

Distribution. Native throughout western Asia and Europe apart from the extreme north and much of the Mediterranean region. It is found in most parts of the British Isles but is less common in northern districts.

Vegetative characteristics. A much-branched, deciduous shrub or small tree with 4-angled twigs and opposite, ovate-lanceolate to elliptical, finely-toothed leaves which often turn reddish in autumn.

Floral formula. K4 C4 A4 $\underline{G}(4)$

Flower and inflorescence. Three to eight hermaphrodite or polygamous flowers are borne in axillary cymes arising from the quadrangular stems. The flowers have yellowish-green petals and appear during May and June. They are followed in September and October by the attractive, 4-lobed, pink fruits, which open to expose the seeds and their pulpy orange arils.

Pollination. The yellowish-green petals are relatively insignificant and play only a small part in the pollination of the flowers, but many small insects, chiefly flies, ants and beetles, are attracted to the nectar that is secreted by the disc surrounding the ovary, and these effect cross-pollination in the course of their visits.

Alternative flowers for study. The Spindle-tree is available in most districts. However, in its absence, other species of *Euonymus* would be suitable, including the evergreen *E. japonicus* which is widely grown as a hedge plant, particularly in seaside gardens. Related genera are rarely in cultivation, though the 5-merous flowers of the genus *Celastrus* may be available for comparison between April and June.

Fig. 46 Celastraceae, *Euonymus europaeus*

A An actinomorphic, hermaphrodite flower. The small, green sepals alternate with the larger, yellowish-green petals. The 4 short stamens are inserted on the flat, nectar-secreting disc which is adnate to the ovary.
Flowers: 8–10 mm in diameter Sepals: 2 mm Petals: 4–5 mm

B A stamen at anther dehiscence – B1 side view, B2 abaxial view. The 2-celled anthers are extrorse and dehisce longitudinally.
Stamen: 3 mm

C L.S. of superior ovary, disc and persistent sepals. The ovary has 4 loculi with 1–2 ovules in each loculus. Each ovule is anatropous and erect on the axile placenta. The stippled portions at the edge of the disc denote the position of the petals and stamens. The ovary is terminated by a single style and slightly 2-lobed stigma.
Receptacle: 4 mm in diameter Ovules: 0.3–0.5 mm
Style + stigma: 2 mm

D T.S. of ovary showing the 4 loculi and ovules. Only one ovule is shown in each loculus though 2 ovules may sometimes occur.
Ovary (at anthesis): *c.* 1.5 mm in diameter

E A single fruit at an early stage of dehiscence. The fruit forms a pink, 4-lobed, loculicidal capsule containing large seeds with orange arils.
Fruit: 11 × 15 mm Seed: 8 × 6 mm

F L.S. of a mature capsule and persistent sepals showing 2 of the large, arillate seeds.

Fig. 46

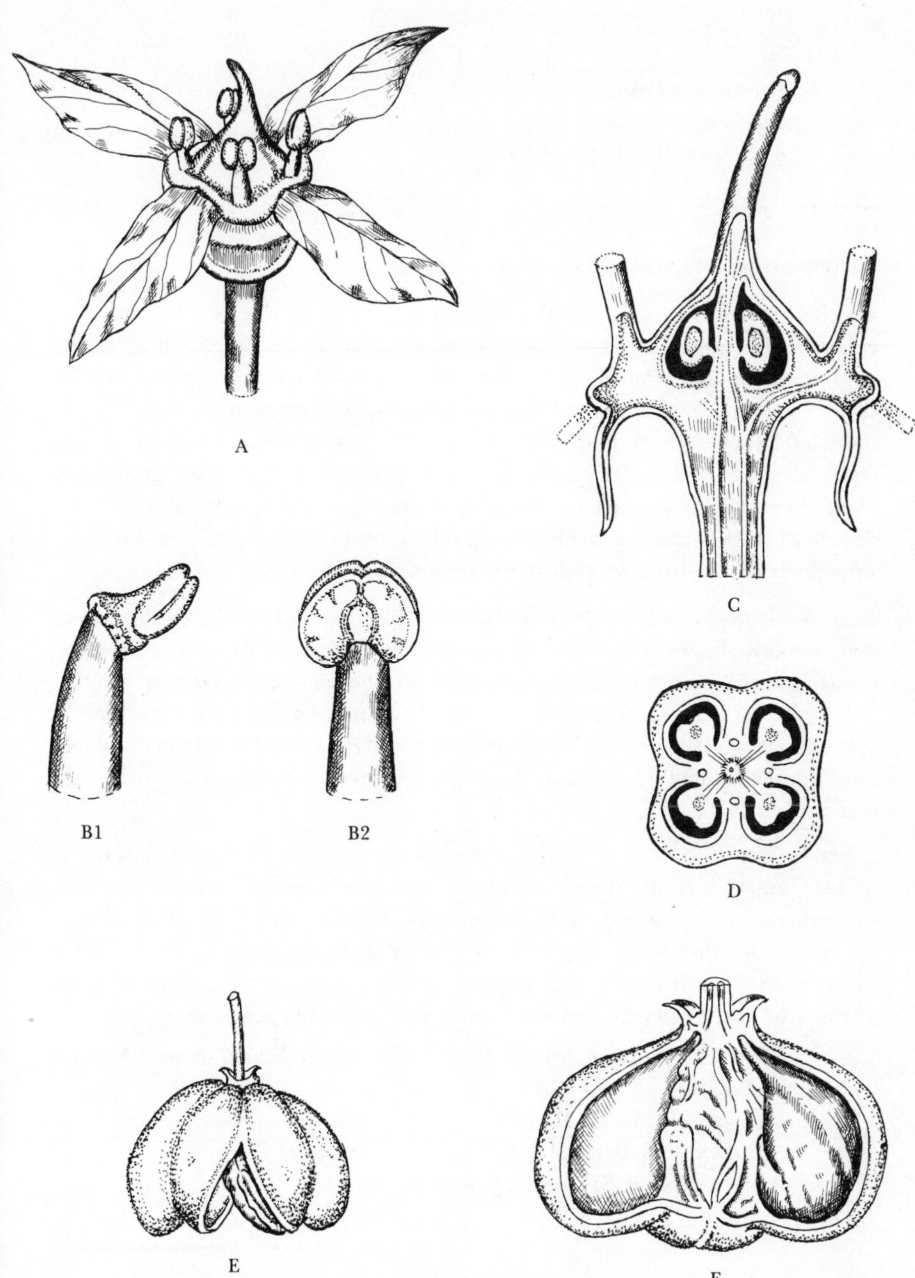

47 Buxaceae Dum.

Box family

4 genera and 100 species

Distribution. Tropical and temperate regions.

General characteristics. Shrubs, more rarely trees or herbs, usually monoecious or dioecious. Leaves evergreen, opposite or alternate, simple, entire or dentate, exstipulate. Flowers actinomorphic, unisexual, rarely hermaphrodite, in dense racemes or heads, or the female solitary. Perianth-segments 2 + 2 or 3 + 3, imbricate. Stamens 4–6, free, with more or less thick filaments and large anthers. Ovary superior, of 3 united carpels, 3-locular, with 1 or 2 pendulous, anatropous ovules in each loculus; placentation axile; styles 3, persistent, with more or less decurrent stigmas. Fruit a loculicidal capsule or drupe; seeds black and shining, sometimes with a caruncle; endosperm fleshy.

Economic and ornamental plants. Large specimens of *Buxus sempervirens*, Box (see below), are of economic importance for their hard, fine-grained wood, though the species is probably better known as a hedge-plant in gardens, often being cut into fantastic shapes. *Sarcococca*, Christmas Box, is of horticultural value for its sweetly scented flowers in winter and *Pachysandra*, a genus of low-growing, semi-woody plants, is useful as ground-cover, particularly in shady places.

Classification. The Buxaceae show possible relationships with the Euphorbiaceae, in which family they were formerly included, and perhaps also with the Celastraceae. The monotypic Californian genus *Simmondsia* has usually been considered as a member of the Buxaceae, but is distinguished by its pentamerous flowers and other characters and is now regarded as sufficiently distinct to warrant a family of its own, the Simmondsiaceae. The 4 remaining genera are:

Buxus (70) Europe, Asia, tropical and S. Africa, N. and Central America, W. Indies.
Sarcococca (16–20) Himalaya to central China, India and S.E. Asia.
Notobuxus (7) tropical and S. Africa, Madagascar.
Pachysandra (3) E. Asia, E. United States.

BUXUS SEMPERVIRENS L.
Box

Distribution. Native in Europe, N. Africa and W. Asia, and probably indigenous to Britain, where it is locally abundant on chalk and limestone in Kent, Surrey, Buckinghamshire and Gloucestershire. It is commonly planted in parks, gardens and churchyards as a clipped shrub or hedge and sometimes becomes naturalised.

Vegetative characteristics. A monoecious, evergreen shrub or small tree 2–4(–10) m in height. The leaves are opposite, leathery, dark glossy green above and paler beneath; ovate, oblong or elliptical, and with a short petiole.

Floral formula. Male: P2+2 A4
Female: P0 $\underline{G}$(3)

Flower and inflorescence. The inflorescence is a cluster of pale green, actinomorphic, unisexual flowers which arise from the axils of leafy shoots. At its centre is the single pistillate flower and this is surrounded by several staminate flowers. If left unclipped, the plant flowers freely in April and May.

Pollination. The flowers of this species are scentless and rather inconspicuous in colouring, but as they open early in the year, when comparatively few other flowers are available, insect visitors, mainly flies and bees, are fairly numerous. They are attracted by the nectar secreted in droplets by 3 small nectaries on the ovary of the female flower, and also apparently by the vestigial ovary of the male flower. The flower-clusters are slightly protogynous, the stigmas of the female flower becoming receptive just before the anthers of the male flowers begin to dehisce, but they remain in that condition until all the pollen has been shed. Although the central pistillate flower can therefore be pollinated easily by the surrounding male flowers, insects usually fly directly to the centre of the cluster, bringing with them pollen from another plant. In this way, cross-pollination takes place.

Alternative flowers for study. Typical *B. sempervirens* is easily obtainable, also a number of cultivars grown for their value in topiary, for their relatively dwarf habit or variegated foliage. The rarer species, *B. balearica*, Balearic Box, has larger leaves and inflorescences. If possible, comparison should be made with the genus *Sarcococca*, which has alternate, entire leaves, 4–6 perianth-segments, 4–6 stamens and a drupaceous fruit, also with *Pachysandra*, which has alternate, toothed leaves and thick, white filaments.

Fig. 47 Buxaceae, *Buxus sempervirens*

A A young, unclipped, leafy shoot from the main stem, showing the clusters of small, green, unisexual flowers borne in the axils of the leaves.

B A single, actinomorphic, male flower, subtended by a bracteole. The 4 well-exserted stamens stand opposite the 4 perianth-segments (3 only shown). Two of the segments, opposite each other, are larger than the others.
Larger perianth-segment: 1.75 × 1.5 mm
Smaller perianth-segment: 1 × 1 mm

C L.S. of the male flower, showing 2 of the 4 stamens with introrse anthers at an earlier stage than in B. Behind the stamens are 2 of the 4 perianth-segments and at their base is the vestigial ovary.

D The upper portion of a stamen, showing the dorsifixed anther, which is 2-celled and dehisces longitudinally.
Young stamens: 2.5–2.75 mm Mature stamens: 3 mm
Anthers: 1 mm

E A single, actinomorphic, female flower from the centre of a cluster. The superior ovary has 3 stout styles, each with a 2-lobed stigma. The perianth of the female flower is not clearly defined, but the ovary is subtended by spirally arranged bracteoles of variable dimensions.
Bracteoles: 1.5–2 × 1.25–2 mm Style: 1 × 1 mm

F L.S. of the superior ovary formed from 3 united carpels, each carpel containing 2 pendulous, anatropous ovules.
Ovary: 2.5 mm

G T.S. of the ovary, showing the 2 ovules in each of the 3 loculi attached to axile placentas. The small circles in the centre are vascular bundles.
Ovary: 1.75 mm in diameter

H The mature fruit, a loculicidal capsule showing 2 of the 3 persistent styles split down the middle as a result of the explosive dehiscence. The capsule encloses a few, shiny, black seeds with a small caruncle. The male flowers have withered away, but the remains of some bracteoles can be seen.
Fruit: 8 × 5 mm

Fig. 47

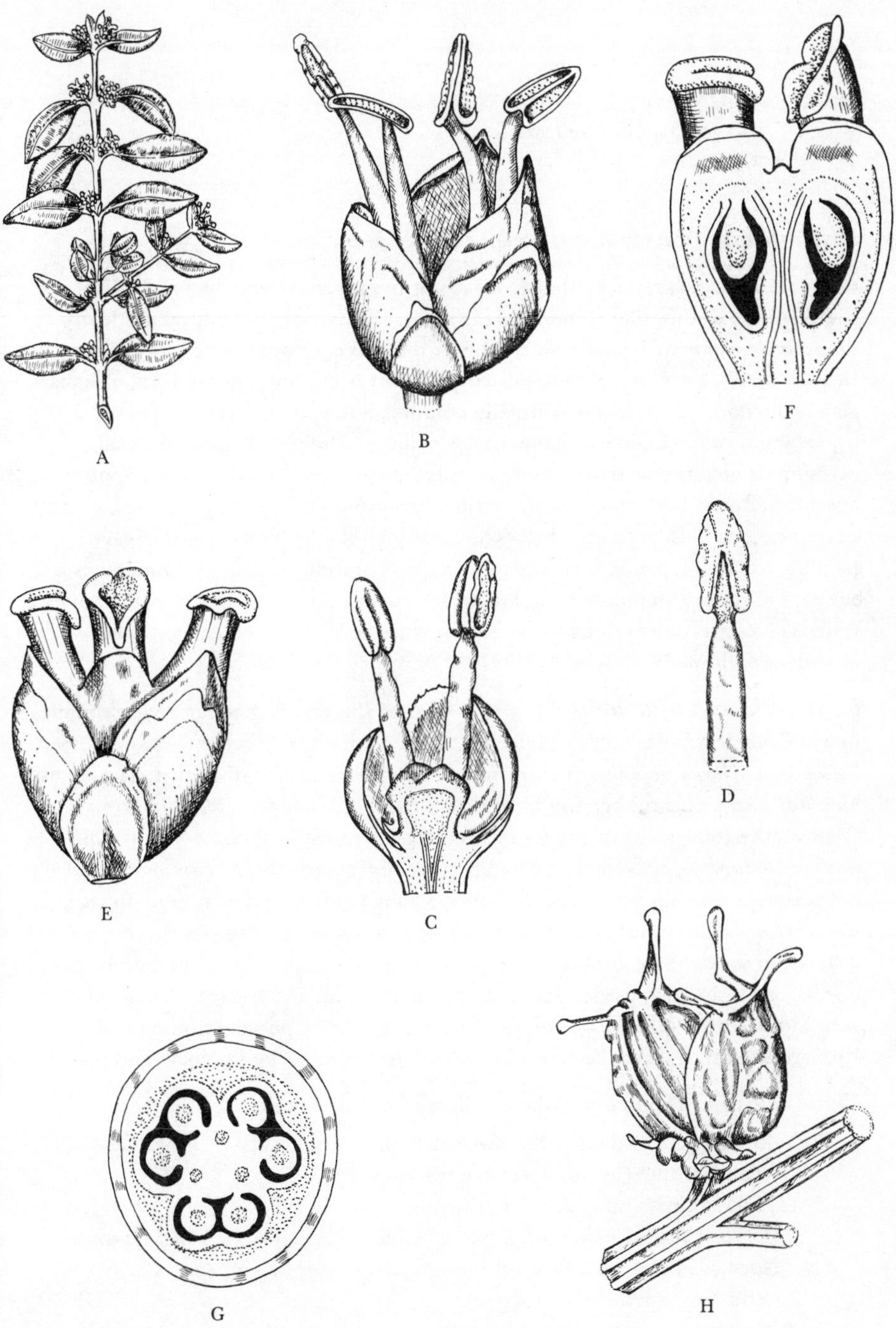

48 Euphorbiaceae Juss.

Spurge family
300 genera and 5000 species

Distribution. Cosmopolitan, except arctic regions.

General characteristics. Mostly shrubs or trees, some (including all British species) herbaceous, plants monoecious or dioecious, latex often present. Many xerophytic, some of ericoid habit, others cactus-like, or possessing phylloclades. Leaves usually alternate. Stipules usually present, sometimes in the form of hairs, glands or thorns. Inflorescence usually complex and varied, often the first branching racemose and subsequent ones cymose. Flowers always unisexual, actinomorphic, hypogynous. Sepals usually 5, sometimes absent. Petals 5, often absent. Stamens 1 to many, free or united in various ways. Ovary superior, usually of 3 united carpels, 3-locular. Styles 3, usually 2-lobed. Ovules 1 or 2 in each loculus, collateral, pendulous, anatropous; placentation axile. Fruit sometimes a drupe, but more commonly a capsule which splits into 3 carpels and at the same time each carpel opens along the ventral surface to let the seed escape. Seed with abundant endosperm.

Economic and ornamental plants. Most of the Euphorbiaceae are poisonous, but nevertheless the family contains a number of plants of economic importance. Latex from *Hevea brasiliensis* is the main commercial source of rubber. The genus *Manihot* also yields rubber and from the roots of *M. esculenta*, Manioc or Cassava, the foodstuff tapioca is prepared. The seeds of *Ricinus communis* and *Croton tiglium* contain castor oil and croton oil respectively. *R. communis* is also cultivated in glasshouses or as a summer bedding plant for its handsome foliage. *Euphorbia pulcherrima*, Poinsettia, is grown as a house plant for its striking red bracts and several hardy species of *Euphorbia* are cultivated in the herbaceous border or on the rock-garden for their coloured glands and foliage. *Codiaeum variegatum*, often incorrectly called 'Croton', is another popular house plant. Numerous cultivars have been produced with leaves of various colours and shapes.

Classification. The family is divided here into 4 subfamilies:

A. 'Platylobeae' (cotyledons much broader than radicle).
- I. Phyllanthoideae (ovules 2 per loculus; no latex).
  - 1. Phyllantheae (male calyx imbricate).
    Phyllanthus, *Glochidion*, *Drypetes*, *Baccaurea*, *Aporusa* (all with many species in tropics and subtropics).
  - 2. Bridelieae (male calyx valvate).
    Bridelia, *Cleistanthus* (both with many species in Old World tropics).

II. Euphorbioideae (Crotonoideae) (ovules 1 per loculus; latex usually present).
3. Crotoneae (stamens bent inwards in bud).
Croton (750) tropics and subtropics.
4. Acalypheae (stamens erect in bud; flowers usually apetalous; male calyx valvate; inflorescence a raceme, spike or panicle, axillary or terminal).
Acalypha, *Mallotus*, *Macaranga*, *Dalechampia*, *Tragia* (all with many species in tropics and subtropics).
Ricinus (1) tropical Africa and Asia.
5. Jatropheae (as in Acalypheae, but inflorescence a dichasial thyrse).
Jatropha (175) tropics and subtropics.
Hevea (12) tropical America.
6. Adrianeae (as in Acalypheae, but inflorescence a simple terminal spike or raceme).
Manihot (170) tropical and subtropical America.
7. Clutieae (male calyx imbricate; male flowers with petals, in groups or cymes, these partial inflorescences axillary or in complex inflorescences).
Clutia (70) Africa, Arabia.
Codiaeum (15) Malaysia, Polynesia, N. Australia.
8. Gelonieae (as in Clutieae, but apetalous, inflorescence leaf-opposed).
Suregada (*Gelonium*) (40) tropics.
9. Hippomaneae (as in Gelonieae, but inflorescence axillary or terminal, spike-like, the partial inflorescences cymes).
Sapium (120) tropics and subtropics.
Excoecarya (40) Old World tropics.
10.Euphorbieae (inflorescence a cyathium).
Euphorbia (2000) cosmopolitan, chiefly in subtropical and warm temperate regions.
Monadenium (47) tropical Africa.

B. 'Stenolobeae' (cotyledons as wide as radicle).
III.Porantheroideae (ovules 2 per loculus; no latex).
Poranthera (10) Australia, New Zealand.
IV.Ricinocarpoideae (ovules 1 per loculus; latex present).
Ricinocarpos (16) Australia, New Caledonia.

EUPHORBIA HELIOSCOPIA L.
Sun Spurge

Distribution. Often found growing as a weed of cultivated ground. It is widespread throughout the British Isles, and is also native in other parts of Europe, N. Africa and Asia.

Vegetative characteristics. A glabrous, annual herb with a slender stem and alternate, obovate leaves which are serrulate towards the apex.

Floral formula. Male: P 0 A 1
Female: P 0 $\underline{G}$(3)

Flower and inflorescence. The cyathium, an inflorescence condensed to simulate a single flower, is characteristic of the genus. These cyathia are borne on a series of branches arising from the axils of the leaves. The 5 primary branches, called rays, have the appearance of an umbel. Subsequent branching into raylets is trichotomous or dichotomous. The plant is monoecious, and the unisexual flowers appear from May to October.

Pollination. The 4 oval glands of the cyathium secrete a shallow layer of nectar which is completely exposed. The inflorescence is strongly protogynous, the 3 bilobed stigmas emerging first from the involucre and becoming dusted with pollen from other flowers. Later, when the pedicel bearing the ovary lengthens and projects far beyond the involucre, the stamens gradually elongate and dehisce. Pollination is effected exclusively by flies, but beetles, wasps and bees are occasional visitors.

Alternative flowers for study. Any species of *Euphorbia* is suitable. Compare with *Ricinus communis*, the Castor Oil Plant, *Codiaeum variegatum* and, where available, genera of other tribes mentioned in the Classification section.

Fig. 48 Euphorbiaceae, *Euphorbia helioscopia*

A One of the 2 or 3 raylets which are borne by each of the 5 primary rays, showing further branches subtended by raylet leaves and terminating in cyathia.
Raylet (1st series): 35 mm
(2nd series): 10 mm
Peduncle: 3.5 mm
Raylet leaves (decreasing in size upwards): 36 × 16 mm, 9 × 7 mm and 7 × 6 mm

B Close-up of a single inflorescence (cyathium). The male flowers are hidden, but the single female flower which is borne on a pedicel arising from the centre of the cyathium is clearly seen.
Cyathium: 3 × 1.75 mm Ovary: 1 × 1 mm Style: 1.25 mm

C The yellow-green cyathium, formed from 4 connate bracts and with 4 oval glands along the upper edge, is opened up to reveal the individual male flowers, each consisting of a single stamen. The stamens in the centre ripen first, the ripening continuing in a centrifugal manner towards the outer wall.

D Male flower, showing the single stamen with its short filament attached to a longer pedicel. A clearly visible line denotes the point of attachment. The pedicel of each male flower is subtended by a feather-like bract.
Anther-lobe: 0.25 × 0.25 mm Bract: 1–1.25(–2) mm

E L.S. of the 3-celled female flower. The ovule in each loculus is pendulous and anatropous. Situated apically are the 3 styles, each with a bifid stigma.
Ovary: 3 × 3.75 mm

F T.S. of the ovary, showing the 3 carpels with well-developed ovules; placentation axile.
Ovary: 4 mm in diameter

G Whole fruit (capsule) in process of dehiscence. The capsule separates into the 3 carpels, each of which splits into 2 valves leaving the central column standing (see H).
Fruit: 4 × 4 mm

H Detail of central column of capsule (also known as a columella).

I A single large seed showing the reticulate surface. Situated near the hilum is a swelling called the caruncle. A number of species of *Euphorbia* have caruncled seeds and it has been observed that these are sought by ants, doubtless for their oil content. The structure is thus an elaiosome, and is of importance in aiding seed-dispersal.
Seed: 2 × 1.75 mm

Fig. 48

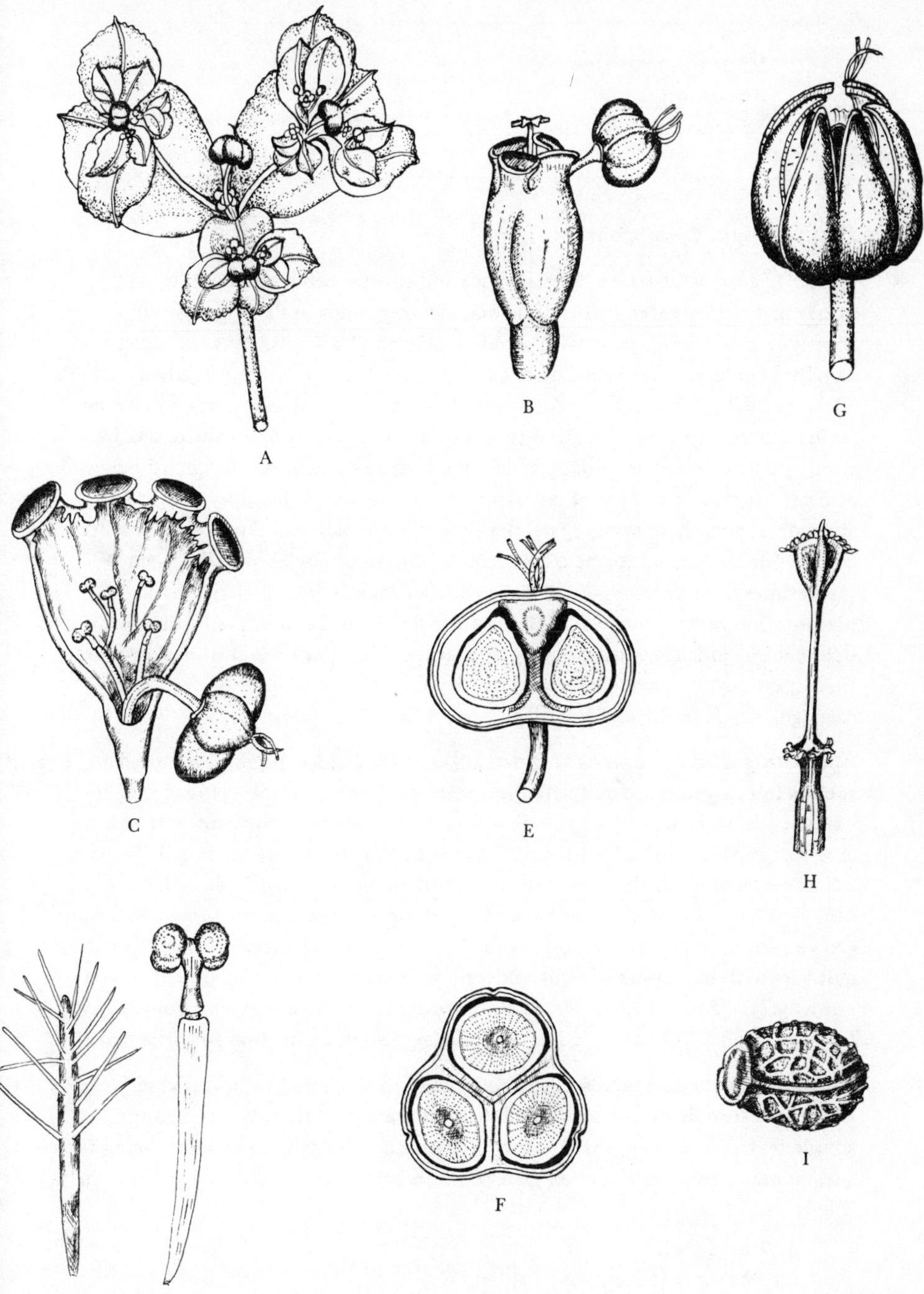

49 Rhamnaceae Juss.

Buckthorn family

58 genera and 900 species

Distribution. Cosmopolitan.

General characteristics. Mostly trees and shrubs, often climbing (by aid of hooks in *Ventilago*, tendrils in *Gouania*, twining stems in *Berchemia*), some thorny, e.g., *Colletia*, some with serial buds in the leaf-axils. Leaves simple, usually stipulate, never lobed or divided. Inflorescence cymose, usually a corymb. Flowers small, hermaphrodite or rarely unisexual, sometimes apetalous. Receptacle hollow, free from or united to the ovary. Calyx of 4 or 5 valvate sepals. Corolla of 4 or 5 petals, usually small, often strongly concave, frequently clawed at base. Stamens 4 or 5, alternate with sepals, usually enclosed by the petals, at any rate at first. Disc usually well-developed, intrastaminal. Ovary superior, free or more or less united to the receptacle, 3-, 2-, (rarely by abortion 1-) locular (sometimes 4- or typically 1-locular), with 1 or rarely 2 ovules with basal placentation in each loculus. Style simple or divided. Fruit dry, splitting into dehiscent or indehiscent mericarps, or a drupe with 1 or several stones, or a nut. Endosperm little or none. Many of the dry fruits show adaptation for wind-carriage, e.g., *Paliurus* and *Ventilago.*

Economic and ornamental plants. *Zizyphus jujuba*, Jujube, and other species in this genus are cultivated in subtropical and tropical regions for their edible fruit. *Hovenia dulcis*, Japanese Raisin-tree, is grown for fruit in the Himalaya eastwards to China and Japan. The genus *Rhamnus* is commercially important as a source of green and yellow dyes and medicinal substances which are obtained from the berries and bark of some species. Perhaps the most commonly grown genus for decorative purposes is *Ceanothus*, of which many hybrids and cultivars with blue, pink or white flowers are now available, and are collectively known as Californian Lilacs. In addition to those genera already mentioned *Paliurus*, *Colletia*, *Berchemia* and *Discaria* are also in cultivation for ornament.

Classification. The family is closely related to the Vitaceae, but is chiefly distinguished from it by the small petals, the receptacle, the endocarp and the simple leaves. It also approaches the Celastraceae, the main distinction being the antipetalous stamens. The chief genera, given below, are placed in 5 tribes as in Engler (4).

1. Rhamneae.
   Rhamnus (160) cosmopolitan.
   Phylica (150) S. Africa, Madagascar, Tristan da Cunha.

Ceanothus (55) N. America.
Hovenia (5) Himalaya to Japan.

2. Zizypheae.
Zizyphus (100) tropical America, Africa, Mediterranean region, Indomalaysia, Australia.
Paliurus (8) S. Europe to Japan.
3. Ventilagineae.
Ventilago (35) India and China to New Guinea, Australia and the Pacific.
4. Colletieae.
Colletia (17) temperate and subtropical S. America.
5. Gouanieae.
Gouania (20) tropics and subtropics.

CEANOTHUS THYRSIFLORUS Esch.
Blue Blossom

Distribution. Native in California and a valuable garden plant for growing against walls, in shrub borders or as a specimen tree.

Vegetative characteristics. An evergreen shrub or small tree with alternate, small, dark green, oblong-ovate leaves with glandular-toothed margins.

Floral formula. K5 C5 A5 $\underline{G}$(3)

Flower and inflorescence. The small, pale blue, actinomorphic, hermaphrodite flowers are borne in round-headed clusters that collectively form showy panicles in May and June.

Pollination. Small insects are attracted by the conspicuous panicles and by the nectar secreted from the intrastaminal disc surrounding the ovary.

Alternative flowers for study. Any other *Ceanothus* would be suitable and usually readily available as the genus is widely cultivated. Floral differences occur within the Rhamnaceae which are demonstrated by the 2 native members of the family. *Rhamnus cathartica*, Buckthorn, has 4-merous, unisexual flowers, while *Frangula alnus*, Alder Buckthorn, has 5-merous, hermaphrodite flowers. Both are in flower during May and June. Interesting fruits occur in *Paliurus spina-christi*, Christ's Thorn, which is native in S. Europe and W. Asia but has proved hardy in the British Isles.

Fig. 49 Rhamnaceae, *Ceanothus thyrsiflorus*

A A complete inflorescence, showing how the numerous small clusters of flowers form a showy panicle.
Inflorescence: 35–75 mm

B A single, actinomorphic, hermaphrodite flower. The 5 concave petals alternate with and expand at right angles to the valvate sepals. The 5 free stamens are opposite the petals.
Petals: 1.75 mm Sepals: 1.75–2 × 2 mm

C The flower with perianth removed to reveal the gynoecium and one of the stamens. The superior ovary is embedded in the nectar-secreting intrastaminal disc which forms the rim of the cup-like receptacle or hypanthium. Surmounting the ovary is the single 3-branched style. The stamens are inserted below the disc and, prior to emergence, are enveloped by the petals. The anthers are 2-celled and dehisce longitudinally.
Stamens: 1.5–3.5 mm Style: 0.75–2 mm

D L.S. of the centre of a flower, showing the ovary, disc and clawed base of 2 of the petals. The ovary, which is formed from 3 united carpels, is 3-locular and has one anatropous ovule with basal placentation in each loculus.
Ovary: 0.8 mm in diameter Ovule: 0.4 × 0.2 mm

E T.S. of ovary, showing the 3 separate loculi each containing one ovule. In transverse section the ovules appear unattached to a placenta owing to the basal placentation.

F The whole fruit prior to dehiscence with its persistent style. The black, dry, globose fruit is 3-lobed and separates into 3 nutlets.
Fruit: 2 × 3 mm

Fig. 49

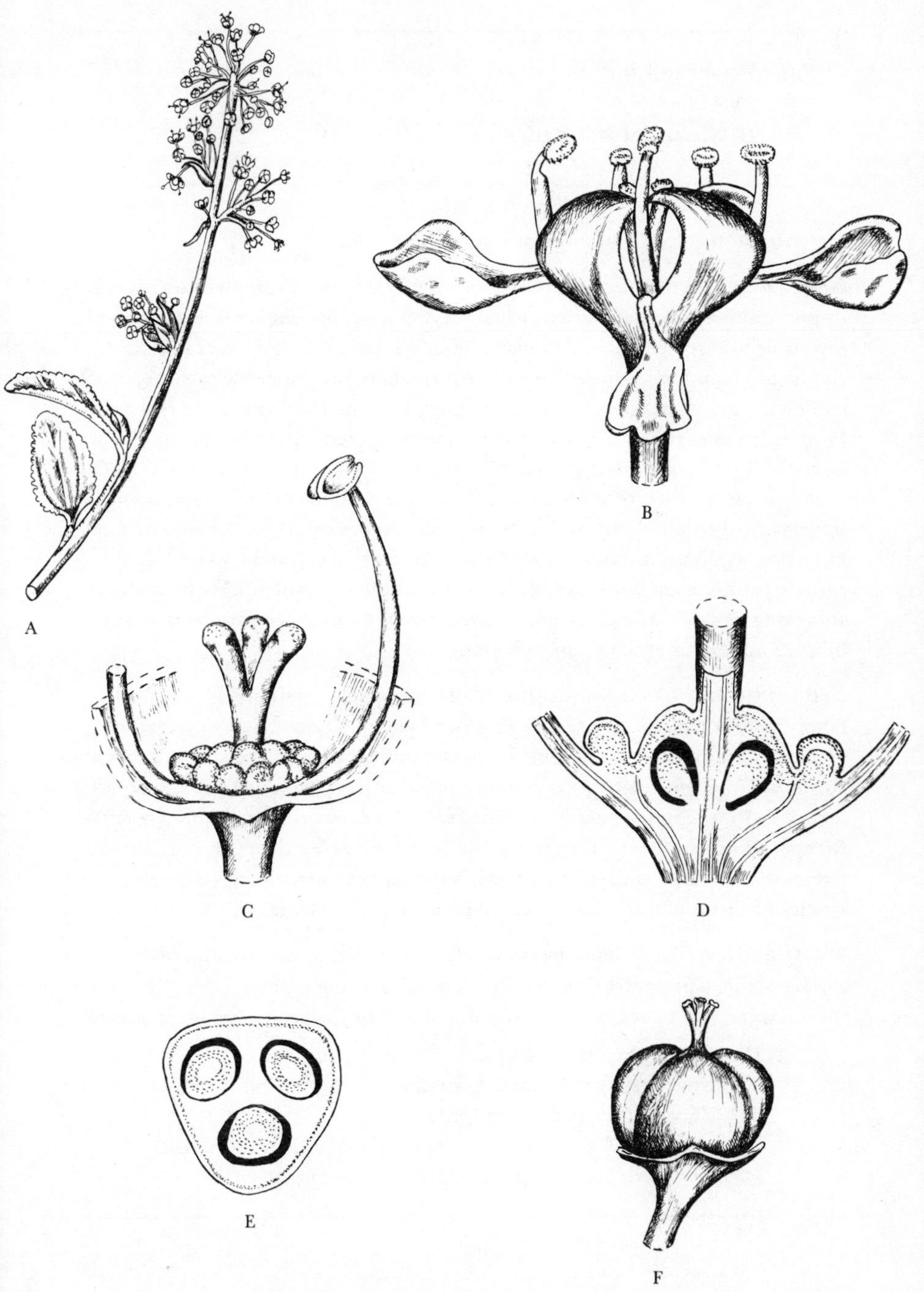

50 Vitaceae Lindl.

Vine family

12 genera and 700 species

Distribution. Mostly tropical and subtropical regions.

General characteristics. Climbing or rarely erect shrubs, sometimes dioecious or polygamo-monoecious; stems usually sympodial, bearing tendrils (modified shoots or inflorescences) which may end in discoid suckers. Leaves alternate, simple or compound, usually distichous, stipulate. Inflorescence cymose, usually complex and leaf-opposed; axis occasionally flattened and expanded (*Pterisanthes*); bracteoles present. Flowers actinomorphic, hermaphrodite or unisexual. Calyx of 4 or 5 small, united sepals, forming a very slightly lobed cup. Corolla of 4 or 5 petals, valvate, often united at the tips and falling off as a hood upon the opening of the bud. Stamens 4 or 5, opposite to the petals, at the base of a hypogynous disc; anthers introrse. Ovary superior, of 2 (rarely 3–6) united carpels, multilocular, with usually 2 anatropous, erect ovules in each loculus; style long or short, stigma inconspicuous, rarely 4-lobed (*Tetrastigma*). Fruit a berry. Endosperm present; embryo straight.

Economic and ornamental plants. The principal economic plant in this family is *Vitis vinifera*, the Grape-vine (see below), whose many varieties are extensively grown for their edible fruit, the source of wines, raisins, sultanas and currants. Other species of *Vitis* produce more or less palatable fruit and are of local importance in their native countries. *Parthenocissus tricuspidata*, Japanese Creeper or Boston Ivy (often referred to in Britain as Virginia Creeper), and *P. quinquefolia*, Virginia Creeper, are cultivated as ornamental vines on walls, while species of *Cissus* and *Rhoicissus* are popular as house plants.

Classification. The Vitaceae is most closely related to the Rhamnaceae, and to the Leeaceae, a palaeotropical family containing a single genus, *Leea*. Division of the Vitaceae into tribes, etc., is difficult as most of the genera are inter-related.

Cissus (350) mainly tropics.
Tetrastigma (90) S.E. Asia, Australia.
Vitis (60–70) northern hemisphere.
Ampelopsis (20) temperate and subtropical Asia and America.
Parthenocissus (15) temperate Asia and America.

VITIS VINIFERA L.
Grape-vine

Distribution. *V. vinifera* is known to have been in cultivation for some 6000 years and it is difficult to ascertain its complete ancestry and correct wild distribution. Two subspecies may be recognised – ssp. *sylvestris* for dioecious plants with unisexual flowers found in central and S.E. Europe, and ssp. *vinifera* for the cultivated vine which has hermaphrodite flowers. It is considered that the cultivated clones probably originated in S.W. Asia and in some areas have become naturalised, forming hybrids with ssp. *sylvestris*. In the past century, many American species of *Vitis* have been introduced into Europe as stocks. The scions, grafted on these, may be cultivars of ssp. *vinifera*, or American species or hybrids, or crosses between the American species and ssp. *vinifera.*

Vegetative characteristics. A deciduous, woody climber, with forked tendrils occurring opposite 2 out of 3 of the palmate, 3- to 5-lobed, coarsely toothed leaves.

Floral formula. K(5) C5 A5 $\underline{G}$(2)

Flower and inflorescence. The inflorescence replaces the tendrils in the upper parts of the stem, and consists of a dense panicle of small, actinomorphic flowers which appears in June.

Pollination. Bees and other insects have been seen visiting the flowers of *V. vinifera* which, although attractively scented, apparently produce little pollen and, at least in the more northerly areas of cultivation, no nectar. However, in warmer regions, the glands situated between the bases of the filaments have been observed to secrete abundant nectar. It is possible that some wind-pollination may take place and even more likely that self-pollination may occur as the stamens and stigma reach maturity at the same time.

Alternative flowers for study. There are numerous cultivars of *Vitis vinifera* with consequent variation in flowers and fruit. If the cultivated Grape is not available, other species of *Vitis* would be suitable or the closely related genus *Parthenocissus*, 2 commonly grown species of which are mentioned above. In this genus, the petals are free, spreading and persist for at least a short time after anthesis. Most members of the Vitaceae grown in this country flower between June and August.

Fig. 50 Vitaceae, *Vitis vinifera*

A The whole inflorescence, a dense panicle which develops in place of a tendril in the upper part of the stem.

B A single flower in bud (B1) and at anthesis (B2). The small, actinomorphic, green flowers have a minute calyx which forms a shallow cup at the base of the flower. The 5 petals are valvate in bud (B1) and apically connate. At anthesis the stamens expand and the petals separate from the base (B2) falling away together as a cap and exposing the other floral parts.
Flower (in bud): 0.8 × 0.7 mm (at anthesis): 1 × 0.7 mm

C A flower after the shedding of the cap of apically connate petals. The 5 free stamens are situated opposite the petals and arise from the glandular disc at the base of the superior ovary.

D The upper portion of a stamen. The anthers are versatile, introrse, 2-celled and dehisce longitudinally.
Filament: 2.25 mm Anther: *c.* 1 mm

E L.S. of the superior ovary bearing a single stigma on a short style. The ovary has 2 (rarely 3) loculi, with 1 or 2 anatropous ovules in each loculus.
Ovary: 3 × 2 mm Ovule: 1 × 0.5 mm

F T.S. of a typical 2-locular (F1) and a rarer 3-locular (F2) ovary with axile placentation.
Ovary: 2 mm in diameter

G Detail of the apex of the short style with its simple, discoid stigma.
Stigma: 0.4 mm in diameter

H The mature fruit, a soft juicy berry containing 1 to 4 seeds with a bony testa. Minor differences occur in the shape, size and colour of the fruits of the various cultivars which are useful diagnostic features in identification.
Fruit: 10 × 9 mm

Fig. 50

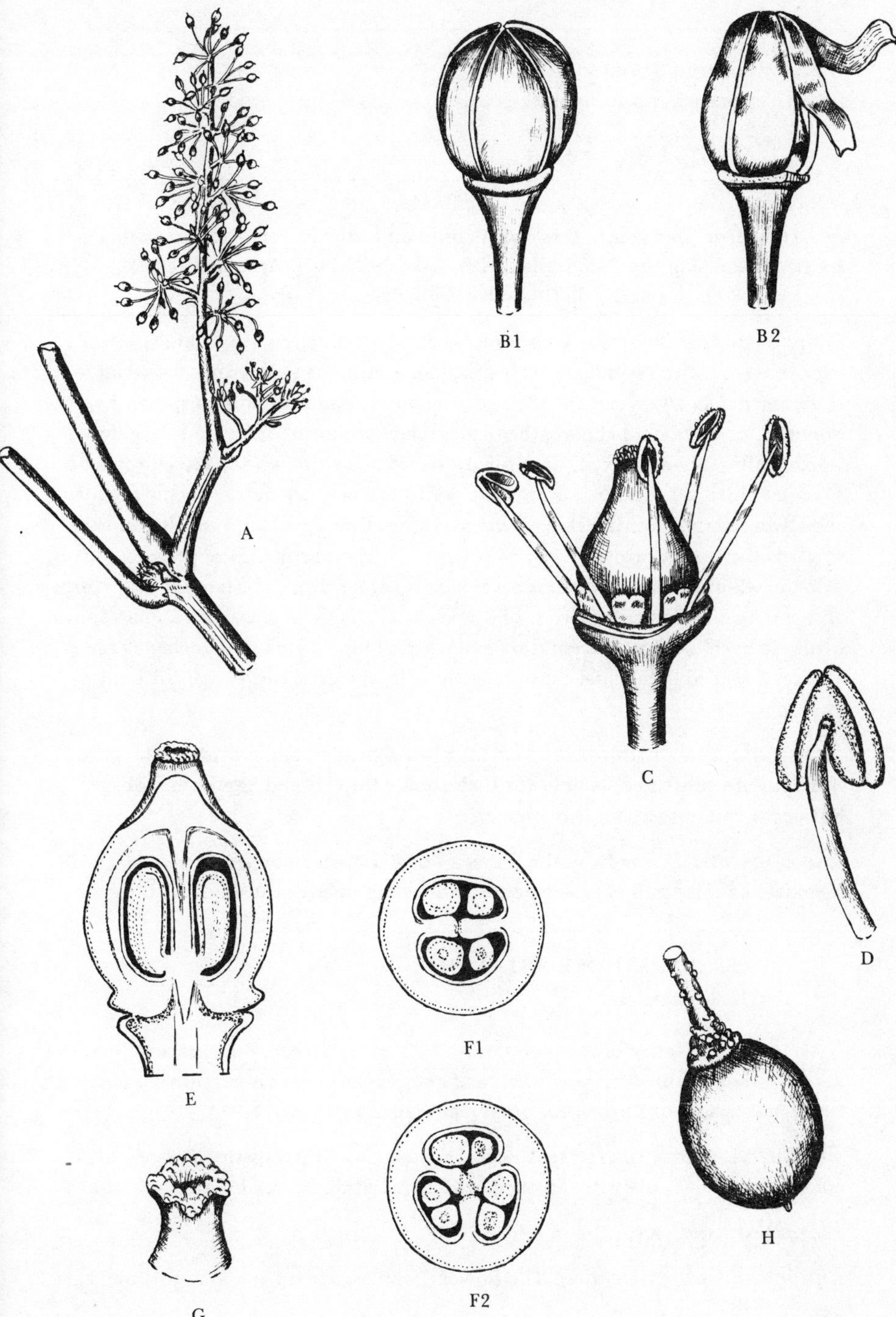

51 Hippocastanaceae DC.

Horse-chestnut family
2 genera and 15 species

Distribution. The genus *Aesculus* is confined to the N. temperate region (1 species in S.E. Europe, 5 in India and E. Asia and 7 in N. America). The 2 species of *Billia* are native in Central and northern S. America.

General characteristics. Trees or large shrubs with large winter buds, sometimes covered with resinous scale-leaves, containing the next year's shoot and inflorescence in a very advanced stage. The bud expands rapidly in spring. Leaves opposite, exstipulate, palmate; the blades when young are hairy and hang downwards. Inflorescence mixed, the primary structure racemose, the lateral branches cymose (cincinni). Upper flowers male, with rudimentary ovary, opening before the lower. Hermaphrodite flowers protogynous. Calyx of 4 or 5 sepals connate towards the base. Corolla of 4 or 5 free petals, zygomorphic. Disc present, outside the whorl of stamens, often developed on one side only. Stamens 5 to 8, free; anthers introrse. Ovary superior, of 3 united carpels, 3-locular, with 2 anatropous ovules in each loculus; placentation axile; style long; stigma unbranched. Fruit a large, 1-seeded loculicidal capsule, opening by 3 valves. Seed large, without endosperm.

Economic and ornamental plants. Several species and hybrids of the genus *Aesculus* are cultivated, mainly for their showy flowers and handsome foliage, but occasionally as a source of timber.

Classification. Apart from the 2 species of *Billia* which are evergreen, all the members of this family are deciduous and belong to the genus *Aesculus*.

AESCULUS HIPPOCASTANUM L.

Horse-chestnut

Distribution. Native in the mountains of Albania, Greece, Bulgaria and Jugoslavia, and naturalised in parts of W. and central Europe. It is extensively planted throughout most of Europe for ornament and as a shade tree.

Vegetative characteristics. A large, broad-crowned tree with palmate leaves composed of 5–7, obovate, toothed leaflets radiating from a long, stout petiole.

Floral formula. K(5) C4–5 A5–8 $\underline{G}(3)$

Flower and inflorescence. The flowers, which appear from May to June, have white petals marked with yellow spots which turn pink with age. There are usually 5 petals, the upper somewhat larger than the lower, and the lowest petal either very small or absent. Each of the lateral branches of the inflorescence bears 3 to 6 zygomorphic flowers in a scorpioid cyme. The group of cymes forms a terminal panicle called a thyrse. The upper flowers in the inflorescence have only rudimentary female organs so are functionally male. Those in the centre are hermaphrodite, while the lowest flowers in the inflorescence are functionally female owing to the incomplete development and early shedding of the anthers.

Pollination. The dense inflorescences of white flowers with coloured markings (nectar-guides) on the 2 upper petals attract bees, mainly bumble-bees, in search of nectar. The nectar-guides, which are pale yellow at first, then deeper yellow and finally carmine in colour, lead the insects to the nectar which is located on the disc lying between the claws of the petals and the stamens. In obtaining the nectar, the bees brush against the stamens and become dusted with the reddish brown pollen which they transfer to other flowers on their foraging trips.

Alternative flowers for study. A. × *carnea*, Red Horse-chestnut, is also widely cultivated in parks and large gardens. It is generally considered to be a garden hybrid between *A. hippocastanum* and an American species *A. pavia*, Red Buckeye.

Fig. 51 Hippocastanaceae, *Aesculus hippocastanum*

A An upper, male flower showing 3 of the 4 petals (the fifth being absent). Two of the 5 hairy sepals, connate at the base, can be seen protruding from behind the petals.
Upper petals: 17 × 11 mm Lower petal: 13 × 9 mm

B A hermaphrodite flower with sepals cut longitudinally to reveal the attachment of the carpel and one of the stamens on the receptacle. The stamens are inserted close to the base of the carpel and within the sometimes partially-developed disc.
Stamen: 18 mm Anther: 1.4 mm Style: 21 mm

C T.S. of ovary showing the 3 loculi. Each loculus contains 2 ovules (see E) on axile placentas.

D T.S. of ovary after fertilisation showing a single seed developing at the expense of all the other ovules. Two of the loculi are displaced in order to accommodate the large seed.

E L.S. of ovary showing the 2 ovules situated one above the other in each loculus.

F Detail of upper portion of style and stigma.

G The mature fruit prior to dehiscence. The spiny, leathery capsule splits longitudinally into 3 valves to release the large shiny seed or 'conker'. Note the large hilum on the seed. The size of the fruit and the number of spines can vary considerably. The fruit illustrated measures 33 × 36 mm

Fig. 51

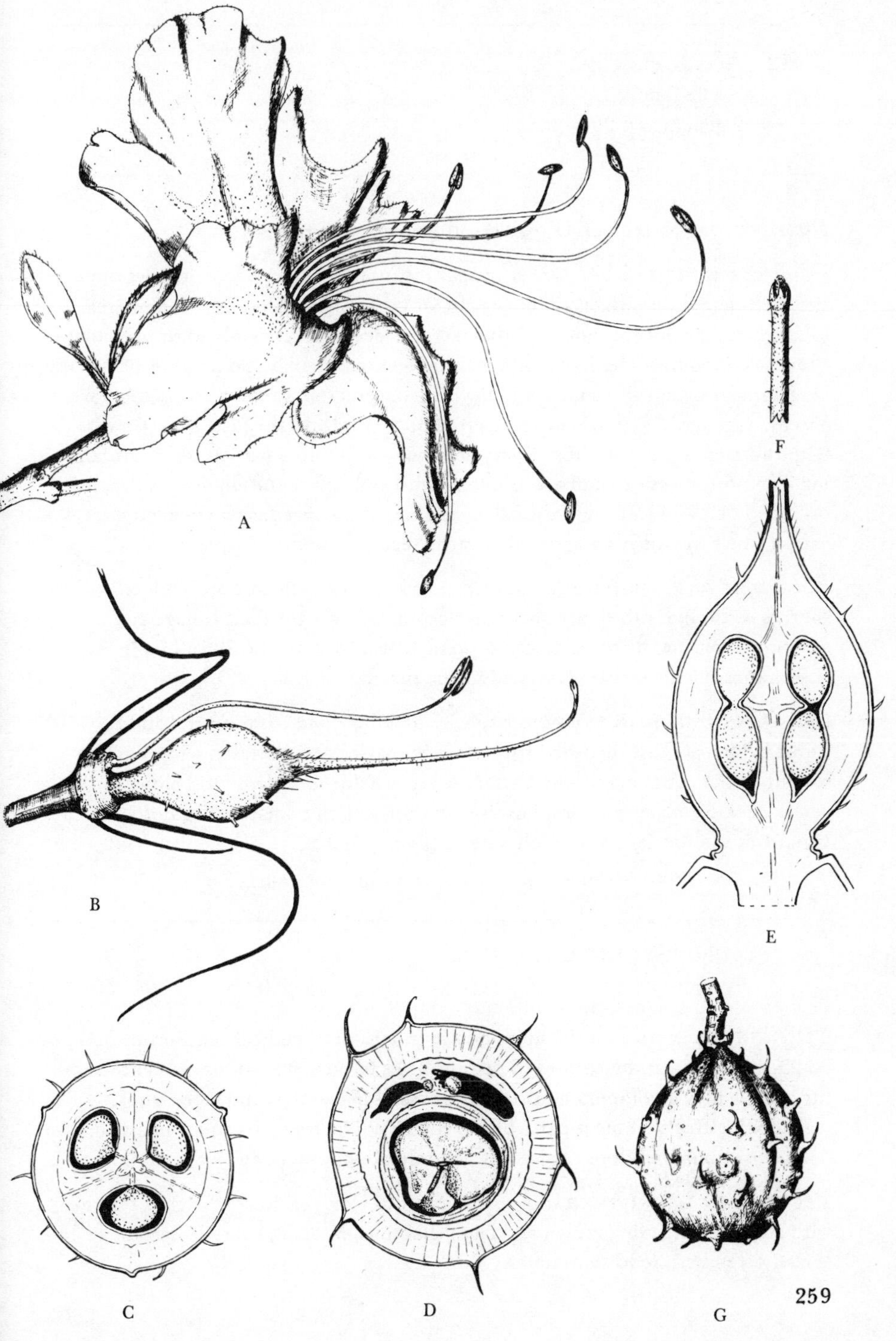

52 Aceraceae Juss.

Maple family

2 genera and 202 species

Distribution. N. temperate region and tropical mountains.

General characteristics. Trees and shrubs, usually monoecious. Leaves opposite, petiolate, exstipulate, simple, entire or more often palmately or pinnately lobed, occasionally compound. Inflorescence racemose, corymbose or paniculate. Flowers actinomorphic, hypogynous or perigynous. Sepals and petals 4 or 5, free. Disc annular or lobed, usually well developed, sometimes reduced to teeth, rarely absent. Stamens 4–10, usually 8 inserted in 2 whorls on the disc. Male flowers with rudimentary gynoecium. Ovary superior, of 2 united carpels, each containing 2 orthotropous or anatropous ovules with axile placentation. Styles 2, free or joined below. Fruit a schizocarp, consisting of 2 winged, single-seeded mericarps (samaras) which separate when ripe. Seeds without endosperm.

Economic and ornamental plants. Some species of maple are valuable sources of timber, others are grown as decorative trees for their foliage and autumn colouring. *Acer saccharophorum* (*A. saccharum*), the Sugar Maple from N. America, yields maple syrup and maple sugar.

Classification. Apart from the 2 species of *Dipteronia* from central and S. China, which are distinguished from *Acer* by having the mericarp winged all round, all the members of the Aceraceae are usually considered to belong to the genus *Acer*. Some authorities, however, recognise a third genus, *Negundo*, which comprises the species with trifoliolate or pinnate leaves.

Acer (200) N. temperate region, tropical mountains.

ACER PSEUDOPLATANUS L.

Sycamore

Distribution. Native on the hills and mountains of S. and central Europe and W. Asia, it was introduced into the British Isles in the fifteenth or sixteenth century and is now a common constituent of many hedgerows and woodlands, regenerating freely in most parts of the country. The tree grows best in a rich, deep, moist but well-drained soil but will tolerate poorer conditions.

Vegetative characteristics. A tall, deciduous tree, reaching a height of 30 m, with a broad, spreading crown and large, glabrous, 5-lobed, coarsely serrate leaves, often with reddish petioles.

Floral formula. K5 C5 A8 $\underline{G}$(2)

Flower and inflorescence. The actinomorphic, yellowish-green flowers are borne in drooping panicles (see Fig. 52.A) that terminate short, leafy branches. Flowering occurs from April to June, with or just after the development of the leaves. The flowers are, in effect, unisexual, as those with a well-developed gynoecium have stamens which do not reach maturity and those with anthers which shed pollen have only a rudimentary ovary.

Pollination. Many kinds of flies and bees are attracted by the easily obtainable nectar secreted by glands on the disc at the base of the ovary. The functionally unisexual nature of the flowers promotes cross-pollination.

Alternative flowers for study. Several cultivars of *A. pseudoplatanus* are grown in gardens, including forms with variegated leaves, and 'Brilliantissimum' which has its young leaves suffused with coral-pink. Alternative species include *A. campestre*, Field Maple, which flowers in May and June and *A. platanoides*, Norway Maple, which flowers in April and May (about 3 weeks earlier than *A. pseudoplatanus*) before the leaves have expanded. *A. palmatum*, Japanese Maple, and *A. saccharophorum*, Sugar Maple, have a corymbose inflorescence, while *A. saccharinum*, Silver Maple, has apetalous flowers. *A. negundo*, Box Elder, is dioecious, the male flowers forming a corymbose inflorescence while the female flowers hang in racemes.

Fig. 52 Aceraceae, *Acer pseudoplatanus*

A A drooping panicle, borne at the end of a short, leafy branch between the final pair of leaves (petiole bases only shown). Two buds can be seen arising from the axils of the long petioles. The flowers near the apex of the inflorescence may be sterile, while those in the centre are functionally male. The functionally female flowers are found towards the base of the panicle.
Inflorescence: 11–12 cm

B A single, actinomorphic, male flower. The 5 free sepals and 5 free petals subtend the fleshy, lobed, nectar-secreting disc which surrounds a hairy prominence representing a rudimentary ovary. The 8 stamens are inserted on the inner surface of the disc and their filaments are hairy towards the base.
Flowers: 5–6 mm in diameter Sepals and petals: 2 mm
Stamens: 5–6 mm

C Detail of a dorsifixed anther at dehiscence with some of the pollen grains. The anther is 2-celled and dehisces longitudinally.
Anther: 1 × 1 mm

D L.S. of a young, actinomorphic, hermaphrodite flower, as indicated by the erect position of the 2 styles. The 2 loculi of the ovary can be seen with one of the 2 pendulous ovules present within each loculus. As in the male flower, there are 8 stamens attached to the nectar-secreting disc, but in this case they do not shed pollen although appearing fully developed. The flower is therefore functionally female. At this stage of development, the perianth envelops the inner organs.
Sepals and petals: 3 mm Ovary: 1 × 1.5 mm
Style: 1 mm Nectar-secreting disc: 3.5 mm in diameter

E T.S. of the superior ovary formed from 2 fused carpels, each with a single loculus containing 2 ovules with axile placentation.
Ovary: 4 mm

F A young fruit with most of the perianth removed. The walls of the carpels are developing into wings. The 2 styles, connate below and divergent above, are beginning to wither now that fertilisation has been completed. Below the ovary is the lobed disc, with one of the sterile stamens still attached to its inner surface.
Young fruits: 3–4 × 5 mm Wings: 4–5 × 2–3 mm
Disc: 4.75 mm in diameter
Styles (connate portion): 3 mm
(free portion): 3 mm

G The mature fruit, a schizocarp that splits into 2 one-seeded portions, each with the pericarp developed into a wing. The winged mericarps are known as samaras. The 2 samaras remain attached to each other as they spin through the air away from the parent tree, but when on the ground they separate before the seeds germinate.
Samara: 3.5–5 cm

Fig. 52

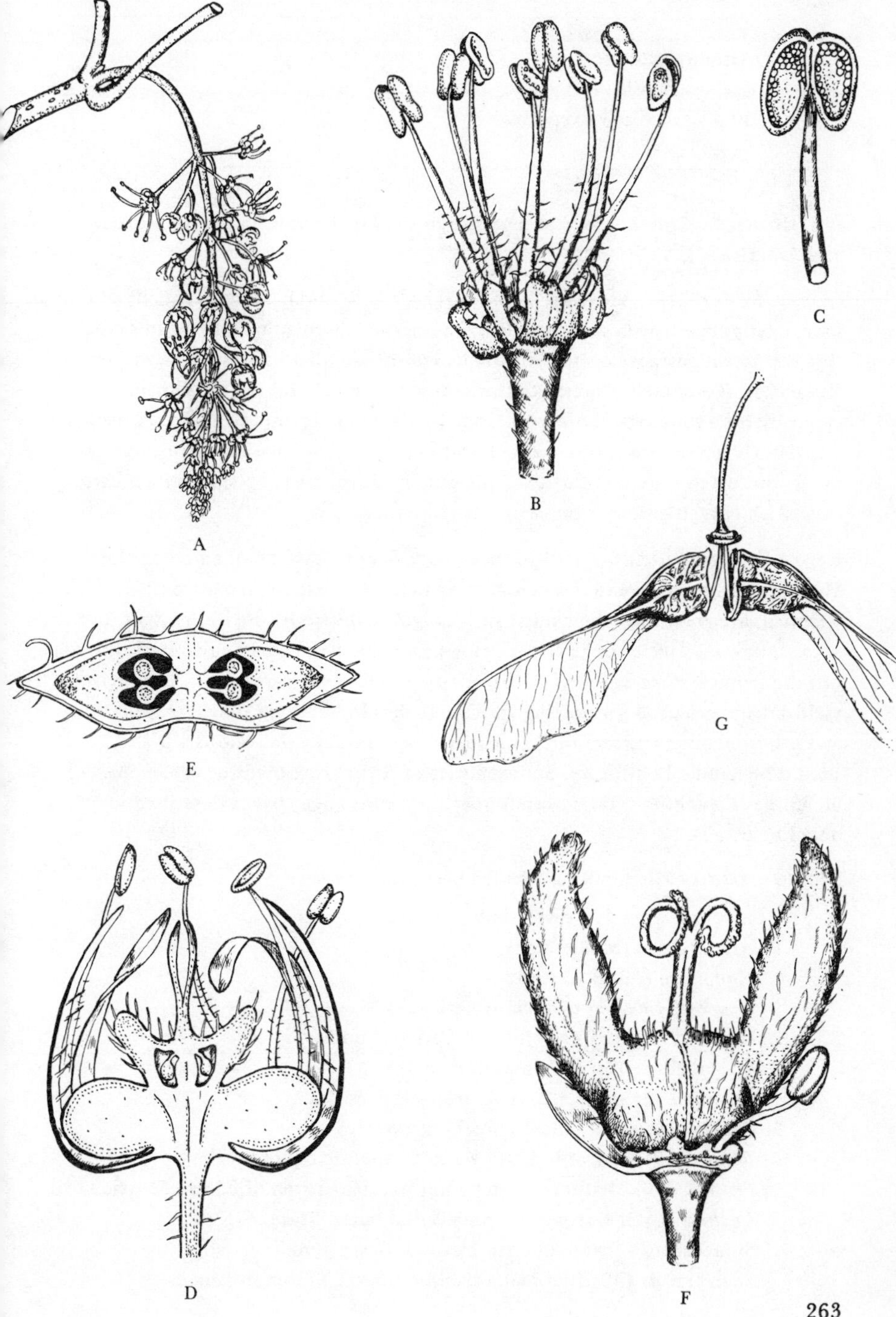

53 Anacardiaceae Lindl.

Cashew family

60 genera and 600 species

Distribution. Chiefly tropical, but also in the Mediterranean region, E. Asia and America.

General characteristics. Trees and shrubs, rarely climbers, with usually alternate, exstipulate leaves, and panicles of numerous, hermaphrodite or unisexual flowers. Resin passages occur, but the leaves are not gland-dotted (in contrast to Rutaceae). Receptacle convex, flat or concave; gynophores occur. Flowers typically 5-merous, actinomorphic, hypogynous to epigynous. Stamens usually 5 or 10. Ovary superior, of usually 3 united carpels, each with a single anatropous ovule, but often only one fertile. Fruit usually a drupe with resinous mesocarp. Seed with curved embryo, but without endosperm.

Economic and ornamental plants. Pistacia vera, Pistachio, native in the Mediterranean region and *Anacardium occidentale*, Cashew, from tropical America, are grown for their nuts, and *Mangifera indica* for its fleshy fruit, the Mango. Several species of *Spondias*, Hog-plum, are also widely cultivated throughout the tropics. *Pistacia lentiscus*, Mastic-tree, and *Rhus verniciflua*, Lacquer-tree, yield substances used for making varnish. It should be noted that *R. verniciflua* and other members of section Toxicodendron contain a poisonous sap and should be handled with care. Some species of *Rhus* are grown for ornament, including *R. typhina*, Stag's-horn Sumach. *Cotinus coggygria* is described in detail below.

Classification. The family is divided here into 5 tribes:

A. Five free carpels, or 1. Leaves simple, entire.

1. Anacardieae (Mangiferae).
   Mangifera (40) S.E. Asia.
   Anacardium (15) tropical America.

B. Carpels united (rarely only 1), leaves rarely simple.

2. Spondiadeae (ovule 1 in each carpel).
   Spondias (10–12) S.E. Asia, tropical America.
3. Rhoeae (ovule in only 1 carpel, ovary free).
   Rhus (250) subtropical and warm temperate regions.
   Pistacia (10) Mediterranean region to Afghanistan, S.E. and E. Asia.
   Cotinus (3) S. Europe to China, S.E. United States.
4. Semecarpeae (ovule in only 1 carpel, ovary sunk in axis).
   Semecarpus (50) Indomalaysia, Micronesia, Solomon Islands.

C. Carpel 1. Female flowers naked. Leaves simple, toothed.

5. Dobineeae.

Dobinea (2) E. Himalaya, S. China.

COTINUS COGGYGRIA Scop. (RHUS COTINUS L.)
Smoke-tree

Distribution. Native of central and southern Europe, the Himalaya and China, and often planted in parks and gardens.

Vegetative characteristics. A deciduous shrub of round bushy habit, with orbicular or obovate leaves which remain long on the plant and turn yellow or red before they fall.

Floral formula. K(5) C5 A5 $\underline{\mathrm{G}}$(3)

Flower and inflorescence. Hermaphrodite or unisexual flowers are borne in loose, terminal panicles in June and July. After flowering, the pedicels lengthen and the silky, spreading hairs give the panicles a hazy, smoke-like appearance.

Pollination. Insects of many kinds are attracted to the large, loose inflorescences and to the nectar which is secreted by the annular disc. As the nectar lies completely exposed it is available even to very short-tongued insects.

Alternative flowers for study. Rhus typhina is also widely cultivated and is an acceptable alternative, though it differs from *C. coggygria* in being consistently dioecious. It too flowers in June and July.

Fig. 53 Anacardiaceae, *Cotinus coggygria*

A The inflorescence, a loose and much-branched panicle. The branches are terminated by hermaphrodite or unisexual flowers whose pedicels lengthen and become silkily hairy as the flowering stage ends and the fruiting stage begins.
Inflorescence: *c.* 20 cm

B A single, actinomorphic hermaphrodite flower, showing the 5 basally connate sepals and 5 free petals attached below the annular, nectar-secreting disc. The 5 free stamens are situated opposite the sepals.
Flowers: 5–6 mm in diameter

C L.S. of flower. The well-developed disc is situated between the gynoecium and the stamens. Two of the 5 stamens, which arise from under the edge of the disc, can be seen. The anthers are 2-celled and dehisce longitudinally.
Disc: 2.5 mm in diameter Anther: 1 mm Filament: 0.75 mm

D L.S. of superior ovary, which is situated at the centre of the concave annular disc. The ovary consists of 3 united carpels surmounted by 3 styles. It is unilocular and contains a single, anatropous, fertile ovule.
Ovary: 0.3 x 0.2 mm

E T.S. of ovary and a portion of the disc. The ovule can be seen at the centre of the loculus.

F The whole fruit, a drupe with resinous mesocarp. Note the irregular shape of the fruit and the 3 persistent styles.
Fruit: 3 x 3 mm

G Detail of the elongated, plumose pedicel of a sterile flower from a fruiting panicle. The mass of plumose pedicels gives the infructescence its characteristic appearance. The fruits finally become detached and are dispersed by the wind.

Fig. 53

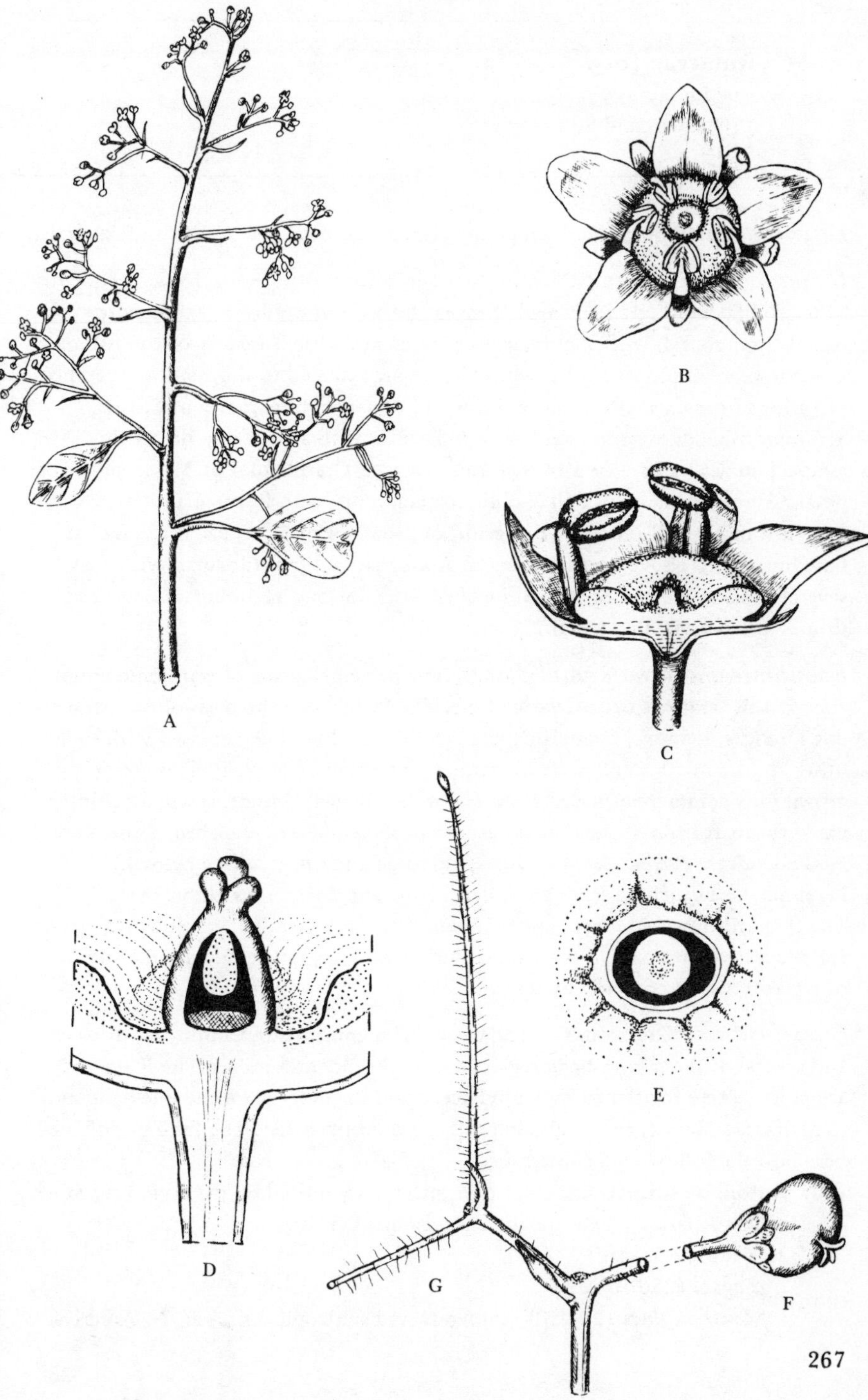

54 Rutaceae Juss.

Rue family

150 genera and 900 species

Distribution. Tropical and temperate regions, especially S. Africa and Australia.

General characteristics. Mostly trees and shrubs, often xerophytic, frequently of heath-like habit (e.g., *Diosma*). Leaves alternate or opposite, exstipulate, usually compound, with glandular dots, often aromatic. In many of the Aurantioideae there are short shoots whose leaves are reduced to thorns. Inflorescence of various forms, usually cymose. Flowers hermaphrodite, rarely unisexual, actinomorphic or zygomorphic, 4- or 5-merous, with a large disc below the gynoecium. Calyx of 4 or 5 free or united sepals. Corolla of 4 or 5 free, imbricate petals. Stamens usually 8 or 10, obdiplostemonous, with introrse anthers. Ovary superior, rarely half-inferior or inferior, of usually 4 or 5 carpels, often free at base but united above by the style (cf. Apocynaceae), multilocular, with 1 to several anatropous ovules in each loculus. Fruit various, including schizocarps, drupes and berries. Endosperm absent.

Economic and ornamental plants. The principal genus of economic importance in this family is *Citrus*, various species of which are the basis of our present-day Oranges, Lemons, Grapefruit, etc. The species have been crossed with each other and also with *Fortunella*, Kumquat, and some of these hybrids are now grown on a commercial scale. *Aegle marmelos*, Bengal Quince, is widely cultivated for its fruit in S. Asia. Amongst the ornamental tree and shrub genera are *Phellodendron*, *Ptelea*, *Zanthoxylum*, *Skimmia* and *Choisya* (see below). Perennial herbs include *Ruta graveolens*, Rue, and *Dictamnus albus*, Dittany. The fragrant-flowered *Murraya paniculata*, Orange Jessamine, is frequently grown for decorative purposes in the tropics and subtropics, but requires the protection of a glasshouse in cooler climates.

Classification. The groups of Rutaceae differ considerably among themselves and several of them have been regarded as independent families. The Rutaceae are most closely related to Zygophyllaceae and Cneoraceae on the one hand and to Meliaceae, Burseraceae and Simaroubaceae on the other. The family is divided here into the following 5 subfamilies:

I. Rutoideae (carpels usually 4 or 5, often only united by the style, and more or less separate when ripe; dehiscence loculicidal).

1. Zanthoxyleae.

   Fagara (250) tropics.

   Zanthoxylum (20–30) temperate and subtropical E. Asia, N. America.

Choisya (6) S. United States, Mexico.

2. Ruteae.

Ruta (7) Canary Islands, Mediterranean region, S.W. Asia.

Dictamnus (6) central and S. Europe to E. Asia.

3. Boronieae.

Boronia (70) Australia.

Eriostemon (32) Australia, New Caledonia.

Correa (11) temperate Australia.

4. Diosmeae.

Adenandra (25) S. Africa.

Diosma (15) S. Africa.

5. Cusparieae.

Cusparia (30) tropical S. America.

Galipea (13) Central and S. America.

Almeidea (5) Brazil, Guiana.

II. Dictyolomatoideae (carpels 5, with several ovules, united only at base; fruit 3- or 4-seeded).

Dictyoloma (2) Peru, Brazil.

III. Spathelioideae (carpels 3, united, each with 2 pendent ovules; drupe winged; oil glands at margins of leaves).

Spathelia (20) W. Indies, northern S. America.

IV. Toddalioideae (carpels (1–)2–5, each with 1 or 2 ovules; fruit a drupe or dry and winged; endosperm present or absent; oil glands in bark and leaves).

Skimmia (7 or 8) Himalaya to Philippines.

Ptelea (3) United States, Mexico.

V. Aurantioideae (fruit a berry, with pulp derived from sappy emergences of the carpel wall; seeds without endosperm, often with 2 or more embryos; oil glands present).

Glycosmis (60) Indomalaysia.

Atalantia (18) tropical Asia, China, Australia.

Citrus (12) S. China, S.E. Asia.

Aegle (3) Indomalaysia.

CHOISYA TERNATA H.B.K.

Mexican Orange Flower

Distribution. A native of Mexico and introduced into Britain in 1825. It is a popular evergreen shrub which is normally quite hardy, but may suffer some damage during severe winters.

Vegetative characteristics. An evergreen shrub of rounded, bushy habit, with opposite, gland-dotted leaves composed of 3 obovate leaflets.

Floral formula. K5 C5 A10 G(5)

Flower and inflorescence. The inflorescence consists of about 6 branches, with 4 or 5 flowers on each branch. The fragrant, white flowers appear during April and May, but if conditions are suitable the flowering period may be extended to September or even to December if the weather is very mild.

Pollination. The clusters of white flowers attract insects which seek out the nectar secreted by the disc at the base of the ovary. The stamens and styles are usually mature at about the same time so that both cross-pollination and self-pollination may occur. It is, however, remarkable that there is a general failure to produce fruit, and therefore it is likely that the stock in cultivation is self-incompatible.

Alternative flowers for study. *Choisya ternata* is easily obtainable, but has the disadvantage, mentioned above, that the cultivated stock does not fruit. Capsules, however, are produced by *Ruta graveolens*, which flowers throughout the summer, and by *Dictamnus albus*, which has zygomorphic flowers in May and June, *Skimmia japonica* provides an example of dioecism, with drupes as fruits on the female plants, while the genus *Citrus* exhibits large fleshy berries with numerous carpels and a leathery epicarp. Flowers of this genus are often available from orange and lemon plants in conservatories.

Fig. 54 Rutaceae, *Choisya ternata*

A A part of the branched inflorescence at an early stage, with only the terminal flower fully expanded.

B A single flower with 2 of the 5 (sometimes 6) free petals removed to reveal the reproductive parts. The stamens are attached to the rim of a disc situated at the base of the ovary.
Sepal: 5 mm Petal: 11 mm

C One of the 10 (sometimes up to 13) free stamens, showing the 2-celled anthers which are introrse and dehisce longitudinally.
Filament: 5 mm Anther: 1.5 mm

D Perianth and stamens have been removed to expose the gynoecium of 5 united carpels, forming a deeply 5-lobed ovary surmounted by 5 united styles with furrowed stigmas.

E L.S. of ovary. The superior ovary is 5-locular, and has 2 superposed ovules with axile placentation in each loculus.
Ovary: 4.5–4.75 mm Loculus: 1.25 × 0.75 mm
Style + stigma: 3 mm

F T.S. of the densely hairy ovary. In wild plants the ovary develops into a 5-valved capsule.
Ovary: 3 mm in diameter

Fig. 54

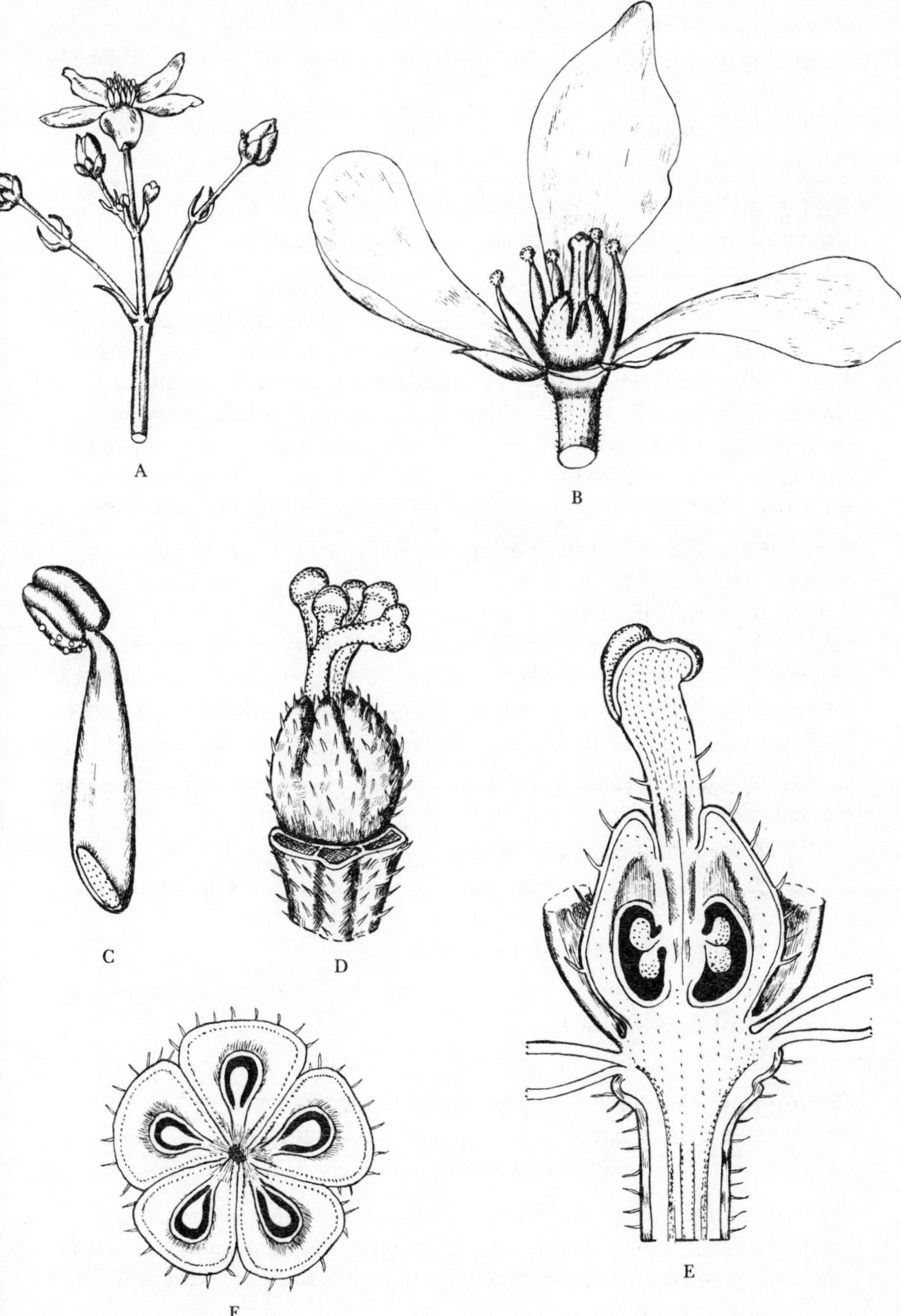

55 Juglandaceae Kunth

Walnut family

7 genera and 64 species

Distribution. N. temperate and subtropical regions, Andes.

General characteristics. Monoecious trees with alternate, exstipulate, pinnate leaves; winter buds brown and hairy. Inflorescence unisexual. Bract and usually 2 bracteoles present. Male flowers appearing as catkins on the twigs of the previous year; perianth typically of 4 segments, but often fewer by abortion; stamens 3–40 (more in the lower flowers). Female flowers with epigynous perianth enclosed in adnate cupule. Ovary inferior, of 2 united carpels, unilocular, with 1 erect, orthotropous ovule; style short, with 2 stigmas. Flowers wind-pollinated. Fruit a drupe or nut, sometimes winged. Seed without endosperm.

Economic and ornamental plants. Several members of the family are of economic importance for their timber and edible nuts, e.g., *Juglans regia*, Walnut, *J. nigra*, Black Walnut, *J. cinerea*, Butternut, *Carya illinoinensis*, Pecan, and various other species of *Carya* known as Hickory. *J. nigra* is often planted as a decorative tree in large gardens in Britain, but other species of *Juglans* are much less common. *Carya* and *Pterocarya*, Wing-nut, are occasionally grown for ornament.

Classification. This small family is considered as having some connection with the Anacardiaceae.

Carya (25) E. Asia, eastern N. America.
Juglans (15) Mediterranean region to E. Asia, N. and Central America, Andes.
Pterocarya (10) Caucasus to Japan.

JUGLANS REGIA L.

Walnut

Distribution. Native in S.E. Europe and central Asia to China. It has been introduced into other parts of Europe and is widely cultivated for its nuts and timber. It is now naturalised in many places, including S. and S.W. England.

Vegetative characteristics. A large, spreading, deciduous tree, reaching a height of 30 m. The bark is grey, smooth and only fissured when old. The alternate, pinnate leaves are composed of usually 7 or 9 obovate or elliptical leaflets.

Floral formula. Male: P4 A6–20
Female: P4 $\overline{\text{G}}$(2)

Flower and inflorescence. J. regia is monoecious and flowers in June. The pendulous, solitary, purplish male catkins are made up of numerous staminate flowers and develop laterally on the stems of the previous year. The female or pistillate flowers develop in pairs or groups of 3 at the end of the current year's growth.

Pollination. The Walnut is wind-pollinated, but individual specimens may be either protandrous or protogynous, which encourages cross-pollination. The pendulous nature of the male catkins with their numerous stamens facilitates the transference of pollen by the wind to the branched, plumose stigmas of the pistillate flowers.

Alternative flowers for study. Juglans regia is a readily available species. The closely related *J. nigra*, Black Walnut, differs in having more numerous, always serrate leaflets, and more globular, pubescent fruit. Comparison should be made with *Carya* and *Pterocarya* which are sometimes represented in the larger gardens and parks.

Fig. 55 Juglandaceae, *Juglans regia*

A End of a branch in June. The solitary, pendulous male catkins grow laterally from the twigs of the previous year. The current year's leafy growth begins just beyond the male catkin and ends with 2 pistillate flowers which develop from the axils of the leaves.
Male catkin: 70 × 8–10 mm

B A single staminate flower attached to the axis of the male catkin. The scars on the axis denote the position of other flowers which have been removed for clarity. Each staminate flower is subtended by a hairy bract and has usually 4 (rarely 3) irregular perianth-segments loosely enveloping a variable number of stamens. Up to 20 stamens may be present in the lower flowers of the catkin but only perhaps 6 in the upper.
Bract: 2 × 1 mm Perianth-segments: 2 × 1.25 mm

C Detail of a pendulous stamen with short filament. C1: before dehiscence. C2: after dehiscence. The anthers are basifixed, 2-celled and dehisce longitudinally.
Anther: 2 × 0.75 mm

D The 2 green, erect, pistillate flowers at anthesis. The female flowers are terminal and arise from the axils of the leaves. Only the bases of the petioles are shown. Foliage at a very early stage of development can be seen in front.

E A single pistillate flower showing 3 of the 4 perianth-segments surrounding the 2-lobed style which terminates the pubescent, inferior ovary.
Perianth-segments: 3–3.5 × 1 mm
Style + stigma: 5–8 × 3 mm Ovary: 4 × 3–4 mm

F L.S. of the pistillate flower. The inferior ovary is unilocular and is formed from 2 fused carpels. The single, erect, orthotropous ovule is attached to the centre of an incomplete partition, although appearing basal in the young flower.

G The fruit in August, an aromatic drupe comprising a glabrous, green, gland-dotted epicarp, a fleshy mesocarp and an endocarp that becomes woody in September or early October and forms the characteristic walnut with its single large seed (the edible 'nut').
Young fruit: 27 × 23 mm

H T.S. of a walnut. The walnut of commerce lacks the fleshy green covering (mesocarp and epicarp) but pickled walnuts are the immature, complete fruit. The woody endocarp ('shell') has a deeply contoured inner surface and encloses a seed with deeply lobed cotyledons covered by a thin, brown seed-coat. Two primary septa grow inwards from the point of union of the 2 carpels and unite the lower part of the fruit to a varying degree, the seed being supported by these septa. 2 secondary septa may also develop in the lower part of the endocarp, causing the seed to become 4-lobed at its base. The endocarp dehisces along the 2 prominent midribs situated opposite the 2 cotyledons.
Endocarp ('shell' of walnut): 40 × 31 mm

Fig. 55

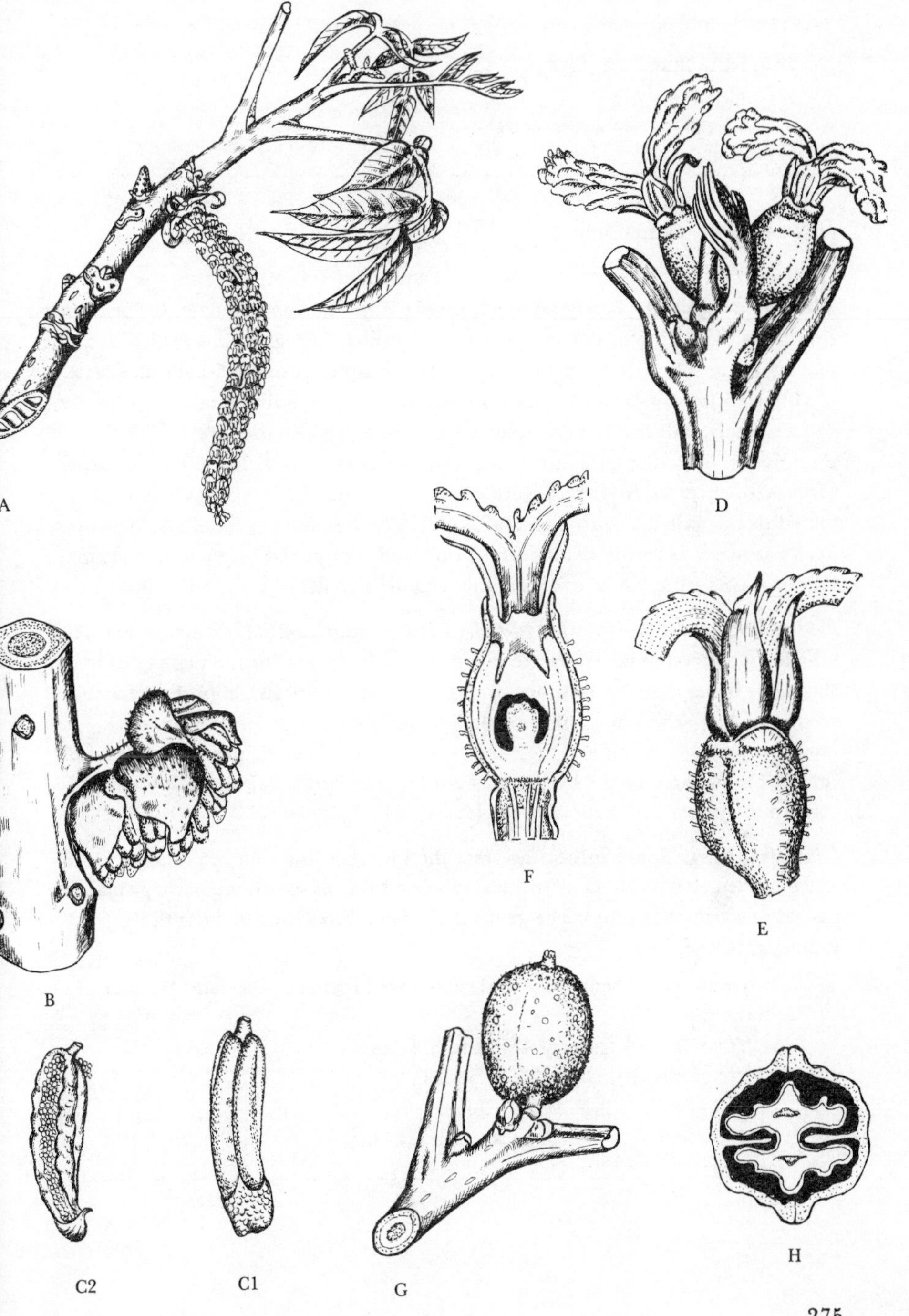

56 Linaceae S.F. Gray

Flax family

12 genera and 290 species

Distribution. Cosmopolitan.

General characteristics. Mostly herbaceous plants, sometimes shrubs. Leaves alternate or opposite, rarely whorled, simple, entire, often stipulate. Inflorescence cymose, a dichasium or cincinnus, the latter usually straightening out and looking like a raceme. Flowers hermaphrodite, actinomorphic, usually 5-merous. Calyx usually of 5 (rarely 4) free or basally connate sepals. Corolla usually of 5 (rarely 4) free petals, imbricate or convolute. Stamens 5, 10 or more, often with staminodes, united at base into a ring. Ovary superior, of 2, 3 or 5 united carpels, often with extra partitions projecting from the midribs of the carpels, but not joined to the axile placentas. Ovules 1 or 2 in each loculus, pendulous, anatropous. Styles as many as loculi, distinct, filiform, each terminated by a capitate stigma. Fruit a septicidal capsule or drupe. Embryo usually straight, in fleshy endosperm.

Economic and ornamental plants. Linum usitatissimum, Common Flax, has long been cultivated in temperate regions as a fibre-plant for the making of linen, the stems being used for this purpose. It is also economically important for its seeds which are the source of linseed oil. A number of species of *Linum* are grown as ornamental plants, the hardy perennial kinds being particularly suitable for the rock-garden or the herbaceous border. The shrub *Reinwardtia indica* is sometimes grown for decoration in glasshouses.

Classification. Some authorities treat the Linaceae as a somewhat larger family and include certain genera of tropical trees and shrubs which are perhaps better placed in separate families. The genus *Anisadenia* links the family with the Plumbaginaceae.

Linum (230) temperate and subtropical regions, especially Mediterranean.
Hugonia (40) tropical Africa to S.E. Asia.
Roucheria (8) tropical S. America.
Reinwardtia (2) N. India, China.
Anisadenia (2) Himalaya to central China.

LINUM PERENNE L.
Perennial Flax

Distribution. This is one of several species comprising the *Linum perenne* aggregate, a taxonomically difficult group of plants showing considerable ecological and genetic variation. *L. perenne* L. has been divided into 5 subspecies, occurring mainly in central and eastern Europe, but extending locally westwards. In Britain the species is represented by the endemic subspecies *anglicum* which is confined almost entirely to the eastern side of the country.

Vegetative characteristics. A glabrous, perennial herb with alternate, linear leaves.

Floral formula. K5 C5 A5 $\underline{G}(5)$

Flower and inflorescence. The conspicuous blue flowers are borne in a loose cyme in June and July.

Pollination. Insects, mainly bees and flies, are attracted by the bright blue petals and the nectar secreted by glands at the base of the stamens. Heterostyly is a feature of many species of *Linum*, including *L. perenne*, and it has been observed that a short-styled flower is infertile both with its own pollen and with that from other short-styled plants, but is fertile with pollen from plants bearing long-styled flowers.

Alternative flowers for study. In the absence of *L. perenne*, any other species of *Linum* would be suitable, including *L. usitatissimum*, mentioned above, which has a similar flowering period.

Fig. 56 Linaceae, *Linum perenne*

A A single flower, with 2 of the 5 petals and sepals removed to reveal the reproductive organs. The sepals are imbricate and the petals are contorted in the bud stage. The petals have a short claw and are alternate with the sepals.
Sepals: 4 mm Petals: 15–15.5 mm

B All petals and stamens have been removed to reveal the gynoecium, composed of 5 united carpels situated opposite the petals. Each of the 5 free styles terminates in a well-developed stigma.

C L.S. of the superior ovary, showing 2 of the pendulous, anatropous ovules. At the apex of the ovary are the remains of 2 of the 5 styles.
Ovary (at anthesis): 2 × 1.25 mm Ovule: 1 × 0.5 mm

D T.S. of ovary. Each of the loculi of the 5-celled ovary is divided into 2 by the growth of a false septum, giving the ovary the appearance of being 10-celled. Placentation is axile.
Ovary: 4 mm in diameter

E The mature fruit, a septicidal capsule. During dehiscence, the capsule splits into the 10 valves formed by the division into 2 portions of each of the 5 carpels.
Fruit: 7 mm

F Two of the 5 stamens, with introrse, 2-celled anthers which dehisce longitudinally. The filaments are fused at the base into a ring which contains small, nectar-secreting glands. Note the small, linear staminode situated between the stamens.
Stamen: 8 mm Anther: 1.5 mm Free portion of stamen: 5 mm

Fig. 56

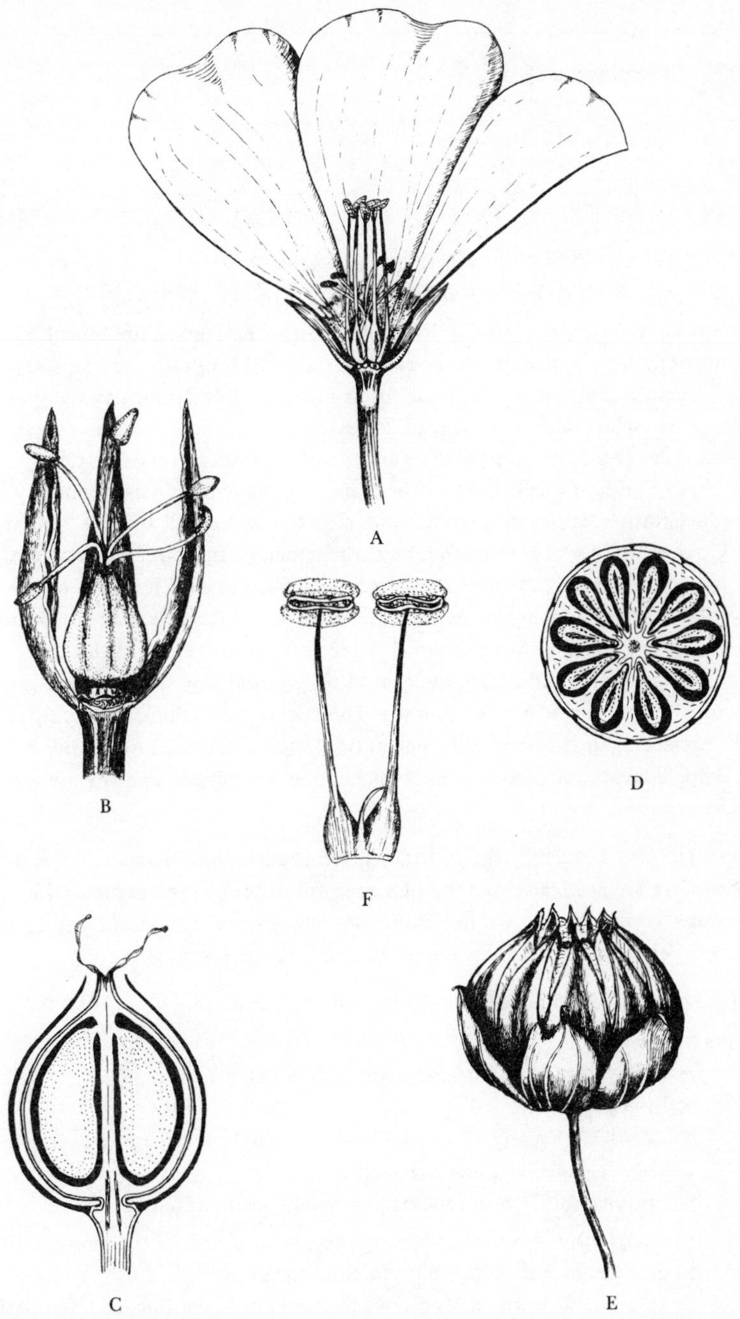

57 Geraniaceae Juss.

Geranium family

5 genera and 790 species

Distribution. Cosmopolitan.

General characteristics. Mostly herbaceous, often hairy. Leaves opposite or alternate, often stipulate, usually lobed or divided. Flowers actinomorphic or zygomorphic, hermaphrodite, hypogynous. Calyx of 5 usually free sepals, persistent. Corolla of 5 free petals. Stamens as many or 2 or 3 times as many as petals, more or less united at base, obdiplostemonous, some sometimes reduced to staminodes (e.g., *Erodium*, Stork's-bill); anthers usually versatile. Ovary superior, of 5 united carpels, with 1 or 2 anatropous ovules in each carpel with axile placentation; styles long, sometimes slightly coalescent, but the 5 stigmas free. Flowers usually protandrous. Fruit a schizocarp, the carpels splitting off from a central column (the persistent style); each takes with it a strip of the tissue of the style, forming an awn. In most species of *Geranium* the awn suddenly curls upwards throwing the seed some distance from the plant. In the other species of *Geranium*, and in *Erodium* and *Pelargonium*, the whole carpel (mericarp) together with the awn is flung off. The stiff apex of the awn usually remains firmly pressed against surrounding vegetation, etc., while the lower, coiled portion, being hygroscopic, extends as it takes up moisture, pushing the mericarp into the ground.

Economic and ornamental plants. Several species and hybrids of *Geranium* and *Erodium* are cultivated for the herbaceous border or rock-garden. The 'geraniums' commonly grown in glasshouses or for summer bedding belong to the genus *Pelargonium* and numerous cultivars are in existence.

Classification. The 5 genera are distinguished as follows:

Flowers zygomorphic.

Pelargonium (250) tropical, especially S. Africa.

Flowers actinomorphic.

Stamens 10, usually all fertile; beak of carpel rolling upwards in dehiscence and releasing the seeds.

Geranium (400) cosmopolitan, especially temperate regions.

Stamens 10, 5 reduced to staminodes; beak of carpel twisting spirally in dehiscence and remaining attached to the seeds.

Erodium (90) Europe, Mediterranean region to central Asia, temperate Australia, S. America.

Stamens fifteen, in 5 groups of 3; plants more or less herbaceous.
Monsonia (40) Africa, S.W. Asia, N.W. India.

Stamens fifteen, monadelphous, stems thick and fleshy, armed with spines formed from persistent petioles.
Sarcocaulon (12) S.W. Africa.

GERANIUM PRATENSE L.
Meadow Crane's-bill

Distribution. Native but local throughout most of the British Isles and rare in N. Scotland and Ireland. Widely distributed in Europe, though rare in the Mediterranean region. Also found in N. and central Asia and naturalised in N. America. *G. pratense* has proved to be a useful plant for the herbaceous border and it is widely cultivated.

Vegetative characteristics. A perennial herb with a stout rhizome. The hairy, long-stalked leaves are divided almost to the base into 5–7 coarsely toothed lobes.

Floral formula. K5 C5 A5+5 $\underline{G}$(5)

Flower and inflorescence. The flowers are arranged in pairs on peduncles which arise from the axils of the leaves, forming a cymose inflorescence. After flowering the pedicels are reflexed, but become erect as the fruit develops. The showy, violet-blue flowers appear from June to September.

Pollination. The flowers are visited by bees and other hymenopterous insects which seek out the nectar secreted by the glands at the base of the inner stamens. To prevent self-pollination the anthers ripen first. The 5 stamens in each row become erect and shed their pollen. The stigmas then expand and ripen in order to receive the pollen from another flower.

Alternative flowers for study. Geranium sylvaticum, Wood Crane's-bill, is a closely related species which differs from *G. pratense* in having petals of more variable length and pedicels which remain erect after anthesis. The flowers of most native species are produced in pairs, but in *G. sanguineum*, Bloody Crane's-bill, the flowers are solitary. Annual members of the genus tend to have smaller flowers than the perennials, e.g., *G. robertianum*, Herb Robert, a familiar plant of walls, hedgebanks and woods. Petals may be emarginate, as in *G. pyrenaicum* or clawed, as in *G. lucidum.*

Fig. 57 Geraniaceae, *Geranium pratense*

A L.S. of the whole flower, showing 3 of the 5 free petals. The separate stamens widen out at the base to form a cup-like structure surrounding the ovary.
Sepals: 12 × 7 mm (arista 3 mm)
Petals: 16–17 × 13–17 mm

B Dorsal view of one of the 10 stamens. The anthers are 2-celled and dehisce longitudinally. The filament broadens out at the base where it is tucked under the ovary. The stamens are obdiplostemonous, i.e., there are twice as many stamens as petals, the outer series of stamens being opposite the petals. The inner row of stamens ripens before the outer row and both rows are protandrous and ripen before the stigmas.
Anthers: 2.5 × 1.5 mm
Filaments: 8–10 mm, 2 mm wide at the base

C T.S. of ovary. The ovary is composed of 5 carpels, with 2 ovules on axile placentas in each carpel. Only one of each pair of ovules develops to form a seed. The ovary increases in width from 2 mm to 3.5 mm when the fruit is formed.

D L.S. of the superior ovary, showing point of attachment to the style. Each ovule is anatropous and pendulous. The central column is known as the carpophore and eventually acts as an important aid to seed dispersal. Situated on the outer side of the bases of the 5 inner stamens are 5 small glands which secrete nectar.

E The petals have been removed in order to expose the female parts and to show their relationship to the filaments and sepals during a late stage of development of the ovary. The central style and carpophore terminates in 5 stigmas. At the apex of each sepal can be seen an awn-like bristle called an arista, thus the sepals are described as being aristate.
Ovary + style + stigma: 14 mm   Spread of stigmas: 3 mm
Fully developed mericarps: 6.5 × 2.5–2.8 mm
Mature carpophore: 28 mm

F The method of seed dispersal. When the carpels ripen, they separate into the one-seeded portions known as mericarps. When the awn suddenly coils up, the seed is ejected through the opening on the ventral side of the mericarp.

Fig. 57

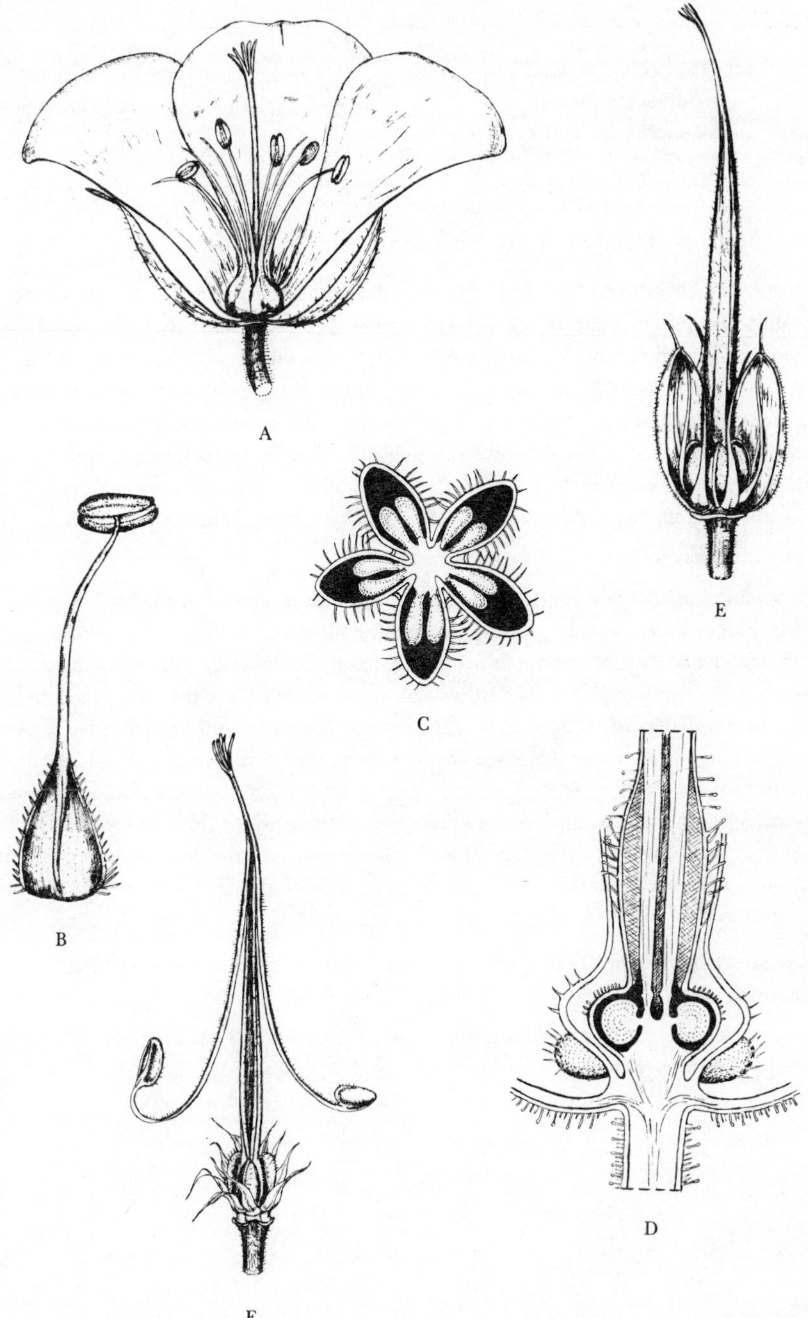

58 Oxalidaceae R. Br.

Wood-sorrel family
8 genera and 900 species

Distribution. Mainly tropical and subtropical regions.

General characteristics. Mostly perennial herbs, with alternate, often compound (pinnate or digitate) exstipulate leaves. Flowers relatively large, usually in cymes, hermaphrodite, actinomorphic. Calyx of 5 sepals, imbricate, persistent, free or slightly united. Corolla of 5 petals, imbricate or contorted, free or slightly united. Stamens 10, obdiplostemonous, united below, with introrse anthers. Ovary superior, of 5 free or united carpels with free styles, 5-locular, with axile placentation. Ovules in 1 or 2 rows in each loculus or few, anatropous, with micropyle facing upwards and outwards. Fruit a capsule. Embryo straight, in fleshy endosperm.

Economic and ornamental plants. The genus *Averrhoa* contains 2 species of evergreen trees, widely grown in tropical countries for their fruit, whose vernacular names, Carambola and Bilimbi, have been taken as the botanical names of the species. The only other economic plant of any importance in the Oxalidaceae is *Oxalis tuberosa*, Oca, which has long been cultivated in its native area of Peru, Ecuador and Bolivia for its tubers, leaves and young shoots. Ornamental plants are represented almost entirely by the genus *Oxalis*, *O. articulata* with its bright pink or more rarely pale pink or even white flowers frequently being grown in warm, dry borders, and other species on the rock-garden or in an alpine house. More tender species, such as *O. deppei* and *O. hedysaroides* require the protection of a heated glasshouse.

Classification. The Oxalidaceae is closely allied to the Linaceae and the Geraniaceae.

Oxalis (800) cosmopolitan, chiefly Central and S. America and S. Africa.
Biophytum (70) tropics.
Averrhoa (2) probably native in Brazil.

OXALIS ACETOSELLA L.
Wood-sorrel

Distribution. A common plant of woodlands, hedgerows and other shady places which have humus-rich soils. It is widely distributed throughout most of Europe, and is also found in parts of N. and central Asia eastwards to Japan.

Vegetative characteristics. A perennial herb with slender, creeping rhizome and long-stalked leaves composed of 3 obcordate leaflets indented at the apex.

Floral formula. K5 C5 A(5+5) $\underline{G}$(5)

Flower and inflorescence. The solitary, actinomorphic flowers have white petals veined with lilac to pale purple or violet and appear on slender stems during April and May, sometimes even until August. The flower-stem is about as long as the petiole, and has 2 small bracteoles at or above the middle.

Pollination. Many foreign species of *Oxalis* are particularly adapted to cross-pollination by the trimorphic heterostyly of their flowers. There are 3 forms of plants, one bearing flowers with long styles, and mid- and short-length stamens, the others with mid-length or short styles and correspondingly long and short or long and mid-length stamens. *O. acetosella*, however, has flowers of one form only and the anthers and stigmas ripen about the same time. Insect visitors, mainly beetles and flies, are apparently few, although nectar is secreted from pits at the base of the petals. In fact, in this species, self-pollination seems to be preferred as cleistogamous flowers occur abundantly in summer after the normal flowering and it is these that produce most of the seed. In appearance they resemble small flower-buds. All their parts are reduced in size, and a smaller quantity of pollen is produced, but this is sufficient since the flowers never open wide and the anthers and stigmas remain in close proximity.

Alternative flowers for study. If *O. acetosella* is not available, *O. corniculata*, Procumbent Yellow Sorrel, is a suitable choice as it is a common weed of gardens, glasshouses and some waste places, also *O. articulata*, which has corymbose cymes of bright rose-pink flowers, and is often cultivated in gardens. Both species have a flowering period extending from June until September or even October.

Fig. 58 Oxalidaceae, *Oxalis acetosella*

A L.S. of flower. Two of the 5 free petals have been removed to reveal the reproductive parts of the flower. At the base is the hairy calyx of 5 imbricate, free sepals. The superior 5-locular ovary is surmounted by 5 free styles and surrounding its base are the 5 long and 5 short stamens.
Sepals: *c.* 5 mm Petals: 10–16 mm

B Sepals and petals have been removed from the flower revealing the free styles arising from the 5 united carpels of the ovary. The stamens are united at their base and form a ring round the ovary.
Filaments: 4–6 mm Stigma + style + ovary: 8–9 mm

C Apical region of a stamen showing the 2-celled, introrse anther dehiscing longitudinally.
Anther: 0.5 × 0.4 mm

D L.S. of ovary, showing 2 carpels, each containing 2 anatropous ovules on axile placentas.
Ovary (at anthesis): 3 × 1.6 mm

E T.S. of ovary, consisting of 5 united carpels, with the developing ovules.
Young ovary: 1.5 mm in diameter

F Upper portion of style and its terminal, shortly-lobed stigma.

G Fruit prior to dehiscence with the styles attached. The fruit, a loculicidal capsule, contains arillate seeds. When it is ripe, the aril surrounding each seed separates suddenly from the seed coat (testa) and expels the seed from the capsule.
Fruits: 8 × 3–4 mm

Fig. 58

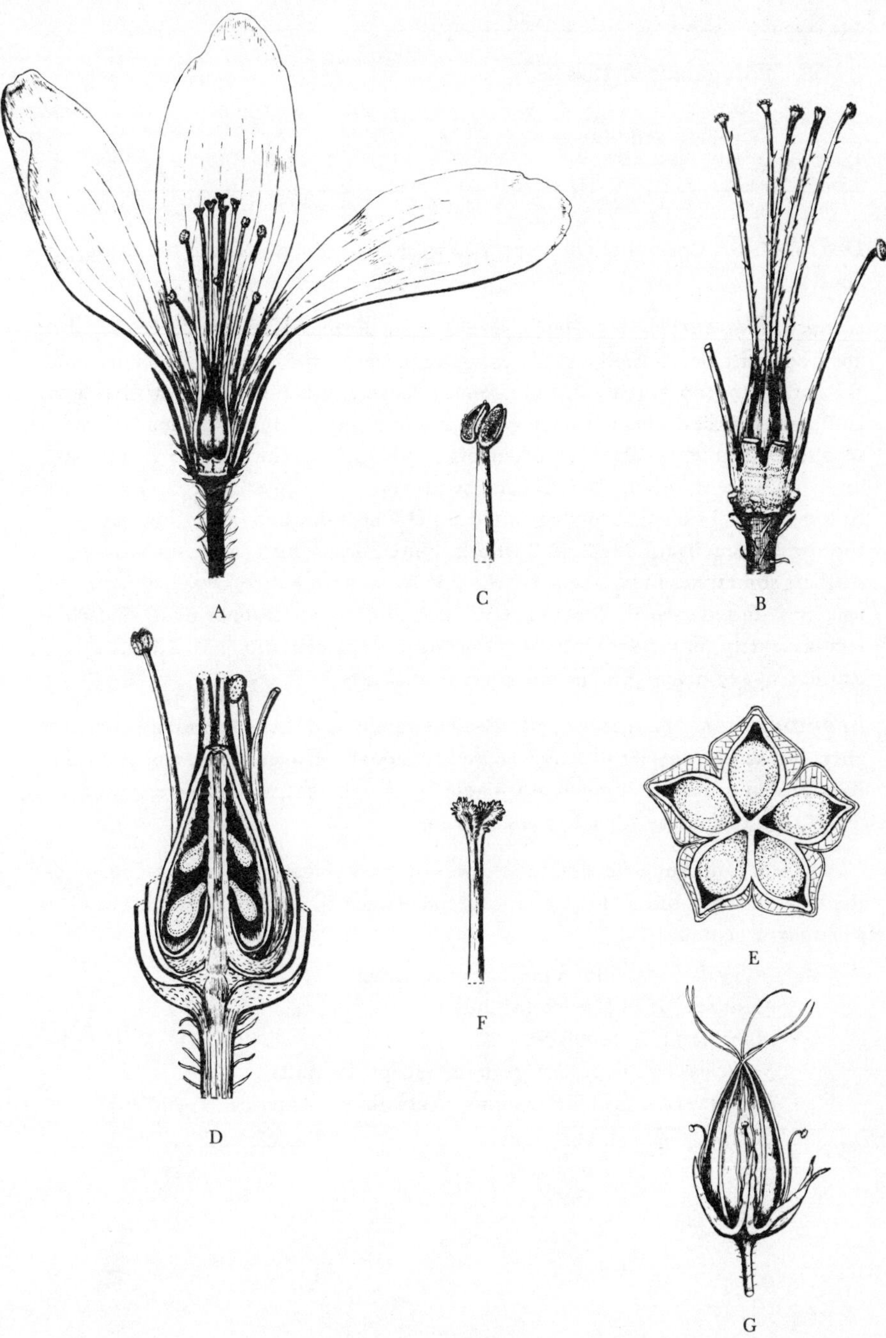

59 Polygalaceae Juss.

Milkwort family

12 genera and 900 species

Distribution. Cosmopolitan, except New Zealand, Polynesia and the arctic zone.

General characteristics. Herbs, shrubs or small trees, with simple, entire, alternate, opposite or whorled, usually exstipulate leaves; the stipules, when present, are usually thorny or scaly. Inflorescence a raceme, spike or panicle, with bracts and bracteoles. Flowers hermaphrodite, zygomorphic. Calyx of 5 usually free sepals, the 2 inner sepals (alae) often large and petaloid. Corolla of 5 free petals, rarely all present, usually only 3 (the lowest and the 2 upper) more or less joined to the staminal tube, the median anterior petal keel-like and often fimbriate at the apex. Stamens usually 8, in 2 whorls, united below into an open tube. Anthers sometimes only 1-celled, dehiscing by an apical pore. Ovary superior, of usually 2 united carpels, 2-locular, with 1 pendulous, anatropous ovule in each loculus (rarely unilocular with numerous ovules); placentation axile. Fruit a capsule, nut or drupe. Endosperm present or absent.

Economic and ornamental plants. The family is of little economic importance, but several species of *Polygala* are occasionally grown for ornament on the rock-garden or in the alpine house. These include the native *P. calcarea* and a dwarf shrub from the Alps, *P. chamaebuxus*.

Classification. Superficially, the flowers of the Polygalaceae resemble many of the Leguminosae, but a closer relationship between the families has not been generally accepted.

Polygala (over 500) almost cosmopolitan.
Monnina (150) Mexico to Chile.
Muraltia (115) S. Africa.
Securidaca (80) tropical regions, except Australia.
Bredemeyera (50) New Guinea, Australia, S. America, W. Indies.

POLYGALA VULGARIS L.

Common Milkwort

Distribution. A common plant of heaths, dunes and grassland throughout Europe including the British Isles, and also found in N. Africa and W. Asia.

Vegetative characteristics. A perennial herb with branched stems and alternate, lanceolate leaves which become narrower towards the inflorescence.

Floral formula. K5 C3 A8 $\underline{G}$(2)

Flower and inflorescence. The zygomorphic, hermaphrodite flowers are blue, pink or white in colour and appear from May to September in dense racemes which elongate in fruit. They are made more conspicuous by the large, petaloid, inner sepals. The corolla is basically 5-merous, but only the median anterior petal and the 2 lateral petals are present.

Pollination. Insects, mainly bees, are attracted to the flowers by the 2 large, petaloid sepals and alight on the median, fimbriated petal. Folds on the upper side of this petal enclose the anthers and the spathulate end of the style. Below this 'spoon' is the sticky, hook-like stigma. The anthers are so placed that at dehiscence the pollen falls into the spoon. When an insect probes for the nectar secreted at the base of the flower, it first encounters the pollen in the spoon, then the stigma. However, it is not until the insect's proboscis has become sticky from touching the stigma that the pollen will adhere to it. Thus the pollen is only collected as the insect leaves the flower, and this arrangement favours cross-pollination. Self-pollination may however occur at the beginning of anthesis, when the amount of pollen in the spoon is so large that some may get pushed on to the stigma as an insect enters the flower. It may also take place in the absence of insect visits, since the stigma is able to bend down and touch the pollen lying in the spoon.

Alternative flowers for study. P. vulgaris is widespread throughout the British Isles and usually easily obtainable. In its absence, the other common native species, *P. serpyllifolia*, which flowers from May to August, would be suitable.

Fig. 59 Polygalaceae, *Polygala vulgaris*

A The whole flower, showing the large, petaloid sepals with branching veins and the small, green, outer sepals.
Flower: *c.* 8 mm Outer sepals: 3 mm Inner sepals: 6–8 mm

B The flower with the 2 petaloid sepals removed to reveal the anterior petal with its fimbriated crest and the 2 lateral (upper) petals. The 3 petals are united at the base and adnate to the staminal tube. Subtending the petals are 2 of the 3 green, outer sepals.

C L.S. of flower showing the 2-locular superior ovary. The stigma and the spathulate end of the single style almost reach the fimbriated portion of the anterior petal.

D Detail of the upper portion of 3 of the 8 stamens prior to dehiscence. The stamens are monadelphous and are united for more than half their length into an open tube adnate to the petals. The anthers are 1-celled, basifixed and dehisce by means of an apical pore. The surface of one of the anthers has been partly cut away to reveal the pollen grains.

E L.S. of the superior, bilocular ovary after fertilisation. The 2 ovules, one in each loculus, are anatropous and pendulous on axile placentas.
Ovary (at anthesis): 1.5 mm
(at maturity): 5 x 4.5 mm
Style: 3 mm

F T.S. of ovary showing the axile placentation of the ovules.
T.S. of mature ovary: 4.5 x 1.5 mm

G Detail of the upper portion of the style, showing the hook-like stigma, with pollen grains attached, and beyond it the spathulate end of the style.

H The fruit prior to dehiscence, a narrowly compressed, marginally winged capsule, which dehisces loculicidally in order to shed the 2 seeds. As in many species of *Polygala* the seeds are hairy and have large, lobed elaiosomes which attract ants by their oil content. These insects carry the seeds away from where they have fallen, but being interested only in the oil-bodies, leave the rest of the seed undamaged and ready to germinate in its new place. In this way the seeds are dispersed over a wider area.
Fruits: 3–5 x 3–4 mm

Fig. 59

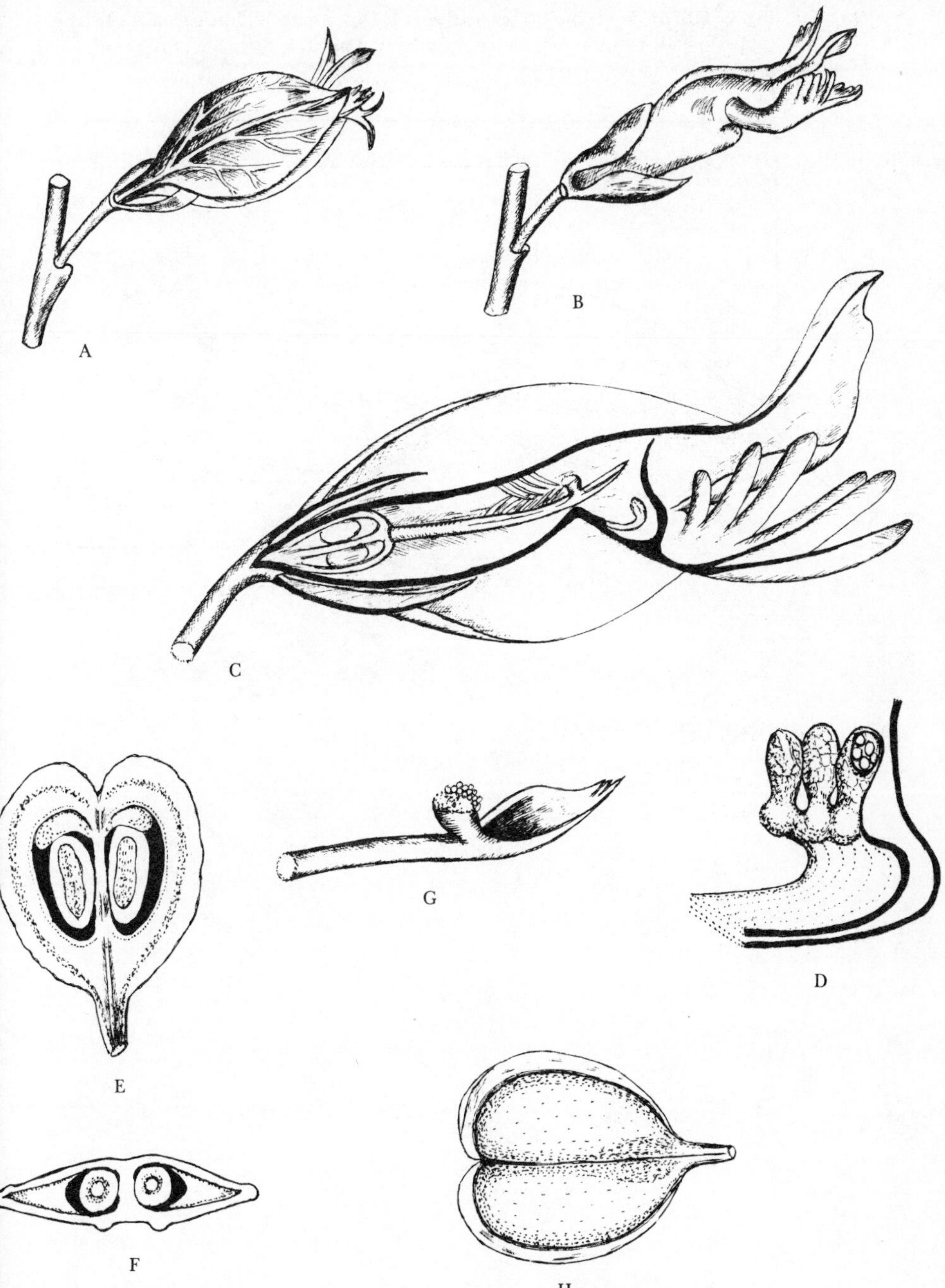

Fig. C. L.S. of flower of *Polygala vulgaris*: 1, Outer sepal; 2, Inner sepal; 3, Lateral (upper) petal; 4, Anther; 5, Anterior fimbriated petal; 6, Stigma; 7, Style; 8, Ovary.

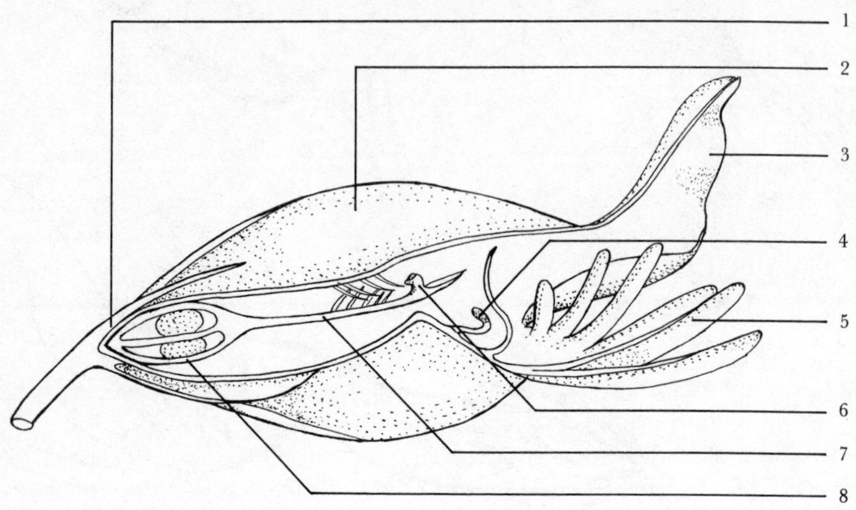

60 Araliaceae Juss.

Ivy family

55 genera and 700 species

Distribution. Chiefly tropical, especially Indomalaysia and tropical America.

General characteristics. Usually trees and shrubs, sometimes palm-like, some twiners and some, e.g., *Hedera*, root-climbers; pubescence often stellate. Leaves alternate, rarely opposite or whorled, often large and compound, with small stipules; seedling leaves often simpler than those of mature plants. Flowers small, in umbels or heads, often massed into compound inflorescences, hermaphrodite or sometimes unisexual, actinomorphic, usually epigynous and 5-merous. Calyx usually of 5 very small sepals. Corolla of 4 to 10 petals, usually 5, free, rarely connate, often valvate. Stamens 3 to many, usually 5. Ovary usually inferior, rarely half-inferior or superior, of 1 to many (usually 5) united carpels, each containing 1 anatropous, pendent ovule with axile placentation. Styles free or united, or sometimes absent, the stigmas then sessile. Fruit a berry or drupe with as many seeds as carpels. Seed with abundant endosperm round a small embryo.

Economic and ornamental plants. The pith of *Tetrapanax papyriferus* is the source of rice-paper and the roots of *Panax schinseng* provide the medicinal substance ginseng. Tender members of the family, including *Dizygotheca* and *Schefflera*, are cultivated indoors as pot plants, while hardier ones, e.g., species of *Aralia* and *Fatsia* can be grown outside in many parts of the country for their ornamental foliage. The native *Hedera helix*, and other species of Ivy, are cultivated as evergreen vines, and numerous cultivars have been developed which show a wide range of leaf-shape and variegation.

Classification. The family is divided here into 3 tribes as follows:

1. Scheffiereae (corolla valvate, with broad base).
   Schefflera (200) tropics and subtropics.
   Hedera (15) Canary Islands to Japan, Queensland.
   Fatsia (2) Japan, Formosa.
   Tetrapanax (1) S. China, Formosa.
2. Aralieae (corolla more or less imbricate, with broad base).
   Aralia (35) Indomalaysia, E. Asia, N. America.
   Panax (8) tropical and E. Asia, N. America.
3. Mackinlayeae (corolla valvate, shortly clawed).
   Mackinlaya (12) E. Malaysia, Queensland.

HEDERA HELIX L.
Ivy

Distribution. Commonly found growing in a wide range of shady situations throughout the British Isles. It is also native in most parts of Europe, in Turkey, south to Israel and east to Iran.

Vegetative characteristics. A woody, evergreen climber or ground-carpeting plant, having stems densely clothed with adhesive roots and bearing glabrous, dark green, palmately 3- to 5-lobed leaves.

Floral formula. K(5) C5 A5 $\overline{\text{G}}$(5)

Flower and inflorescence. The hermaphrodite, yellowish-green flowers are in subglobose umbels which are often arranged racemosely to form a terminal panicle.

Pollination. The well-developed disc situated on top of the ovary secretes nectar which attracts many kinds of insects, mainly flies and wasps. The flowers have been described both as protandrous and as homogamous. Where they are protandrous, the insects visiting them in the first (male) stage become dusted with pollen which is subsequently transferred to the stigmas of those in the second (female) stage. In the case of homogamous flowers cross-pollination is again promoted, for the insects alight on the central stigma and do not touch the radiating stamens until afterwards.

Alternative flowers for study. Ivy flowers abundantly from September to November or December and its fruits persist for long after this time, making an alternative almost unnecessary. However, *Fatsia japonica*, the 'aralia' of commerce, which is sometimes grown for its attractive foliage outside in gardens and perhaps more frequently as a pot plant for decoration indoors, may be used, but it should be noted that some flowers of this plant may prove to be unisexual. *F. japonica* has a similar flowering period to *Hedera helix*.

Fig. 60 Araliaceae, *Hedera helix*

A Close-up view of an individual actinomorphic flower with an inferior ovary. The 5 hooded, yellowish-green petals are attached to a broad disc. This disc is nectar-secreting, covers the ovary and is terminated by a sessile stigma. The 5 stamens, bearing versatile anthers, alternate with the petals.
Petals: 3–4 × 2 mm Disc: *c.* 3 mm in diameter

B The entire inflorescence after petal-fall showing the umbel with its tomentose pedicels and receptacles. Note the atypical foliage leaf below the inflorescence.
Peduncle: *c.* 15 mm Pedicels: 5–8 mm
Diameter of inflorescence: 25 mm enlarging to 30 mm in the early fruiting stage

C Close-up of a single flower at the stage of seed formation. The petals and stamens have fallen away to reveal the minute, deltoid sepals attached to the nectar-secreting disc.

D The black, globose berry which contains 5 seeds (see G).
Fruits: 6–8 mm

E L.S. of the inferior ovary. Each ovule is pendulous and anatropous. Some members of the family have 5 free styles, but in *Hedera* they have become reduced and fused to form a single, sessile stigma on top of the nectar-secreting disc. Attached to one side is a section of one of the tiny, triangular sepals.
Ovary: 3 mm in diameter Stigma + style: 0.75 mm

F T.S. of ovary, with a single ovule on the axile placenta in each loculus.
Ovule: 0.75 × 0.5 mm

G One of the 5 seeds after it has been removed from the fruit.

H A whole stamen. The stamens have versatile, 2-celled anthers which dehisce longitudinally.
Filaments: 3.5 mm Anthers: 1.75–2 × 0.75 mm

Fig. 60

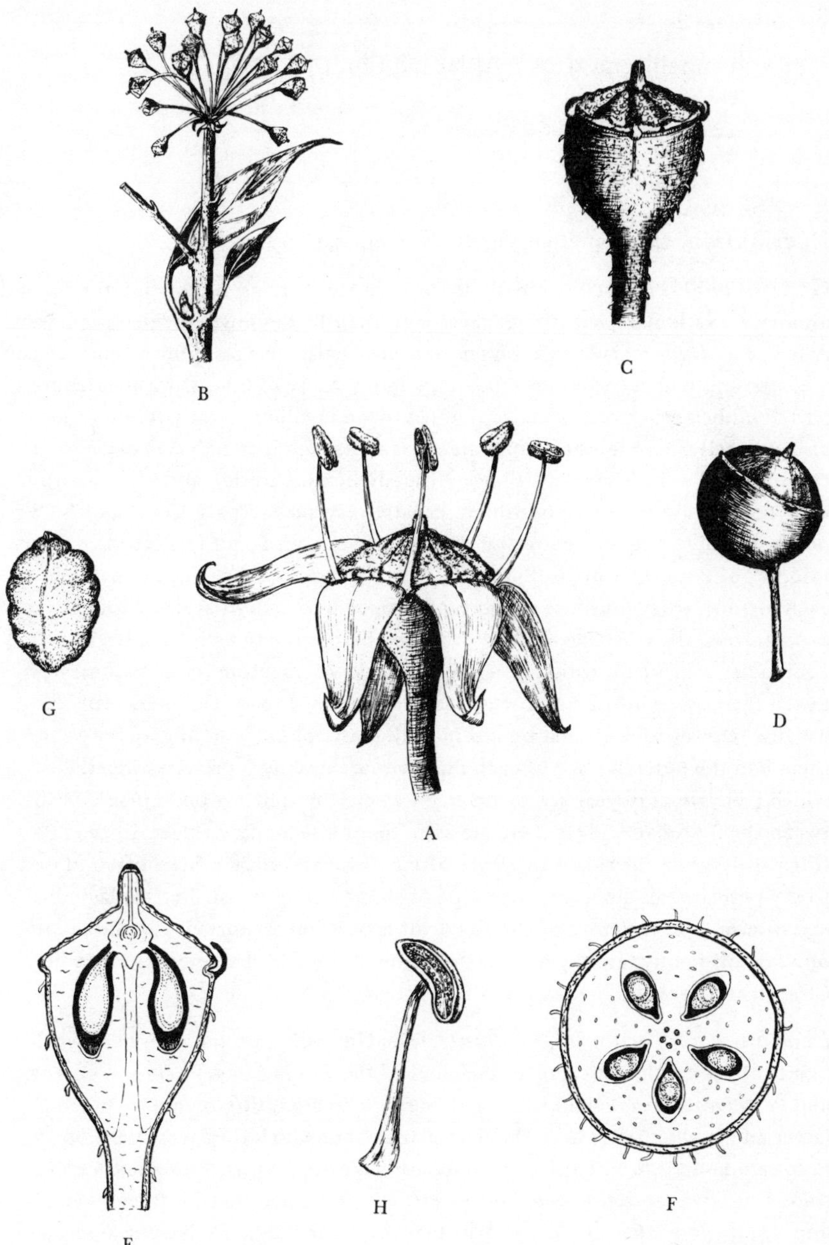

61 Umbelliferae Juss. (Apiaceae Lindl.)

Umbellifer family

275 genera and 2850 species

Distribution. Cosmopolitan, chiefly N. temperate region.

General characteristics. Mostly herbs with stout stems, hollow internodes and alternate, exstipulate, sheathing leaves with their blades much divided pinnately. A few, e.g., *Hydrocotyle* and *Bupleurum*, have entire leaves. Inflorescence usually a compound umbel (often subtended by an involucre of bracts) formed from partial umbels (each one often subtended by an involucre of bracteoles termed an involucel), sometimes a simple umbel as in *Astrantia* and *Hydrocotyle*, or a cymose head as in *Eryngium*. Flowers usually hermaphrodite and actinomorphic, epigynous. Calyx of 5 free or united, usually very small sepals. Corolla of 5 free petals, usually white or yellow, rarely absent. Stamens 5, anthers introrse. Ovary inferior, of 2 united carpels, 2-locular, with 1 pendulous, anatropous ovule in each loculus; epigynous disc present, prolonged into 2 short styles. Flowers very protandrous, the male stage usually finished before the female stage begins. Fruit a dry schizocarp which generally splits down a septum (commissure) between the carpels into 2 mericarps, each containing 1 seed. The mericarps are usually held together at first by a thin stalk (carpophore) running up between them. On the outer surface of each mericarp are usually 5 primary ridges, 2 of which (the lateral ridges) are at the edges where the splitting takes place. Between the 5 primary ridges there are sometimes 4 secondary ridges. In the furrows between the ridges there are often oil-canals (vittae). Seed often united to the pericarp, albuminous, with a small embryo in oily, usually cartilaginous endosperm. The structure of the ripe fruit is of prime importance in the taxonomy of this family, as may be seen from the diagnostic characters given in the Classification section below.

Economic and ornamental plants. The Umbelliferae contain a number of plants of economic importance, the chief of these being *Daucus carota*, Carrot, and *Pastinaca sativa*, Parsnip. The cultivars which are now grown have much larger and fleshier roots than the wild plants from which they were developed. *Petroselinum crispum*, Parsley, *Foeniculum vulgare*, Fennel, *Anthriscus cerefolium*, Chervil, and *Anethum graveolens*, Dill, are cultivated for their leaves, and *Apium graveolens*, Celery, for its blanched leaf-stalks. The young stems and leaf-stalks of *Angelica archangelica*, Angelica, are crystallised with sugar and used in confectionery. Dill and Fennel are also grown for their fruits, as are *Carum carvi*, Caraway, *Pimpinella anisum*, Anise, *Coriandrum sativum*, Coriander, and *Cuminum cyminum*, Cumin, all of these being used for flavouring. Plants of this family cultivated solely for decorative purposes are few. They include species of *Eryngium*, which are suitable for the herbaceous border and *Astrantia*, which require a damp but not wet situation, also *Heracleum mantegazzianum*, Giant Hogweed, whose tall stems and large inflorescences produce a bold effect when planted in clumps in the wild garden. Caution should be exercised in handling this species, since subsequent exposure to sunlight may cause dermatitis.

Classification. The subfamily Hydrocotyloideae is raised in rank by some authorities and treated as a separate family. The following classification into 3 subfamilies and 12 tribes is based on that of Engler (4).

I. Hydrocotyloideae (stipules present; free carpophore absent; endocarp woody; vittae absent or in main ribs).

1. Hydrocotyleae (fruit with narrow commissure, laterally compressed).
   Hydrocotyle (100) tropical and temperate regions.
   Azorella (70) Andes, temperate S. America, Falkland Islands, Antarctic islands.
2. Mulineae (fruit with flattened or rounded back).
   Bowlesia (14) America.

II. Saniculoideae (stipules absent; endocarp soft; style long with capitate stigmas, surrounded by a ring-like disc; vittae various).

3. Saniculeae (ovary 2-locular; fruit 2-seeded; commissure broad; vittae distinct).
   Eryngium (230) tropical and temperate regions, except tropical and S. Africa.
   Sanicula (37) cosmopolitan, except Australasia.
   Astrantia (10) central and S. Europe, W. Asia.
4. Lagoecieae (ovary unilocular; fruit 1-seeded; vittae indistinct).
   Lagoecia (1) Mediterranean region.

III. Apioideae (stipules absent; endocarp soft; style on apex of disc; secondary ridges usually absent; vittae various).

5. Echinophoreae (secondary umbels each with one central, hermaphrodite flower surrounded by male flowers; fruit enclosed by hardened stalks of male flowers).
   Echinophora (10) Mediterranean region to Iran.
6. Scandiceae (parenchyma around carpophore with crystal layer).
   Chaerophyllum (40) N. temperate region.
   Anthriscus (20) Europe, temperate Asia.
   Torilis (15) Canary Islands, Mediterranean region to E. Asia.
7. Coriandreae (parenchyma without crystal layer; fruit usually ovoid, nut-like).
   Coriandrum (2) Mediterranean region.
8. Smyrnieae (commissure narrow, mericarps rounded outwards).
   Smyrnium (8) Europe, Mediterranean region.

Conium (4) temperate Eurasia, S. Africa.

9. Ammieae (primary ridges all alike; seed semicircular in section).
   Bupleurum (150) Europe, Asia, Africa, N. America.
   Pimpinella (150) chiefly Eurasia and Africa, a few in America.
   Seseli (80) Europe to central Asia.
   Ligusticum (60) northern hemisphere.
   Oenanthe (40) temperate Eurasia, mountains of tropical Africa.
   Carum (30) temperate and subtropical regions.
   Foeniculum (5) Europe, Mediterranean region.
   Petroselinum (5) Europe, Mediterranean region.
   Apium (1) Europe to India, N. and S. Africa.
10. Peucedaneae (lateral ridges forming wings; seed narrow in section).
    Peucedanum (120) temperate Eurasia, tropical and S. Africa.
    Angelica (80) N. hemisphere, New Zealand.
    Heracleum (70) N. temperate regions, tropical mountains.
    Pastinaca (15) temperate Eurasia.
11. Laserpitieae (secondary ridges very marked, often extended into broad, undivided or wavy wings).
    Laserpitium (35) Canary Islands, Mediterranean region to S.W. Asia.
    Thapsia (6) Mediterranean region.
12. Dauceae (secondary ridges with spines).
    Daucus (60) Europe, Africa, Asia, America.

HERACLEUM SPHONDYLIUM L.
Hogweed, Cow Parsnip

Distribution. Native in W. and N. Asia, N. Africa, and throughout Europe except the extreme north and much of the Mediterranean region. It is common in the British Isles, and is found growing by hedges, woods, roadsides and other grassy places on a wide range of soils.

Vegetative characteristics. A stout, erect biennial or short-lived perennial, reaching a height of 2.5 m or more. The hollow, ridged stem may be glabrous or hairy. The leaves, which are palatable to herbivorous animals, are simply pinnate, variously lobed and hairy on both surfaces. The lower leaves are stalked.

Floral formula. K5 C5 A5 $\overline{G}(2)$

Flower and inflorescence. The inflorescence is typical of the family and consists of an umbel composed of partial umbels. These compound umbels may be terminal or axillary. The white or pinkish flowers appear between June and September. The outer, larger flowers of each partial umbel open first, followed shortly afterwards by the smaller, inner flowers, so that in any one umbel all the flowers are at about the same stage.

Pollination. The large, flat-topped inflorescences afford a good landing-ground for a wide variety of insect visitors. The orders represented include Hymenoptera (bees, wasps, ichneumons and sawflies), Coleoptera (longhorn and soldier beetles), Diptera (drone and hover flies) and Hemiptera (capsid bugs). Many of these visit the flowers for their pollen and for the readily accessible nectar secreted by an epigynous disc, others use the flowers merely as a convenient resting-place. The arrival of one of the larger insects often causes even more movement among the visitors and probably assists the pollination process. The anthers of the flowers are incurved at first, but then become erect to shed their pollen. Only when they have all dehisced and most of them have fallen do the stigmas mature. The strongly protandrous nature of the flowers prevents self-pollination.

Alternative flowers for study. H. sphondylium is so readily obtainable that an alternative is scarcely necessary. Comparison should be made with members of the other subfamilies and tribes, most of which are represented by plants native or frequently cultivated in the British Isles.

Fig. 61 Umbelliferae, *Heracleum sphondylium*

A A partial umbel removed from the compound inflorescence. The outer flowers are larger and open slightly before the smaller, inner flowers.
Compound umbel: 20 cm in diameter
Partial umbels: 25–35 mm in diameter

B Two hermaphrodite flowers taken from different parts of the inflorescence. At the centre of the flowers are the 2 united carpels, one pointing towards the centre of the inflorescence, the other pointing outwards. B1: an outer or marginal flower with petals and stamens of different lengths making it distinctly zygomorphic. The large, deeply notched outer petals of the marginal flowers form a conspicuous border to the partial inflorescence (see A). B2: one of the inner, more or less actinomorphic flowers with petals and stamens of similar lengths. The petals of both types of flower are bilobed to a varying extent, with an incurved point between the 2 lobes. The stamens, which alternate with the petals, are inflexed within the flower-bud, but become erect at anthesis.
Petals: 3–10 × 2–3 mm Filaments: 3–10 mm

C Detail of the upper part of a stamen. The introrse anther is dorsifixed, 2-celled and dehisces longitudinally. The filament broadens gradually towards the base.
Anther: 1.25 mm

D A single flower, with 2 petals and 3 stamens removed to reveal the gynoecium. The hairy, inferior ovary is terminated by 2 diverging styles, their enlarged bases forming a stylopodium. The stamens and petals are attached to the upper rim of the ovary together with a barely visible calyx represented by 5 minute teeth.

E L.S. of the gynoecium. The inferior ovary is composed of 2 united, unilocular carpels. In each of the 2 loculi there is a solitary, pendulous, anatropous ovule. Crowning the ovary are the 2 styles with their enlarged and thickened bases.
Ovary: 2.25 × 1.5 mm Stylopodium: 1 × 2 mm
Style + stigma (at early anthesis): 1 mm
(at late anthesis): 2.25 mm

F1 T.S. of the ovary, hairy at anthesis, showing the 2 united, unilocular carpels, each loculus containing a single ovule attached to an axile placenta.
Ovary: 1.5 × 1.25 mm

F2 T.S. of the fruit at an early stage of development, its ridged pericarp now glabrous. The carpels have grown larger and have become dorsally compressed, and the 2 ovules are developing into flattened seeds. The septum or commissure, which is visible along the line of the placentas, will eventually form the plane of separation of the 2 carpels.
T.S. of young fruit: 6 × 2 mm

G Detail of the receptive stigma, situated on the upper, inner surface of the style.
Stigma: 0.4 × 0.2 mm

H1 The mature, dry fruit before dehiscence, showing the 4 dark lines that indicate the presence of oil-canals or vittae, which are embedded in the fruit-wall between the narrow ribs. At the top can be seen the remains of the 2 persistent styles.
Fruits: 6.5–9 × 6–7 mm

Fig. 61

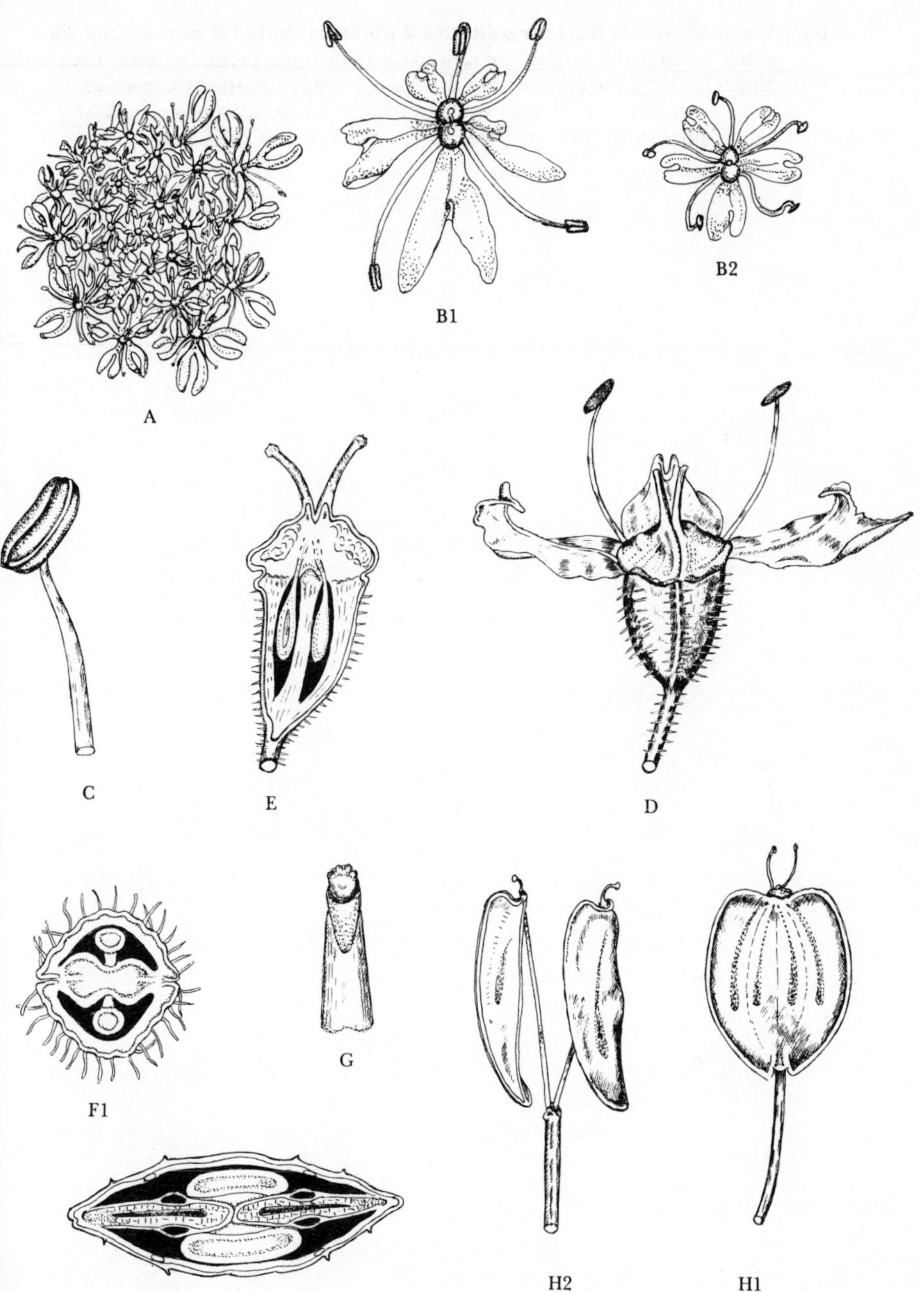

H2 The fruit, which has now split into 2 portions ready for dispersal by the wind. At first the carpels are united by their inner faces, but later they separate along the commissure into two 1-seeded portions known as mericarps. The 2 mericarps are held in position by a hair-like stalk or carpophore until they are detached by the force of the wind.

62 Loganiaceae Mart.

Buddleia family

30 genera and 700 species

Distribution. Mainly tropical, a few in warm temperate regions.

General characteristics. Trees, shrubs and herbs, some climbers. Leaves usually opposite, simple, stipulate, the stipules often much reduced. Inflorescence usually cymose. Flowers with bracts and bracteoles, usually actinomorphic, hermaphrodite. Disc small or absent. Calyx of 4 or 5 united sepals, imbricate. Corolla of 4 or 5 united petals. Stamens 4 or 5, rarely 1, epipetalous. Ovary superior, rarely half-inferior, of usually 2 united carpels, usually 2-locular; style simple; ovules usually numerous, amphitropous or anatropous, with usually axile placentation. Fruit usually a septicidal capsule, rarely a berry or drupe. Endosperm present.

Economic and ornamental plants. Several species of *Strychnos* are of importance as drug plants, including *S. nux-vomica*, the source of strychnine, and *S. toxifera* which provides curare. *S. spinosa* has edible fruit. Various species and varieties of *Buddleia* are grown for decorative purposes and a number of cultivars are now available. The genera *Gelsemium* and *Spigelia* are sometimes found in cultivation.

Classification. The family is considered here in a broad sense. Some authorities, however, restrict the Loganiaceae to 7 genera, placing the others, including *Buddleia*, in separate families, each with only a few genera.

Strychnos (200) tropics.
Buddleia (100) tropics and subtropics, especially E. Asia.
Geniostoma (60) Madagascar to New Zealand.
Spigelia (50) warm regions of America.
Nuxia (40) tropical Africa, Madagascar.
Fagraea (35) Indomalaysia, N. Australia, Pacific.
Mitrasacme (35) Indomalaysia, Australasia.
Logania (25) Australasia.

BUDDLEIA DAVIDII Franch.
Butterfly-bush

Distribution. Native of central and W. China, and introduced into British gardens about 1900. It is naturalised in many parts of W. and central Europe and is commonly found growing in waste places, particularly in urban areas. The shrub grows well near the coast.

Vegetative characteristics. A deciduous shrub of variable habit with 4-angled branchlets and lanceolate, finely-toothed leaves which are white-felted beneath.

Floral formula. K(4) C(4) A4 $\underline{G}$(2)

Flower and inflorescence. The actinomorphic, hermaphrodite flowers appear from June to October forming dense terminal panicles made up of many-flowered cymes. The colour of the corolla varies from pale lilac to deep violet with an orange-yellow ring at the mouth of the tube.

Pollination. Butterflies of various kinds are attracted by the colour of the inflorescence and by its characteristic fruity smell. The butterfly lands on the flat surface formed by the spreading corolla-lobes. It then uncoils its long proboscis in order to reach the nectar that is situated at the base of the long, funnel-shaped corolla-tube. The stamens present no obstacle as the filaments are entirely adnate to the corolla-tube and the anthers lie flat against its surface.

Alternative flowers for study. In the absence of *B. davidii* any other species of *Buddleia* would be acceptable. *B. davidii*, however, is usually readily available, being represented by numerous varieties and cultivars ranging in colour from white, through lilac and purplish-red to deep violet. In some species the cymes forming the inflorescence are more widely spaced, as in the drooping inflorescences of *B. alternifolia* and the erect ones of *B. globosa.*

Fig. 62 Loganiaceae, *Buddleia davidii*

A The complete inflorescence, a narrowly cylindrical panicle made up of many-flowered cymes and terminating a shoot of the current year's growth. The rachis, peduncles and pedicels are more or less tomentose.
Inflorescence: 8–22.5 cm

B A single flower. The corolla consists of a long, funnelform tube with 4 or sometimes 5 more or less imbricate lobes. Surrounding the base of the corolla is the sparingly pubescent 4- or sometimes 5-lobed calyx, subtended by a pubescent bract.
Corolla: 7 mm across lobes Corolla-lobes: 2.5–3 mm wide
Corolla-tube: 12 × 1.75 mm

C Upper portion of a typically 4-lobed corolla cut to show a facial view of the lobes. Note the irregular margins.

D Middle section of corolla-tube opened out to reveal the 4 dehiscing stamens whose filaments are completely adnate to the inner surface of the tube. The anthers are 2-celled, introrse and dehisce longitudinally. Several unicellular hairs are present in the region of the filaments.
Anther (at maturity): 1 mm

E Detail of anther prior to dehiscence.
Young anther: 0.7 mm

F L.S. of lower portion of flower showing the gynoecium. The ovary is superior and terminates in a single style with a bilobed stigma.
Style + stigma (at anthesis): 4.5 mm
Ovary (at anthesis): 1.5 × 0.75–1 mm

G T.S. of the superior ovary, which is bilocular and contains numerous ovules attached to the well-developed axile placenta.
Ovary: 0.75 mm in diameter

H Fruit prior to dehiscence, a bivalved capsule with septicidal dehiscence. The valves part from each other, exposing numerous small seeds attached to the axis formed by the persistent placenta. The seeds are light and are readily dispersed by the wind.
Fruit: 10 mm

Fig. 62

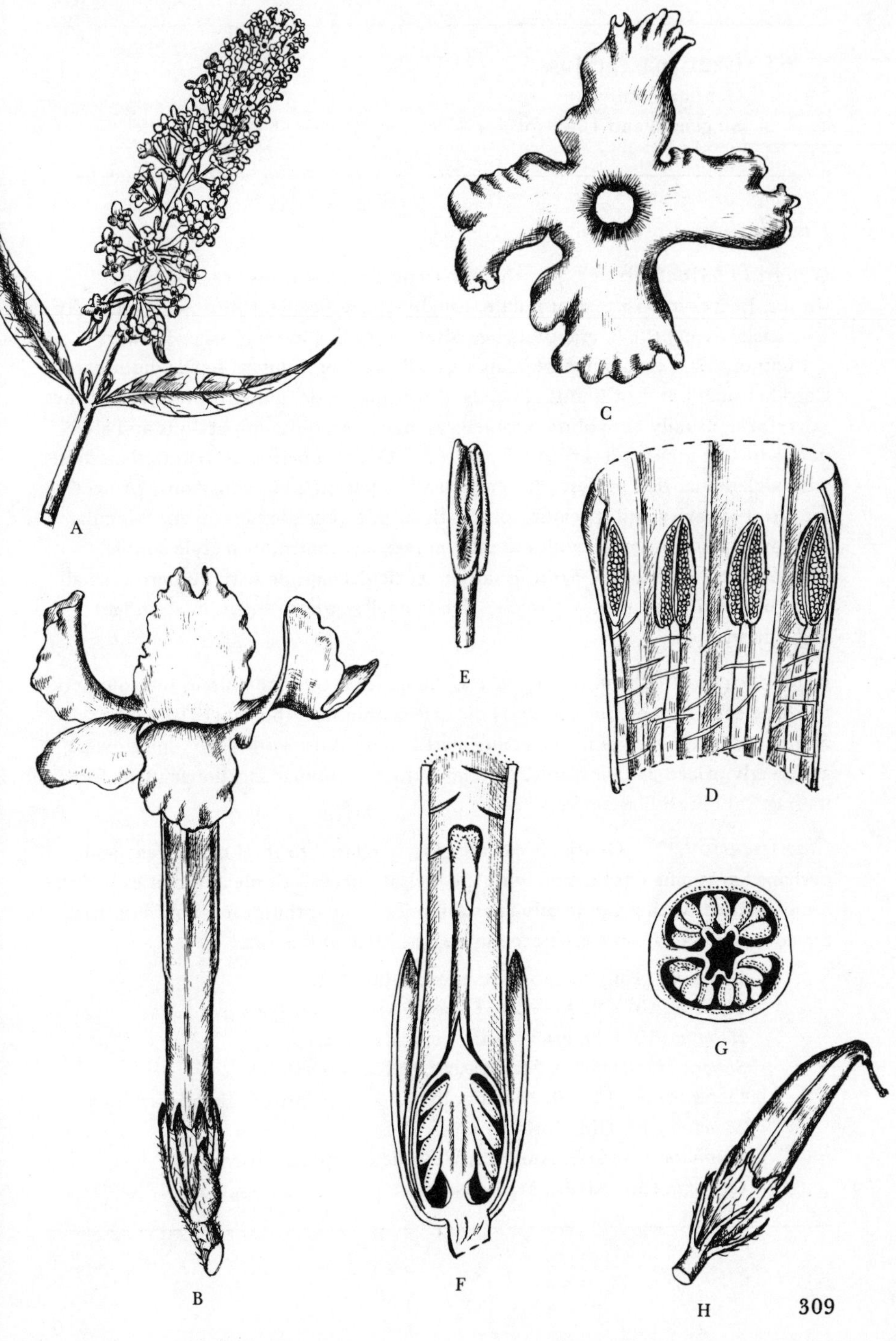

63 Gentianaceae Juss.

Gentian family

80 genera and 1000 species

Distribution. Cosmopolitan.

General characteristics. Mostly herbaceous, rhizomatous perennials, rarely shrubs. Leaves opposite, exstipulate, usually entire, sessile. Inflorescence usually a dichasial cyme, the lateral branches often becoming monochasial. Flowers actinomorphic, hermaphrodite. Calyx usually of 4 or 5 united sepals, imbricate. Corolla usually of 4 or 5 united petals, campanulate or funnelform, or sometimes salverform, usually convolute. Stamens as many as petals, epipetalous and alternate with the lobes; anthers usually introrse. Ovary superior, of 2 united carpels, with a glandular disc at base. Placentas usually parietal, but commonly projecting far into the cavity and spreading out at their ends (occasionally ovary 2-locular with axile placentation); ovules usually numerous, anatropous; style simple, stigma simple or 2-lobed. Fruit usually a septicidal capsule with numerous small seeds, rarely a berry (e.g., *Chironia*). Seeds small; embryo small, in abundant endosperm.

Economic and ornamental plants. Apart from *Gentiana lutea*, the source of the drug gentian root, the family is of little economic importance. However, a number of other species in the genus, particularly those with deep blue flowers, are greatly prized garden plants. *Exacum affine* is popular as a house plant for its fragrant bluish-lilac flowers.

Classification. The Gentianaceae are closely related to the Loganiaceae and perhaps have some connection with the Melastomaceae. Some authorities include 5 more genera within the family (as subfamily Menyanthoideae), but here these are considered to form a separate family, the Menyanthaceae.

Gentiana (400) cosmopolitan, excluding Africa.
Swertia (100) N. America, Eurasia, Africa, Madagascar.
Halenia (100) America, India, central and E. Asia.
Sebaea (100) Africa, Madagascar, India, Australasia.
Centaurium (40–50) cosmopolitan, excluding tropical and S. Africa.
Exacum (40) Old World tropics.
Leiphaimos (40) S. America, W. Indies, tropical Africa.
Chironia (30) Africa, Madagascar.

GENTIANA SEPTEMFIDA Pallas

Distribution. Native in W. and central Asia; a very variable species, tolerant of many soils, and frequently cultivated in the rock-garden or border.

Vegetative characteristics. A perennial with several more or less erect stems, on which pairs of sessile, ovate leaves are set at short intervals.

Floral formula. K(5) C(5) A5 $\underline{G}$(2)

Flower and inflorescence. The actinomorphic, hermaphrodite flowers are borne in terminal clusters in late summer. Their colour ranges from deep blue to deep purplish-blue, with paler spots within the corolla-tube.

Pollination. Like many other gentians, this species is pollinated by bumble-bees, which are attracted by the deep blue flowers and the nectar secreted at the base of the ovary. The flowers open for several days in succession and are protandrous, a device favouring cross-pollination. In many species of gentian the corolla-tube is whitish inside and the interior remains light when an insect enters the flower. This is often a feature of flowers with tubular or bell-shaped corollas and assists in luring the insect further into the corolla. The translucent effect is present also in *G. septemfida*, which has pale spotted markings within the corolla-tube.

Alternative flowers for study. All our native gentians are rare and in the interests of conservation it is inadvisable to collect any wild species for study. Apart from *G. septemfida*, the most readily available are probably *G. asclepiadea*, Willow Gentian, and *G. sino-ornata*, which are comparatively easy to cultivate in suitable soils. A few British members of the family are widespread and even common in some areas, e.g., *Centaurium erythraea*, Common Centaury, which is in flower from June onwards and *Gentianella amarella*, Felwort, which flowers in August and September. *Exacum affine*, mentioned above, has anthers with poricidal dehiscence.

Fig. 63 Gentianaceae, *Gentiana septemfida*

A A single, actinomorphic, hermaphrodite flower, subtended by 2 bracteoles. The 5-lobed calyx is imbricate in bud. The campanulate corolla is plicate between the lobes. In the sinuses between the corolla-lobes are fimbriate appendages to the plicae.
Calyx-tube: 10 mm Calyx-lobes: 8–10 mm
Corolla: 32–35 × 7–8 mm

B Corolla-tube opened out to reveal the essential floral parts. The fimbriate appendages in the sinuses of the corolla-lobes are clearly visible. The 5 long stamens are adnate to the corolla-tube for about half their length and stand alternate with the corolla-lobes. The fusiform, superior ovary is surmounted by a bilobed stigma.
Gynoecium (at anthesis): 26 × 2 mm Stamen: 23 mm

C Upper (C1) and middle (C2) portions of a stamen, showing respectively the anther and the point where the filament becomes free from the corolla. The 2-celled anther is introrse, versatile and dehisces longitudinally.
Anther: 3 mm

D L.S. of gynoecium, subtended by portions of the calyx and the corolla. The ovary is unilocular and contains numerous ovules. It is terminated by a relatively short style and a bilobed stigma.

E T.S. of ovary, showing the numerous anatropous ovules with parietal placentation. Evidence of placental intrusion into the loculus is not apparent in this species.
Ovary: 1.2–2 mm in diameter.

F Detail of the stigma showing the receptive surface of the 2 lobes.
Stigma-lobe: 4.25 mm

G The entire fruit, a septicidal capsule which dehisces by 2 valves to liberate the numerous small seeds.
Fruit: 32–35 × 5 mm

Fig. 63

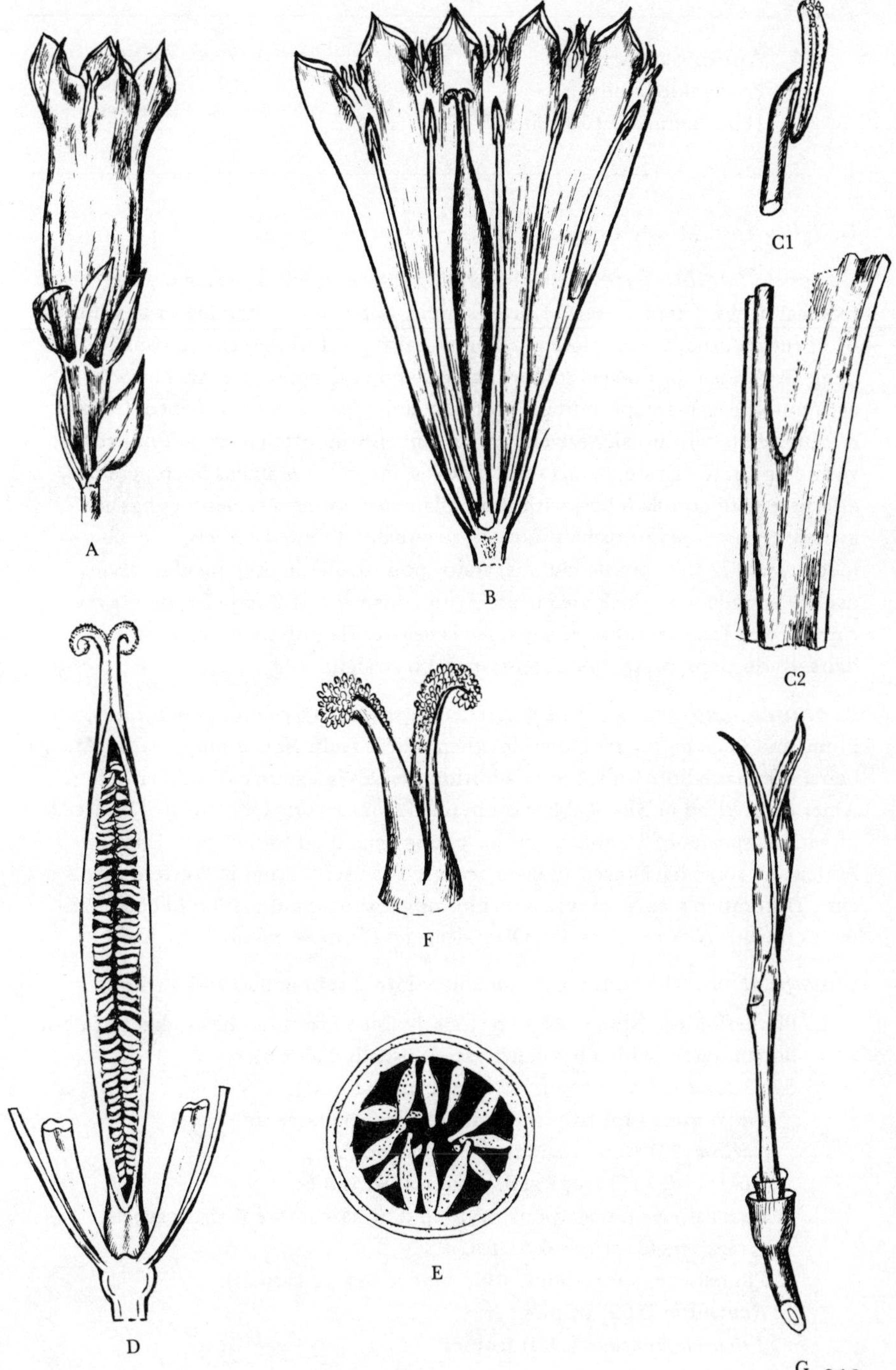

64 Apocynaceae Juss.

Periwinkle family

180 genera and 1500 species

Distribution. Mostly tropical, a few in temperate regions.

General characteristics. Usually twining shrubs, rarely erect; many large tropical lianes. Latex present. Leaves simple, opposite or alternate or in whorls of 3, entire, usually with close parallel lateral veins. Inflorescence a panicle, cyme or raceme, or flowers solitary; bracts and bracteoles present. Flowers actinomorphic, hermaphrodite. Calyx of 5 united sepals, deeply lobed. Corolla of 5 united petals, usually salverform or funnelform, often hairy within, convolute or rarely valvate, petals sometimes asymmetric. Stamens 5, epipetalous, alternate with corolla-lobes, with short filaments; anthers sometimes basally awned. Ovary superior or half-inferior, of usually 2 united carpels, 1- or 2-locular, with 2 to many pendulous, anatropous ovules in each loculus; style usually simple, with thickened head. Fruit consisting of 2 follicles, or a berry, capsule or 2 indehiscent mericarps; seeds usually flat, often with a crown of hairs. Endosperm present or absent; embryo straight.

Economic and ornamental plants. *Carissa macrocarpa* (*C. grandiflora*), Natal Plum, and *C. carandas* are grown for their edible fruit. *Hancornia speciosa*, Mangabeira, also has edible fruit, but in addition the sap is a source of rubber in S. America. African or Silk Rubber is obtained from *Funtumia elastica*. The seeds of various species of *Strophanthus* have long been used for arrow poisons in Africa and some have recently been accepted as useful drugs in Western medicine. Decorative plants for warm regions or glasshouse cultivation in cooler climates include *Nerium oleander*, Oleander, and *Plumeria rubra*, Frangipani.

Classification. The family is divided here into 2 subfamilies and 5 tribes:

I. Plumerioideae (stamens free or loosely joined to stylar head; thecae full of pollen, rarely with a basal awn; seeds usually hairless).
   1. Arduineae (syncarpous, style not split at base).
      Landolphia (55) tropical and S. Africa, Madagascar.
      Carissa (35) warm regions of Africa and Asia.
      Allemanda (15) tropical S. America, W. Indies.
   2. Pleiocarpeae (apocarpous, style split at base, more than 2 carpels).
      Pleiocarpa (3) tropical Africa.
   3. Plumerieae (apocarpous, style split at base, 2 carpels).
      Rauwolfia (100) tropics.
      Tabernaemontana (100) tropics.

Alyxia (80) Madagascar, Indomalaysia.
Aspidosperma (80) tropical and S. America, W. Indies.
Ochrosia (30) Madagascar to Australia and Polynesia.
Amsonia (25) Japan, N. America.
Plumeria (7) warm regions of America.
Vinca (5) Europe, N. Africa, W. Asia.

II. Apocynoideae (stamens firmly joined to stylar head, thecae empty at base and with a basal awn, seeds hairy).

4. Apocyneae (anthers included).
Mandevilla (114) Central and tropical S. America.
Strophanthus (60) tropical Africa, Madagascar, Indomalaysia.
Dipladenia (30) tropical S. America.
Apocynum (7) N. America, Mexico.
Nerium (3) Mediterranean region to Japan.

5. Parsonsieae (anthers exserted).
Parsonsia (100) S.E. Asia, Australasia, Polynesia.
Prestonia (65) Central and tropical S. America, W. Indies.
Forsteronia (50) Central and tropical S. America, W. Indies.

VINCA MINOR L.
Lesser Periwinkle

Distribution. Native in W. Asia and in Europe from Denmark southwards. Doubtfully native in the British Isles, but found locally in woodlands, copses and hedgebanks. It is a popular ground-cover plant in gardens.

Vegetative characteristics. An evergreen, suffruticose perennial with trailing stems rooting at intervals, short erect flowering stems and glabrous, lanceolate or elliptical leaves.

Floral formula. K(5) C(5) A5 $\underline{G}$(2)

Flower and inflorescence. The flowers appear from March until May, and arise singly or rarely in pairs on long pedicels from the axils of the leaves. The corolla is tubular with spreading lobes, usually blue-violet in colour, but occasionally pink or white.

Pollination. Long-tongued bees and bee-flies are attracted by the colour of the flowers and by the nectar secreted by the pair of nectaries situated at the base of the ovary. The reproductive organs are highly specialised to prevent self-pollination. The filaments are bent abruptly (see Fig. 64.B) and the anthers are poised over the style. The top of the style, which bears a ring of white hairs, is non-receptive, but below this is a disc with a viscid edge forming the true receptive stigma (see Fig. 64.E). The anthers, which dehisce introrsely, shed their pollen in coherent masses on to the hairs at the top of the style. As the visiting insect probes for nectar, its proboscis becomes sticky through touching the edge of the stigma. On withdrawing its proboscis, pollen from the apical tuft of hairs adheres to it and is subsequently transferred to the stigmatic surface of another flower.

Alternative flowers for study. *Vinca minor* is easily obtainable from wild localities or from gardens where cultivars may be found with varying flower-colour and sometimes variegated foliage. The closely related *V. major*, Greater Periwinkle, which is naturalised in similar habitats throughout the British Isles, differs mainly in its larger flowers and ciliate calyx-lobes. The genus *Vinca* is somewhat unusual within the family on account of the solitary flowers.

Fig. 64 Apocynaceae, *Vinca minor*

A L.S. of flower which has a funnelform corolla-tube terminating in 5 asymmetrical lobes. The tube is fluted and has an inner zone of hairs in the region of the anthers and the head of the style. The 5 stamens are adnate to the corolla-tube and are alternate with the lobes. The calyx is glabrous and has 5 lanceolate lobes. The ovary is superior and is surmounted by the highly specialised style and stigma.
Corolla: 28 mm in diameter Corolla-tube: 15–16 mm
Calyx-lobes: 6–7 mm Ovary to apex of style: 8–10 mm

B Portion of corolla-tube showing a stamen and the surrounding hairs. The short filament is sharply bent or kneed near the base. The anthers are 2-celled, introrse, and have expanded, hairy connectives which meet over the top of the style. The position of the anthers ensures that the granular pollen is transferred only to the non-receptive apical region of the style.
Filament (from knee to anther): 2 mm
Anther (including connective): 2.25 mm

C L.S. of lower part of flower. The superior ovary consists of 2 distinct carpels united by the long, single style. A few anatropous ovules are present within each loculus.
Corolla-tube (at base): 2.5 mm in diameter
Carpel: 1.5 mm in diameter

D T.S. of paired carpels. The ovules are attached to the ventral sutures of the carpels. The 2 nectar-secreting glands (not shown) are situated between the carpels.
T.S. of ovary: 0.8–1.25 × 1–1.5 mm

E The upper region of the style. The apex of the style is a non-receptive area and has a plume of white hairs for the purpose of receiving the pollen shed by the overhanging anthers. Below the tuft of hairs, round the thickened head of the style, is a viscid band which forms the receptive stigma.
Base of stigma to apex of style: 2–2.5 mm
Stigma: 0.9 mm in diameter

F A single fruit after dehiscence. The ripened carpel forms a follicle which dehisces down its ventral suture to disperse the long, narrow seeds. The seeds have a glabrous, warted surface and are blackish in colour. Fruit is rarely set in the British Isles.
Fruit: 20 × 8 mm

Fig. 64

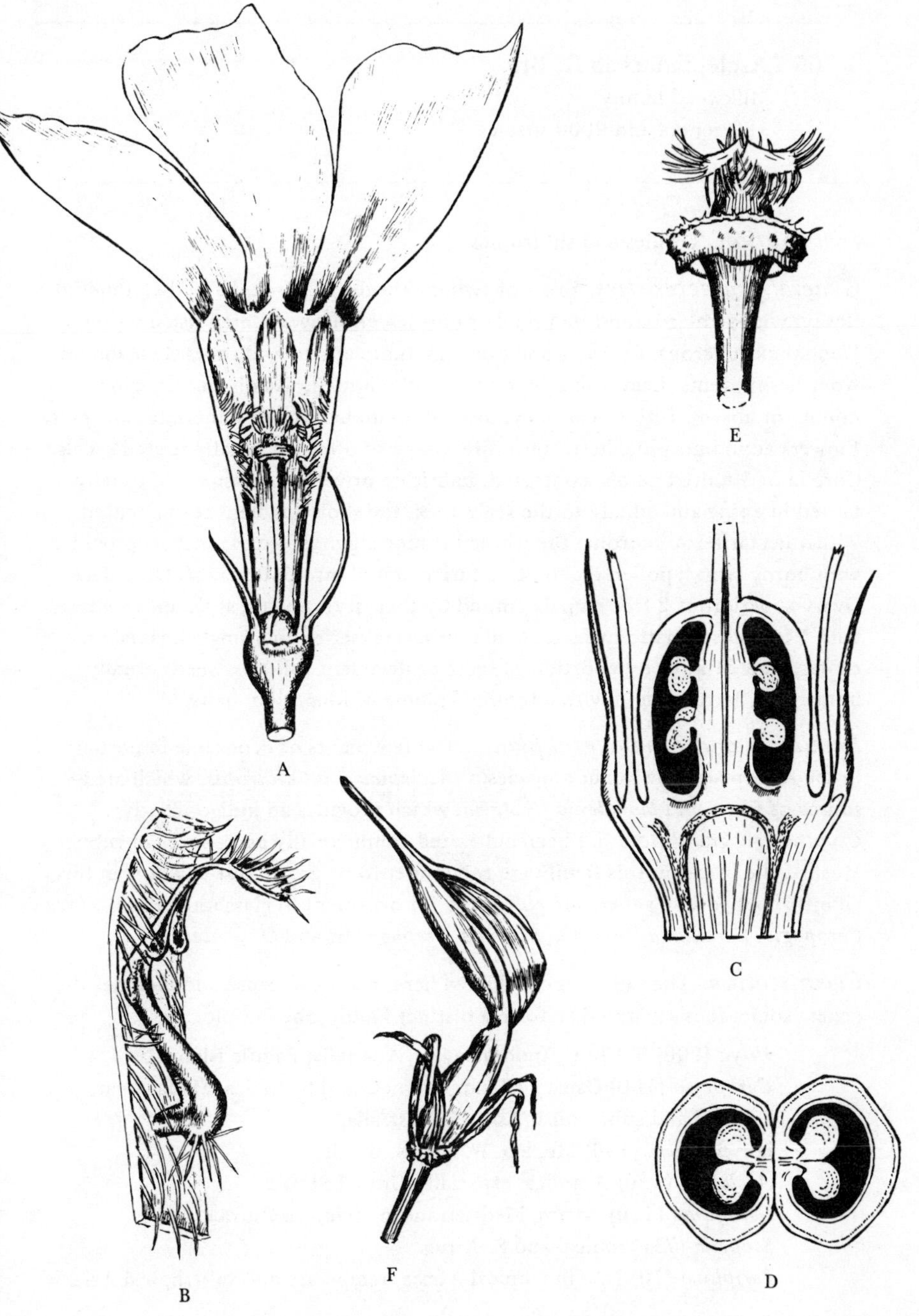

65 Asclepiadaceae R. Br.

Milkweed family

130 genera and 2000 species

Distribution. Tropics and subtropics.

General characteristics. Erect or twining shrubs or perennial herbs; sometimes fleshy, with reduced non-functional or obsolescent leaves. Latex usually present. Rootstock tuberous, fleshy, woody or sometimes absent and roots then annual from fleshy stems. Leaves usually opposite or whorled, usually entire, stipules minute or absent. Inflorescence cymose, often umbelliform, sometimes racemose. Flowers actinomorphic, hermaphrodite. Calyx of 5 free or basally united sepals. Corolla of 5 united petals, contorted, imbricate or valvate. Stamens 5, usually united in a ring and adnate to the style apex, the short filaments ornamented with a nectariferous corona, the whole forming a gynostegium; anthers provided with horny wings; pollen grains at maturity united into pollinia (cf. Orchidaceae). Ovary superior, of 2 free carpels, united by their style apices; style apex peltate, with 5 lateral stigmatic surfaces; ovules in several series on a single adaxial placenta. Fruit of 2 (or by abortion 1) erect or divergent follicles. Seeds usually numerous, flattened and with a terminal plume of long, silky hairs.

Economic and ornamental plants. The few plants of economic importance in the Asclepiadaceae include species of *Asclepias* and *Calotropis*, which are a source of fibre, and *Marsdenia tinctoria*, which provides an indigo-like dye. *Cryptostegia grandiflora* has been cultivated commercially as a source of rubber. Most of the plants in this family are too tender to be grown out of doors in this country, but several genera are cultivated for ornament in glasshouses, e.g., *Hoya*, *Ceropegia*, *Caralluma*, *Huernia*, *Stapelia*, *Stephanotis* and *Oxypetalum*.

Classification. The family is considered here in a broad sense and includes the genera sometimes separated to form a distinct family, the Periplocaceae.

Hoya (200) S. China, Indomalaysia, Australia, Pacific islands.
Ceropegia (160) Canary Islands, tropical and S. Africa, Madagascar, tropical and subtropical Asia, N. Australia.
Oxypetalum (150) Mexico, W. Indies, Brazil.
Asclepias (120) America, especially United States.
Caralluma (110) Africa, Mediterranean region to Burma.
Stapelia (75) tropical and S. Africa.
Periploca (10) N. and tropical Africa, temperate and subtropical Asia.

ASCLEPIAS CURASSAVICA L.
Blood-flower

Distribution. Native in tropical America and naturalised in the southern United States. It was introduced into Britain in 1692 and is often grown in glasshouses for its attractive inflorescence.

Vegetative characteristics. An almost glabrous perennial, woody near the base and reaching a height of 1 m in its native habitat. The leaves are opposite, oblong to oblong-lanceolate, acuminate and shortly petioled.

Floral formula. K(5) C(5) A(5) $\underline{G}$(2)

Flower and inflorescence. The flowers, which have a highly specialised structure, are actinomorphic and hermaphrodite, and are borne in terminal and lateral umbels from June until October. The small calyx is clearly visible at the bud stage but is later hidden by the reflexed lobes of the purplish red corolla. A conspicuous feature of the flower is the bright orange column terminating in the horned hoods of the corona.

Pollination. The chief pollinating agents of this species are butterflies which visit the brightly coloured flowers for the nectar secreted by glands in the stigmatic cavities and stored in the large pouches arranged round the top of the column. When a butterfly alights on a flower to extract nectar, its legs tend to slip down the slits in the column and hook on to the band connecting a pair of pollinia. As it flies away, this connecting band twists sharply, bringing the pollinia close together. On arrival at another flower, the pollinia are in such a position that they lodge in a slit in the column and break off, distributing the pollen grains over the receptive surface of the stigma.

Alternative flowers for study. The genus *Asclepias* includes several species that are comparatively easy to obtain and to dissect. Dissection, however, is much more difficult in the case of the readily available *Hoya carnosa*, Wax Plant, whose fleshy flowers are borne in a cymose inflorescence. The glasshouse climber *Stephanotis floribunda* has fragrant, white flowers with a leafy calyx and a salver-form corolla, while the genus *Stapelia*, Carrion Flower, has foetid flowers, with a rotate or campanulate corolla, barred or mottled with dark or dull colours.

Fig. 65 Asclepiadaceae, *Asclepias curassavica*

A A lateral umbel at anthesis, borne on a more or less erect peduncle. Two buds can be seen in the axils of the shortly petioled leaves.
Peduncle: 40 mm Pedicels: *c.* 25 mm

B A flower-bud, showing the small calyx with 5 basally connate sepals.
Bud: 8 × 5 mm

C The actinomorphic, hermaphrodite flower. The corolla, which consists of 5 basally connate petals, is strongly reflexed. The central column is crowned by the 5 cuculli or corona-hoods.
Corolla-lobes: 10 × 4 mm Column: 4 × 2 mm
Corona-hood: 3 × 0.75 mm

D L.S. of central part of flower. The tubular column forms a ring round the 2 free carpels. It is enlarged at the apex into a corona consisting of 5 horned cuculli which act as nectar-pouches. The 2 free styles are terminated by stigmas which are united to each other and to the upper part of the column, forming a gynostegium.

E Detail of part of the gynostegium. The connate stigmas have 5 grooves, alternating with the 5 anthers. The translator connecting a pair of pollinia is poised within each of the grooves. Part of the stigmatic tissue has been cut away to reveal a pollinium attached to its translator. The second pollinium of the pair is just visible through the stigmatic tissue. The circles lower down show where 2 of the nectar-pouches (cuculli) have been removed for clarity.

F A pair of pollinia connected by the translator. There are 5 translators in each flower, alternating with the 5 anthers. Each anther has 2 pollinia. The translator, which consists of a horny clip (corpusculum) and 2 bands (retinacula), connects 2 pollinia, one from each of 2 adjacent anthers.
Pollinium: 1 mm Retinaculum: 0.5 mm
Corpusculum: 0.3 mm

G L.S. of the lower part of the 2 carpels. The numerous anatropous ovules are imbricate and are attached to marginal placentas. Surrounding the carpels is the column and below them the bases of the petals and sepals can be seen.

H T.S. of the unilocular ovary with a marginal placenta on its ventral surface bearing numerous ovules.
T.S. of ovary: 0.75 × 0.6 mm

I The mature fruit opened up prior to dehiscence to show the imbricate arrangement of the brown seeds. It should be noted that the fruit is a follicle, and in nature it usually dehisces along one suture only.
Fruit: 43–60 × 10 mm

J A seed, showing the tuft of long, silky hairs attached to the micropylar end, aiding dispersal by the wind. (cf. Compositae and Valerianaceae).
Seed: 5.5 × 3–3.5 mm Hairs: 15–18 mm

Fig. 65

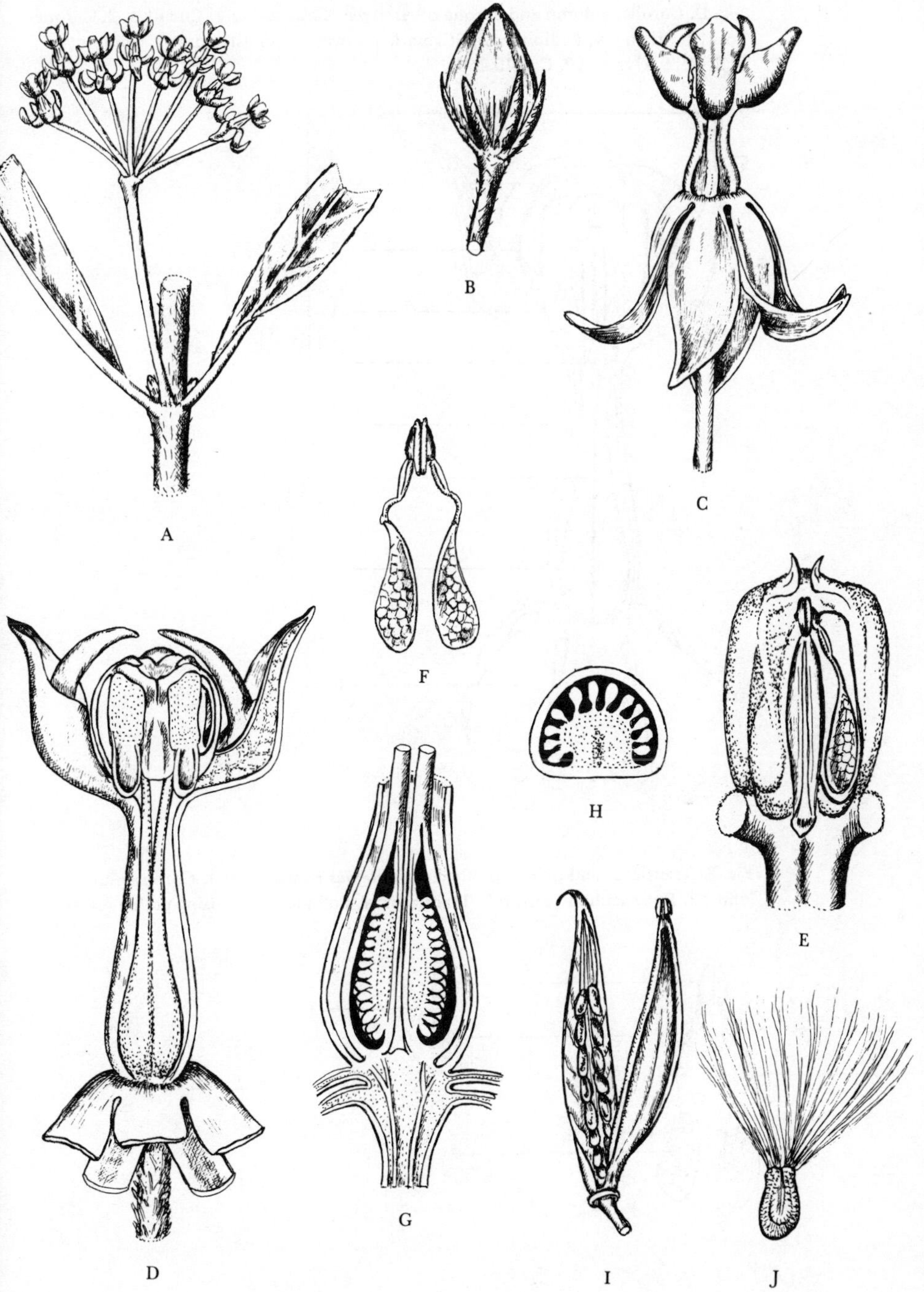

Fig. D. Corolla, column and corona of *Asclepias curassavica*: 1, Cucullus; 2, Corona-horn; 3, Stigma; 4, Pollinium; 5, Stigmatic groove; 6, Gynostegium; 7, Column; 8, Style; 9, Ovary; 10, Corolla-lobe.

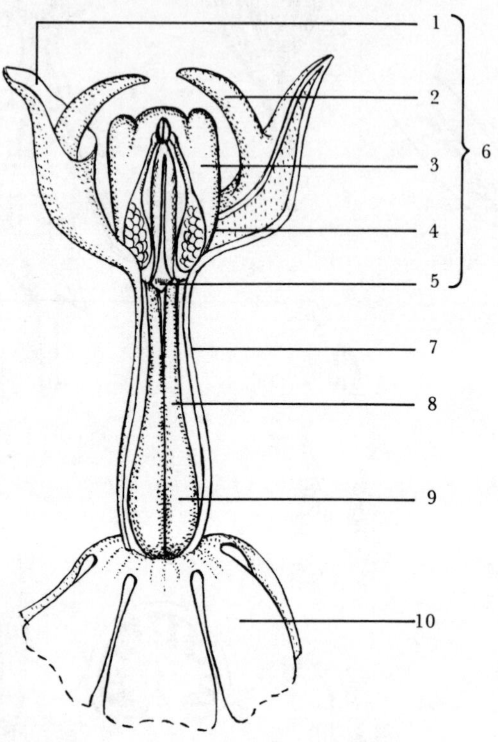

Fig. E. Translator and pair of pollinia of *Asclepias curassavica*: 1, Corpusculum (clip); 2, Retinaculum (band); 3, Translator; 4, Pollinium (enclosing pollen mass).

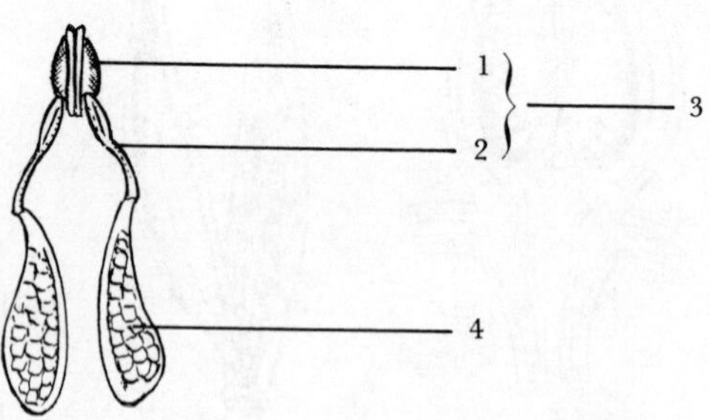

66 Oleaceae Hoffmgg. & Link

Olive family

29 genera and 600 species

Distribution. Cosmopolitan, especially temperate and tropical Asia.

General characteristics. Trees or shrubs, sometimes climbing or rambling. Leaves usually opposite, simple or pinnate, often entire, exstipulate. Inflorescence racemose or cymose, often bracteolate. Flowers usually hermaphrodite, actinomorphic, 2- to 6-merous, sometimes polypetalous or apetalous (e.g., *Fraxinus*). Calyx of usually 4 united sepals, valvate. Corolla of usually 4 united petals, valvate or imbricate, rarely convolute. Stamens 2, epipetalous, alternate with the carpels; anthers 2-celled, the cells back to back. Disc absent. Ovary superior, of 2 united carpels, 2-locular, with 2 anatropous ovules in each loculus; placentation axile; stigma 2-lobed on simple style. Fruit a berry, drupe, capsule or schizocarp, with 1–4 seeds. Seeds with or without endosperm; embryo straight.

Economic and ornamental plants. *Olea europaea*, Olive, has been cultivated in the Mediterranean region since ancient times as a source of food and an edible oil. *Fraxinus excelsior*, Ash, and other species are commercially important for their timber, and the flowers of various species of *Jasminum*, Jasmine, are the source of an oil used in perfumery. *J. nudiflorum* and *J. officinale*, known respectively as Winter and Summer Jasmine, are popular flowering shrubs for the garden. Others include *Syringa*, Lilac, with a wide range of cultivars, *Forsythia*, a conspicuous shrub in early spring on account of its abundant yellow flowers, and evergreen shrubs such as *Osmanthus*, *Phillyrea* and *Ligustrum*, Privet, the latter frequently represented in gardens by the species *L. ovalifolium*, generally used as a hedging plant.

Classification. Following Engler (4) the family is divided into 2 subfamilies.

I. Jasminoideae (Ovules generally erect).

Jasminum (300) Old World tropics and subtropics.

Forsythia (7) mainly E. Asia, 1 species in S.E. Europe, *F. europaea*.

II. Oleoideae (Ovules pendulous).

Linociera (80–100) tropics and subtropics.

Fraxinus (70) northern hemisphere.

Ligustrum (40–50) Europe, Asia, E. Australia.

Syringa (30) S.E. Europe to E. Asia.

Olea (20) Mediterranean region, Africa, E. Asia to New Zealand.

Osmanthus (15) E. and S.E. Asia.

SYRINGA VULGARIS L.
Common Lilac

Distribution. Native on the mountains of E. Europe from Romania to Albania and N.E. Greece and introduced into Britain in the sixteenth century. It has proved to be a successful shrub for sunny gardens and is suitable for most soils, particularly those that contain chalk. It occasionally escapes from cultivation and has become naturalised in W. and central Europe.

Vegetative characteristics. A large, deciduous shrub or small tree reaching a height of 7 m and suckering at the base. Its leaves are opposite, entire, ovate and subcordate to broadly cuneate at the base.

Floral formula. K(4) C(4) A2 $\underline{G}$(2)

Flower and inflorescence. The sweet-scented, actinomorphic, hermaphrodite flowers are borne in axillary, paniculate inflorescences just after leaf emergence in May. The panicle has an indeterminate growth along the main axis though the lateral axes are determinate. This type of inflorescence is termed a thyrse. The flowers are lilac or more rarely white in the species, but there are over 500 named cultivars with single or double flowers and colours ranging from white and creamy yellow to red and various shades of purple. The inflorescences are usually paired and grow from the apical axillary buds, the terminal bud being abortive.

Pollination. Many kinds of insects, mainly bees and butterflies, are attracted by the heavy, sweet scent, the colour of the flowers and the presence of nectar. The genus *Syringa* is usually homogamous and self-pollination can take place in the absence of insect visitors.

Alternative flowers for study. Syringa vulgaris is so readily available that there is little need for any alternative during the month that it is in flower. The closely related genus *Ligustrum* includes *L. vulgaris*, Common Privet, which flowers in June and July. Its floral structure is similar to that of *Syringa* but the fruit forms a berry instead of a capsule. Heterostyly occurs in *Forsythia* and *Jasminum*, and *J. nudiflorum*, Winter Jasmine, has a 5- or 6-lobed corolla. *Fraxinus excelsior*, Common Ash, provides an example of polygamous, apetalous, wind-pollinated flowers.

Fig. 66 Oleaceae, *Syringa vulgaris*

A A pyramidal, paniculate inflorescence which has developed from an apical axillary bud.
Inflorescence: 12–20 cm

B L.S. of a single, actinomorphic, gamopetalous flower. The small-lobed calyx protects the base of the corolla which has a more or less cylindrical tube and hooded lobes. The 2 stamens are adnate to the corolla-tube and are situated just above the apex of the style.
Calyx: 2.9 mm Corolla-tube: 9 mm Corolla-lobe: 5 mm

C Portion of the corolla-tube with adnate stamen. The stamens alternate with the 2 united carpels. The anthers are 2-celled and dehisce longitudinally.

D L.S. of the gynoecium, showing the superior ovary surrounded by the 4-lobed calyx. The ovary, composed of 2 fused carpels, is terminated by a single style with a 2-lobed stigma. The anatropous ovules are pendulous from an axile placenta.
Style + stigma (at beginning of anthesis): 5 mm
Ovary: 1 mm

E T.S. of the 2-carpelled ovary. Each carpel has 2 loculi, containing 2 ovules in each loculus.

F The fruit prior to dehiscence, a loculicidal capsule containing 2 seeds in each loculus.

G The fruit after dehiscence. The 2 valves of the capsule have opened allowing the brown, winged seeds (only one of which is shown) to be dispersed.
Fruit: 14 × 5 mm Seed: 8 × 4 mm

Fig. 66

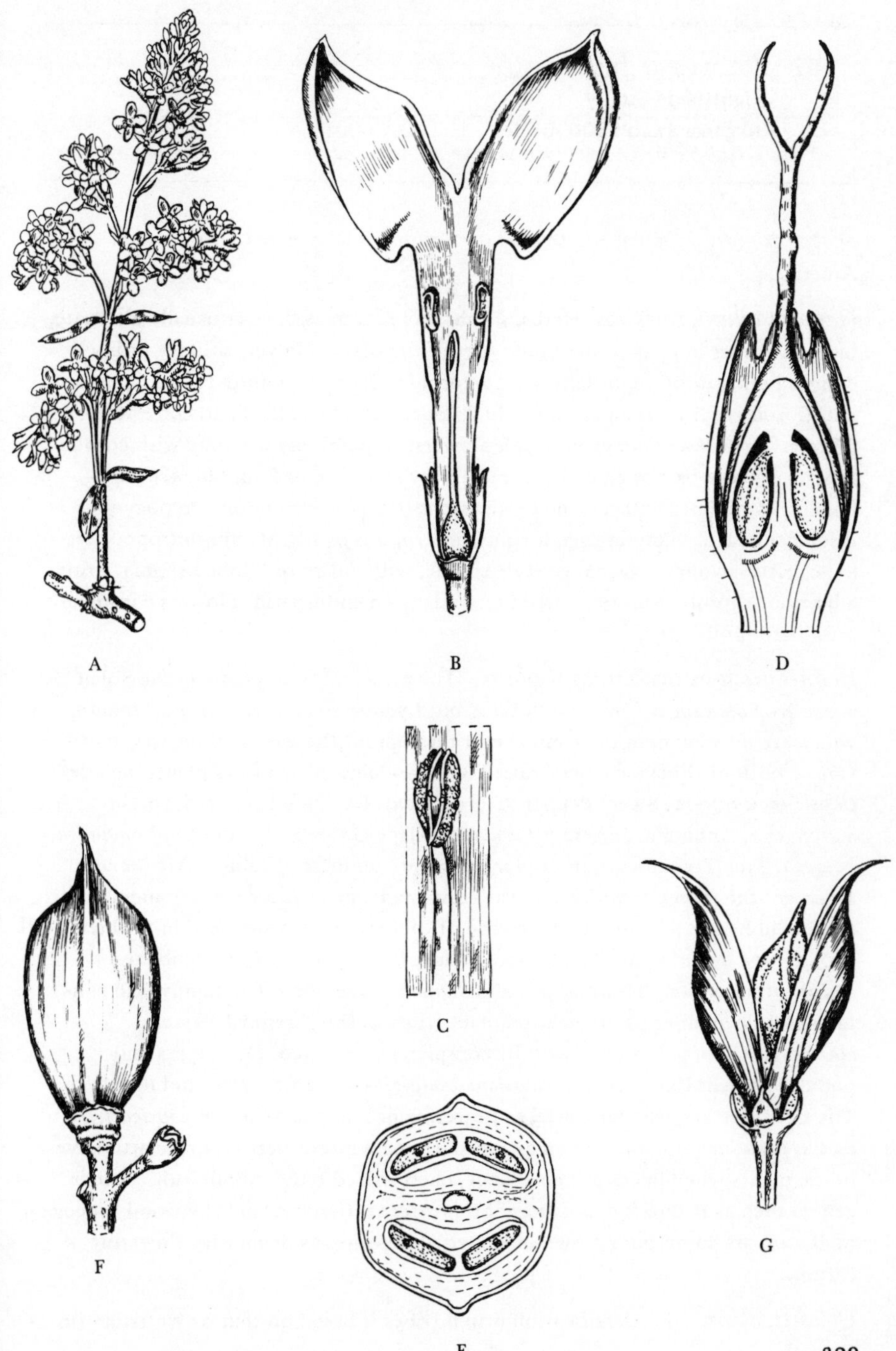

67 Solanaceae Juss.

Nightshade family

90 genera and 2300 species

Distribution. Tropical and temperate regions, chief centre Central and S. America.

General characteristics. Herbs, shrubs or small trees. Leaves usually alternate, but sometimes in pairs in the upper part of the plant. Flowers solitary or in cymes, hermaphrodite, usually actinomorphic. Calyx of 5 united sepals, persistent. Corolla of 5 united petals, rarely 2-lipped, usually folded and convolute. Stamens 5, or fewer in zygomorphic flowers, epipetalous, alternate with corolla-lobes; anthers often opening by pores. Ovary superior, of 2 united carpels, 2-locular, sometimes with secondary divisions (e.g., *Datura*) upon a hypogynous disc; ovules 1 to many in each loculus, anatropous or slightly amphitropous on axile, often swollen, placentas; style simple, with entire to 2-lobed stigma. Fruit a berry or capsule. Embryo curved or straight, in endosperm. Flowers conspicuous, insect-pollinated.

Economic and ornamental plants. The principal food plants in the Solanaceae are *Solanum tuberosum*, Potato, and *Lycopersicon esculentum*, Tomato, which are now economically important throughout the world. Numerous cultivars of both of these have been raised. Other widely grown food plants include *Capsicum annuum*, Sweet Pepper, *C. frutescens*, Cayenne Pepper, *Solanum melongena*, Aubergine, *Physalis peruviana*, Cape Gooseberry, and *Cyphomandra betacea*, Tree Tomato. Another plant of great commercial value is *Nicotiana tabacum*, the leaves of which are dried and made into tobacco. Many members of the family are poisonous and several of these provide drugs used in medicine, e.g., *Atropa belladonna*, Deadly Nightshade, *Hyoscyamus niger*, Henbane, *Datura stramonium*, Thorn-apple, and *Scopolia carniolica*. The family also contains a large number of ornamental plants such as the perennial *Physalis alkekengi*, Chinese Lantern, with its conspicuous, inflated, red calyces and popular annuals like *Petunia*, *Nicotiana*, *Salpiglossis*, *Schizanthus* and *Browallia*. The last three are also cultivated as flowering pot-plants. *Solanum capsicastrum* and *S. pseudocapsicum*, known as Jerusalem or Winter Cherries, make attractive house plants when bearing their orange, cherry-sized but inedible fruit. Tender genera such as *Brunfelsia*, *Solandra*, *Streptosolen*, *Cestrum* and the woody species of *Datura* are sometimes grown in heated glasshouses as decorative flowering shrubs.

Classification. The classification into 5 tribes is based on that of Wettstein (in Ref. 9).

A. Embryo curved through more than a semicircle. All 5 stamens fertile, equal or only slightly different in length.

1. Nicandreae (ovary 3- to 5-locular, the walls of the loculi dividing the placentas irregularly).
   Nicandra (1) Peru, *N. physalodes*.
2. Solaneae (ovary 2-locular).
   Solanum (1700) tropical and temperate regions.
   Physalis (100) cosmopolitan, especially America.
   Lycium (80–90) temperate and subtropical regions.
   Capsicum (50) Central and S. America.
   Lycopersicon (7) Pacific S. America.
   Atropa (4) Europe, Mediterranean region to central Asia and Himalaya.
3. Datureae (ovary 4-locular, the walls dividing the placentas equally).
   Datura (10) tropical and warm temperate regions.
   Solandra (10) Mexico to tropical S. America.

B. Embryo straight or slightly curved.

4. Cestreae (all 5 stamens fertile).
   Cestrum (150) warm regions of America, W. Indies.
   Nicotiana (66) warm regions of America, Australia, Polynesia.
   Petunia (40) warm regions of America.
5. Salpiglossideae (2 or 4 stamens fertile, of different lengths).
   Salpiglossis (18) S. America.
   Schizanthus (15) Chile.

SOLANUM DULCAMARA L.

Bittersweet, Woody Nightshade

Distribution. Native in Europe, N. Africa and Asia. It is common throughout most of the British Isles and is found in hedges, woodlands and waste ground.

Vegetative characteristics. A scrambling, woody perennial, 30–200 cm, with glabrous to pubescent or even tomentose stems and leaves. The leaves are alternate and ovate, acute to acuminate at the apex, truncate, cordate or hastate at the base, and may be entire or with 1–4 lobes or small, stalked, basal pinnae.

Floral formula. K(5) C(5) A5 $\underline{G}$(2)

Flower and inflorescence. Lax cymes of 10–25 or more hermaphrodite, actinomorphic flowers, the bright yellow stamens contrasting strongly with the purple corolla, appear from June until September. The flowers have pedicels which are erect during flowering but become recurved in fruit.

Pollination. The flowers are visited mainly by bees for the collection of pollen, since little nectar is produced. The insect clings to the cone of exserted anthers and by vibrating its wings rapidly causes pollen to be drawn out of the apical pores on to its body. Although the flowers are homogamous, cross-pollination is achieved because the style protrudes considerably beyond the anther-cone (see Figs. 67.A and B).

Alternative flowers for study. Any member of the genus *Solanum* would be a suitable alternative, including *S. tuberosum*, Potato, which has larger, white or purple flowers and larger fruit, sometimes formed from 3 carpels. *Lycopersicon esculentum*, Tomato, differs mainly in having anthers which dehisce by longitudinal slits instead of apical pores. Examination of the T.S. of a Tomato fruit will show clearly the typical structure of a berry (see caption to Fig. 67.F), also the subdivision of a fruit by false septa. Comparison with other genera will reveal differences in corolla-shape, number of stamens and type of fruit. In the native *Hyoscyamus niger*, Henbane, the fruit is a capsule, the upper portion of which forms a lid which falls off when the seeds are ripe.

Fig. 67 Solanaceae, *Solanum dulcamara*

A The hermaphrodite, actinomorphic flower attached to an erect pedicel. The 5 lobes of the dark purple corolla are recurved and have 2 green spots at their base. The small calyx, hidden behind the corolla, has 5 rounded lobes.
Calyx: 3 mm in diameter Calyx-lobes: 0.5–1 mm
Corolla: 18–20 mm in diameter Corolla-lobes: 9 mm

B L.S. of the flower showing the essential organs. The superior ovary supports a long style that is exserted beyond the stamens. The stamens, which are inserted on the corolla-tube, alternate with the corolla-lobes. The 5 anthers form a cone round the style.
Ovary: 1.8 mm Style + stigma: 6–7.5 mm
Anther: 5 mm

C Detail of the capitate stigma.
Stigma: 0.13 × 0.375 mm

D Two of the 5 stamens whose very short filaments are inserted at the top of the corolla-tube. The long anthers appear to be coherent but are not strictly connate. They dehisce by a terminal pore to liberate the pollen.

E L.S. of the superior ovary which is formed by the fusion of 2 carpels. Surrounding the ovary are portions of the calyx, corolla and stamens.

F T.S. of the 2-locular ovary, showing the ovules attached to axile placentas. At maturity, the ovary develops into a typical berry with a pericarp divided into 3 distinct layers – the inner, juicy layer (endocarp) with embedded seeds, a fleshy layer (mesocarp) and an outer layer of 'skin' (epicarp). The much larger fruit of the Tomato illustrates even more clearly the general structure.
Ovary: 1.4 mm in diameter

G The fruit at maturity, a red berry with the persistent calyx at its base. The pedicel that is erect at anthesis becomes recurved at the fruiting stage.
Fruit: 10 × 9–10 mm

Fig. 67

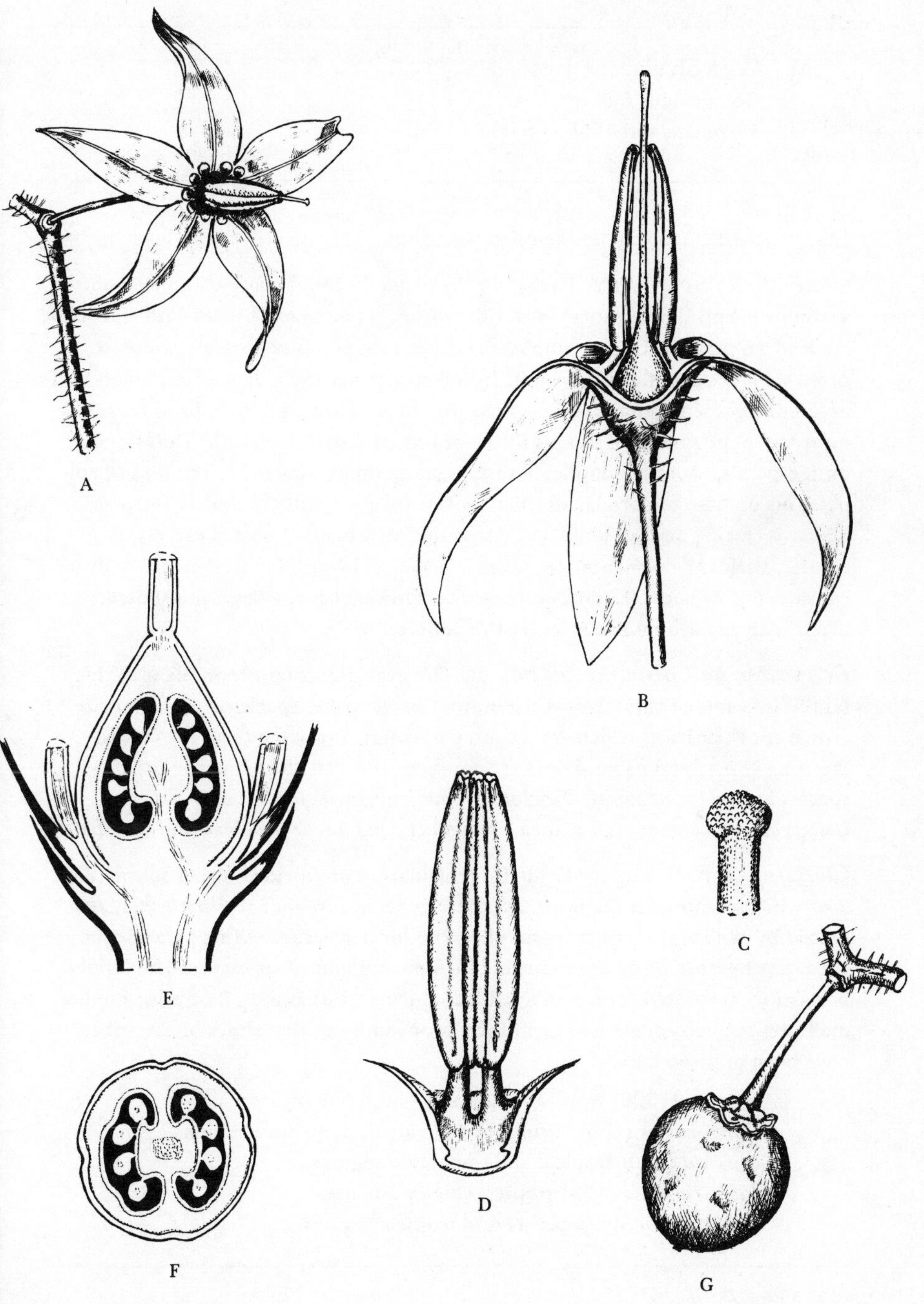

68 Convolvulaceae Juss.

Convolvulus family

55 genera and 1650 species

Distribution. Tropical and temperate regions.

General characteristics. Herbs, shrubs or rarely trees, many climbing, some xerophytic and thorny; some with tuberous roots or stems, others with rhizomes; latex often present. Leaves simple, alternate, usually stalked, rarely stipulate, often with accessory axillary buds. Inflorescence usually a dichasium; bracts and bracteoles present, sometimes close to the calyx. Flowers hermaphrodite, actinomorphic, hypogynous. Calyx of 5 free or united sepals, imbricate. Corolla of 5 united petals, usually induplicate-valvate, sometimes convolute. Stamens 5, epipetalous on base of corolla, alternate with the lobes, anthers usually introrse; nectar-secreting disc present. Ovary superior, of usually 2 united carpels, 2-locular, with axile placentation. Ovules usually 2 in each loculus, erect. Fruit a capsule, nut or berry. Endosperm present. Flowers usually large and showy, often with extrafloral nectaries on the petiole.

Economic and ornamental plants. The most important food plant in this family is *Ipomoea batatas*, Sweet Potato. Certain other species of *Ipomoea* are grown for their roots which are a source of drugs, and some (e.g., *I. tricolor*, Morning Glory, and its cultivars) are climbers of horticultural merit. Other species grown for decoration include *Convolvulus tricolor*, *C. cneorum*, *Calonyction aculeatum* (*Ipomoea bona-nox*), *Quamoclit coccinea* and *Q. lobata.*

Classification. In Engler (4) the Convolvulaceae are divided into 4 subfamilies, 2 of which comprise a single genus. These genera, *Cuscuta* and *Humbertia*, are placed by Willis (5) in separate families, the former because of its parasitic habit, the latter because of its zygomorphic flowers and numerous ovules. The family in its more restricted form is divided into 8 tribes, but as only 2 of these are normally represented in gardens or in the European flora, the names of the tribes have been omitted here.

Ipomoea (500) tropical and warm temperate regions.
Convolvulus (250) cosmopolitan, mostly temperate regions.
Cuscuta (170) tropical and temperate regions.
Jacquemontia (120) tropics, chiefly America.
Calystegia (25) temperate and tropical regions.

CONVOLVULUS ARVENSIS L.
Field Bindweed

Distribution. Native throughout the temperate regions of both hemispheres. It is common in waste places, beside roads and railways and on grassland especially near the coast, and is a persistent weed of cultivated ground.

Vegetative characteristics. A perennial herb with stout rhizomes and stems which scramble or climb by twisting in an anti-clockwise direction round the stems of other plants. The alternate leaves are ovate or oblong, hastate or sagittate, with the petiole shorter than the blade.

Floral formula. K(5) C(5) A5 $\underline{G}$(2)

Flower and inflorescence. The actinomorphic, hermaphrodite, pink or white funnelform flowers, often with 5 purplish stripes on the outside, are borne singly or in groups of 2 or 3 on long peduncles arising from the leaf-axils. They appear from June to September and are short-lived.

Pollination. Bees and various kinds of flies visit the flowers in search of the nectar which is secreted from a well-developed disc (see Fig. 68.E) surrounding the base of the ovary. In the centre of the flower is a short column formed by the 5 stamens which closely surround the style. Between the bases of the filaments are 5 narrow passages leading to the nectar. On entering the corolla an insect bearing pollen from another flower will readily effect cross-pollination as it comes into contact with the 2-lobed stigma that projects beyond the surrounding stamens. After the transference of pollen, the visiting insect moves further down the corolla and becomes dusted with pollen shed by the anthers which dehisce outwards.

Alternative flowers for study. Owing to the frequent occurrence of the Field Bindweed an alternative is hardly necessary. However, *Calystegia sepium*, Hedge Bindweed, is similar in many respects and is also readily obtainable during the summer months.

Fig. 68 Convolvulaceae, *Convolvulus arvensis*

A The 1–3 actinomorphic, hermaphrodite, gamopetalous flowers are borne on a peduncle longer than the leaves. Small scale-like bracteoles are attached to the peduncle well below the 5-lobed calyx. The mid-vein of each corolla-lobe is strengthened and prior to anthesis the corolla is folded inwards, the strengthened portions only being exposed.
Corolla: 20 mm Spread of lobes: 20–25 mm
Calyx: 4 mm in diameter Lobes (at fruiting stage): 4 mm

B The 5-lobed, funnel-shaped corolla opened out to reveal its pleated form and the point of attachment of the stamens. The 5 stamens are adnate to the base of the corolla-tube and alternate with the lobes.
Corolla-lobes: *c.* 10 mm in width

C Anterior view of upper portion of stamen prior to dehiscence. The anthers are 2-celled and dehisce longitudinally. The pollen grains are ellipsoidal, papillose and white.

D Posterior view of stamen showing the dorsifixed anther.
Entire stamen: 11 mm Anther: 4 × 1.5 mm

E Lower portion of perianth cut away exposing the nectar-secreting disc that surrounds the superior ovary.

F L.S. of lower part of flower showing 2 of the 4 erect, anatropous, sessile ovules which develop within the ovary. The lobe-like structures visible on each side of the ovary are sections of the nectar-secreting disc.
Ovary: 1–1.5 × 1.5 mm Ovule (at anthesis): 0.75 × 0.5 mm

G T.S. of ovary formed from 2 united carpels and surrounded by the nectar-secreting disc. The ovary consists of 2 loculi, each containing 2 ovules. Placentation is technically axile, but the erect position of the ovule (see F) gives an impression of basal placentation.
Ovary: 1 mm in diameter.

H Detail of the 2-lobed stigma which terminates the long, filiform style.
Stigma-lobe: 3 mm

I The fruit, a 2-valved, loculicidal capsule, surrounded by the persistent, 5-lobed calyx.
Fruit: 8 × 6 mm

Fig. 68

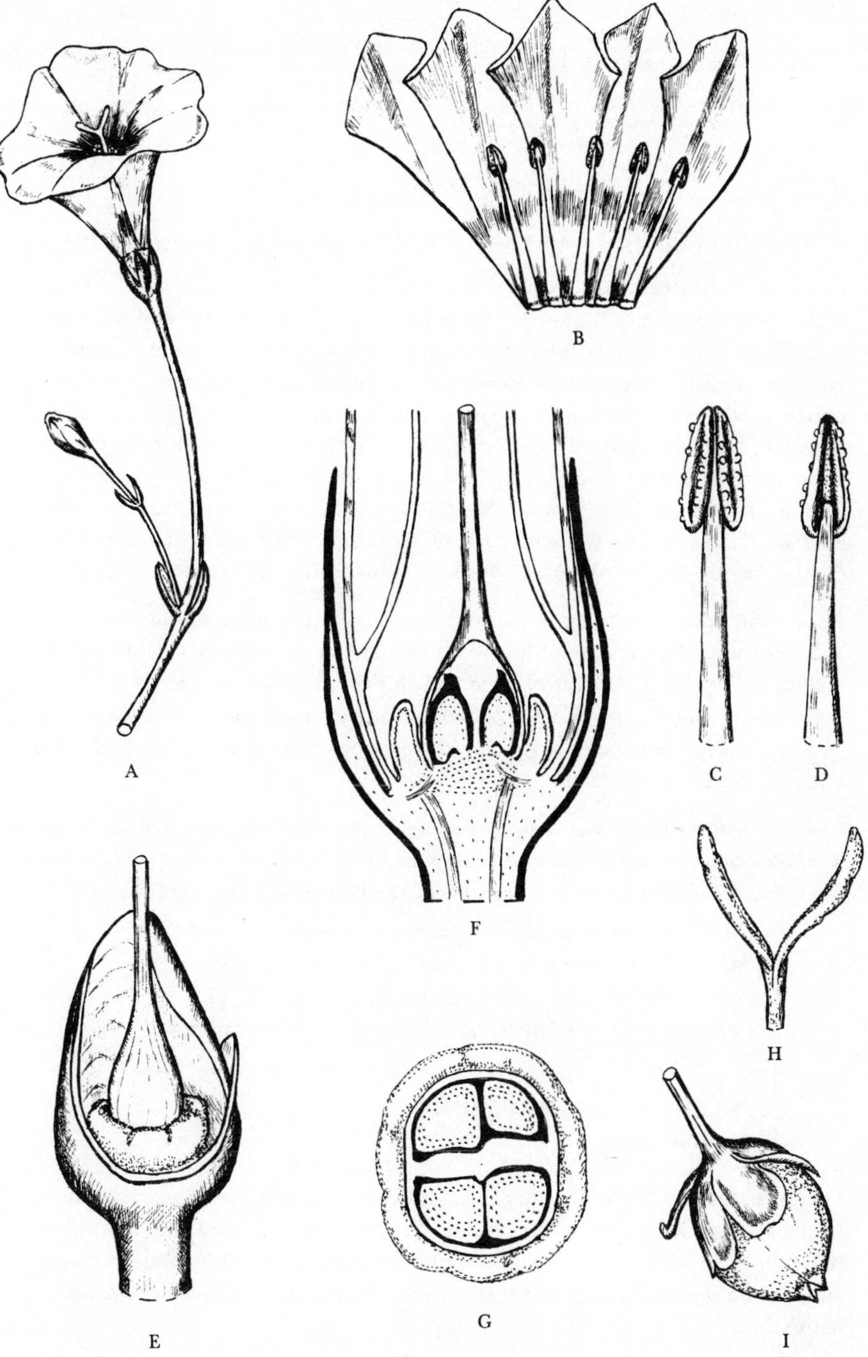

69 Polemoniaceae Juss.

Phlox family

16 genera and 318 species

Distribution. Chiefly N. America, a few in S. America, Europe and N. Asia.

General characteristics. Annual or perennial herbs, rarely shrubs or twining vines, with alternate or opposite, exstipulate leaves (stipules large and foliaceous in *Cobaea*). Inflorescence cymose to capitate. Flowers hermaphrodite, actinomorphic or slightly zygomorphic, with or without bracteoles. Calyx of 5 usually united sepals, valvate or imbricate, persistent. Corolla of 5 united petals, campanulate, funnelform or salverform, usually convolute. Stamens 5, epipetalous, alternate with corolla-lobes. Ovary superior, of usually 3 united carpels, 3-locular, with simple style more or less 3-lobed at tip; ovules 1 to many in each loculus, anatropous, sessile, with axile placentation. Fruit usually a loculicidal capsule (septicidal in *Cobaea*). Endosperm usually abundant (absent in *Cobaea*).

Economic and ornamental plants. The family is of no economic importance, but several genera are of horticultural value in the herbaceous border. These include *Polemonium*, *Gilia*, *Collomia* and *Phlox*, especially the annual *P. drummondii* and the perennial *P. paniculata* and its cultivars. *Cobaea scandens*, with its large, purple, bell-shaped flowers, may be grown as a half-hardy climber out of doors or in the glasshouse.

Classification. The genus *Cobaea* is sometimes considered intermediate between the Polemoniaceae and the Bignoniaceae, and is then placed in a separate family, the Cobaeaceae. Here, the broader view of the Polemoniaceae is maintained.

Gilia (120) temperate and subtropical America.
Phlox (67) N. America, Mexico, 1 species in N.E. Asia.
Polemonium (52) N. temperate region, Chile.
Collomia (15) western N. America, temperate S. America.

POLEMONIUM CAERULEUM L.

Jacob's Ladder

Distribution. Native in N. and central Europe, extending southwards to the Pyrenees, Jugoslavia and S. Russia; also in the Caucasus, Siberia and N. America. It is found in rocky and grassy places, often on limestone hills and mountains, and is sometimes cultivated in gardens from which it may escape and become naturalised.

Vegetative characteristics. A perennial herb, with erect, hollow, angled stems bearing alternate, pinnate leaves, the lower petiolate, the upper subsessile. The leaves are composed of a terminal and 6–12 pairs of lateral leaflets, lanceolate to oblong in shape and acuminate at the apex.

Floral formula. K(5) C(5) A5 $\underline{G}$(3)

Flower and inflorescence. The numerous, hermaphrodite, actinomorphic flowers are borne in terminal or axillary cymes. The flowers are blue, or rarely white, have short pedicels and bloom in June and July.

Pollination. Hover-flies and bumble-bees are attracted by the clusters of conspicuous, blue flowers which secrete nectar from a fleshy ring at the base of the ovary. Cross-pollination is achieved by the protandry of the flowers and by the relative positions of stamens and style.

Alternative flowers for study. Polemonium caeruleum is the most easily obtainable member of the genus, though any other species would be an acceptable substitute. Corolla-shape within the family varies considerably, from salver-form in *Phlox* to campanulate in *Cobaea.*

Fig. 69 Polemoniaceae, *Polemonium caeruleum*

A L.S. of the actinomorphic, gamopetalous flower. The 5-lobed, sparsely glandular-hairy calyx protects the base of the corolla, 3 of whose 5 lobes can be seen. Three of the 5 exserted stamens are shown. They are adnate to the corolla-tube and are alternate with the corolla-lobes. The ovary is superior and is terminated by a long, 3-branched style.
Calyx: 5–6.5 mm Corolla: 10–15 mm

B L.S. of the upper part of the flower showing the inner surface of the corolla-tube and the point of attachment of 2 of the stamens. The numerous hairs at the bases of the filaments form a ring inside the corolla-tube, almost closing its throat. The anthers are 2-celled and dehisce longitudinally.
Stamen: 12 mm

C L.S. of ovary surrounded by part of the hairy calyx and the hairy-throated corolla. The superior ovary has a well-developed axile placenta bearing sessile, anatropous ovules.
Ovary (at anthesis): 1.7 × 1.5 mm

D T.S. of ovary showing axile placentation. There are 3 loculi, corresponding to the number of fused carpels, with several ovules in each loculus.
Ovary: 1.5 mm in diameter

E Detail of the 3-branched stigma which terminates the long style. The branches are only receptive on the upper surface.
Spread of stigma: 3.5 mm

F The fruit after dehiscence, a loculicidal capsule which dehisces by 5 teeth to liberate the angular, rugose and shortly winged seeds.
Fruit: 8 × 4 mm

Fig. 69

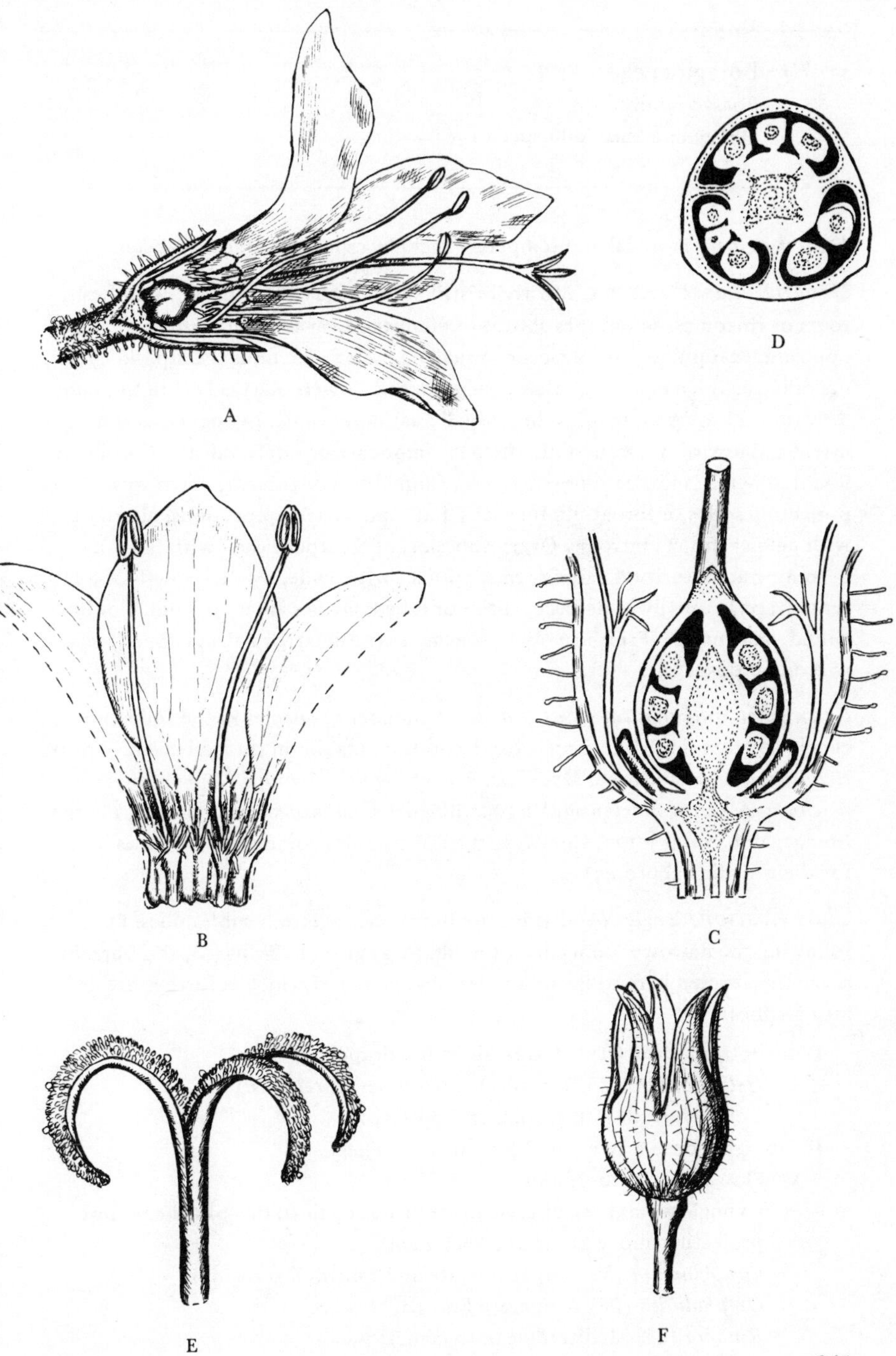

70 Boraginaceae

Borage family

100 genera and 2000 species

Distribution. Tropical and temperate regions, especially Mediterranean.

General characteristics. Mostly bristly hairy, herbaceous plants, with fleshy roots or rhizomes, sometimes shrubs or climbers. Leaves usually alternate, rarely opposite, exstipulate. Inflorescence usually a coiled cincinnus ('scorpioid cyme'), uncoiling as flowers open so that newly opened flowers always face in the same direction. Flowers hermaphrodite, usually actinomorphic, hypogynous and 5-merous. Calyx of 5 usually united sepals, imbricate or rarely valvate. Corolla of 5 united petals, imbricate or convolute, funnelform or tubular, often with projecting scales in throat, limb usually flat. Stamens 5, epipetalous, alternate with petals; anthers introrse. Ovary superior, of 2 carpels, each with 2 ovules, becoming at maturity 4 carpels, each with a single ovule, by the growth of a false septum; style usually gynobasic, entire or lobed. Ovules erect, ascending or horizontal, anatropous. Fruit usually of 4 achenes (nutlets), or a drupe. Seed usually without endosperm.

Economic and ornamental plants. A number of species in the following genera are commonly cultivated for decoration, mainly in the herbaceous border: *Heliotropium*, Heliotrope, *Myosotis*, Forget-me-not, *Echium*, Viper's Bugloss, *Anchusa*, Alkanet, *Mertensia*, Virginia Bluebells, *Pulmonaria*, Lungwort, *Borago*, Borage, and *Symphytum*, Comfrey, the last two also sometimes grown as herbs for their medicinal properties.

Classification. Engler (4) divides the Boraginaceae into 5 subfamilies, but following the narrower concept of the family as given in Willis (5), the Boraginaceae is restricted here to 2 subfamilies, the second of which is further divided into 5 tribes:

I. Heliotropoideae (style terminal; fruit a drupe).
  - *Heliotropium* (250) tropical and temperate regions.
  - *Tournefortia* (150) tropics and subtropics.

II. Boraginoideae (style gynobasic; fruit achenes).
  - A. (Flowers actinomorphic).
    1. Cynoglosseae (base of style more or less conical; tips of achenes not projecting above point of attachment).
       - *Cynoglossum* (50–60) temperate and subtropical regions.
       - *Omphalodes* (28) temperate Eurasia, Mexico.
       - *Rindera* (25) Mediterranean to central Asia.

2. Eritricheae (as for Cynoglosseae, but tips projecting above point of attachment).
   Cryptantha (100) Pacific America.
   Eritrichium (65) N. temperate region.
   Lappula (55) temperate Eurasia, Australia, N. America.
3. Boragineae (base of style flat or slightly convex, achenes with concave attachment surface).
   Anchusa (50) Europe, N. America, W. Asia.
   Alkanna (25–30) S. Europe and Mediterranean region to Iran.
   Symphytum (25) Europe and Mediterranean region to Caucusus.
   Pulmonaria (10) temperate Eurasia.
   Borago (3) S. Europe and Mediterranean region.
4. Lithospermeae (as for Boragineae, but surface of attachment flat).
   Lithospermum (60) temperate regions.
   Myosotis (50) temperate Eurasia, mountains of tropical Africa, S. Africa, Australasia.
   Arnebia (25) Mediterranean region, tropical Africa, Himalaya.
   Cerinthe (10) Europe and Mediterranean region.

B. (Flowers somewhat zygomorphic).

5. Echieae.
   Echium (40) Canary Islands, Azores, N. and S. Africa, Europe, W. Asia.

SYMPHYTUM OFFICINALE L.
Common Comfrey

Distribution. This species is found in damp places, especially beside rivers and streams, throughout the British Isles, though it is less common in the north. Its distribution extends eastwards across Europe to W. Asia.

Vegetative characteristics. An erect, hispid perennial with a thick, fleshy root, stout stem and large, ovate-lanceolate leaves.

Floral formula. K(5) C(5) A5 $\underline{G}$(2)

Flower and inflorescence. Pendulous, cream to purple or pink flowers appear from May to July in terminal scorpioid cymes.

Pollination. Only certain kinds of bumble-bees and other insects with long tongues are able to reach, in a legitimate way, the nectar at the base of the corolla. Due to the narrower lower portion of the corolla-tube and the ring of scales between the stamens, even a long-tongued insect is obliged to push its proboscis past the anthers in order to obtain the nectar. The proboscis becomes dusted with pollen which is subsequently transferred to other flowers and in this way cross-pollination is effected. However, some of the shorter-tongued bees make holes in the base of the corolla-tube and steal the nectar without contributing to the pollination of the flower.

Alternative flowers for study. Symphytum × uplandicum, Russian Comfrey, is a hybrid between *S. officinale* and *S. asperum* and is the commonest *Symphytum* of roadsides, hedgebanks and woods, but usually absent from the waterside habitats occupied by *S. officinale.* Examples of most tribes of the Boraginaceae can be found readily, either as native species or as popular garden plants. A feature of many members of this family is the striking colour-change from pink in bud to blue when the flowers are fully open.

Fig. 70 Boraginaceae, *Symphytum officinale*

A L.S. of the actinomorphic flower. The 5-lobed corolla-tube is surrounded by a persistent, hairy calyx with 5 tooth-like lobes. The introrse stamens are joined to the corolla-tube. The style arises from below the ovary and protrudes beyond the corolla.
Corolla-tube 14 mm Style: 15 mm Sepals: 6–8 mm

B The calyx with 2 of its 5 lobes removed to expose 2 of the 4 fertilised carpels seated on a nectar-secreting disc on the surface of the receptacle. The 2 original carpels divide into 4 as they mature (see E). Like other members of the subfamily Boraginoideae the style is gynobasic.
The group of 4 carpels is 2 mm across.

C The detached corolla-tube cut open to show the 5 stamens situated between the corolla-lobes and alternating with the 5 triangular, hairy scales which form the corona.

D Detail of one stamen. The dorsifixed, introrse anther has 2 cells which dehisce longitudinally to liberate the pollen.
Anther: 4.5 mm

E T.S. of ovary showing the 4 carpels, each containing one ovule with sub-basal placentation, which surround the central style. At maturity the carpels develop into glossy black nutlets.
Carpel: 1 mm in diameter

F L.S. of 2 mature carpels, showing how the obliquely orientated seeds are attached to the receptacle. Between them can be seen the basal portion of the style.

G Detail of the simple stigma at the top of the single style.

Fig. 70

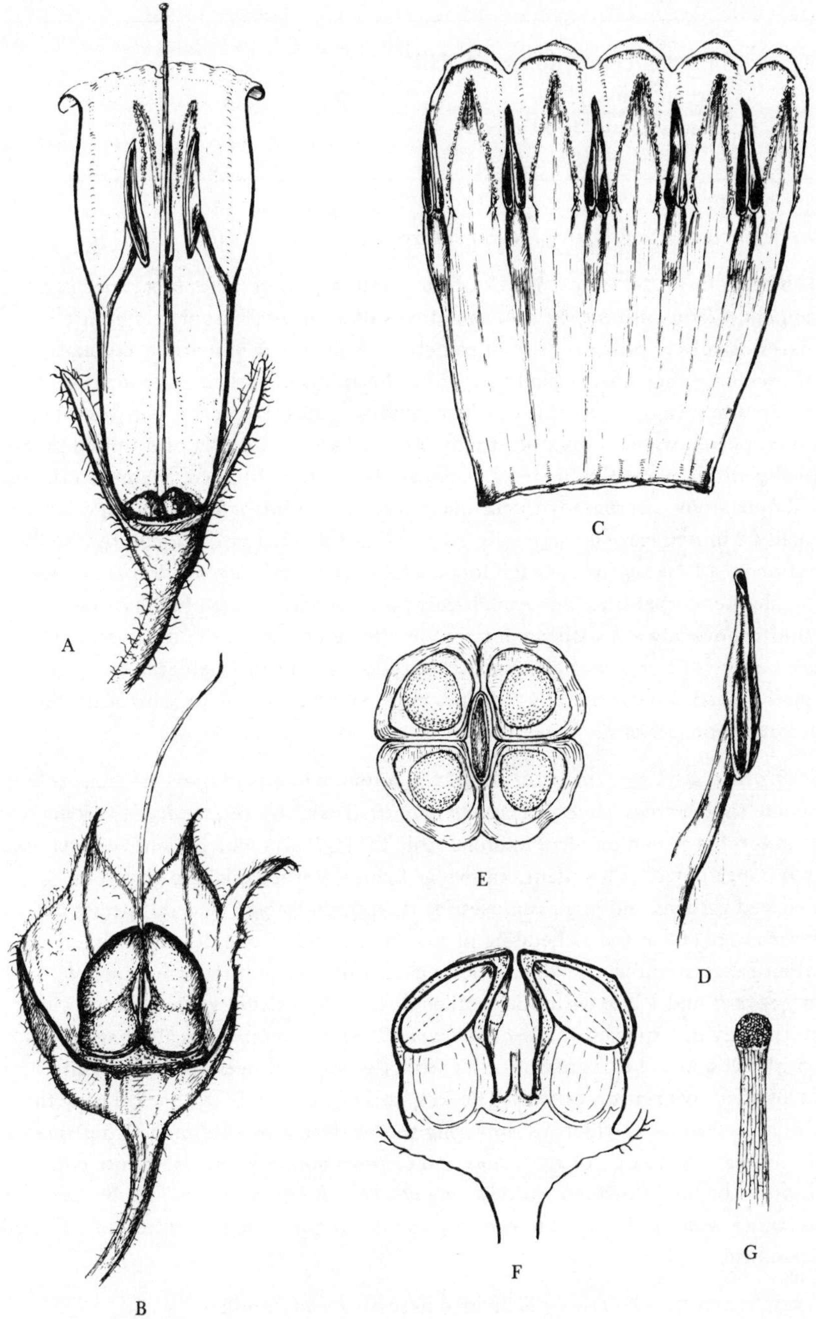

71 Verbenaceae Jaume St.-Hil.

Verbena family

75 genera and 3000 species

Distribution. Chiefly tropics and subtropics.

General characteristics. Herbs, shrubs or trees, often lianes (e.g., species of *Lantana*, *Clerodendrum*, *Vitex*), sometimes of a thorny, xerophytic nature. Leaves usually opposite, entire or sometimes palmate or pinnate, exstipulate. Inflorescence racemose, usually a spike or head, often with an involucre of coloured bracts, or consisting of dichotomous cymes. Flowers usually hermaphrodite, zygomorphic. Calyx of usually 5 united sepals. Corolla of 5 united petals, usually with a narrow tube, rarely campanulate, often 2-lipped. Stamens usually 4, didynamous, alternate with corolla-lobes; anthers introrse. Ovary superior, of usually 2 united carpels, originally 2-locular but divided into 4 loculi by the formation of a false septum in each loculus (cf. Labiatae); placentation axile, with 2 ovules per carpel (i.e., 1 in each loculus after septation); ovules anatropous to orthotropous, always with the micropyle directed downwards; style terminal, rarely more or less sunk between the ovary-lobes (contrast Labiatae); stigma usually lobed. Fruit generally a drupe, more rarely a capsule or schizocarp. Seed without endosperm.

Economic and ornamental plants. Economic plants in the Verbenaceae include timber trees, such as *Tectona grandis*, Teak. *Aloysia triphylla* (*Lippia citriodora*) is grown on a commercial scale for its leaves which yield verbena oil, used in perfumery. This plant, known as Lemon Verbena, is also grown in sheltered gardens and in glasshouses for its scented foliage. Several species of *Verbena* are cultivated as bedding plants and a number of cultivars with a wide colour-range are now available. Woody genera of horticultural value include *Caryopteris* and *Vitex* with blue or lilac flowers, and *Callicarpa* which has violet fruits. Only one species of *Clerodendrum*, *C. trichotomum*, is really hardy in the British Isles, and this is grown for its attractive white flowers and blue fruits surrounded by crimson calyces. The less hardy *C. bungei* is cut back during the winter, but sends up vigorous flowering shoots during the summer. Other species of *Clerodendrum*, e.g., *C. splendens* and *C. thomsoniae*, need glasshouse conditions. The blue-flowered, woody vine *Petrea volubilis* and the shrubby *Lantana camara* (a widespread weed in the tropics) also require the protection of a heated glasshouse.

Classification. The family is divided here into 4 subfamilies.

A. Inflorescence racemose or spicate.

I. Verbenoideae.

Verbena (250) chiefly America, a few species in the Old World.

Lippia (220) tropical America, Africa.

Lantana (150) tropical America, W. Indies, tropical and S. Africa.

Citharexylum (115) S. United States to Argentina.

Stachytarpheta (100) America.

Petrea (30) tropical America, W. Indies.

B. Inflorescence cymose, the cymes often united into panicles or corymbs, or, if axillary, often reduced to 1 flower.

II. Viticoideae (fruit drupaceous; flowers sometimes actinomorphic).

Clerodendrum (400) tropics and subtropics.

Vitex (250) tropical and temperate regions.

Callicarpa (140) tropics and subtropics.

Tectona (3) Indomalaysia.

III. Nyctanthoideae (fruit a 2-locular, 2-valved, 2-seeded capsule; flowers actinomorphic or almost so).

Nyctanthes (2) India to Java.

IV. Caryopteridoideae (fruit capsule-like and 4-valved, or 1-celled and indehiscent)

Caryopteris (15) Himalaya to Japan.

VERBENA RIGIDA Spreng.
(V. VENOSA Gill. & Hook.)

Distribution. Native in Brazil and Argentina and grown in British gardens as a border or bedding plant since 1830. It sometimes escapes from cultivation and is found on roadsides and waste places.

Vegetative characteristics. A hispid perennial with erect or ascending, 4-angled stems reaching a height of 60 cm. The leaves are opposite, sessile, amplexicaul, oblong-lanceolate, irregularly dentate and acute at the apex.

Floral formula. K(5) C(5) A4 $\underline{G}$1

Flower and inflorescence. The inflorescence is in the form of a lax, terminal corymb, with each branch ending in a dense spike (see Fig. 71.A). The spikes consist of reddish purple, hermaphrodite, weakly zygomorphic flowers that appear in July and August.

Pollination. Although details of the pollination of *Verbena rigida* are lacking, it is probable that, like other members of the genus, it is entomophilous, and that insects are attracted by the dense spikes of purple flowers and the nectar secreted at the base of the ovary. As in the native *V. officinalis*, there are upward-pointing hairs at the mouth of the corolla-tube, presumably to exclude unwanted visitors.

Alternative flowers for study. Any other members of the genus *Verbena* would be suitable alternatives, several of these being readily obtainable as they are popular garden plants. They include the lilac-flowered *V. bonariensis*, the red-flowered *V. peruviana*, and a hybrid with the latter species in its parentage, *V.* × *hybrida.* In *Verbena* and *Caryopteris* the fruits are dry, but in *Callicarpa*, *Clerodendrum*, *Vitex* and most of the other genera, the fruit is a drupe.

Fig. 71 Verbenaceae, *Verbena rigida*

A A spike of flowers terminating a flowering branch. Each individual flower is subtended by a bract that exceeds the calyx.
Spike: 23 mm

B A single, hermaphrodite, weakly zygomorphic flower subtended by a lanceolate-subulate, ciliate bract. The tubular calyx is covered in small hairs and is 5-ribbed, the ribs ending in small lobes or teeth. The reddish-purple corolla has a tube at least twice as long as the calyx, opening out into a weakly 2-lipped limb consisting of 5 more or less unequal lobes.
Bract: 5.5–6 mm Calyx: 4 mm
Corolla-tube: 7.5 × 2 mm Corolla-lobes: 2 × 2 mm
Corolla-limb: *c.* 6 mm in diameter

C The corolla, cut down the centre and opened out to reveal the 4 stamens that are attached about half-way up the tube. The stamens are in pairs of different lengths, which can be seen more clearly in D. Upward-pointing hairs clothe the inner surface of the corolla-tube and are particularly dense at its mouth.

D L.S. of lower portion of flower. The didynamous stamens are adnate to the corolla-tube. The superior ovary is terminated by a long style with a bilobed stigma. Surrounding the base of the corolla-tube is the calyx, subtended by the long bract; all these parts are hairy.
Ovary (at anthesis): 1.75 mm Style + stigma: 3 mm

E Free portion of stamen removed from corolla. The anthers are introrse, 2-celled and dehisce longitudinally. The stamens are attached about 3 mm up the corolla-tube.
Anther: 0.75 × 0.4 mm

F L.S. of ovary showing 2 of the 4 erect, anatropous ovules.
Ovary: 2 × 1.75 mm Ovule: 1.5 × 0.5 mm

G T.S. of ovary after fertilisation. Placentation is axile. The ovary is initially unilocular, but later becomes 4-locular by the growth of false septa. The fruit divides into 4 separate portions, each portion containing a single nutlet.
Ovary (at anthesis): 1.25–1.5 mm in diameter

H Apical region of style showing the 2-lobed stigma.
Stigma-lobe: 0.2 mm

I The persistent calyx has been cut away to reveal 2 of the 4 mature, single-seeded nutlets. A portion of the bract subtends the calyx-tube.
Nutlets: *c.* 2.5 × 0.75 mm

Fig. 71

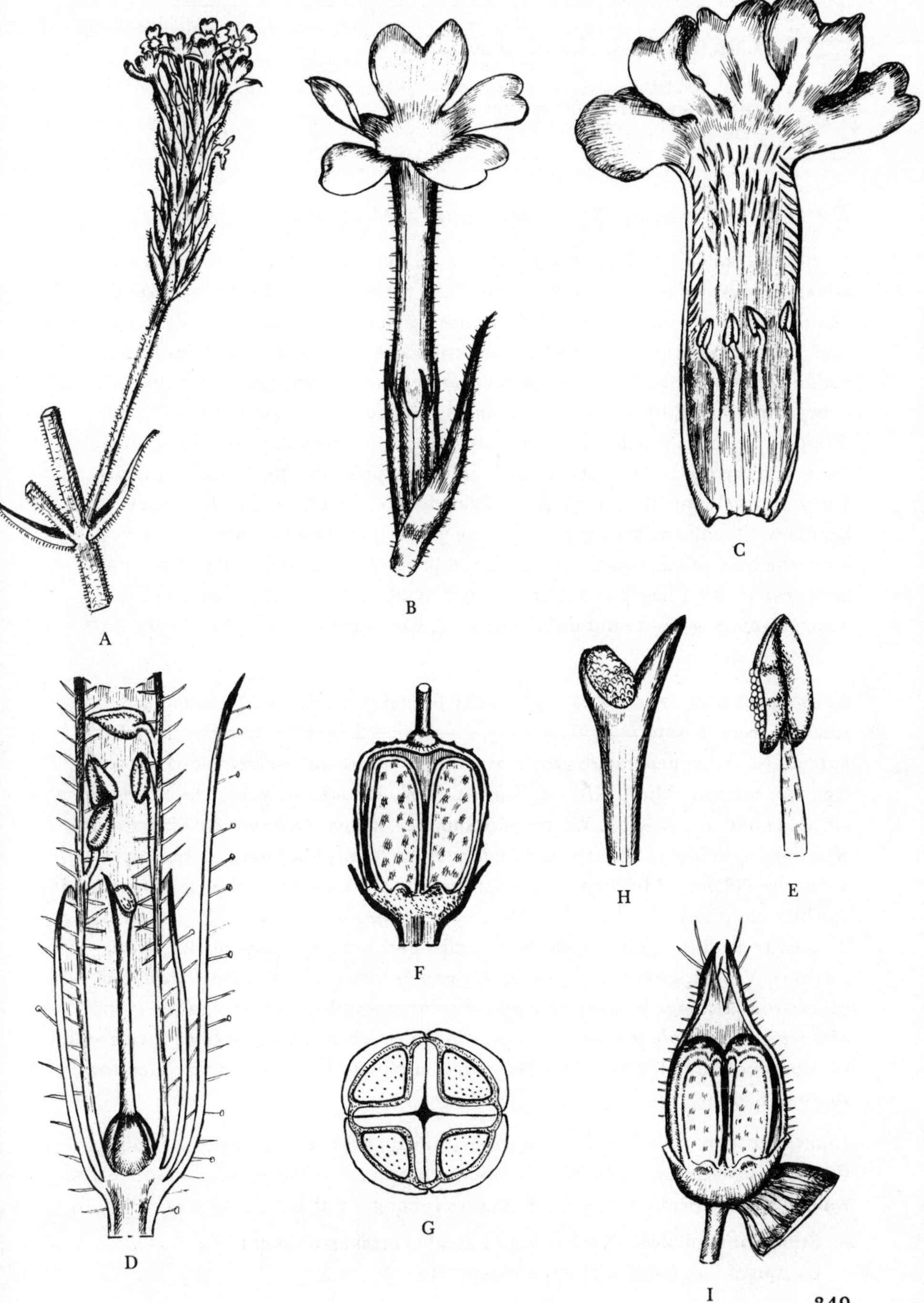

72 Labiatae Juss. (Lamiaceae Lindl.)

Mint family

180 genera and 3500 species

Distribution. Cosmopolitan, chief centre the Mediterranean region.

General characteristics. Mainly terrestrial herbs or undershrubs with square stems. Leaves decussate, simple, exstipulate, often hairy and with epidermal glands secreting volatile oils. Inflorescence racemose or cymose, the cymes at the nodes condensed into a false whorl or verticillaster. Flowers usually hermaphrodite, zygomorphic. Calyx of 5 sepals, united into a campanulate or funnelform tube, sometimes 2-lipped, persistent in fruit. Corolla of 5 united petals, usually 2-lipped. Stamens usually 4 and didynamous, sometimes 2, epipetalous, with introrse anthers. Ovary superior, of 2 united carpels, on a nectar-secreting disc. Early in development each carpel is divided into 2 loculi, so that the ovary becomes 4-locular as it matures. In some genera the style is gynobasic, arising from the base of the ovary between the 4 portions. Stigma 2-lobed; placentas axile, each with 1 basal, erect, anatropous ovule. Fruit usually a group of 4 achenes (nutlets) each containing one seed, sometimes a drupe. Seed with little or no endosperm.

Economic and ornamental plants. Oils and perfumes are obtained by distillation from *Rosmarinus*, Rosemary, *Lavandula*, Lavender, and *Pogostemon*, Patchouly. Other genera are grown as culinary herbs and are used for flavouring, e.g., *Mentha* spp., Mint, *Salvia officinalis*, Sage, *Thymus vulgaris*, Thyme, *Satureja montana* and *S. hortensis*, Winter and Summer Savory, *Ocimum basilicum*, Basil, *Hyssopus officinalis*, Hyssop, and *Majorana hortensis*, Marjoram. Other species of *Salvia* are cultivated for ornament, particularly the red-flowered *S. splendens* and the blue-flowered *S. patens*, also *Monarda didyma*, Bergamot, *Molucella laevis*, Bells of Ireland (so called from the conspicuous, broadly campanulate, green calyces), *Nepeta*, *Physostegia*, *Perovskia* and *Phlomis. Thymus serpyllum* and other aromatic, mat-forming species are sometimes planted between the stones of paved walks. *Coleus blumei*, a name which covers an assemblage of forms with variously coloured leaves, is popular as a house plant and may also be used for summer bedding.

Classification. The Labiatae are closely allied to the Verbenaceae as well as to the Boraginaceae and are similar in minor characters to the Scophulariaceae. The following classification into 8 subfamilies is based on that by Briquet (in Ref. 9).

A. Style not gynobasic. Nutlets with lateral-ventral attachment.

   I. Ajugoideae (seed without endosperm).

1. Ajugeae (corolla various, upper lip if present rarely concave; stamens 4 or 2; anthers 2-celled; nutlets more or less wrinkled).
   Teucrium (300) cosmopolitan, especially Mediterranean region.
   Ajuga (40) Old World temperate region.
2. Rosmarineae (corolla strongly 2-lipped, upper lip very concave and arched; stamens 2; anthers 1-celled; nutlets smooth).
   Rosmarinus (3) Mediterranean region.

II. Prostantheroideae (endosperm present).
   Prostanthera (50) Australia.

B. Style gynobasic. Nutlets with basal attachment.

III. Prasioideae (nutlet drupaceous).
   Gomphostemma (40) India, E. Asia.

IV. Scutellarioideae (nutlet dry; seed more or less transversal; embryo with curved radicle).
   Scutellaria (300) cosmopolitan.

V. Lavanduloideae (nutlet dry; seed erect; embryo with straight radicle; disc-lobes opposite to ovary-lobes).
   Lavandula (28) Atlantic islands and Mediterranean region to India.

VI. Lamioideae (as V, but disc-lobes, when distinct, alternate with ovary-lobes; stamens ascending or spreading and projecting straight forwards).
   Salvia (700) tropical and temperate regions.
   Thymus (300–400) temperate Eurasia.
   Stachys (300) tropics and subtropics.
   Lamium (40–50) Europe, Asia, Africa.

VII. Ocimoideae (as VI, but stamens descending, lying upon lower lip or enclosed by it).
   Hyptis (400) warm regions of America, W. Indies.
   Ocimum (150) tropical and warm temperate regions.

VIII. Catopherioideae (nutlet dry; seed erect; embryo with curved radicle).
   Catopheria (3) tropical America.

LAMIUM ALBUM L.
White Dead-nettle

Distribution. Native in most of Europe, the Himalaya and Japan, but introduced into Ireland. It is commonly found growing in hedges, waste places and near roadsides.

Vegetative characteristics. A hairy, perennial herb with creeping rhizome and a 4-angled, erect stem which reaches a height of between 20 and 60 cm. Its leaves are petiolate, ovate, coarsely and often doubly serrate or crenate-serrate, acuminate at the apex and cordate at the base.

Floral formula. K(5) C(5) A4 $\underline{G}$(2)

Flower and inflorescence. The hermaphrodite, zygomorphic, white flowers appear from May onwards, sometimes until December if the weather is mild. They are borne in cymes in the axils of opposite bracts but having very short pedicels, the flowers appear to be grouped closely round the stem, forming the verticillaster or false whorl characteristic of many of the Labiatae.

Pollination. As the white flowers are in dense groups, they appear more conspicuous to pollinating insects, generally long-tongued bees such as bumble-bees, which seek out the nectar secreted by the disc (see Fig. 72.H) situated at the base of the ovary. Sometimes short-tongued insects steal the nectar by biting through the base of the corolla-tube. The upper lip of the corolla forms a hood, protecting the stamens and stigma from possible damage (see Fig. 72.A), while the lower lip acts as a landing-stage for pollinating insects. On arrival, the insect grasps the 2 lateral lobes of the flower, causing the hooded upper lip to descend and press first the style and then the stamens on to its back. As the style projects beyond the stamens it picks up the pollen brought in from another flower before pollen from its own flower is deposited on the insect's back. In this way cross-pollination is usually achieved although, as the flowers are homogamous, there appears to be no barrier to self-pollination.

Alternative flowers for study. If *L. album* is not available, any other species of *Lamium* would be suitable, including the equally common *L. purpureum*, Red Dead-nettle. Comparison should be made with genera in other subfamilies (see Classification section) and also with members of the Verbenaceae, Boraginaceae and Scrophulariaceae.

Fig. 72 Labiatae, *Lamium album*

A L.S. of the hermaphrodite, zygomorphic flower with its 2-lipped corolla surrounded at the base by the tubular 5-lobed calyx. The hooded upper lip, formed from 2 united lobes, protects the stamens and style which lie close to its inner surface. The 4 stamens (only 2 of which are shown) are didynamous, the outer pair being longer than the inner. The stamens are adnate to the lower part of the corolla-tube and alternate with the corolla-lobes, the upper lip counting as a single lobe. Arising from the base of the 4-partite ovary is the long style, 2-branched at the apex, and projecting beyond the stamens.
Calyx-tube: 5–6 mm Corolla-tube: 7–8 mm
Upper lip of corolla: 12.5 mm Lower lip: 6 mm

B The persistent, tubular calyx, with 5 more or less equal lobes or teeth. At the base of the tube can be seen the fruit consisting of 4 nutlets.
Calyx-lobes: 4–5 mm

C Ventral and lateral views of the hairy anthers, which are introrse, 2-celled and dehisce longitudinally.
Long stamen: 6 mm Short stamen: 4 mm
Anther: 1.5 mm

D L.S. of one of the lobes of the 4-lobed, superior ovary. The single ovule is anatropous, erect and attached to a basal placenta derived from an axile type.
Ovary: 1.5 mm

E T.S. of the ovary formed from 2 united carpels. At a later stage of development there is a further division of the ovary resulting in 4 separate loculi, each containing a single ovule.

F T.S. of the mature fruit showing the division into 4 separate portions, each portion becoming a 1-seeded nutlet.
Fruit: 1.75 mm in diameter

G Basal portion of style with one of the ovary-lobes and part of the hypogynous, nectar-secreting disc. The style is gynobasic and arises from the base of the ovary, between the 4 lobes.

H The 4-lobed ovary at maturity showing the gynobasic style. The hypogynous disc is more developed on the anterior side of the ovary.
Nutlet: 2 × 1.25 mm

I Apical portion of the 2-branched style. The receptive stigmas are at the tips of the branches.
Style (from base to fork): 20 mm Style-branches: 1.5 mm

Fig. 72

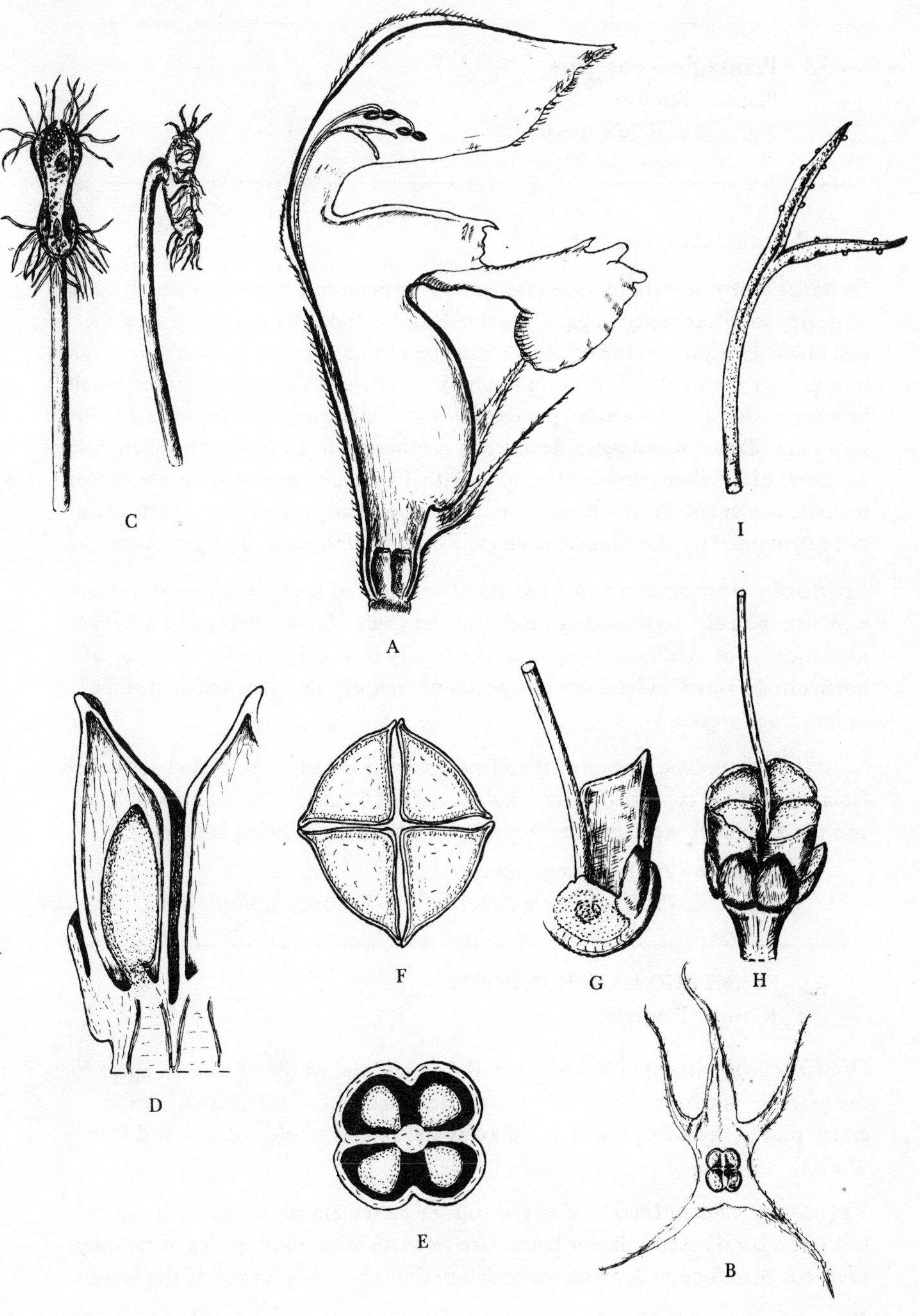

73 Plantaginaceae Juss.

Plantain family

3 genera and 269 species

Distribution. Cosmopolitan.

General characteristics. Scapose, annual or perennial herbs. Leaves all basal or nearly so, often narrow, parallel-veined, and without distinction into stalk and blade, exstipulate. Inflorescence usually a head or spike. Flowers inconspicuous, usually hermaphrodite, actinomorphic, bracteate. Calyx of 4 united sepals, imbricate. Corolla of 4 united petals, imbricate, membranous. Stamens 4, with very long filaments and versatile anthers containing much powdery pollen. Ovary superior, of 2 united carpels, 2-locular, with 1 to many semi-anatropous ovules on axile placentas. Fruit a membranous, circumscissile capsule or sometimes a nut surrounded by the persistent calyx. Embryo straight, in fleshy endosperm.

Economic and ornamental plants. The leaves of some species of *Plantago* have occasionally been used as food, and the seeds of *P. psyllium* and others are used locally for medicinal purposes. The family is usually considered to be of no horticultural value, indeed several species of *Plantago* are regarded as troublesome garden weeds.

Classification. The flower of the Plantaginaceae is usually regarded as derived from a 5-merous type in the same way as that of *Veronica* (Scrophulariaceae) and the 2 families are therefore sometimes considered as being allied.

Plantago (265) cosmopolitan.
Littorella (3) 2 species in America, 1 in Europe, *L. uniflora* (*L. lacustris*)

PLANTAGO LANCEOLATA L.

Ribwort Plantain

Distribution. Native in N. and central Asia and the whole of Europe apart from the extreme north. It is generally distributed throughout the British Isles in grassy places, including lawns and playing fields, and is tolerant of a wide range of soils.

Vegetative characteristics. A glabrous or pubescent perennial with leaves forming a basal rosette, linear-lanceolate to ovate-lanceolate, entire or remotely toothed, 3- to 5-nerved, sessile or with a petiole up to the length of the lamina.

Floral formula. K(4) C(4) A4 $\underline{G}$(2)

Flower and inflorescence. A deeply furrowed scape, usually twice as long as the leaves, supports a dense flower-spike (see Fig. 73.A). The numerous actinomorphic flowers appear from April to August. Male sterility is not infrequent, so that some of the flowers in an inflorescence may be functionally only female. The spike then consists partly of hermaphrodite and partly of female flowers and the plant is described as gynomonoecious.

Pollination. The species is primarily adapted to wind-pollination as is indicated by the numerous small flowers with inconspicuous corollas, well-exserted stamens with versatile anthers and a long feathery stigma. It is protogynous, but not always strongly so, and self-pollination may occur. Insect-pollination may also take place, since honey-bees have been seen to visit the flowers to collect the pollen which they smear with nectar to render more adhesive.

Alternative flowers for study. Several commonly occurring species may be used instead of *P. lanceolata.* These vary mainly in the size of the inflorescence and the number of seeds in the capsule. *P. media*, Hoary Plantain, has a 4-seeded capsule, in *P. major*, Greater Plantain, it is 8- to 16-seeded, and in *P. coronopus*, Buck's-horn Plantain, the capsule contains 3 or 4 seeds. All these species have approximately the same flowering period.

Fig. 73 Plantaginaceae, *Plantago lanceolata*

A The top of the furrowed scape that bears a dense spike of actinomorphic, usually hermaphrodite flowers. The apical flowers are at the female stage and have their feathery stigmas exserted. The central flowers are at the male stage, with stamens fully extended and styles beginning to wither. The basal flowers are at the end of anthesis and both stamens and styles have withered.
Inflorescence: 40 × 10 mm

B L.S. of the top of a young inflorescence, showing the arrangement of the flower-buds.
Buds: 2.25 × 1.5 mm

C A single flower at an early stage subtended by a glabrous, scarious bract with a green keel. The 4 sepals are similar to the bract but their keels are slightly hairy. The 2 anterior sepals are connate. The feathery stigma protrudes from the closed corolla.
Flower (at early stage): 4 × 2 mm Bract: 3 × 2 mm

D The flower at a later stage with its subtending bract. The corolla has 4 glabrous lobes (3 shown) each with a brown midrib. The 4 stamens, with versatile anthers on long filaments, are inserted on the upper part of the corolla-tube between the corolla-lobes. The style, which is only slightly shorter than the stamens, arises from the apex of the superior ovary.
Stamen: 5 mm

E Detail of a versatile anther which is 2-celled and dehisces longitudinally.
Anther: 1.75 × 1.25 mm

F L.S. of the superior ovary surrounded by a portion of the corolla-tube. The 2-locular ovary is formed from 2 fused carpels and has a single, semi-anatropous ovule in each loculus.
Ovary: 0.8 × 0.4 mm

G T.S. of ovary at early anthesis, showing the 2 ovules attached to axile placentas.
T.S. of ovary: 0.6 × 0.4 mm

H H1: the young fruit, a circumscissile capsule (pyxis). At dehiscence, the lid (operculum) lifts off, allowing the seeds to be dispersed. H2: shows the operculum of a ripe capsule containing the 2 seeds still attached to their shrivelled placentas. The seeds and placentas often come away with the operculum and can be seen protruding below its rim.
Complete fruit: 4 × 2 mm Operculum: 3 mm

Fig. 73

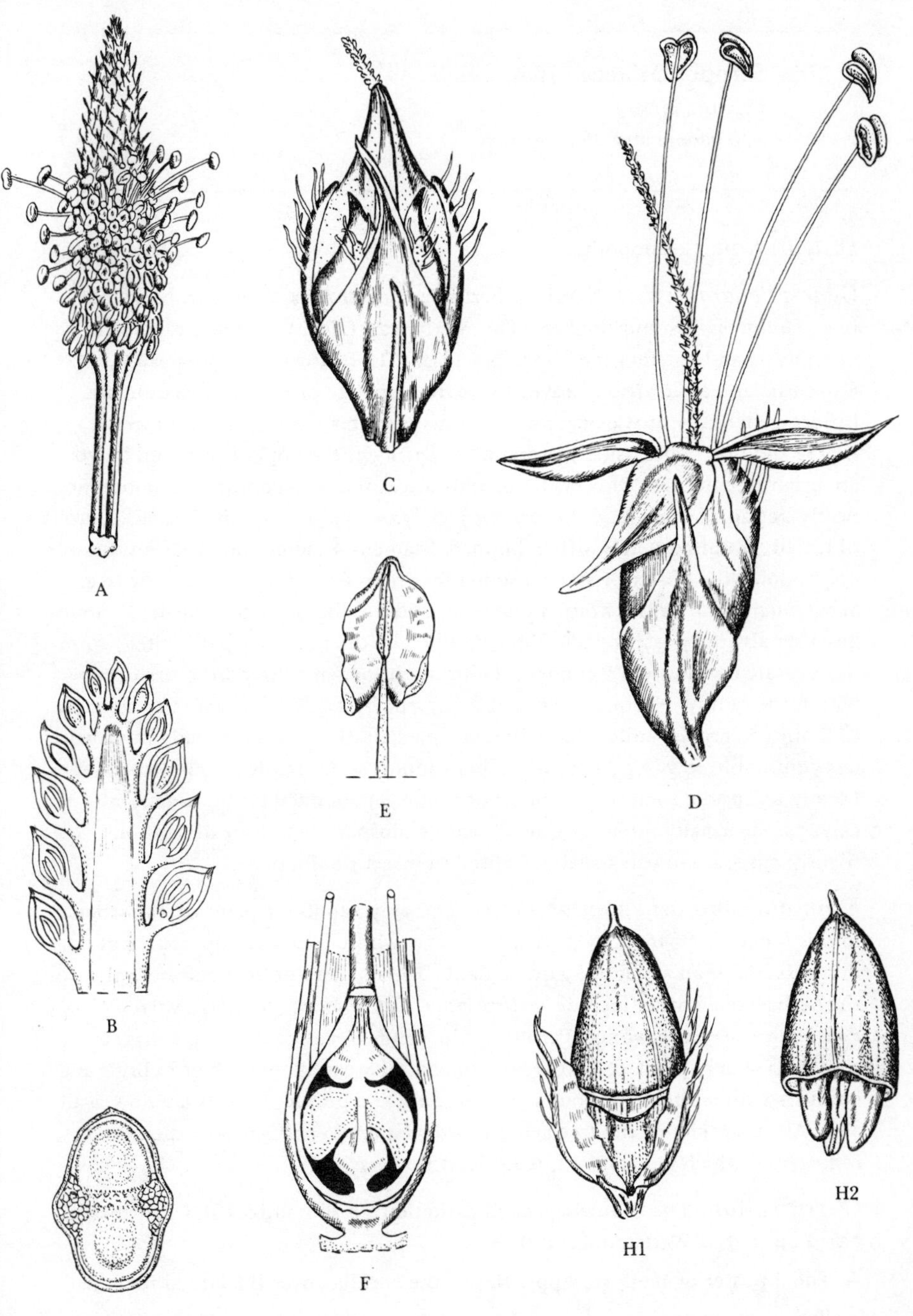

74 Scrophulariaceae Juss.

Figwort family

220 genera and 3500 species

Distribution. Cosmopolitan.

General characteristics. Mostly herbs or undershrubs, a few shrubs or trees (e.g., *Paulownia*), some climbers. The 'Veronicas' (*Hebe*) of New Zealand are xerophytic and several genera in tribes 11 and 12 below are semi-parasitic, e.g., *Euphrasia* and *Pedicularis*. Leaves alternate, opposite or whorled, exstipulate. Inflorescence racemose or cymose, or flowers solitary in axils (e.g., *Linaria*). Bracts and bracteoles usually present. In *Castilleja*, the upper leaves and bracts are brightly coloured. Flowers hermaphrodite, usually zygomorphic, sometimes nearly actinomorphic (e.g., *Verbascum*). Calyx of usually 5 united sepals. Corolla of usually 5 united petals, often 2-lipped. Stamens 4 (sometimes 2), didynamous, epipetalous, the posterior stamen sometimes represented by a staminode (e.g., *Scrophularia* and *Penstemon*). *Verbascum* and its allies have 5 stamens. *Veronica* has 4 sepals (the posterior one absent), 4 petals (the posterior pair united) forming a rotate corolla, and 2 stamens. Other variations in flower structure occur within the family. Nectar-secreting disc present below the ovary. Ovary superior, of 2 united carpels, unilocular, with axile placentation; ovules usually numerous, less commonly few (e.g., *Veronica*), anatropous; style simple or bilobed. Fruit usually a capsule, sometimes a berry or drupe, surrounded by the persistent calyx; seeds usually numerous, small, with endosperm; embryo straight or slightly curved. Flowers usually adapted to insect-pollination.

Economic and ornamental plants. The only economic plant of importance in this family is *Digitalis purpurea*, Foxglove, whose leaves are the source of drugs. It is also an attractive garden plant. Other genera of horticultural value for the herbaceous border include *Antirrhinum*, *Calceolaria*, *Mimulus*, *Nemesia*, *Penstemon*, *Verbascum* and *Veronica*. The genus *Hebe*, comprising evergreen shrubs, is widely planted, particularly in coastal areas. A number of hybrids and cultivars with white, or various shades of pink and purple, flowers are now available. A few species of the tree genus *Paulownia* are in cultivation, especially *P. tomentosa*, which has fragrant, pale violet flowers.

Classification. The following classification into 3 subfamilies and 12 tribes is based on that of Wettstein (in Ref. 9).

A. The 2 posterior teeth (or upper lip) of the corolla cover the lateral teeth in bud.

   I. Verbascoideae (all leaves usually alternate; 5 stamens often present).

1. Verbasceae (corolla with very short tube or none, rotate or shortly campanulate).
   Verbascum (including *Celsia*) (360) temperate Eurasia.
2. Aptosimeae (corolla with long tube).
   Aptosimum (42) tropical and S. Africa.

II. Scrophularioideae (at least lower leaves opposite; 5th stamen wanting or staminodial).

3. Calceolarieae (corolla 2-lipped, the lower lip concave and bladder-like).
   Calceolaria (300–400) Mexico to S. America.
4. Hemimerideae (corolla spurred or saccate at base, with no tube).
   Alonsoa (6) tropical America.
5. Antirrhineae (as tribe 4, but with tube).
   Linaria (150) northern hemisphere.
   Antirrhinum (42) Pacific N. America, W. Mediterranean region.
6. Scrophularieae (corolla not spurred or saccate; inflorescence cymose, compound).
   Scrophularia (310) chiefly temperate Eurasia.
   Penstemon (250) chiefly N. America.
   Paulownia (17) E. Asia.
7. Manuleae (corolla as in tribe 6, but inflorescence not cymose, usually simple; anthers finally 1-locular).
   Sutera (130) tropical and S. Africa, Canary Islands.
   Zaluzianskya (36) chiefly S. Africa.
8. Gratioleae (as tribe 7, but anthers finally 2-locular).
   Mimulus (100) cosmopolitan, especially America.
9. Selagineae (fruit a drupe or indehiscent, few-seeded capsule).
   Selago (150) tropical and S. Africa.

B. The 2 posterior teeth (or upper lip) of the corolla are covered in bud by one or both of the lateral teeth).

III. Rhinanthoideae

10. Digitalideae (2 upper corolla-lobes often erect; not parasitic).
    Veronica (300) chiefly N. temperate region.
    Digitalis (20–30) Europe, Mediterranean region, Canary Islands.
11. Gerardieae (corolla-lobes all flat, divergent; often parasitic).
    Gerardia (60) America, W. Indies.
12. Rhinantheae (2 upper corolla-teeth form a helmet-like upper lip; often parasitic).
    Pedicularis (500) N. hemisphere.
    Euphrasia (200) chiefly N. temperate region.
    Castilleja (200) chiefly N. and S. America.
    Rhinanthus (50) temperate Eurasia, N. America.
    Melampyrum (35) N. temperate region.

ANTIRRHINUM MAJUS L.
Snapdragon

Distribution. Native of S.W. Europe eastwards to Sicily, and often grown in gardens from which it sometimes escapes and becomes naturalised, mainly on old walls.

Vegetative characteristics. A perennial, sometimes treated as an annual in gardens, reaching between 30 and 80 cm in height in the British Isles. The erect or straggling herbaceous stem becomes woody near the base. The entire leaves, opposite below and alternate above, are lanceolate or linear-lanceolate and cuneate at the base.

Floral formula. K(5) C(5) A4 $\underline{G}$(2)

Flower and inflorescence. The hermaphrodite, zygomorphic flowers form a terminal raceme, each individual flower arising from the axil of a short, sessile bract. The flowering period is from July to September. The corolla of the wild types is usually pink or purple with a yellow palate. The numerous cultivars grown in gardens have colours ranging from pure white to yellow, orange, pink, red and deep purple.

Pollination. A conspicuous corolla, and the presence of pollen and nectar, are attributes of entomophilous flowers, but in *Antirrhinum* and certain related genera the corolla is so designed that only long-tongued bees can penetrate the flower. In fact the mouth of the corolla is completely closed by the upper and lower lips and is termed 'personate'. The flowers of *A. majus* are homogamous. The lower lip of the corolla has a palate with 2 swellings which fit accurately into 2 depressions in the upper lip. The 2 lips can only be prised open by large, strong insects such as bumble-bees. On arriving at a flower, the bee alights on the lower lip, and as it forces its way into the corolla it touches the anthers, which lie close to the upper lip (see Fig. 74.1.B) and the pollen, aggregated into 2 rounded masses, adheres to the insect's back. The nectar, which is secreted by the base of the ovary, collects round the bases of the filaments and does not normally flow into the short, broad spur or pouch situated at the base of the corolla-tube. As the bee probes for nectar, its proboscis is guided by the stiff hairs which form lines along the lower lip of the corolla. When the insect leaves, it carries with it the pollen masses, which are removed by the receptive stigma of another flower.

Alternative flowers for study. A. majus and its numerous cultivars are commonly available and no alternative is necessary. Comparison should be made, not only with the representatives of the other subfamilies described here, but if possible with genera from other tribes of the Scrophularioideae.

Fig. 74.1 Scrophulariaceae, *Antirrhinum majus*

A The hermaphrodite, zygomorphic flower with personate corolla (see section on Pollination) produced abaxially at the base into a short pouch. The flower is borne on a short pedicel which arises from the axil of a bract. The calyx is deeply 5-lobed and considerably shorter than the corolla.
Corolla: 35–37 mm Corolla-tube: *c.* 20 × 12 mm
Calyx-lobes: 6–7 × 5–6 mm

B L.S. of the flower showing the essential organs. The corolla is prominently 2-lipped, the upper lip 2-lobed and the lower 3-lobed. Stiff, capitate hairs are arranged in interrupted lines on the anterior inner surface of the corolla-tube. These hairs extend to the base of the filaments and act as nectar-guides. A small pouch (appearing wrinkled due to sectioning) is situated on the abaxial side of the superior ovary. The ovary terminates in a long, curved style with a capitate stigma. Two of the 4 stamens which arise from under the base of the ovary can be seen.
Ovary (at anthesis): 3.5 mm Style + stigma: 17–18 mm

C L.S. of the flower, cut horizontally to show the 2-lobed upper lip behind the 4 stamens. The stamens are in 2 pairs of different lengths. The anthers of the 2 longer stamens are fully developed while those of the shorter stamens are less mature. The stiff, capitate hairs can be seen at the base of the filaments and on the calyx.
Longer stamens: 22 and 24 mm Shorter stamens: 14 and 18 mm

D Detail of the upper portion of a stamen prior to dehiscence. D1: anterior view. D2: posterior view. The anthers are introrse, 2-celled and dehisce longitudinally.
Anther: 3 × 3 mm

E L.S. of the lower part of the flower. The ovary protrudes abaxially into the basally saccate portion of the corolla. Nectar is secreted at the base of the ovary but does not usually flow into the pouch. The basal part of 2 of the filaments can be seen with the stiff, capitate hairs. Hairs are also present on the ovary and the style. Numerous anatropous ovules are attached to the axile placenta.
Ovary: 3.5 × 4 mm

F T.S. of ovary composed of 2 united carpels showing the 2 loculi and the ovules attached to the well-developed placenta.
T.S. of ovary: 4 × 2 mm

G Detail of the capitate stigma.
Stigma: 1.5 mm in diameter

H The fruit after dehiscence, showing the persistent calyx and the withered style. The 2-celled capsule opens by 3 small pores. The adaxial cell is longer, narrower above and opens by a single apical pore. The abaxial cell is shorter, wider above and opens by 2 apical pores. The pores are covered by small teeth that curl back at dehiscence to liberate the numerous small seeds.
Fruit: 13–14 × 7–8 mm

Fig. 74.1

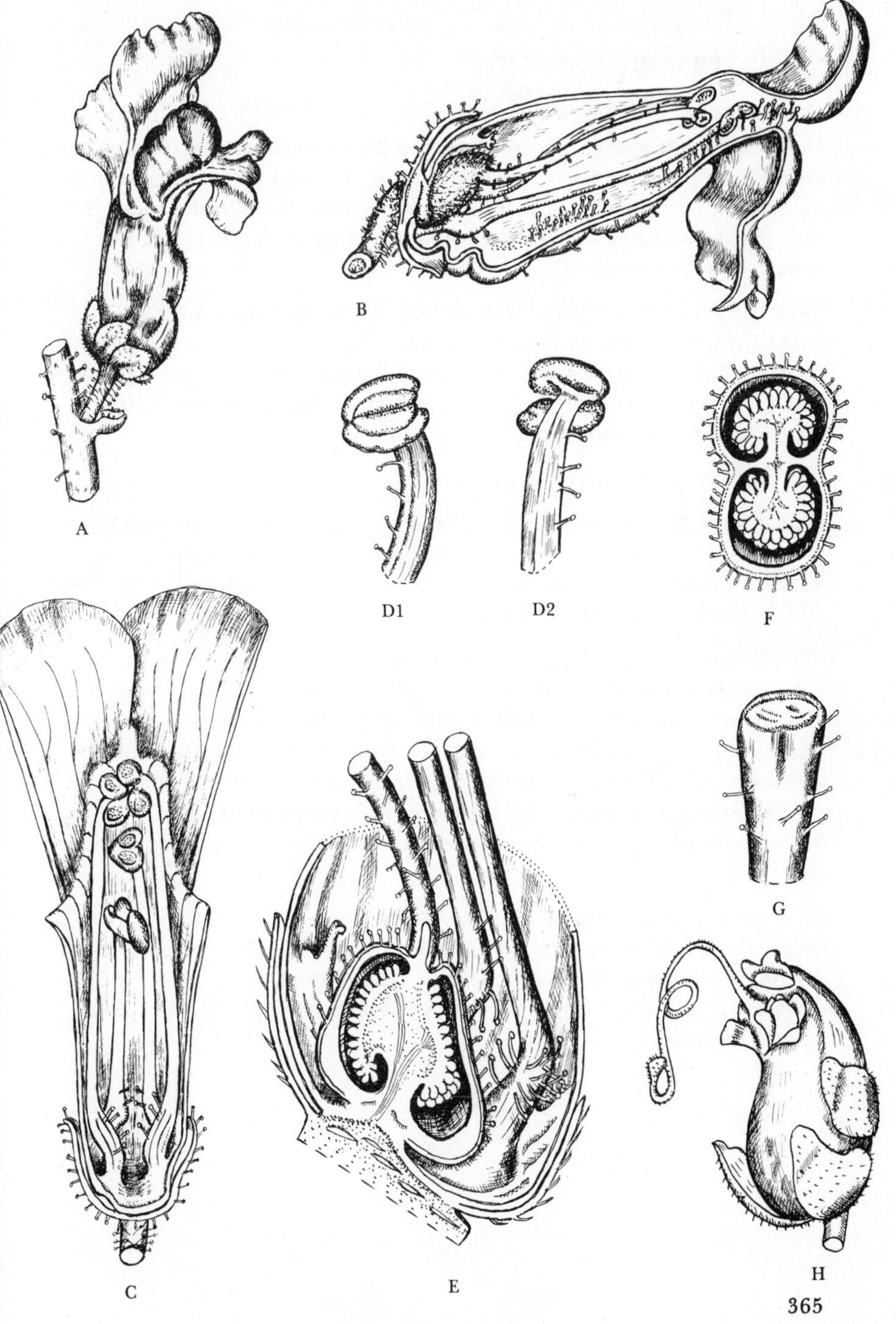

VERBASCUM THAPSUS L.
Aaron's Rod, Great Mullein

Distribution. Native in Asia as far south as the Caucasus and the Himalaya, and eastwards to W. China, also found wild in most of Europe, though absent from the extreme north and much of the Balkan Peninsula. In the British Isles it is widespread and locally frequent on dry, grassy banks and waste places. It is naturalised in N. America.

Vegetative characteristics. An erect biennial, 30–200 cm in height, with stem and leaves densely clothed with soft, whitish hairs. The basal leaves form a rosette and the cauline leaves are alternate. All the leaves are elliptical to obovate-oblong, obtuse, entire or finely crenate, and the cauline leaves have the base decurrent as wings on to the round stem.

Floral formula. K(5) C(5) A5 $\underline{G}$(2)

Flower and inflorescence. The inflorescence usually consists of a dense, terminal raceme, and axillary racemes are rarely present. The numerous yellow, weakly zygomorphic flowers appear from June to August on very short pedicels arising from the axils of softly hairy bracts.

Pollination. The conspicuous yellow flowers forming the long raceme remain open even in wet weather, allowing bees and other insects to collect pollen without difficulty. No nectar is produced. The pollen-collecting insect alights on the lowest corolla-lobe, which is larger than the others. The filaments are clothed with long hairs and these enable the insect to maintain a firm hold while it collects pollen, at the same time transferring pollen from another flower on to the receptive stigma. In the absence of cross-pollination, self-pollination can take place.

Alternative flowers for study. There are a number of minor floral differences between species of *Verbascum* but any other member of the genus may be used if *V. thapsus* is not available. In *V. nigrum*, Dark Mullein, the hairs on the filaments are purple, and in the *V. daenzeri* group from the Balkans only 4 fertile stamens are present.

Fig. 74.2 Scrophulariaceae, *Verbascum thapsus*

A The flower in bud. The calyx is covered in soft, whitish hairs and has 5 more or less equal lobes. Hairs are also present on the outer surface of the corolla.
Flower-bud: 6 mm Calyx-lobes: *c.* 3 × 1.5 mm

B L.S. of the hermaphrodite, weakly zygomorphic flower. The anterior (lower) lobe of the corolla is slightly larger than the others. Three of the 5 stamens are shown, the filaments of the 3 upper (2 shown) are always hairy, while those of the 2 lower (1 shown) may be with or without hairs. The superior ovary is terminated by a long style with a capitate stigma. Numerous anatropous ovules are attached to the axile placentas.
Stamens: *c.* 7 mm Ovary: 2 × 1.5–2 mm
Style + stigma: 10 mm

C Detail of the anther from a posterior (upper stamen). The anther is reniform and is attached at its centre to the filament. The anterior (lower) stamens may be similar to the posterior stamens, or the anther may lie against and appear to recline on the apex of the filament.
Anther: 1.75 × 1 mm

D Detail of the capitate stigma, showing papillae.

E T.S. of ovary. The ovary is formed from 2 fused carpels and has 2 loculi and axile placentation. The fruit develops into a 2-celled, septicidal capsule containing numerous seeds.
Ovary (at anthesis): 1.5–2 mm in diameter

Fig. 74.2

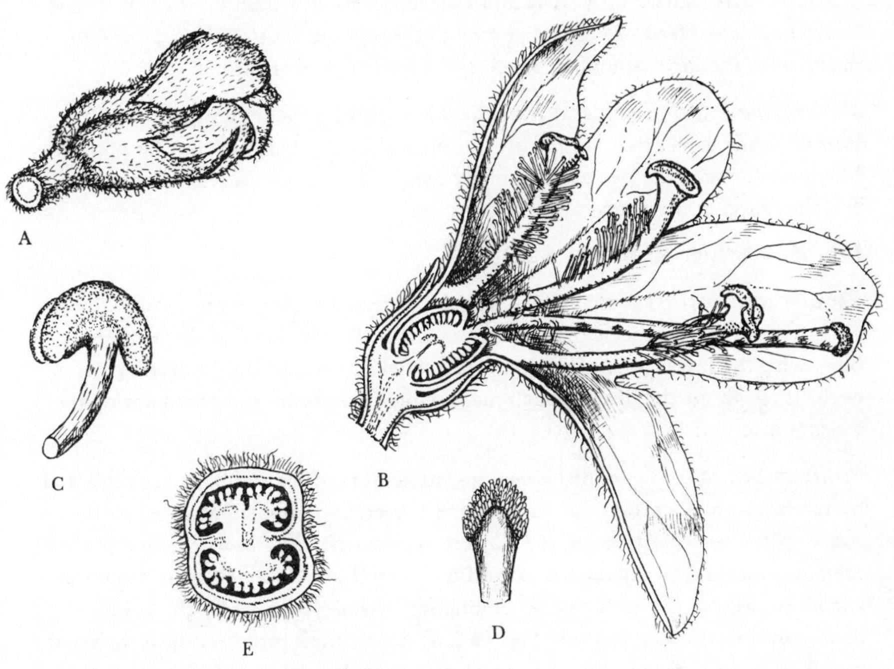

VERONICA PERSICA Poir.
Common Field Speedwell

Distribution. Native in W. Asia and first recorded in Britain in 1825, it is now a very common weed of cultivated ground throughout most of Europe and in many areas the most abundant species of *Veronica* in this habitat.

Vegetative characteristics. An annual herb, with procumbent, pubescent stems bearing leaves that are mostly alternate though the lower may be opposite. The leaves are shortly petiolate, broadly ovate with a more or less truncate base and the margins are coarsely crenate-serrate.

Floral formula. K(4) C(4) A2 $\underline{G}$(2)

Flower and inflorescence. The solitary, hermaphrodite, zygomorphic flowers are borne on long pedicels that arise from the leaf-axils. The corolla is bright blue, with the lower lobe often somewhat paler or even white. Flowering can occur throughout the year, though many more flowers are produced during the warmer months.

Pollination. Many types of insects, including hover-flies and bees, are attracted by the bright blue corolla and the nectar secreted by a disc situated below the ovary. If the weather is warm the flowers open in the early morning, and as they are homogamous, the filaments extend to a lateral position well away from the stigma to prevent self-pollination. Pollinating insects, directed by the guide-marks on the corolla-lobes (see Fig. 74.3.B), insert their proboscis into the short corolla-tube in order to seek out the nectar which is protected from rain by the short hairs near its base. On arriving at a flower the insect clings to the stamens and becomes dusted with pollen. At the same time its underside, which may be covered with pollen from another flower, brushes against the stigma. In this way cross-pollination is achieved. On damp mornings the stamens do not spread out so far, and as the anthers remain closer to the stigma they may deposit their pollen on to it, effecting self-pollination.

Alternative flowers for study. In the absence of *V. persica* any other species of *Veronica* would be suitable, or any member of the closely related genus *Hebe*, which embraces all the shrubby Veronicas. In *V. spicata*, Spiked Speedwell, and related garden plants, the inflorescence is a terminal raceme, and in *V. chamaedrys*, Germander Speedwell, the racemes are axillary.

Fig. 74.3 Scrophulariaceae, *Veronica persica*

A The flower in bud. The calyx has a short tube and is deeply divided above into 4 unequal, ciliate lobes.
Calyx-lobes (at anthesis): *c.* 4 × 2.5 mm

B Anterior view of the flower. The corolla has a very short tube with 4 unequal lobes, the upper lobe being the largest and the lower lobe the smallest. Lines, darker in colour than the rest of the corolla, converge towards the centre of the flower and act as nectar-guides.
Corolla: 7.5 mm in diameter
Upper lobe: 3 × 4.5 mm
Lower lobe: 3 × 1.5 mm
Lateral lobes: 2 × 4 mm

C L.S. of the corolla, showing the ring of nectar-protecting hairs at the base of the tube. One of the 2 stamens is shown. The anthers are introrse, 2-celled and dehisce longitudinally. The filaments narrow considerably towards the base.
Filament: 3 mm   Anther: 1 mm

D L.S. of the ovary after anthesis, subtended by the calyx. The ovary is hairy, and is flattened at right angles to the septum. Several anatropous ovules are attached to the axile placentas.
Ovary: 2.5 × 3.75 mm   Style + stigma: 2 mm

E Detail of the upper portion of the style and the stigmatic papillae.
Stigma: 0.2 × 0.3 mm

F T.S. of ovary. The ovary is formed by the fusion of 2 carpels and has 2 loculi and axile placentation. The fruit develops into a flattened, 2-celled, loculicidal capsule. The seeds are broadly elliptical, concave on one face.

Fig. 74.3

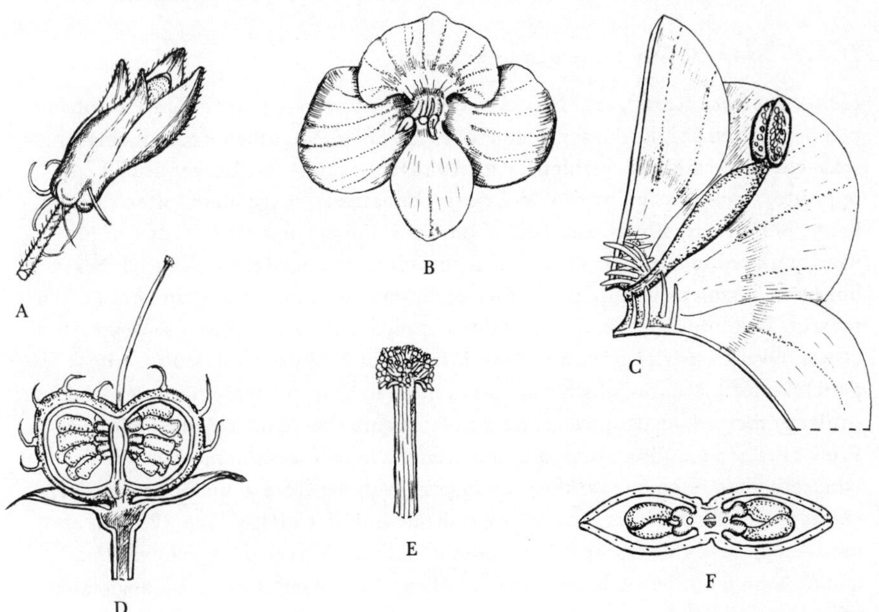

75 Gesneriaceae Dum.

Gloxinia or Gesneria family
120 genera and 2000 species

Distribution. Mostly tropics and subtropics.

General characteristics. Herbaceous or slightly woody plants, rarely shrubs or climbers, often epiphytic; some tuberous e.g., *Sinningia*, others reproducing by scale-covered, catkin-like stolons, e.g., *Kohleria*, *Achimenes*. Leaves usually opposite, simple, entire or toothed, rarely pinnatisect, exstipulate, often densely hairy. Flowers usually cymose, hermaphrodite and zygomorphic. Calyx of 5 usually united sepals. Corolla of usually 5 united petals, often 2-lipped, lobes imbricate. Stamens usually either 2 or 4, didynamous, rarely 5; staminodes often present. Disc annular, cupular, one-sided or represented by up to 5 separate glands. Ovary superior or more or less inferior, of 2 carpels, unilocular, with 2 parietal bifid placentas which sometimes meet so that the ovary becomes 2-locular; ovules numerous, anatropous; style simple; stigma simple or variously bilobed. Fruit usually a loculidical capsule, sometimes a berry; seeds numerous, small, with reticulate or spiral markings; endosperm varying from abundant to absent; embryo straight. Flowers mostly protandrous and often fairly large (25 mm or more long); many probably insect-pollinated, but others (e.g., *Aeschynanthus* and *Columnea*) have the large, curved, red corolla and copious nectar associated with bird-pollination.

Economic and ornamental plants. Plants of economic importance are absent from the Gesneriaceae, but a number of genera are of horticultural value, particularly as house plants. These include *Saintpaulia*, African Violet, *Achimenes*, *Streptocarpus*, *Episcia*, *Columnea*, *Hypocyrta* and *Sinningia*, the latter usually referred to commercially as *Gloxinia*, but which botanically is a separate genus. The small genera *Ramonda* and *Haberlea*, both from S. European mountains, are suitable for cultivation in the rock-garden or alpine house.

Classification. The family Gesneriaceae is divided geographically into 2 sub-families and subsequently into numerous tribes. It is closely allied to the Scrophulariaceae and Bignoniaceae and is separated from those families with difficulty.

I. Cyrtandroideae (almost entirely confined to the Old World; ovary superior).
   - *Cyrtandra* (350) Malaysia, Polynesia.
   - *Streptocarpus* (132) tropical and S. Africa, Madagascar.
   - *Aeschynanthus* (80) Indomalaysia, China.
   - *Chirita* (80) Indomalaysia, S. China.
   - *Saintpaulia* (20) tropical E. Africa.

II. Gesnerioideae (confined to the New World and S.E. Australasia).
(ovary superior).
Columnea (200) tropical America.
Episcia (40) tropical America, W. Indies.
(ovary more or less inferior).
Achimenes (50) tropical America.
Kohleria (50) tropical America.
Gesneria (50) tropical America, W. Indies.

SAINTPAULIA IONANTHA Wendl.
African Violet

Distribution. Native in shady, wooded mountain regions of tropical E. Africa and introduced into the British Isles in 1893. It is now widely cultivated as a decorative, indoor pot plant and is readily propagated by seed or leaf-cuttings in warm conditions.

Vegetative characteristics. A stemless, herbaceous perennial, with long-petiolate, basal leaves which are orbicular to oblong-ovate in shape, densely hairy on both surfaces and often purplish beneath.

Floral formula. K(5) C(5) A2 $\underline{G}$(2)

Flower and inflorescence. One to six hermaphrodite, slightly zygomorphic flowers with conspicuous yellow anthers are borne in cymes on hairy peduncles (see Fig. 75.1.A). The species and its cultivars flower more or less continuously throughout the year, their colour ranging from the typical violet-blue, through maroon and pink to pure white.

Pollination. Wild plants are mainly bird-pollinated, pollen being shed as the bird inserts its beak into the corolla-tube. Later on, the style bends over in front of the tube so that pollen from another flower is transferred to the receptive stigma.

Alternative flowers for study. S. ionantha is readily available and there is little need for an alternative. The colour of the flower is of little consequence here, but it should be noted that some cultivars have semi-double flowers. The genus *Saintpaulia* differs from most other genera in the family in having a corolla-tube considerably shorter than the lobes. However, the small genus *Ramonda*, native in the mountains of S. Europe, also shares this character, but has 4 or 5 stamens. The popular house plants *Achimenes* and *Sinningia* have the typical 4 stamens but exhibit a half-inferior ovary. *S. speciosa* is the plant commonly known as Gloxinia.

Fig. 75.1 Gesneriaceae, *Saintpaulia ionantha*

A A long-peduncled cyme, bearing 5 hermaphrodite flowers. Two of the flowers are in bud, each showing 3 of the 5 hairy calyx-lobes which are united only at the base. The mature flowers can be seen to be slightly zygomorphic, the short corolla-tube opening out into a bilobed upper lip and a 3-lobed lower lip. The somewhat unequal corolla-lobes are elliptical and imbricate.
Peduncle: 45 mm Pedicels: *c.* 24 mm
Flower-bud: 6 × 6 mm Calyx-lobes: 4.75 × 2 mm
Corolla: 36–38 × 30–32 mm

B L.S. of the calyx and corolla exposing the essential organs. The zygomorphic nature of the flower is demonstrated by the corolla, the lower lip being somewhat longer than the upper. The rather fleshy, short corolla-tube is subtended by the deeply-divided, hairy calyx. The 2 stamens, which are borne on the corolla-tube, have 2-celled anthers which dehisce longitudinally. The anther-cells are confluent at the apex. The hairy, superior ovary tapers into a single style which may be bent to left or right. The stigma is shown here at an early stage of development, but when mature it is bilobed and papillose. The disc, which forms a ring round the base of the ovary, is not visible in the drawing.
Filaments: 3.5–4 mm Anther: 3 × 2 mm
Style + stigma: 6 mm Ovary: 2 mm

C One of the papillae which are found inside the corolla-tube. These structures, usually 2 or 3 in number, are regarded as rudimentary stamens (staminodes).
Staminodes: up to 1.5 mm

D T.S. of the unilocular ovary formed from 2 fused carpels. The parietal placentas intrude into the loculus and their margins, which bear numerous anatropous ovules, curl back towards the ovary-wall. In cultivation the flowers have to be pollinated by hand in order to produce seed, but in nature the flowers are pollinated in the normal way and the ovary develops into an oblong, 2-valved, loculicidal capsule.
Ovary (at anthesis): 2 mm in diameter

Fig. 75.1

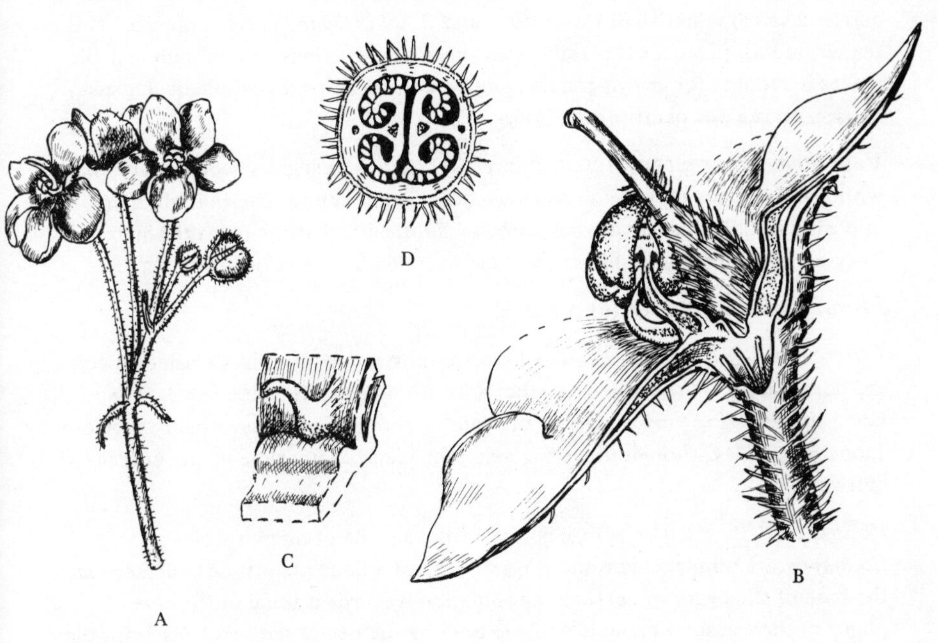

COLUMNEA × BANKSII Lynch

Distribution. An artificial hybrid which was raised by G.H. Banks in 1918 at the University Botanic Garden, Cambridge. It is the result of a cross between *C. oerstediana* (♂), a native of Costa Rica, and *C. schiedeana* (♀) from Mexico. This tender, indoor plant roots readily from short stem-cuttings and its trailing habit makes it suitable for growing in hanging baskets or on peat-clad pillars in glass-houses, or as a pot plant for the house.

Vegetative characteristics. A fleshy, herbaceous or slightly woody perennial with trailing stems that are more or less hairy when young. The sparsely hairy, opposite, petiolate leaves are more or less elliptical and are acute at the apex. They are dark green and shiny above and have reddish veins beneath.

Floral formula. K(5) C(5) A4 $\underline{G}$(2)

Flower and inflorescence. The hermaphrodite, zygomorphic, scarlet flowers are borne singly on short pedicels that arise from the axils of the leaves. They occur in increasing profusion near the ends of the trailing stems, usually between January and May, though flowering may take place at any time of the year in cultivation.

Pollination. The red or orange colour of the corolla of many species of *Columnea*, its tubular shape and the secretion of a large quantity of nectar near the base of the ovary indicate that pollination by birds is usual in the case of plants in their natural habitat. While extracting the nectar with its beak from the base of the corolla, the bird's head becomes dusted with pollen from the block of anthers borne at the ends of the curved filaments and protected from rain by the hooded upper lip. The flowers of *Columnea* are protandrous, and as the stamens wither, the filaments draw the anthers away from the stigma, partly by swinging in a vertical arc but also by a coiling movement. In this way self-pollination is avoided.

Alternative flowers for study. The genus *Columnea* contains several species that would be suitable alternatives to *C.* × *banksii*, e.g., *C. gloriosa*, a commonly grown, winter-flowering species, and *C. glabra*, an erect shrub that flowers in April.

Fig. 75.2 Gesneriaceae, *Columnea* × *banksii*

A A zygomorphic, hermaphrodite flower removed from the axil of a leaf. The corolla-tube is uniformly scarlet, but the 4-lobed upper lip and the single-lobed lower lip have indistinct yellow veins. The 2 lateral lobes of the upper lip and the single lobe comprising the lower lip are reflexed. The corolla is hairy on both the inner and outer surfaces, and is subtended by a hairy, deeply-divided calyx whose lobes broaden considerably towards the base.
Corolla: 78 mm Upper lip: 36 mm
Lower lip: 16–17 mm
Corolla-tube (at middle): 4–5 mm in diameter
Calyx-lobe: 15 mm

B The rectangular block of 4 stamens which are coherent by their anthers. The anthers are dorsifixed, 2-celled and dehisce by longitudinal slits to liberate the pollen.
Block of anthers: *c.* 5 × 4 mm Individual anthers: 2–2.5 × 1.5–2 mm
Filaments: *c.* 65 mm

C L.S. of the basal part of a flower. The corolla is strongly pouched on the upper side at the base and readily accommodates the ovary and the dorsal nectary. However, the pouch is larger than is strictly necessary for this purpose and it is likely that its chief task is to contribute to the structural stability of the long, narrow corolla. The 4 stamens, 2 of which are visible, are attached to the base of the corolla-tube. Part of the narrow disc can be seen near the nectary at the base of the ovary. The ovary is terminated by a long style which is about as thin as the filaments.

D L.S. of the unilocular, superior ovary, formed from 2 fused carpels, which contains numerous anatropous ovules within the single loculus. The nectary can be seen lying close to the outer wall of the ovary and part of the hairy calyx is also visible. The corolla has been removed for clarity. In suitable conditions the ovary develops, after fertilisation, into a white, depressed-globose berry about 10 mm in diameter.
Ovary: 2.5 × 1.75 mm

E T.S. of the ovary, showing the parietal placentas which intrude into the single loculus. Ovules at various stages of development are attached to the well-developed placental tissue.
T.S. of ovary: 1.75 × 1.5 mm

F Detail of the bilobed stigma that terminates the solitary style. The lower part of the style is glabrous, but the upper portion bears some short hairs.
Style + stigma: 75 mm Stigma-lobe: 2.25 mm

Fig. 75.2

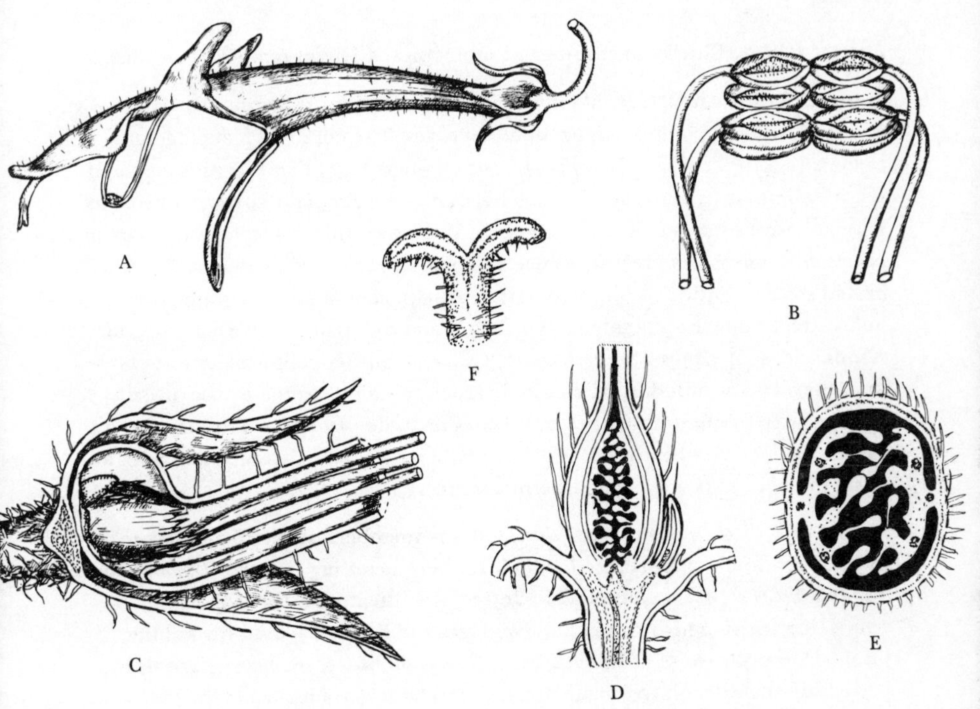

76 Bignoniaceae Juss.

Bignonia family

120 genera and 650 species

Distribution. Chiefly in tropics and subtropics, a few in temperate regions.

General characteristics. Mainly trees and shrubs, most commonly lianes, with opposite, usually compound, exstipulate leaves. The climbing plants include twiners (e.g., *Tecomaria*, *Pandorea*), root-climbers (e.g., *Campsis radicans*) and tendril-climbers (most genera); the tendrils may be simple, hooked, or provided with adhesive discs (cf. *Parthenocissus* in Vitaceae). Inflorescence usually cymose; bracts and bracteoles present. Flowers hermaphrodite, zygomorphic. Calyx of 5 united sepals. Corolla of 5 united petals, usually campanulate or funnelform, imbricate. Stamens epipetalous, typically 4 and didynamous, with a posterior staminode, sometimes 2 stamens (e.g., *Catalpa*); anther-cells usually one above the other. Ovary superior, of 2 united carpels, on a hypogynous disc, usually 2-locular, with numerous, anatropous ovules on axile placentas. Fruit usually a capsule, sometimes fleshy and indehiscent; seeds usually flattened and with a large, membranous wing. Endosperm absent.

Economic and ornamental plants. Some members of the Bignoniaceae are of local economic importance for their timber, including *Catalpa*, *Tecoma* and *Tabebuia*. *Crescentia cujete*, Calabash-tree, is cultivated in the tropics for its gourd-like fruits, which, when hollowed out and dried, are used for holding liquids. Several species of *Catalpa* (e.g., *C. bignonioides*, see below) are also prized for their decorative value in parks and large gardens, while the genus *Jacaranda* is widely grown in the warmer regions as a street-tree. Species of *Campsis*, Trumpet Vine, and *Eccremocarpus scaber* are sometimes cultivated as ornamental climbers in suitably sheltered places, and the herbaceous genus *Incarvillea* is occasionally represented in gardens as a border plant.

Classification. The classification into 4 tribes is based on that of Schumann (in Ref. 9).

1. Bignonieae (ovary 2-locular, compressed parallel to septum, or cylindrical; capsule septicidal, with winged seeds; tendrils usually present).
   Arrabidaea (70) tropical America.
   Adenocalymma (40) tropical America.
   Anemopaegma (30) tropical America.
2. Tecomeae (ovary 2-locular, compressed at right angles to septum, or cylindrical; capsule loculicidal, with winged seeds; tendrils rarely present).
   Tabebuia (100) Mexico to N. Argentina, W. Indies.

Jacaranda (50) Central and S. America, W. Indies.
Tecoma (16) Florida, W. Indies, Mexico to Argentina.
Incarvillea (11–14) central and E. Asia, Himalaya.
Catalpa (11) E. Asia, America, W. Indies.

3. Eccremocarpeae (ovary 1-locular; capsule splits from below up; seeds winged; tendrils present).
Eccremocarpus (5) western S. America.
4. Crescentieae (ovary 1- or 2-locular; fruit a berry or dry and indehiscent; seeds not winged; plant usually erect).
Phyllarthron (13) Madagascar, Comoro Islands.
Parmentiera (8) Mexico to Colombia.
Crescentia (5) tropical America.

CATALPA BIGNONIOIDES Walt.
Indian Bean-tree, Southern Catalpa

Distribution. Native in the S.E. United States and naturalised in adjoining parts of that country. It was introduced into the British Isles in 1726 and is often grown as an ornamental tree in parks and large gardens. It prefers a sunny, sheltered position on deep, moist loam.

Vegetative characteristics. A deciduous, wide-spreading tree, reaching a height of 12–18 m. The leaves, usually in whorls of 3 but sometimes opposite on weaker shoots, are long-petiolate, glabrous above, pubescent beneath, broadly ovate, with a cordate base and abruptly acuminate apex. They are normally light green, but may be purplish at first on young trees.

Floral formula. K(2) C(5) A2 $\underline{G}$(2)

Flower and inflorescence. The broadly pyramidal panicles are borne terminally on the current year's growth. Each panicle is composed of a number of zygomorphic, white flowers with orange and purple markings, which make an attractive show in late July or early August.

Pollination. The form of the flower resembles that of *Mimulus* in the closely related family Scrophulariaceae and the pollination process is also similar. The white, scented flowers attract the larger insects, e.g., various kinds of bees, which land on the lower lip of the corolla and, following the nectar-guides of orange stripes and purple dots, proceed towards the nectar-secreting disc situated at the base of the ovary. As the insect enters the corolla-tube, it first touches the 2 spreading lobes of the stigma, depositing on their inner surface any pollen which is adhering to its back. The sensitive lobes then close together, reducing the chances of self-pollination which might be brought about by the insect's foraging movements. As the insect moves further into the flower it comes into contact with the anthers and is dusted with their pollen. When the insect backs out of the corolla, covered with pollen from that flower, it contacts only the non-receptive outer surface of the closed stigma.

Alternative flowers for study. Several species and hybrids of *Catalpa* are grown as decorative trees and any of these would be suitable alternatives to *C. bignonioides.* Both species of *Campsis* are in cultivation and *Eccremocarpus scaber* is even more readily available but has much smaller flowers. *Campsis* differs in having a 5-partite calyx and *Eccremocarpus* in having a unilocular ovary. The herbaceous genus *Incarvillea,* which has alternate leaves, is sometimes grown in gardens.

Fig. 76 Bignoniaceae, *Catalpa bignonioides*

A A single, hermaphrodite, zygomorphic flower, its campanulate corolla-tube opening out into 5 frilled lobes. Three of the lobes form the larger, lower lip, while the other 2 form the upper lip. The sepals are connate, splitting unevenly at anthesis into 2 lips.
Lips of calyx: *c.* 10 × 8 mm Corolla-tube: 14 × 10 mm
Lower lip of corolla: 14–20 × 10–16 mm
Upper lip: 10–15 × 18–20 mm

B L.S. of flower, showing the essential organs enclosed within the corolla-tube. On the lower surface of the tube is a deep groove, bordered by irregular orange stripes. These, together with the numerous purple dots covering the inner surface of the corolla-tube and the lower lobe, act as nectar-guides for insects. One of the 2 stamens can be seen attached to the base of the tube. The superior ovary tapers gradually upwards into a long, curved style terminated by a bilobed stigma. Part of the 2-lipped calyx surrounds the base of the corolla-tube.
Ovary: 4 mm Style + stigma: 13 mm
Stamens: 13–16 mm

C Part of the corolla-tube folded back to reveal 2 staminodes attached to its inner surface (in some flowers 3 staminodes are found). The 2 stamens are made up of long filaments, each bearing a 2-celled anther that dehisces longitudinally, exposing the spherical pollen grains. The 2 stripes on the lower surface of the corolla and a number of the dots can be seen.
Anther-cells: 3–3.5 × 2 mm Staminodes: 5–7 mm

D The upper portion of a stamen taken from a flower-bud so as to show the 2-celled anther before dehiscence. The 2 cells are divergent, and each splits down a longitudinal groove to liberate the pollen.
Anther-cell: 3 mm

E L.S. of the superior ovary, which has 2 loculi each containing numerous, anatropous ovules on axile placentas. At the base of the ovary is a nectar-secreting disc and surrounding this is part of the calyx, the corolla having been removed. The ovary is green, but becomes whitish as it tapers upwards into the style.
Ovary: 4 × 1–1.75 mm

F T.S. of ovary, showing the 2 loculi and the numerous ovules attached to axile placentas. The placentas change shape in the course of the ovary development.
Ovary: 1 mm in diameter

G The upper portion of the style with the 2-lobed stigma which is partially closed in the drawing but opens out when an insect visits the flower. A few pollen grains can be seen on the receptive inner surface of the stigmatic lobes.
Stigmatic lobe: 1.75 mm

H The fruit ('bean') prior to dehiscence, a capsule that is green at first but becomes dark brown and woody in late autumn. The 2 valves of the ripe capsule separate from the longitudinal septum (cf. the siliqua in the Cruciferae) and release the densely packed, winged seeds.
Fruit: 15–24 × 0.8 cm

I The seed is light in weight and has membranous wings with tufts of

Fig. 76

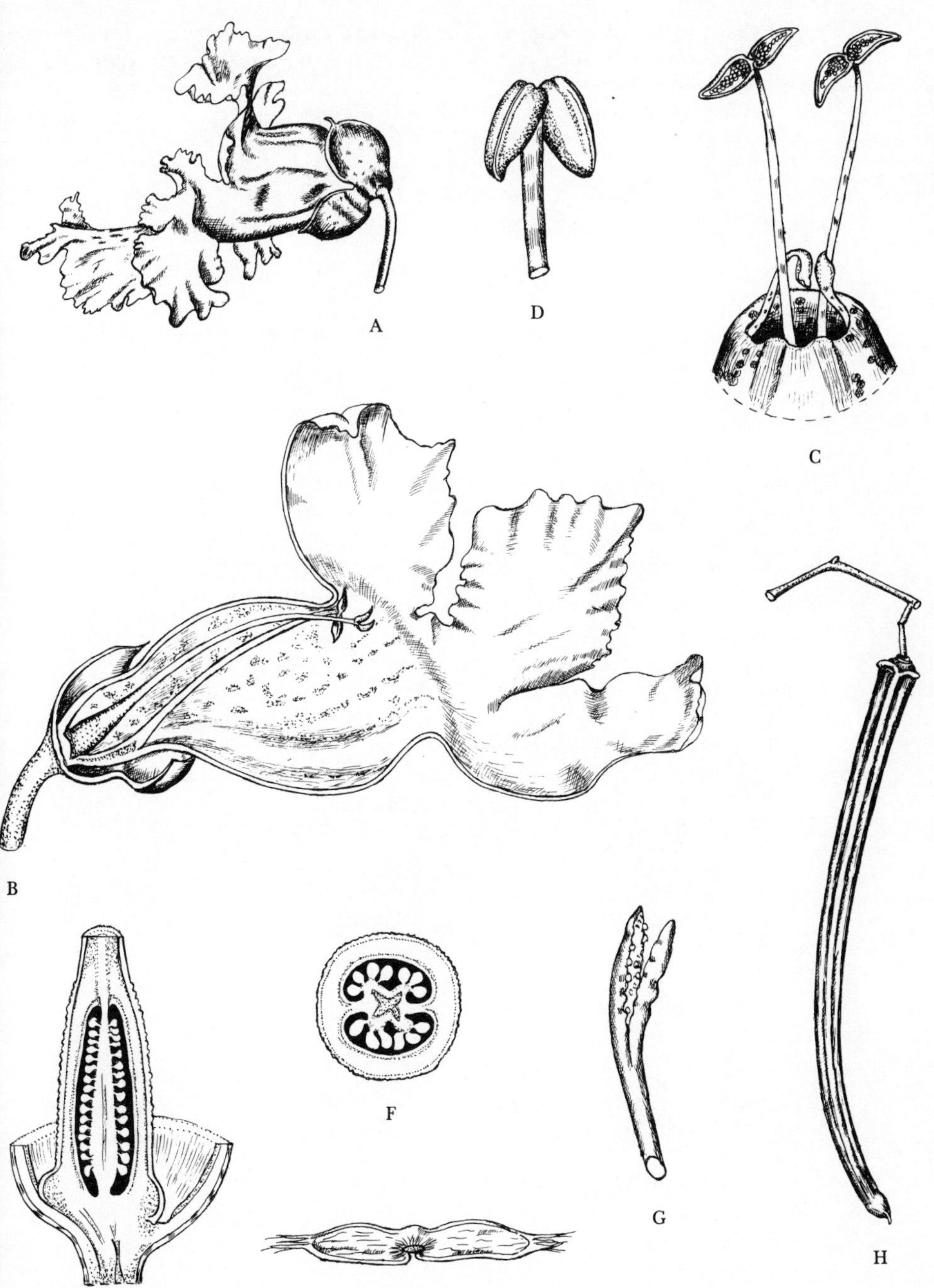

hairs at each end enabling it to be dispersed easily by the wind. The seeds germinate readily but the seedlings may need some protection during the winter.

Seeds (excluding tufts of hairs): 20–25 × 5 mm

77 Acanthaceae Juss.

Acanthus family

250 genera and 2500 species

Distribution. Mainly tropics, but also Mediterranean region, United States and Australia. The 4 chief centres of distribution are Indomalaysia, Africa, Brazil and Central America.

General characteristics. Mostly shrubs or herbs, rarely trees, some climbers, very many in damp places in tropical forests. Leaves opposite, usually decussate, entire, thin, exstipulate. Cystoliths, visible as streaks or protuberances, usually common on leaves and stems. Inflorescence most commonly a dichasial cyme, frequently condensed in the leaf-axils (cf. Labiatae), sometimes racemose, or flowers solitary. Bracts and bracteoles usually present, often coloured, the latter frequently large, more or less enclosing the flower. Flowers hermaphrodite, zygomorphic, usually with a nectar-secreting disc below the ovary. Calyx of 4 or 5 united sepals. Corolla of 4 or 5 united petals, commonly 2-lipped (upper lip sometimes not developed, e.g., *Acanthus*). Stamens rarely 5, usually 4 or 2, epipetalous, usually exserted; 1–3 staminodes frequently present; anthers often with one lobe smaller than the other, or abortive; connective often long (cf. *Salvia* in Labiatae). Ovary superior, of 2 united carpels, 2-locular, with axile placentas, each with 2 to many, usually anatropous ovules in 2 rows; style usually long, with 2 stigmas, the posterior often smaller. Fruit usually a capsule, loculicidal to the very base. In the Acanthoideae the carpels spring apart at maturity leaving a persistent central column; the seeds are thrown out largely by the aid of hook-like outgrowths from their stalks. Seeds usually without endosperm.

Economic and ornamental plants. Very few economic plants occur in the Acanthaceae. One of these is *Strobilanthes cusia* (*S. flaccidifolius*), cultivated in S.E. Asia for its leaves which are the source of a blue dye. The family is richer in ornamental plants, several species of *Acanthus* being sufficiently hardy for British gardens. Decorative house plants include *Beloperone guttata*, Shrimp Plant, *Pachystachys lutea*, *Sanchezia nobilis*, *Thunbergia alata*, and species of *Aphelandra*, *Ruellia*, *Crossandra* and *Fittonia*.

Classification. The family is divided here into 4 subfamilies. Subfamily I closely approaches the Scrophulariaceae and its genera are sometimes placed in that family. Subfamilies II and III are intermediate between the Acanthaceae and the Bignoniaceae and are sometimes considered as separate families. Subfamily IV contains the Acanthaceae *sensu stricto*.

I. Nelsonioideae (ovules many; fruit a capsule).

Staurogyne (80) tropics, especially W. Malaysia.

II. Mendoncioideae (ovules 4; fruit a drupe).

Mendoncia (60) Central and tropical S. America, tropical Africa, Madagascar.

III. Thunbergioideae (ovules 4; fruit a capsule).

Thunbergia (200) Old World tropics.

IV. Acanthoideae (ovules 2 to many; fruit a capsule).

Justicia (300) tropics and subtropics.
Strobilanthes (250) Madagascar, tropical Asia.
Barleria (230) tropics.
Aphelandra (200) warm regions of America.
Dicliptera (150) tropics and subtropics.
Blepharis (100) Old World tropics, Mediterranean region, S. Africa, Madagascar.
Beloperone (60) warm regions of America, W. Indies.
Acanthus (50) S. Europe, tropical and subtropical Asia, Africa.
Crossandra (50) tropical Africa, Madagascar, Arabia.
Eranthemum (30) tropical Asia.
Ruellia (5) tropical and subtropical America.
Fittonia (2) Peru.

ACANTHUS MOLLIS L.
Bear's Breech

Distribution. Native in the W. and central Mediterranean region, including Portugal, where it grows in cool, shady places and on roadsides. It was introduced into the British Isles in 1548, and is a handsome plant for the herbaceous border. It has become naturalised in Cornwall and the Scilly Isles.

Vegetative characteristics. A strong-growing, perennial herb with stout, erect stems reaching a height of 30–80 cm. The radical leaves are ovate in outline, pinnatifid and have long petioles. The cauline leaves are opposite, more or less ovate, spinose-dentate and almost sessile.

Floral formula. K(4) C(3) A4 $\underline{G}$(2)

Flower and inflorescence. The hermaphrodite, zygomorphic flowers are borne in dense, terminal, bracteate spikes from June until August or early September. The corolla is whitish with purple veins and is partially enclosed by the dark purple calyx. Only the 3-lobed lower lip of the corolla is present.

Pollination. The large, trilobed lower lip of the corolla forms a landing-stage for pollinating insects, mainly bumble-bees, which visit the flowers for their pollen and for the nectar which is secreted from a ring at the base of the ovary. The 4 stamens surrounding the style (see Fig. 77.D1) have strong filaments which can only be forced apart by large and powerful insects. When this occurs, pollen is shed from the anthers on to the insect's body. The flowers are protandrous, and the immature style lies close to the upper side of the flower. When the stigma becomes receptive, the style bends downwards causing pollen-carrying insects to touch it as they enter the flower and so effect cross-pollination.

Alternative flowers for study. A. mollis is the most readily available species of the genus *Acanthus* and an alternative is hardly necessary. Comparison between genera will reveal some variation in floral structure, e.g., the 5-partite calyx of *Crossandra*, the strongly 2-lipped corolla of *Aphelandra*, the presence of only 2 stamens in *Beloperone*, and a corolla of more or less equal lobes in *Thunbergia* and *Ruellia.*

Fig. 77 Acanthaceae, *Acanthus mollis*

A Lateral view of the hermaphrodite, zygomorphic flower, sessile on the flowering stem and subtended by a large, deeply-incised bract. The largest of the 4 basally connate sepals protects the rest of the flower from above. One of the 2 narrow, lateral sepals can be seen but the anterior sepal is almost entirely hidden by the bract. The scar on the stem indicates where the next flower below has been removed for clarity.
Bract: 40 × 18 mm Flower: 50 mm

B Anterior view of flower, showing the lower sepal with a bifid apex, and behind it the apex of the 3-lobed lower lip of the corolla (the upper lip is absent in *Acanthus*). The lateral sepals are shorter and much narrower.
Posterior sepal: 50 × 25–27 mm Anterior sepal: 45 × 19 mm
Lateral sepals: 35 × 3 mm Corolla: 48 × 40 mm

C L.S. of a flower exposing one of the 4 stamens and the gynoecium with its superior ovary. The posterior sepal protects the long, hairy anthers and the curved style. The inner surface of the corolla is densely hairy.
Gynoecium: 40–43 mm

D D1: the perianth has been cut away to reveal the superior ovary and the long style surrounded by the 2 pairs of stamens. The hairy anthers fit closely round the style. D2: a 1-celled anther which dehisces longitudinally, depositing its pollen on the numerous white hairs. The dorsifixed anthers are attached to strong, rigid filaments.
Filaments: 20–23 mm Anthers: 10 mm

E L.S. of one half of the ovary, showing one of the 2 loculi containing 2 anatropous ovules arranged one above the other on axile placentas.
Ovary: 5 × 5 mm Ovule: 2.25 × 1 mm

F T.S. of the 2-locular ovary, showing one ovule in each loculus. The other 2 ovules are situated immediately below them and are therefore not visible.
T.S. of ovary: 5 × 4 mm

G Detail of apex of style with its 2 stigmas.
Stigma: 1–1.25 mm

H The mature fruit, a loculicidal capsule with 2 large seeds in each of the 2 loculi. At dehiscence the 2 valves of the woody capsule spring back suddenly and eject the seeds some distance from the parent plant. The explosion is caused by the gradual hardening of the funicle which develops into a jaculator.
Fruit: 36 × 17 mm Seed: 14 × 11 mm

Fig. 77

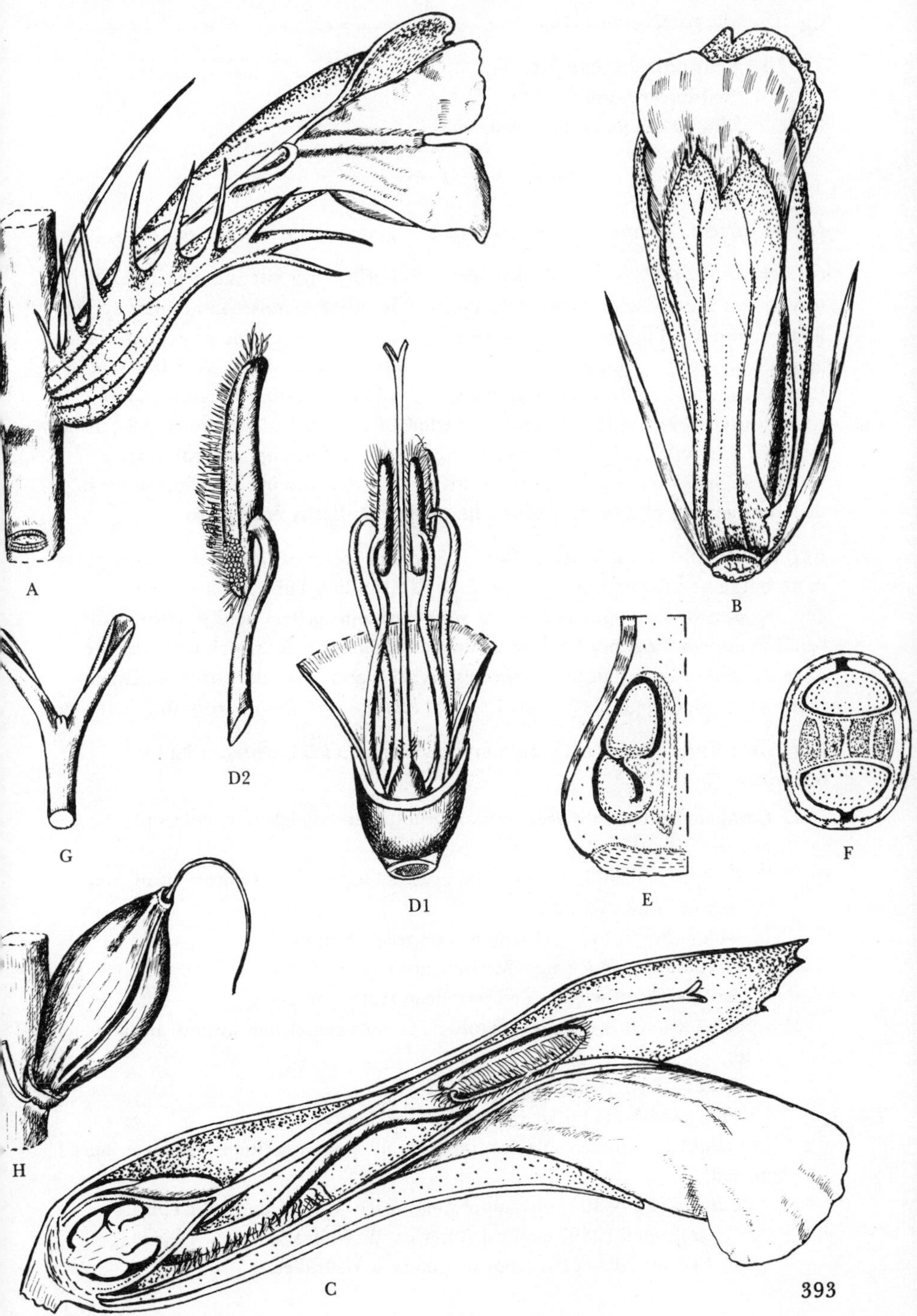

78 Campanulaceae Juss.

Bellflower family

60–70 genera and 2000 species

Distribution. Temperate and subtropical regions, and tropical mountains.

General characteristics. Mostly perennial herbs, with alternate, exstipulate leaves, and usually with latex. Inflorescence usually racemose, sometimes cymose (e.g., *Canarina*). Flowers usually hermaphrodite, actinomorphic or zygomorphic, epigynous, rarely hypogynous (e.g., *Cyananthus*). Calyx of usually 5 free sepals. Corolla of usually 5 united petals, valvate. Stamens 5; anthers introrse, sometimes united. Ovary inferior, rarely superior, of usually 2–5 united carpels, usually multilocular, with axile placentas bearing numerous anatropous ovules; style simple, stigmas as many as carpels. Fruit a capsule, dehiscing in various ways in different genera, or a berry. Seeds with abundant, fleshy endosperm.

Economic and ornamental plants. Several species of *Lobelia* have been used in herbal remedies and one of these, *L. inflata*, Indian Tobacco, has been employed in Western medicine as the source of a drug. The family is horticulturally important mainly because of the genus *Campanula*, which is represented in gardens by a large number of species, hybrids and cultivars. Other genera containing ornamental plants include *Lobelia*, *Platycodon*, *Codonopsis* and *Phyteuma*.

Classification. The classification into 3 subfamilies is based on that by Schönland (in ref. 9).

I. Campanuloideae (flowers actinomorphic, rarely slightly zygomorphic; anthers usually free).

Campanula (300) N. temperate region, especially Mediterranean, and tropical mountains.

Wahlenbergia (150) chiefly S. temperate region.

Phyteuma (40) Europe, Mediterranean region, Asia.

Jasione (20) Europe, Mediterranean region, W. Asia.

II. Cyphioideae (flowers zygomorphic; stamens sometimes united; anthers free).

Cyphia (50) Africa, especially S. Africa.

Nemacladus (10) S.W. United States, Mexico.

III. Lobelioideae (flowers zygomorphic, rarely almost actinomorphic; anthers united).

Lobelia (200–300) cosmopolitan, mostly tropics and subtropics.

Centropogon (230) tropical America, W. Indies.

Siphocampylus (215) tropical America, W. Indies.

CAMPANULA ROTUNDIFOLIA L.

Harebell (Bluebell in Scotland)

Distribution. Native throughout N. temperate regions including the British Isles. It is locally common in dry, grassy places and on fixed dunes, often on poor, shallow soils.

Vegetative characteristics. A slender perennial with ascending to erect stems reaching a height of 15–40(–60) cm. The basal leaves, which are sometimes present at anthesis, are ovate to suborbicular, crenate and long-petioled. The cauline leaves are alternate, lanceolate, petiolate and remotely serrate on the lower part of the stem, and linear, sessile and entire on the upper part.

Floral formula. K(5) C(5) A5 $\overline{\text{G}}$(3)

Flower and inflorescence. The blue (rarely white) hermaphrodite, actinomorphic, nodding, bell-shaped flowers appear solitarily or in a lax, branched panicle from July until September.

Pollination. The usual pollinators of the pendulous flowers are bees, which are sufficiently agile to reach the nectar hidden by the expanded bases of the stamens (see Fig. 78.D). The flowers are protandrous and when still in bud the anthers shed their pollen on to the hairy, upper part of the style (see Fig. 78.E1). At anthesis, the filaments have already shrivelled and the style has extended (see Fig. 78.E2). As a bee, foraging for nectar, crawls into the corolla, it picks up pollen from the style. A few days later, the 3 style-branches separate, exposing the stigmatic surface (see Fig. 78.C). An insect now visiting the flower is likely to transfer pollen from another flower on to the mature stigmas and so effect cross-pollination. In the absence of insect visits, self-pollination can take place, since, before withering, the style-branches curve back and allow the stigmas to touch any pollen remaining on the style.

Alternative flowers for study. Any other species of *Campanula* would be a suitable alternative, including the popular garden plant *C. medium*, Canterbury Bell, which flowers in May and June and is one of the group with a 5-branched style. Differences in floral structure between genera include zygomorphic flowers in *Lobelia* (placed by some authorities in a separate family, Lobeliaceae) a capitate inflorescence in *Jasione* and many species of *Phyteuma*, and a 5-celled capsule opening by apical valves in *Platycodon*.

Fig. 78 Campanulaceae, *Campanula rotundifolia*

A A flower-bud, showing the corolla subtended by a calyx with 5 linear lobes. Below the calyx is the ribbed, inferior ovary.
Bud: 12 × 3 mm

B The actinomorphic, hermaphrodite flower at anthesis, shown as erect though it is nodding on the plant. The campanulate corolla has 5 lobes that are shorter than the tube.
Corolla: 14 × 13 mm Calyx-lobe: 5 mm

C L.S. of flower, showing the stamens and style. The 5 stamens alternate with the 5 corolla-lobes and the expanded bases of the filaments form a dome over the nectar-secreting disc at the base of the style. The anthers have shrivelled and the 3 branches of the style have begun to separate in order to expose the stigmatic surface.

D A complete stamen. The anther is introrse, 2-celled and dehisces longitudinally. It is attached by its base to a short filament which is considerably enlarged at the base and edged with hairs.

E The reproductive organs in the male and female stages.
E1: The style and 2 of the 5 stamens as they appear inside the flower-bud. At this stage, the style is only as long as the stamens and the anthers are able to deposit their pollen on to the hairy upper part of the style.
Filament: 2.5 mm Anther: 5 mm
Style + stigma: 7.5 mm
E2: The style and 2 of the stamens at early anthesis. The style has lengthened, extending its pollen-covered portion well beyond the stamens. The stamens have begun to wither after shedding their pollen. (See C for next stage of style development.)
Style + stigma: 10 mm

F L.S. of the inferior ovary. Numerous anatropous ovules are borne on axile placentas. Above the ovary is the nectar-secreting disc, protected by the expanded bases of the filaments.
Loculus of ovary: 2 × 0.5 mm

G T.S. of ovary formed from 3 united carpels, showing the 3 loculi with numerous ovules in each.
Ovary: 1.75 mm in diameter

H The fruit before dehiscence. The withered corolla and calyx enclose a subglobose capsule which opens by basal pores.
Capsule: 5 × 2.5 mm

Fig. 78

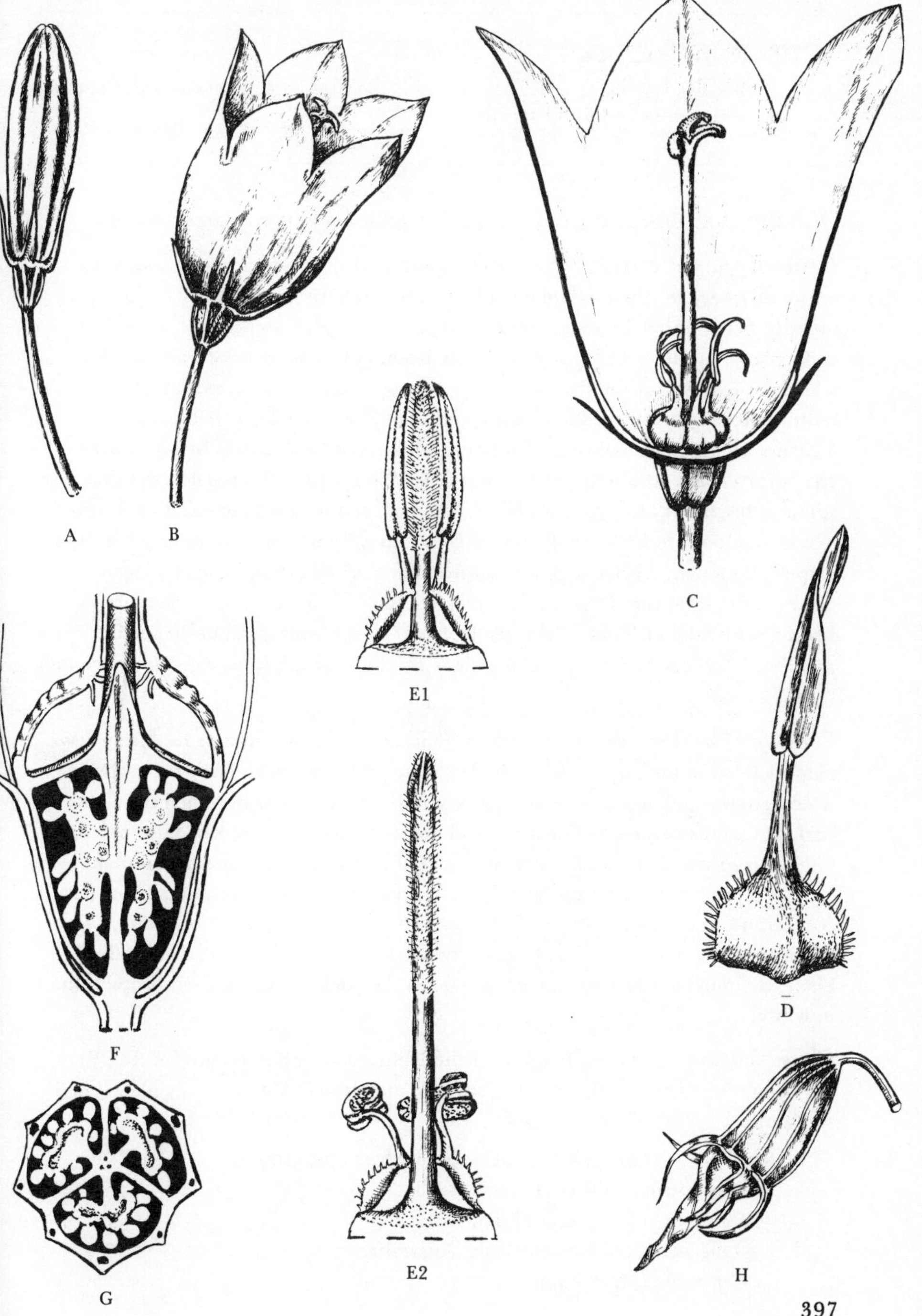

79 Rubiaceae Juss.

Madder family

500 genera and 6000 species

Distribution. Chiefly tropics, but some in temperate or even arctic regions.

General characteristics. Trees, shrubs and herbs. Leaves decussate, simple, entire or rarely toothed, stipulate. The stipules exhibit great variety of form, frequently being united to one another and to the petioles, sometimes foliaceous and as large as the ordinary leaves. In the latter case, the plants appear to have whorls of leaves, but the true leaves are distinguished by the presence of axillary buds. Inflorescence typically cymose and usually a much branched panicle. Flowers usually hermaphrodite, actinomorphic. Calyx of 4 or 5 free or united sepals, often almost absent, sometimes (e.g., *Mussaenda*) one sepal larger than the rest and brightly coloured. Corolla of 4 or 5 united petals. Stamens 4 or 5, epipetalous, alternate with corolla-lobes. Ovary usually inferior, of usually 2 united carpels, 2-locular, with 1 to many anatropous ovules in each loculus; placentation axile; style simple or bifid, stigma capitate. Epigynous disc often present. Fruit a septicidal or loculicidal capsule, berry or schizocarp. Embryo small, usually in rich endosperm. Several genera are myrmecophilous, e.g., *Myrmecodia*, *Hydnophytum*.

Economic and ornamental plants. This large family contains relatively few plants of economic importance. The best known is *Coffea arabica*, the seeds of which are roasted and ground to produce coffee. Quinine is obtained from the bark of various species of *Cinchona* and ipecacuanha from the rhizomes of *Cephaelis ipecacuanha* and *C. acuminata.* The most familiar ornamental plant is *Gardenia jasminoides*, noted for its strong scent. In Britain it requires glasshouse conditions, as do most species of *Manettia*, *Bouvardia*, *Pentas*, *Rondeletia*, *Ixora* and *Coprosma*, which are less frequently cultivated. *Nertera granadensis*, Bead Plant, is almost hardy and may be grown on the rock-garden if given protection in winter.

Classification. The classification of the Rubiaceae by Schumann (in Ref. 9) has been revised by Verdcourt (14) and the numerous tribes are now placed in 3 subfamilies.

I. Rubioideae (raphides present; seeds with endosperm).

Psychotria (700) warm regions.

Galium (400) cosmopolitan.

Asperula (200) Europe, Asia, Australia.

Cephaelis (180) tropics.

Coprosma (90) Malaysia, Australasia, Polynesia, Chile.
Hydnophytum (80) S.E. Asia.
Rubia (60) Europe, temperate Asia, Africa, Central and S. America.
Pentas (50) Africa, Madagascar.
Myrmecodia (45) Malaysia, tropical Australia, Fiji.
Nertera (12) S.E. Asia, Australasia, S. America.

II. Cinchonoideae (raphides absent; seeds with endosperm).
Ixora (400) tropics.
Gardenia (250) Old World tropics.
Mussaenda (200) Old World tropics.
Rondeletia (120) warm regions of America, W. Indies.
Bouvardia (50) tropical America.
Coffea (40) Old World tropics.
Cinchona (40) Andes.

III. Guettardoideae (raphides absent; seeds without endosperm or with traces only).
Guettarda (80) tropical America, New Caledonia.

GALIUM VERUM L.

Lady's Bedstraw

Distribution. Native in W. Asia and throughout most of Europe including the British Isles where it is abundant in grassland, hedgebanks, open woodlands and sand-dunes.

Vegetative characteristics. A perennial, stoloniferous herb with erect to decumbent 4-angled stems which are glabrous to sparsely pubescent. The leaves, arranged in whorls of 8–12, are linear, mucronate, dark green and rough above, and pale and pubescent beneath.

Floral formula. K(4) C(4) A4 $\overline{\mathrm{G}}$(2)

Flower and inflorescence. The panicles of actinomorphic, hermaphrodite flowers appear in July and August, the bright yellow corolla-lobes lying flat when the flowers are fully open.

Pollination. Many kinds of insects are attracted to the dense panicles of yellow, coumarin-scented flowers which produce small amounts of nectar from a disc situated above the ovary and surrounding the bifid style. The flowers are markedly protandrous. In the first stage, the 4 stamens bend back so far that the lower parts of the filaments lie between the lobes of the expanded corolla. The upper parts curve upwards so that the anthers may come into contact with insect visitors. In the second stage, the style-branches, so far united, separate and grow, raising the mature stigmas to the level of the anthers. Self-pollination may take place if the filaments of a flower bend over and touch the stigmas of the same flower. Also, because of the proximity of the flowers, pollen may fall from one flower to another, or the wind may act as a pollinating agent.

Alternative flowers for study. Several other widespread, native species would be suitable alternatives to *Galium verum*. *G. aparine*, Cleavers or Goosegrass, is in flower from June to August, while *G. cruciata*, Crosswort, and *G. odoratum*, Sweet Woodruff, begin flowering in May. The closely related genus *Asperula* is distinguished by a longer corolla-tube and *Rubia* by a 5-merous corolla and fleshy fruit. The only other genus found wild in Britain is monotypic and is represented by *Sherardia arvensis*, an annual which bears lilac flowers from May to October.

Fig. 79 Rubiaceae, *Galium verum*

A One of the inflorescences of numerous, hermaphrodite, actinomorphic flowers that arise from the axils of the leaves.
Inflorescence: 15 mm

B A single flower with its 4 corolla-lobes spread out wide to become more conspicuous to visiting insects. The calyx is in the form of an annulus or ring.
Flower: 3 mm in diameter Corolla-lobe: 0.5 mm wide

C The corolla removed from the flower at anthesis. The corolla consists of a short tube which opens out into 4 widely spread lobes. The 4 stamens are inserted in the throat of the tube and alternate with the lobes.
Stamen: 2.5 mm

D The stamen at an early stage showing the short filament inserted between 2 of the corolla-lobes. As the flower-bud opens, the filaments lengthen and bend outwards (see B) before the 2 style-branches separate.
Filament: 0.4 mm Anther: 0.4 mm

E The gynoecium, consisting of an inferior ovary terminated by a bifid style with a globular stigma at the end of each branch. At the base of the style is a nectar-secreting disc.
Ovary: 1.1 × 0.7 mm Style: 0.5 mm
Stigma: 0.14 mm

F L.S. of the inferior ovary formed from 2 fused carpels. The ovary is 2-locular with a single, anatropous ovule in each loculus attached to an axile placenta.

G T.S. of ovary, showing the 2 loculi each containing one ovule.

H The glabrous fruit consisting of two 1-seeded mericarps. The fruit turns black when fully ripe and the style falls away.
Fruit: 1.25 × 2 mm

Fig. 79

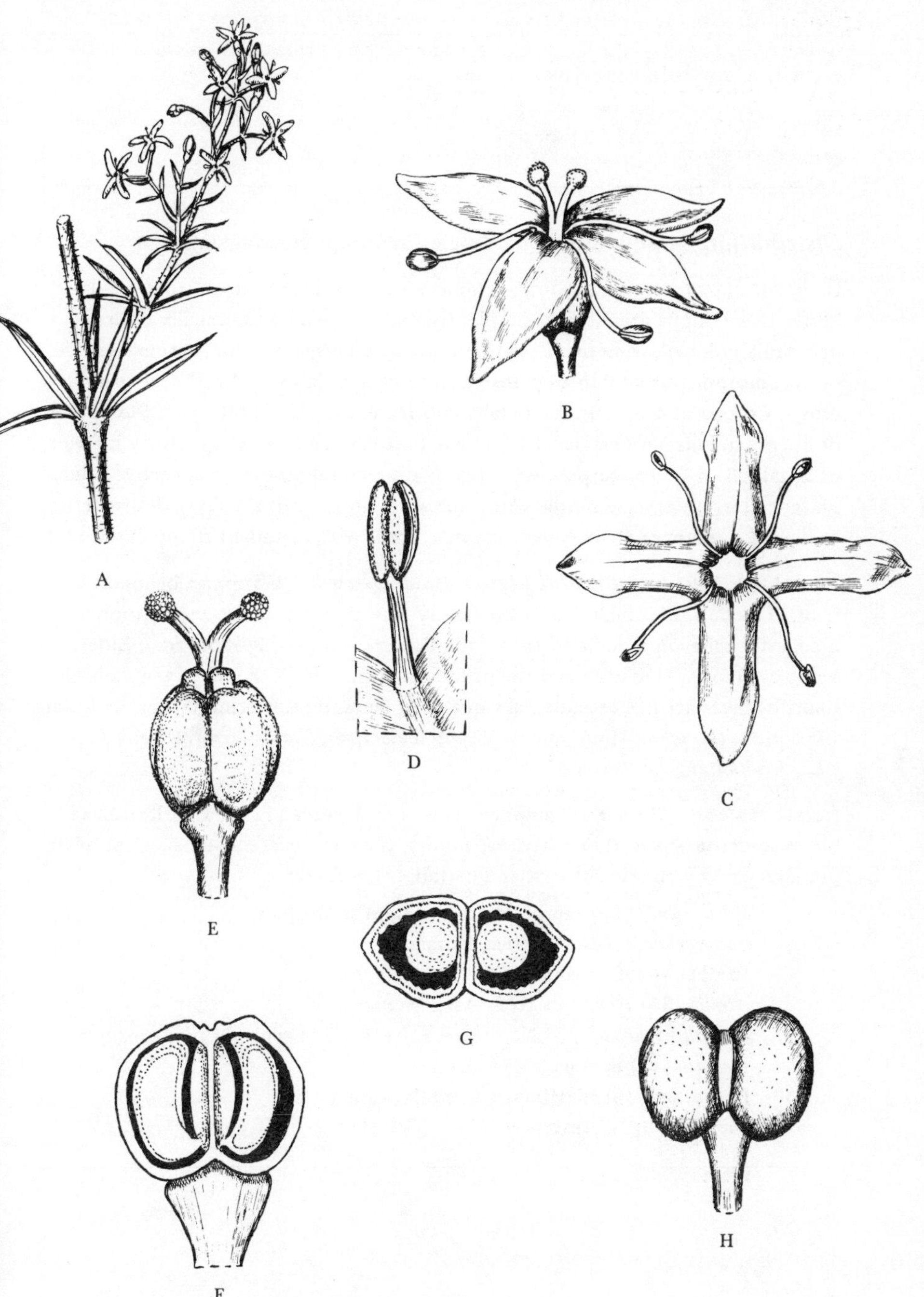

80 Caprifoliaceae Juss.

Honeysuckle family

13 genera and 530 species

Distribution. Mostly N. temperate region and tropical mountains.

General characteristics. Shrubs (sometimes climbing) or small trees, rarely herbs. Leaves opposite, usually simple, (pinnate in *Sambucus*) usually entire, occasionally lobed, sometimes stipulate. Flowers hermaphrodite, actinomorphic or zygomorphic, usually in a cymose inflorescence. Calyx of 4 or 5 free or united sepals. Corolla of 4 or 5 united petals, imbricate, sometimes bilabiate. Stamens 4 or 5, epipetalous; anthers usually introrse (extrorse in *Sambucus*). Ovary inferior, of usually 3–5 united carpels, with 1 to many pendulous ovules in each loculus; placentation axile; style simple with capitate stigma. Fruit usually a fleshy berry or drupe, sometimes an achene or capsule; seeds with abundant fleshy endosperm.

Economic and ornamental plants. Some species of *Viburnum* bear edible fruit. If *Sambucus* is included in the family, this genus may also be mentioned as being of local economic importance, the flowers and berries of *S. nigra*, Elder, being used in making wine and the pith in microscopic work. Many genera in the Caprifoliaceae are of horticultural value as ornamental flowering shrubs, including *Viburnum*, *Lonicera*, Honeysuckle, *Symphoricarpos*, Snowberry, *Weigela, Diervilla*, *Kolkwitzia*, *Leycesteria* and *Abelia.*

Classification. The genus *Sambucus* is usually included in the Caprifoliaceae, but is sometimes placed in a separate family, the Sambucaceae, on account of its pinnate leaves, extrorse anthers, and certain other features.

Viburnum (200) temperate regions and subtropics.
Lonicera (200) N. America, Eurasia.
Sambucus (40) cosmopolitan.
Abelia (30) Himalaya to E. Asia, Mexico.
Symphoricarpos (18) N. America, 1 species in central China, *S. sinensis.*
Weigela (12) E. Asia.
Leycesteria (6) W. Himalaya to S.W. China.
Diervilla (3) N. America.

VIBURNUM OPULUS L.
Guelder-rose

Distribution. Native in Europe, N. and W. Asia, and N. Africa. It is common throughout most of the British Isles in woods, hedgerows and scrub, particularly on damp soils.

Vegetative characteristics. A deciduous shrub, reaching a height of 2–4 m, with opposite, stipulate leaves which are glabrous above and sparingly pubescent beneath. The leaf-blade has 3–5 irregularly dentate lobes and usually has an attractive red tint in autumn.

Floral formula. K(5) C(5) A5 $\overline{\mathrm{G}}$(3)

Flower and inflorescence. The inflorescence consists of a rather flat-topped cluster of corymbose cymes, and the actinomorphic, white flowers appear in June and July, to be followed in September and October by conspicuous groups of bright red fruits. The small, inner flowers of the inflorescence are fertile, but the larger, marginal ones are sterile (see Fig. 80.1.A).

Pollination. *V. opulus* is a good example of differentiation of function within an inflorescence, since the small, central flowers are for normal reproductive purposes, while the outer ones, several times larger, serve to render the cluster more conspicuous to pollinating insects but are sterile and have only vestigial reproductive organs. The hermaphrodite flowers are homogamous and an abundance of pollen and nectar is produced at the same time. Both are easily available to short-tongued insects, so that bees, flies and beetles visit the flowers and, in landing upon the inflorescence, effect cross-pollination. Because of the proximity of the receptive stigma to the mature anthers self-pollination may also occur.

Alternative flowers for study. The other species of *Viburnum* found wild in Britain is *V. lantana*, Wayfaring-tree, which is common on calcareous soils in the south. *V. tinus*, Laurustinus, is a free-flowering, evergreen species from the Mediterranean region which is frequently grown in parks and gardens. Most Viburnums have entirely hermaphrodite flowers, but a few (like *V. opulus*) have some sterile flowers, and more rarely all the flowers are sterile, e.g., *V. opulus* 'Roseum' ('Sterile'), Snowball-tree, in which the inflorescence consists of a globular cluster of large, white flowers. The closely related genus *Sambucus* is represented in the native British flora by *S. nigra*, Elder, a common shrub in hedgerows and thickets, which differs from *Viburnum* in having extrorse anthers. Comparison should be made with other genera, e.g., *Weigela* and *Diervilla*, which are often grown in gardens and are examples of members of the family with capsular fruits.

Fig. 80.1 Caprifoliaceae, *Viburnum opulus*

A The inflorescence, a flat-topped cluster of small, fertile flowers surrounded by much larger, sterile flowers.
Sterile flowers: 17–18 mm in diameter
Fertile flowers: 4–5 mm in diameter

B Detail of a portion of the inflorescence, showing one flower in bud and the other fully open. The bases of 2 other flowers can be seen. There are often small bracts subtending the flowers and at the bases of the pedicels. The flower is actinomorphic, with a campanulate tube and 5 lobes which are imbricate in bud. The 5 stamens are well exserted from the tube.
Corolla-lobes of fertile flowers: 1.25 × 1.25–2.25 mm

C The 5-lobed corolla, showing the 5 introrse stamens adnate to the base of the tube and alternating with the lobes. The anthers are 2-celled and dehisce longitudinally.
Stamens: 5 mm

D Detail of a versatile anther prior to dehiscence.
Anther: 1 × 1 mm

E The corolla and stamens have been removed from the flower to reveal the gynoecium. The 5 calyx-teeth are arranged round the top of the inferior ovary and subtend the 3-lobed, almost sessile stigma. Below the ovary is a portion of the pedicel.
Stigma: 0.75 × 0.6 mm Calyx-teeth: 0.3 × 0.2 mm

F L.S. of the gynoecium. F1: at early anthesis, showing 2 of the 3 pendulous ovules on axile placentas. All 3 ovules are only visible in the early stages of ovary development. Each ovule is in a separate loculus. F2: at late anthesis, showing the one remaining ovule within the ovary. The ovary is composed of 3 fused carpels, but in the course of development, 2 of the carpels are normally suppressed.
Ovary: 2 × 1.25 mm
Ovule in F1: 0.25 × 0.125 mm
Ovule in F2: 0.7 × 0.125 mm

G T.S. of the ovary at late anthesis, shown by the single loculus, surrounded at this stage by a mass of cellular tissue (stippled in drawing). A small space at the centre, close to the surviving loculus, denotes the remains of an aborted carpel. The 5 small circles inside the margin are vascular bundles. The ovule has been teased into a horizontal position within the loculus in order to show more clearly how it is attached to the placenta.
T.S. of ovary: 1.5 × 1 mm

H One of the red, subglobose fruits attached to its stalk. The fruit is a drupe containing a single seed or 'stone'. For clarity, the fruits were removed from the 2 lateral branches. The scars on the stalk indicate where pedicels have fallen away.
Fruit: 10 × 7 mm

Fig. 80.1

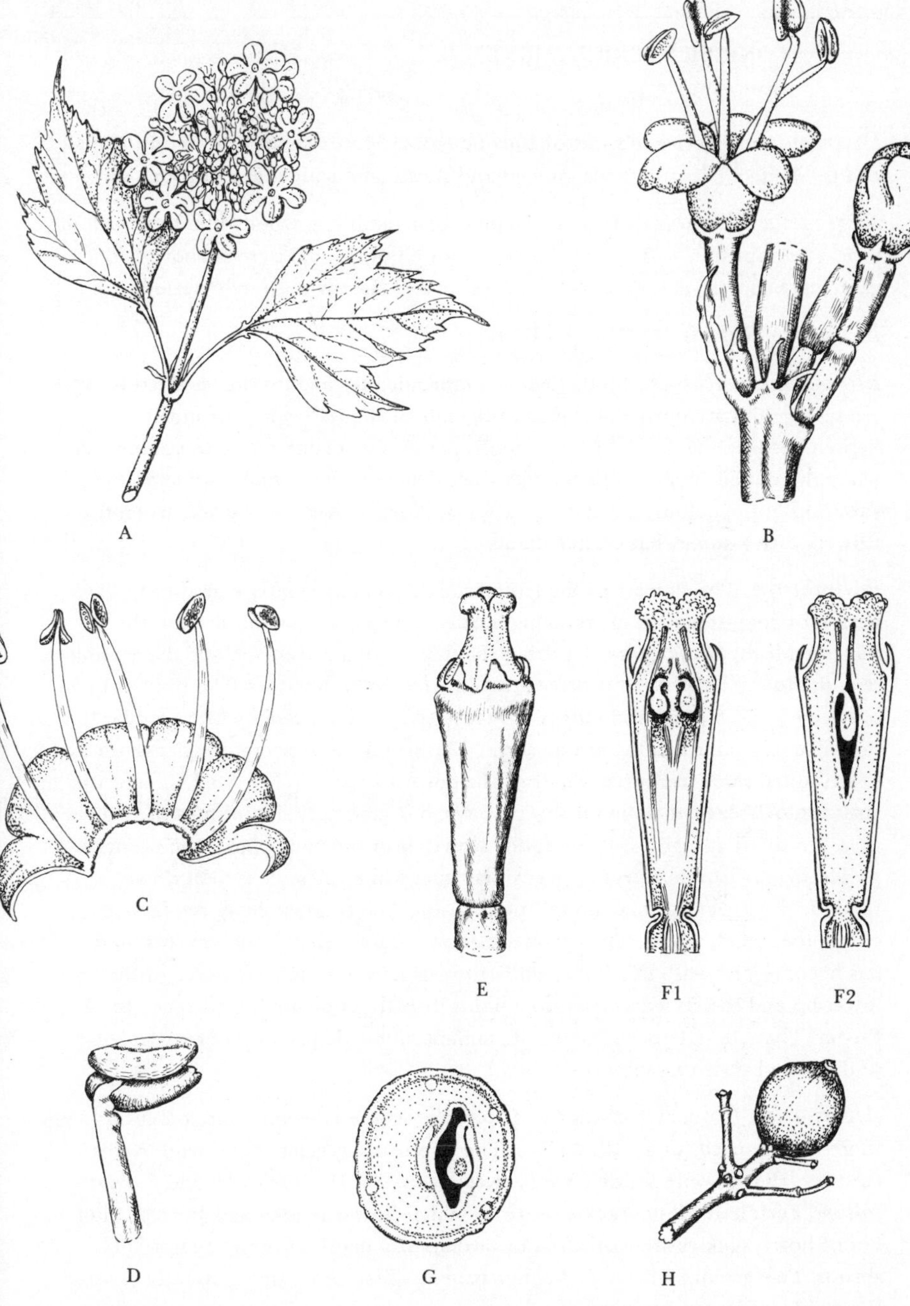

LONICERA PERICLYMENUM L.
Honeysuckle

Distribution. Native in most of Europe and in Morocco, and common throughout the British Isles in woods, hedges and scrub on a wide range of soils.

Vegetative characteristics. A deciduous, scrambling, woody climber reaching 6 m. The opposite, entire leaves are ovate or elliptical, dark green above and glaucous beneath, the upper ones subsessile and the lower shortly petiolate.

Floral formula. K(5) C(5) A5 $\overline{\mathrm{G}}$(3)

Flower and inflorescence. The zygomorphic, hermaphrodite, scented flowers are sessile, and are borne in heads at the ends of shoots from June until September (see Fig. 80.2.A). The corolla is usually creamy white to yellow inside, while the outside may have a reddish tinge. There is, however, considerable variation in the colour, particularly after pollination has taken place, when the flowers turn a somewhat darker shade.

Pollination. The flowers of the Honeysuckle are particularly suited to pollination by long-tongued, night-flying moths, especially hawk-moths and the Silver-Y Moth, *Plusia gamma.* During the day visiting insects include the bumblebee, *Bombus hortorum*, but these are not effective pollinators. The flowers open between 7 and 8 p.m., and moths are attracted by the creamy white corolla, the presence of nectar and the strong scent. The secretion of nectar by glands on a longitudinal ridge in the basal half of the corolla-tube (see Fig. 80.2.C), and the emission of scent continue for several days, but gradually diminish as the flower ages. On the first night of opening the flower is in the male phase, the stamens projecting horizontally from the corolla-tube, while the style is bent down towards the lower lip to avoid self-pollination. The following day the flower enters the female phase. The corolla-tube now curves slightly downwards and has become yellowish in colour, while the stamens have moved closer to the lower lip and the style has risen to a point directly opposite the entrance to the flower. The alternate prominence of stamens and style promotes cross-pollination by visiting insects.

Alternative flowers for study. L. periclymenum is readily obtainable and there is little need for an alternative. The two other species found wild in this country, though only locally, are *L. xylosteum*, Fly Honeysuckle, and *L. caprifolium*, Perfoliate Honeysuckle, both of which flower in May and June. A number of honeysuckles are cultivated in gardens and many of these are upright shrubs. This group includes *L. fragrantissima*, which bears small, sweetly scented flowers from December until March and *L. involucrata*, whose pairs of flowers appear in May, subtended by broad, reddish bracts which deepen in colour as the plant approaches the fruiting stage. *L. nitida* is a common, evergreen species, useful in hedging and topiary, but it should be noted that some clones rarely produce flowers, so it cannot be regarded as a suitable alternative.

Fig. 80.2 Caprifoliaceae, *Lonicera periclymenum*

A The inflorescence, a head of sessile flowers at the end of a leafy shoot. Some of the zygomorphic, hermaphrodite flowers are still in bud while others are fully open.

B L.S. of the corolla, with the inferior ovary and 5-toothed calyx left intact. The corolla is 2-lipped, the upper lip being 4-lobed and the lower one entire. The well-exserted stamens are joined to the corolla-tube just inside the mouth, and the long style, terminating in a 3-lobed stigma, projects even beyond the stamens. Part of the long, narrow nectary (shown as a wavy line) can be seen on the lower inner surface of the corolla-tube. Both corolla and calyx have small glandular hairs on their outer surface.
Calyx-teeth: 1.5 x 1 mm Corolla-tube: 20–25 mm
Lobed upper lip: 18–20 x 12 mm Lobes: 4–8 x 3 mm
Lower lip: 17–20 x 4 mm Style + stigma: 50 mm

C Detail of a portion of the corolla when opened out. The upper part of the drawing shows how 4 of the 5 stamens are attached to the corolla-tube (the central portion is the base of the lower lip). The lower part shows the longitudinal ridge in the basal half of the corolla-tube that bears the nectar-secreting glands.
Nectary: 7–11 mm

D Detail of the upper part of a stamen. The introrse anthers are versatile, 2-celled and dehisce longitudinally.
Filament: 20 mm Anther: 5 mm

E L.S. of the inferior ovary, with the glandular calyx and base of the corolla-tube. Within the tube is the style with a small swelling at its base. The ovary is composed of 3 fused carpels, each with a single loculus containing 2 pendulous ovules attached to axile placentas. On one side of the ovary is part of a glandular bract.
Ovary: 3 mm

F T.S. of the ovary, showing the 3 loculi with 2 ovules in each loculus. The attachment of the ovules to the placentas is not visible in this T.S.
T.S. of ovary: 2.75 x 2.5 mm

G The inferior ovary at an early stage of development into the fruit, a red, few-seeded berry. The ovary is subtended by small, glandular bracts and crowned by the persistent, glandular calyx.

Fig. 80.2

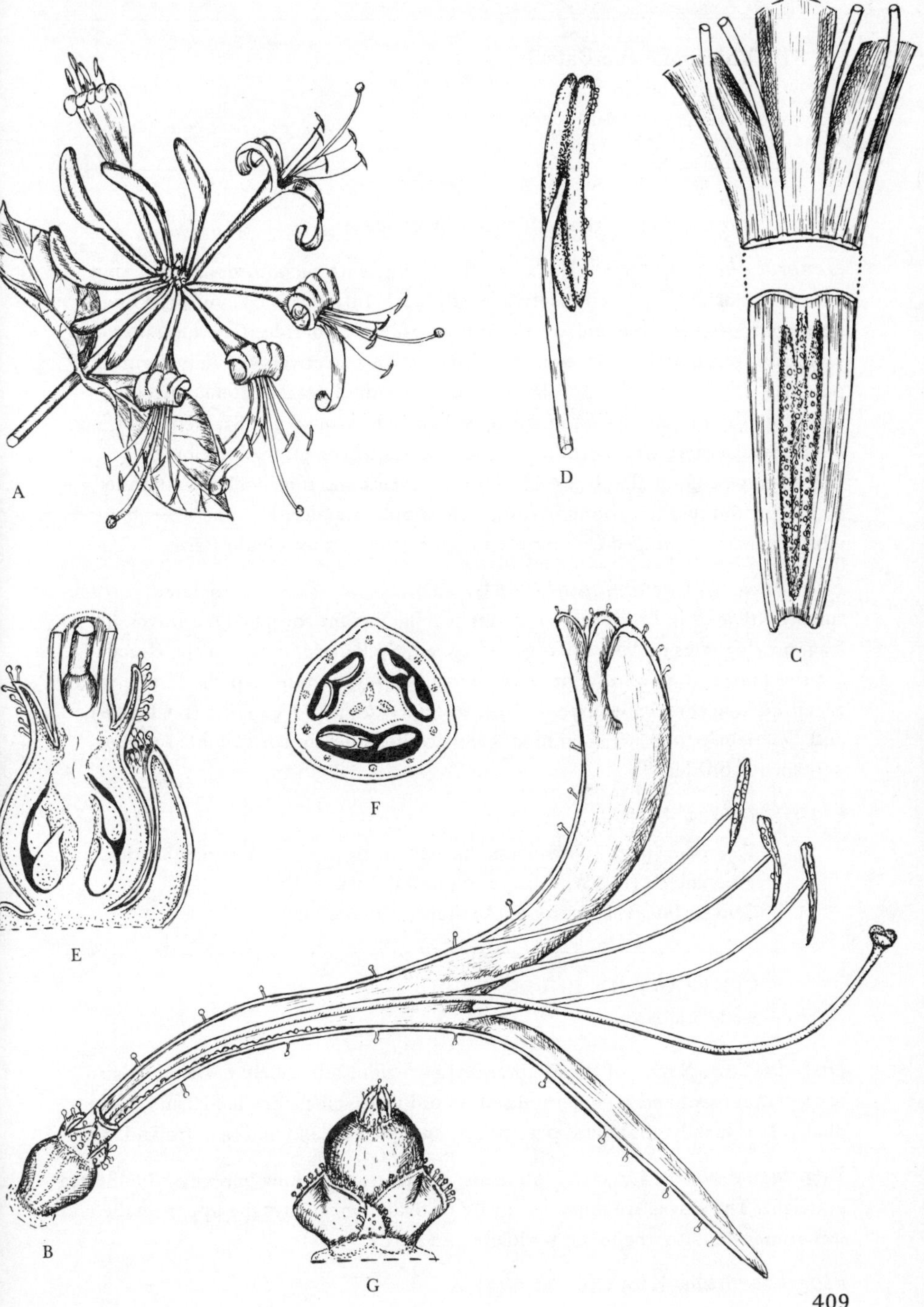

81 Valerianaceae Batsch

Valerian family
13 genera and 400 species

Distribution. Europe, Asia, Africa and America.

General characteristics. Herbs, rarely shrubs, with dichotomous branching. Leaves opposite or in basal rosettes, exstipulate. Inflorescence cymose. Flowers hermaphrodite or unisexual, zygomorphic, usually 5-merous. Calyx annular or variously toothed, little developed at anthesis, but afterwards usually forming a pappus (cf. Compositae). Corolla of usually 5 united petals, often spurred at base, lobes imbricate, sometimes 2-lipped (e.g., *Centranthus*). Stamens 1–4, epipetalous, alternate with corolla-lobes; anthers introrse. Ovary inferior, of 3 united carpels, 3-locular, but often only 1 developing, this loculus containing a single, pendulous, anatropous ovule; style simple, slender. Fruit an achene, often with a persistent winged or plumose pappus. Seed without endosperm.

Economic and ornamental plants. Valerianella locusta, Corn Salad, is sometimes cultivated in European countries as a salad plant for its edible leaves. The fragrant rhizomes of *Nardostachys jatamansi*, Spikenard, native in the Himalaya, are the source of an oil used in perfumery. The few members of the family which are of horticultural value include *Centranthus ruber*, Red Valerian (see below), and *Valeriana officinalis*, Common Valerian, both perennials suitable for the herbaceous border.

Classification.

Valeriana (over 200) Eurasia, S. Africa, tropical N. America, Andes.
Valerianella (80) W. Europe to central Asia.
Centranthus (12) Europe, Mediterranean region.

CENTRANTHUS RUBER (L.) DC.

Red Valerian

Distribution. Native of the Mediterranean region, but widely cultivated as an ornamental plant and now naturalised on old walls, cliffs, dry banks and waste places. It is locally abundant, particularly in S.W. England and S.E. Ireland.

Vegetative characteristics. An erect, glaucous, branching perennial 30–80 cm in height. The leaves are opposite, ovate or ovate-lanceolate, the upper sessile and sometimes dentate, the lower petiolate and entire.

Floral formula. K(5) C(5) A1 $\overline{G}(3)$

Flower and inflorescence. Masses of red, or more rarely white, hermaphrodite, zygomorphic flowers are borne in terminal, compound cymes from June until late August or early September.

Pollination. Long-tongued insects, mainly butterflies, are attracted by the conspicuous inflorescences of scented flowers and by the nectar which is secreted in a spur at the base of the corolla-tube (see Fig. 81.B). The flowers are markedly protandrous. At first, the single stamen projects straight out of the corolla-tube, but after dehiscence it becomes reflexed, and its place is taken by the style which elongates to a similar extent. Automatic self-pollination is therefore effectively prevented. The corolla-tube is longitudinally divided into two by a thin septum, with the style and stamen on one side. The other side is unobstructed, and it is along this portion, lined with downward-pointing hairs, that the proboscis of the visiting insect must be directed in order to reach the nectar-secreting spur at its base.

Alternative flowers for study. C. ruber is a popular garden plant and well naturalised so it is usually readily available. The related genus *Valeriana* contains 2 species native to Britain, in one of which, *V. dioica*, the plants are dioecious. The number of stamens differs between genera, there being 3 in *Valeriana* and *Valerianella*, and 2 (or 3 of which 2 are connate) in *Fedia*, a small genus from the Mediterranean region.

Fig. 81 Valerianaceae, *Centranthus ruber*

A The inflorescence, a terminal, compound cyme of hermaphrodite, actinomorphic flowers.

B A single flower. The 5-lobed corolla has a nectar-secreting spur which projects from one side of the base of the long tube. Above the inferior ovary is the calyx in the form of a ring. The single style and stamen are well exserted from the corolla.
Flower: 5 mm in diameter Corolla-tube: 8 mm
Spur: 5 mm

C The zygomorphic corolla, showing the 5 lobes and the relative positions of stamen and style. The style is terminated by a simple stigma.

D L.S. of upper portion of the corolla-tube, showing the septum that divides it lengthwise into 2 parts. The section shown contains the stamen and style, while the section removed leads to the nectar-spur and is lined with downward-pointing hairs.
Exserted portion of stamen: 3–5 mm
Exserted portion of style: 4.5–5 mm

E Detail of anther at dehiscence. The anther is introrse, 2-celled, and dehisces longitudinally.
Anther: 1.75 × 1 mm

F L.S. of the inferior ovary at anthesis. The ovary has a single loculus, but is considered to have evolved from a 3-locular ovary, the other 2 loculi having aborted. The solitary ovule is anatropous and pendulous. Above the ovary are the lower portions of the style and the corolla-tube, and surrounding the base of the latter is the inrolled, annular calyx.
Ovary: 3 × 1 mm Ovule: 0.75 × 0.5 mm

G The mature fruit, an achene crowned by a plumose pappus which was small and ring-like at anthesis. The pappus enables the seed to be carried a considerable distance by the wind.
Achene: 4 × 1.75 mm Rays of pappus: 6 mm

Fig. 81

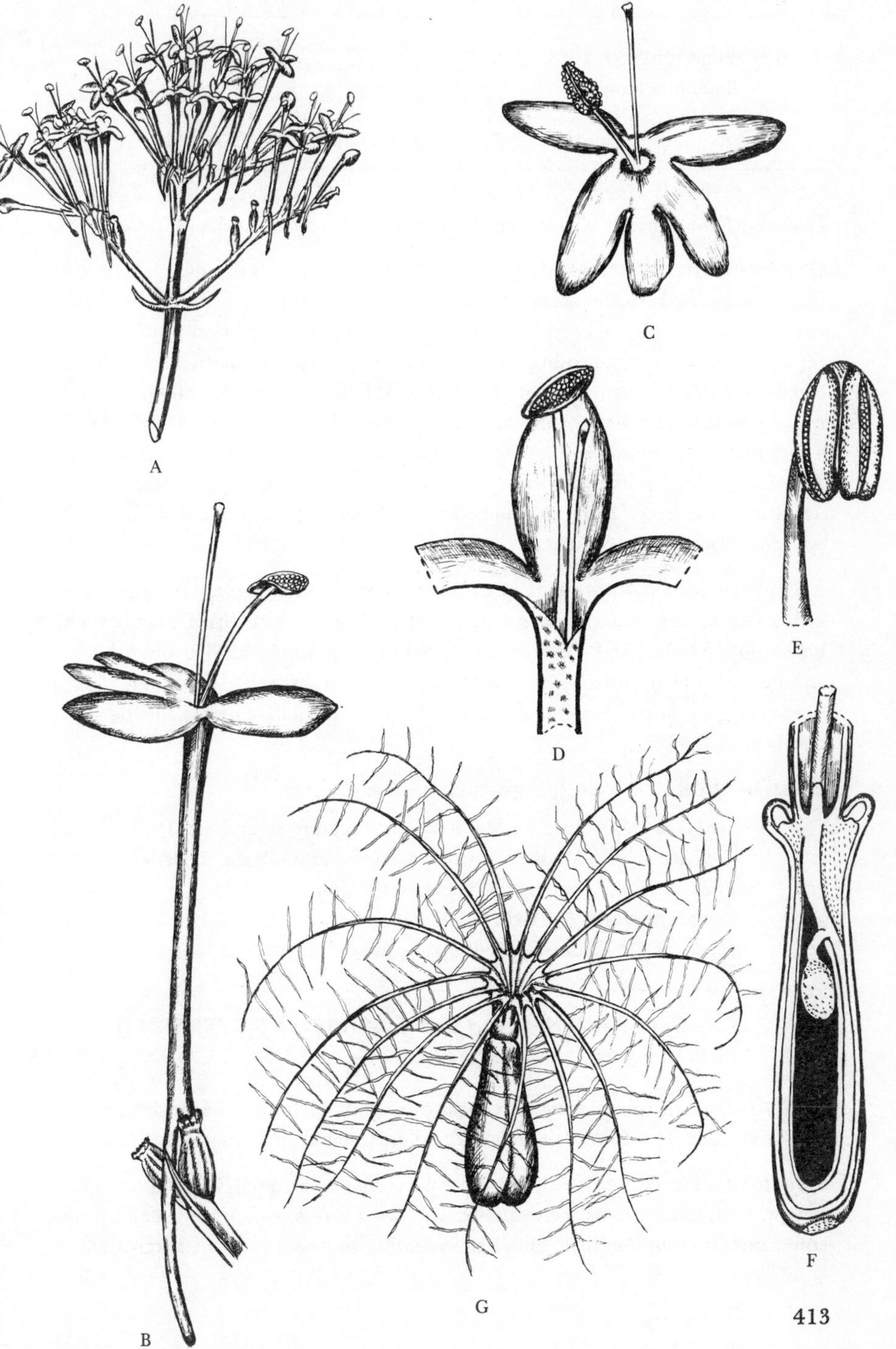

82 Dipsacaceae Juss.

Scabious family

8 genera and 260 species

Distribution. Chiefly N. temperate Eurasia, and tropical and S. Africa.

General characteristics. Mostly herbs, with opposite, exstipulate leaves (connate in *Dipsacus*). Inflorescence cymose, consisting of dense heads of flowers, the outer ones with the corolla often extended abaxially (cf. Compositae, Cruciferae). Flowers hermaphrodite, zygomorphic or rarely asymmetrical. Epicalyx, usually regarded as composed of 2 united bracteoles, often present. Calyx of usually 5 united sepals. Corolla of usually 5 united petals. Perianth sometimes 4-merous by the union of 2 members. Stamens 4, epipetalous. Ovary inferior, of 2 united carpels, unilocular, with one pendent, anatropous ovule. Flowers usually protandrous. Fruit an achene (cf. Compositae), usually enclosed in the epicalyx. Endosperm present.

Economic and ornamental plants. The only economic plant of importance in the Dipsacaceae is *Dipsacus sativus*, Fuller's Teasel, the fruiting heads of which have hooked bracts and are used for raising the nap on cloth. This and other species of *Dipsacus* are also cultivated for ornament, as well as *Scabiosa*, Scabious, *Morina* and *Cephalaria*, all of which are attractive plants for the herbaceous border.

Classification. The principal genera in the family are:

Scabiosa (100) Eurasia, Mediterranean region, E. and S. Africa.
Cephalaria (65) Mediterranean region to central Asia, S. Africa.
Knautia (50) Europe, Mediterranean region.
Pterocephalus (25) Mediterranean region to W. China, tropical Africa.
Dipsacus (15) Eurasia, Mediterranean region, tropical Africa.

KNAUTIA ARVENSIS (L.) Coult. (SCABIOSA ARVENSIS L.)

Field Scabious

Distribution. Native throughout most of Europe including the British Isles, and in W. Asia. It is found on well-drained grass banks, pastures and fields.

Vegetative characteristics. An erect perennial or biennial, 25–100 cm in height, with more or less terete stems. The basal leaves, which are usually simple, sometimes lyrate-pinnatifid, entire or crenately toothed, form an overwintering rosette at the base of the old flowering stem. The stem leaves are usually deeply pinnatifid, though some of the uppermost are more or less entire.

Floral formula. K(8) C(4) A4 $\overline{\text{G}}$(2)

Flower and inflorescence. The inflorescence, which is borne on a long peduncle, is in the form of a flattened capitulum (see Fig. 82.A) with an involucre of leafy bracts. The bluish-lilac, zygomorphic flowers make an attractive show from July until late September. Most plants have capitula consisting of hermaphrodite flowers, but some may have smaller capitula composed solely of female flowers.

Pollination. Various kinds of bees, beetles and flies have been observed visiting the flowers which are made more conspicuous by being grouped into heads. Nectar is secreted on the upper surface of the ovary and is readily accessible to short- as well as long-tongued insects. Pollen is also easily obtainable. In the first (male) stage, the stamens protrude from the corolla, exposing the anthers. These subsequently fall off and the filaments shrivel. The style then extends and in its turn protrudes from the corolla, exposing the stigma. Insect-visitors to the capitulum at either stage of its development may visit several flowers at one time and cross-pollination is usually achieved.

Alternative flowers for study. In the absence of *K. arvensis*, any member of the closely related genus *Scabiosa*, including the large, blue-flowered Garden Scabious, *S. caucasica*, would be suitable. The genus *Scabiosa* differs from *Knautia* in having a 5-toothed calyx that is persistent in fruit and a 5-lobed corolla, while *Knautia* has a 8- to 16-toothed, deciduous calyx and a usually 4-lobed corolla. Comparison should be made with the native genus *Dipsacus*, which includes the familiar Fuller's Teasel and with the Himalayan *Morina longifolia* that has pink to crimson flowers arranged in whorls. The latter is sometimes grown as an interesting, summer-flowering perennial in gardens.

Fig. 82 Dipsacaceae, *Knautia arvensis*

A The inflorescence, a flattened capitulum of zygomorphic flowers surrounded by an involucre of leafy bracts, and borne on a hairy peduncle. The flowers at the margin of the capitulum are larger than those in the centre.
Inflorescence: 28–30 mm in diameter Bracts: 9–13 × 5.5 mm

B A single flower from the margin of the capitulum. The corolla has 4 unequal lobes and the bristly, cup-shaped calyx has 8 teeth. Below the calyx is the inferior ovary, bristly like the calyx. The style terminates in a slightly 2-lobed stigma, and the stamens are well exserted from the corolla, a distinguishing feature when comparing this family with the Compositae.
Corolla-tube: 7 mm Corolla-lobes: 6–10 × 2–5 mm

C A single flower from the centre of the capitulum. The corolla is shorter and somewhat less zygomorphic than that of the marginal flowers, and the stamens appear more exserted than those illustrated in B.
Corolla-tube: 5 mm Corolla-lobes: 3–4 × 1.5 mm

D L.S. of the calyx and corolla with the inferior ovary left intact. The stamens are attached to the lower part of the corolla-tube and alternate with the corolla-lobes. A long style arises from the top of the ovary, terminating in a 2-lobed stigma.
Filament: 8 mm Style + stigma: 9 mm

E Detail of the anther which is introrse, 2-celled, and dehisces longitudinally.
Anther: 2–2.5 × 1 mm

F L.S. of the inferior ovary formed from 2 united carpels, one of which has been suppressed so that the ovary is unilocular. A single, anatropous ovule is attached to the apex of the loculus.
Ovary: 1.5 × 0.8 mm Ovule: 0.8 × 0.4 mm

G T.S. of young fruit, showing the seed formed from the single, pendulous ovule. The 2 shaded structures at the centre are the developing cotyledons or seed-leaves.
T.S. of fruit: 3 × 2 mm

H The young fruit attached to a portion of the hairy, hemispherical receptacle. The calyx that surmounts the fruit is deciduous in *Knautia* but persistent in *Scabiosa*. At maturity the fruit is dry and indehiscent, and is termed an achene.
Young fruit: 3 × 2 mm Mature fruit: 5 × 3 mm
Calyx: 2–3 mm

Fig. 82

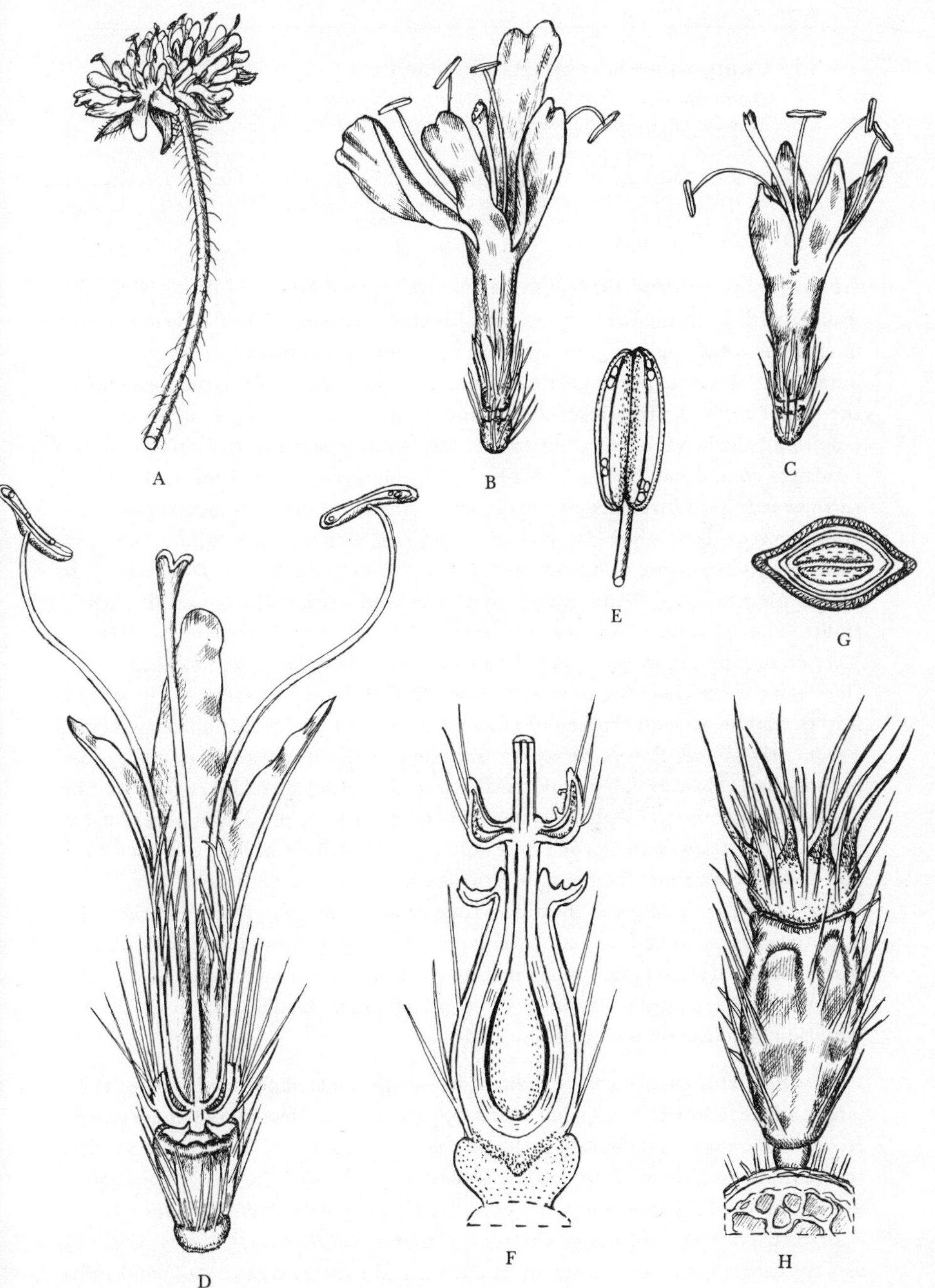

83 Compositae Giseke (Asteraceae Link)

Daisy family

900 genera and up to 20000 species

Distribution. Cosmopolitan.

General characteristics. Mainly herbaceous plants, rarely shrubs or trees, usually with a taproot, sometimes with tubers (e.g., *Dahlia*). Leaves usually alternate, sometimes opposite, rarely whorled, simple or compound, frequently in basal rosettes, usually exstipulate. Oil passages usually present. Latex present in the tribe Lactuceae. Inflorescence racemose, the flowers arranged in heads (capitula), the heads often grouped into larger inflorescences. In *Echinops* the heads are compound, but each smaller head contains only one flower. Head surrounded by an involucre of usually green bracts. Flowers arranged upon a common receptacle, the enlarged end of the axis, which may be smooth or hairy. Scaly bracts belonging to individual flowers may be present (e.g., *Helianthus*). In the simplest case, the flowers (florets) of a single head are all alike and hermaphrodite. They may be all actinomorphic (tubular) or all zygomorphic (ligulate). Very commonly, as in *Bellis* and *Helianthus*, there is a disc of actinomorphic florets and a marginal ring of zygomorphic florets. In, e.g., *Centaurea* the marginal florets may be actinomorphic but larger than the central, and completely sterile. Frequently the ray-florets are female and the disc-florets hermaphrodite (gynomonoecism). Flowers epigynous, usually 5-merous. Calyx usually represented by a pappus, sometimes a slightly 5-lobed rim on the top of the ovary, occasionally absent (e.g., *Ambrosia*). Corolla of 5 united petals, valvate in bud. Stamens 5, epipetalous with short filaments, alternate with the lobes; anthers introrse, cohering by their edges and forming a tube around the style. Ovary inferior, of 2 united carpels, unilocular, containing one erect, anatropous ovule with basal placentation; style simple, branching into 2 stigmas. Fruit an achene, often crowned by the persistent pappus which enlarges after fertilisation to aid distribution. Endosperm absent; embryo straight.

Economic and ornamental plants. Although the Compositae is one of the largest plant families, it contains relatively few species of economic importance. *Helianthus annuus*, Sunflower, is the source of an edible oil, extracted from the seed, and the tubers of *H. tuberosus*, Jerusalem Artichoke, are grown for food, as are the roots of *Tragopogon porrifolius*, Salsify, and *Scorzonera hispanica*, Scorzonera or Spanish Salsify. The root of *Cichorium intybus*, Chicory, is used as a substitute for or an adulterant of coffee, and the leaves of *C. endivia*, Endive, are used in salads. Another, more widely cultivated salad plant is *Lactuca sativa*, Lettuce, while the aromatic leaves of *Artemisia dracunculus*, Tarragon, are used for flavouring. The leaf-stalks of *Cynara cardunculus*, Cardoon, and the young flower-heads of *C. scolymus*, Globe Artichoke, are popular as vegetables, particularly in France. The dried flower-heads of *Chrysanthemum cinerariifolium* and *C. coccineum* are the source of the insecticide pyrethrum. Many genera are highly decorative and include a number of well-known garden plants, e.g., *Aster*, Michaelmas Daisy, *Callistephus*, China Aster, *Echinops*, Globe Thistle, *Onopordon*, Scotch Thistle, *Solidago*, Golden Rod, *Tagetes*, Marigold, *Calendula*, *Chrysanthemum*, *Coreopsis*, *Cosmos*, *Dahlia*, *Gaillardia*, *Rudbeckia* and *Zinnia*. *Senecio* × *hybridus*, the florist's Cineraria, is a popular house plant, with a wide range of flower colour.

Classification. The Compositae are well marked in their characters, and are unlikely to be confused with any other family. They are probably related to the Goodeniaceae, Stylidiaceae and Campanulaceae, and have a superficial likeness to the Dipsacaceae. The traditional classification, employed here, is a division into 2 subfamilies, one with a number of tribes, the other containing only one.

I. Tubuliflorae (disc-florets not ligulate; latex usually absent).

1. Heliantheae.
   Helianthus (110) America.
   Tagetes (50) warm regions of America.
   Helenium (40) western America.
   Dahlia (27) Mexico, Guatemala.
   Zinnia (20) S. United States to Brazil and Chile.
2. Astereae.
   Aster (500) America, Eurasia, Africa.
   Baccharis (400) America.
   Erigeron (200) cosmopolitan, especially N. America.
3. Anthemideae.
   Artemisia (400) N. temperate region, S. Africa, S. America.
   Chrysanthemum (200) Eurasia, Africa, America.
4. Arctotideae.
   Arctotis (65) tropical and S. Africa, Australia.
   Gazania (40) tropical and S. Africa.
5. Inuleae.
   Helichrysum (500) warm regions of Old World.
   Inula (200) Eurasia, Africa.
   Gnaphalium (200) cosmopolitan.
6. Senecioneae.
   Senecio (2000–3000) cosmopolitan.
   Doronicum (35) temperate Eurasia, N. Africa.
7. Calenduleae.
   Calendula (20–30) Mediterranean region to Iran.
8. Eupatorieae.

Eupatorium (1200) mostly America, a few in Eurasia and Africa.
Mikania (250) tropical America, W. Indies.
Ligularia (150) temperate Eurasia.

9. Vernonieae.
Vernonia (1000) America, Africa, Asia, Australia.

10. Cynareae.
Centaurea (600) Eurasia, N. Africa, America.
Cousinia (400) E. Mediterranean region to central and E. Asia.
Cirsium (150) N. temperate region.
Carduus (100) Eurasia, N. Africa.

11. Mutisieae.
Gerbera (70) Africa, Madagascar, Asia.
Mutisia (60) S. America.

II. Liguliflorae (all florets ligulate; latex present).

12. Lactuceae.
*Hieracium** (1000) temperate regions and tropical mountains.
Crepis (200) N. hemisphere, tropical and S. Africa.
Lactuca (100) temperate Eurasia, tropical and S. Africa.
*Taraxacum** (60) mostly N. temperate region.

*These genera are apomictic. In *Hieracium*, a total of 5000 microspecies have been recognised and in *Taraxacum*, about 1200 from Europe alone.

DORONICUM PLANTAGINEUM L.
Plantain-leaved Leopard's-bane

Distribution. Native of S.W. Europe and often grown in gardens for its early flowering.

Vegetative characteristics. A herb, perennating by numerous stout, hypogeal stolons which are tuberised at their tips producing rosettes of long-stalked, ovate-elliptical leaves with prominent, curving lateral veins.

Floral formula. Kpappus C(5) A(5) $\overline{\mathrm{G}}(2)$

Flower and inflorescence. The inflorescence is usually solitary on a long stalk, and consists of a head of yellow florets surrounded by 2 or 3 rows of green bracts which comprise the involucre. The florets are of 2 kinds, those forming the central disc being hermaphrodite and tubular, while those in the single marginal row are female and ligulate. This species flowers in June and July.

Pollination. Various kinds of insects are attracted by the conspicuous head of florets and by the nectar secreted round the base of the style. As the florets are massed together, many can be visited by a single insect in a short time, and their structure permits all but the shortest-tongued insects to obtain the nectar. Since the florets are protandrous, the 2 stigmas are still pressed tightly against each other while the pollen is shed. As the style grows, it gradually pushes the pollen out of the end of the tube formed by the anthers, where it can be picked up by visiting insects. When the style emerges from the tube, the stigmas separate and become receptive to pollen from other flowers. In some cases the stigmas curl back so far that they touch the pollen on their own style and self-pollination takes place. This may not necessarily result in fertilisation as the florets may be self-incompatible.

Alternative flowers for study. Alternatives to *Doronicum* are numerous. Native British plants include *Tussilago farfara*, Colt's-foot, which flowers in March and April, and *Senecio jacobaea*, Ragwort, which is in flower from June until October. Garden plants such as the single forms of *Aster*, *Dahlia* and *Chrysanthemum* could also be used.

Fig. 83.1 Compositae, *Doronicum plantagineum*

A L.S. of the capitulum, consisting of actinomorphic, hermaphrodite disc-florets in the centre and irregular, female ray-florets at the margin. Attached to the receptacle is an involucre of bracts in 2 series.
Receptacle: 10 mm in diameter
Inflorescence: 6.2–6.4 cm in diameter
Bracts: 10–12 × 2 mm

B A single disc-floret, showing the inferior ovary and the 5-lobed corolla-tube with protruding stigmas. The corolla of each disc-floret is surrounded by a pappus of long hairs, considered to be a modified form of calyx.
Corolla: 5.25 mm Pappus: 3.5 mm

C L.S. of a disc-floret revealing 2 of the 5 stamens. The filaments are short and are borne separately on the corolla-tube. The anthers are 2-celled, introrse and connate into a tube. The single style is branched into 2 stigmas in the case of the disc-florets, but is a united structure in the ray-florets.
Filament: 3 mm Anther: 2.65 mm
Style: 6.65 mm Stigma: 0.4 mm

D Lower portion of a ray-floret. The corolla-tube surrounds the unbranched style. The pappus is absent from the ray-florets.

E L.S. of ovary with its single, basal, anatropous ovule.
Ovary: 1.5 × 0.7 mm Ovule: 1 mm

F T.S. of ovary showing the single ovule and the fluted external wall of the ovary formed from 2 united carpels.

G A fruit, crowned by the pappus, which has developed from one of the disc-florets. As it came from a garden-plant it is probably sterile. Fertile fruits are ribbed and more or less cylindrical.
Achene: 1.5 mm

Fig. 83.1

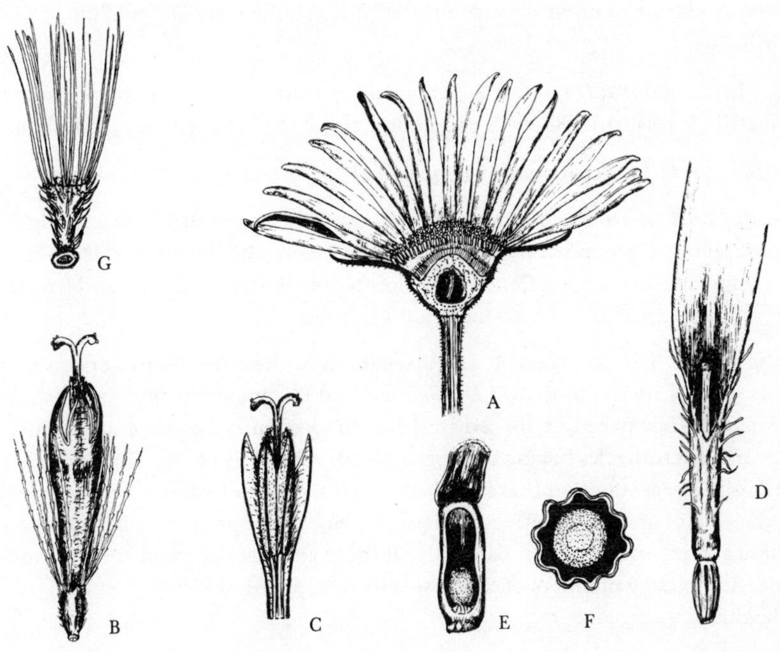

TARAXACUM OFFICINALE Weber
Dandelion

Distribution. Common throughout the N. hemisphere and abundant in the British Isles.

Vegetative characteristics. A perennial herb with a basal rosette of runcinate-pinnatifid leaves from which rises a hollow scape bearing a solitary capitulum.

Floral formula. Kpappus C(5) A(5) $\overline{\mathrm{G}}$(2)

Flower and inflorescence. The solitary inflorescence consists of a head of yellow florets surrounded by 2 rows of green bracts, the inner erect, the outer erect, spreading or even reflexed. The florets are all hermaphrodite and ligulate. The flowering period is from March until October.

Pollination. The arrangement and mechanism of the reproductive organs in *Taraxacum* is similar to that of *Doronicum*, but like many members of the genus *Hieracium*, Hawkweed, in the British Isles, *Taraxacum officinale* is apomictic, over 100 microspecies having been recognised. Apomixis covers the various methods of reproduction that have been developed as substitutes for the sexual production of seeds. The effect is that the progeny of apomictic plants are genetically identical to their parents, and those genera concerned are commonly divided into large numbers of 'microspecies' composed of almost identical individuals.

Alternative flowers for study. The Dandelion is so common that an alternative is hardly necessary. Other native British members of the tribe Lactuceae would be suitable, including *Lapsana communis*, Nipplewort, and *Hypochoeris radicata*, Cat's-ear, which are widely distributed throughout the British Isles. *Cichorium intybus*, Chicory, which is often cultivated, could also be used. All these flower during the summer and early autumn.

Fig. 83.2 Compositae, *Taraxacum officinale*

A L.S. of a capitulum, made up of florets which are all hermaphrodite and zygomorphic. The reflexed outer bracts are clearly visible.
Receptacle: 10 mm in diameter   Inflorescence: 4.5 cm in diameter
Bracts: 10–14.5 × 2–4 mm

B A single floret, showing the inferior ovary, and the corolla surrounded by the hairy pappus, The corolla is extended on one side into a long strap-like structure and is termed ligulate.
Floret: 21 mm   Pappus: 7 mm   Beak: 1 mm   Achene: 2 mm

C The upper part of the corolla has been rolled back to reveal more clearly the reproductive organs. The 5 short filaments are borne on the corolla-tube and are separate, but the anthers are connate into a tube round the style. The style extends far beyond the stamens and branches into 2 stigmas covered with small protuberances (papillae) which help the pollen to adhere.
Filament: 1.5 mm   Anther: 4 mm   Style: 12 mm   Stigma: 2 mm

D L.S. of young ovary, showing the single, basal, anatropous ovule.

E T.S. of ovary. Both ovary and ovule are somewhat compressed.
Ovary (at anthesis): 0.9 × 0.5 mm

F A mature fruit with the pappus attached to the achene by a long beak (absent in the genus *Doronicum*). This type of structure, common within the Compositae, provides a very efficient method of fruit dispersal.
Achene: 3.5 mm   Beak: 12 mm   Pappus: 7 mm

G Enlarged view of the achene, showing the longitudinal ribs.

Fig. 83.2

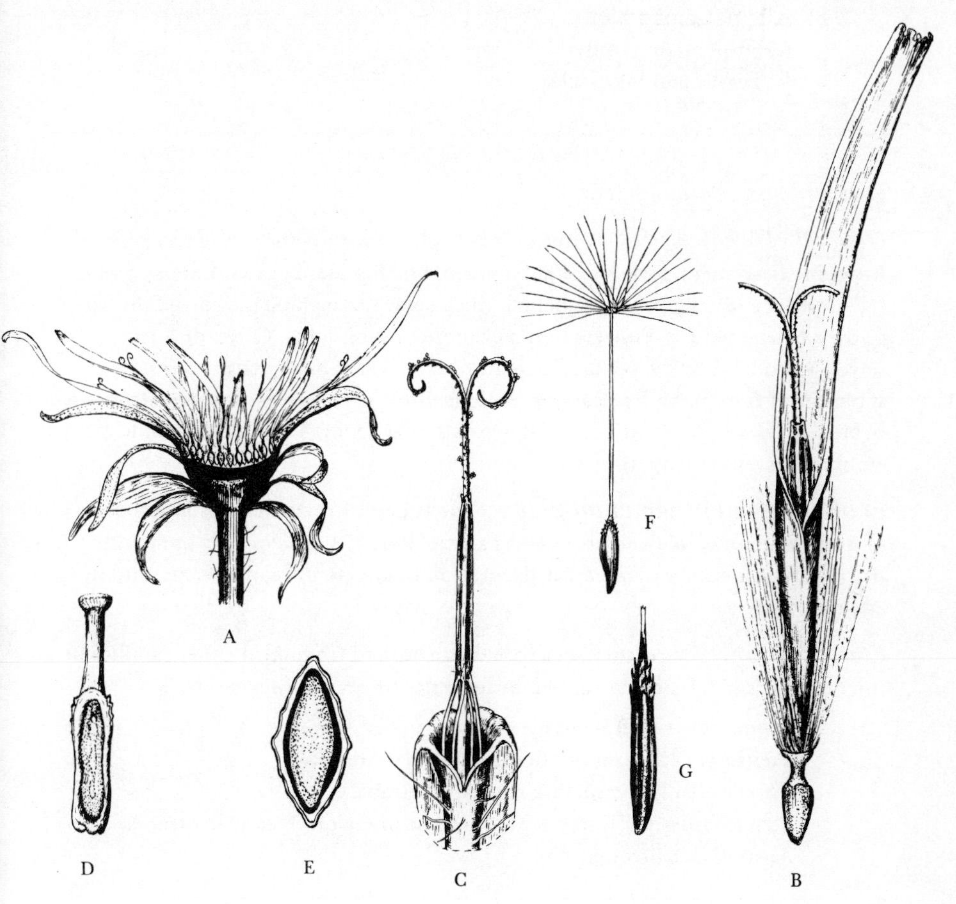

84 Alismataceae Vent.

Water-plantain family
13 genera and 90 species

Distribution. Cosmopolitan.

General characteristics. Water or marsh plants with perennating rhizomes. Leaves radical, erect, floating or submerged. Small scales in axils. Latex present. Inflorescence usually much branched, the primary branching racemose, the secondary often cymose. Flowers hermaphrodite or unisexual. Calyx of 3 free sepals. Corolla of 3 free petals. Stamens 6 to many; anthers extrorse. Ovary superior, of 6 to many free carpels, each having 1 (rarely 2 or more) anatropous ovule with basal placentation. Fruit a group of achenes. Seed without endosperm; embryo horseshoe-shaped.

Economic and ornamental plants. The tuberous rhizomes of some species of *Sagittaria* are edible and are used as a vegetable. The genera *Sagittaria* and *Alisma* are sometimes planted for decoration beside ponds and streams, and in aquaria.

Classification. The Alismataceae are distinguished from most other families in the order Alismatales by the green, outer series of perianth-segments.

Echinodorus (30) America, Africa.
Sagittaria (20) cosmopolitan, especially America.
Alisma (10) N. temperate region, Australia.
Damasonium (5) Europe, Mediterranean region, W. and central Asia, Australia, California.

ALISMA PLANTAGO-AQUATICA L.

Water-plantain

Distribution. Native in Europe, including the British Isles, N. Africa and temperate Asia, and introduced into America, tropical and S. Africa, Indonesia and Australia. It is found growing in shallow water or damp ground near ponds, lakes, ditches and slow-flowing rivers.

Vegetative characteristics. An erect, glabrous perennial, 20–100 cm in height, with acrid juice. The leaves are all basal but vary in shape. Those which are submerged are ribbon-shaped, while the aerial leaves are ovate, rounded to subcordate at base, and have long petioles. Floating leaves sometimes occur when the plant is growing in water.

Floral formula. K3 C3 A6 $\underline{G}\infty$

Flower and inflorescence. The inflorescence consists of a panicle of whorled branches bearing actinomorphic, hermaphrodite, white or pale lilac flowers that appear from June until August. Flowering is most successful when the plant is growing in shallow water or damp ground and produces aerial leaves. When growing in deep water, however, the inflorescence may be reduced or even entirely absent.

Pollination. Hover-flies, short-tongued bees and various kinds of flies are attracted by the conspicuous panicle of flowers, and the nectar which is secreted in 12 small drops by the inner side of a fleshy ring formed by the coherent bases of the 6 stamens. The insects are directed to the nectar by a yellow nectar-guide situated at the base of each petal. The flowers are homogamous, but as the stamens are directed obliquely upwards and outwards, and are well away from the erect styles clustered at the centre of the flower (see Fig. 84.C), cross-pollination is favoured. Self-pollination may occur if the insect alights on a petal first, instead of flying directly to the centre of the flower.

Alternative flowers for study. A. plantago-aquatica is a common, native representative of the family and an alternative is scarcely necessary. Comparison between British genera will show some important variation in floral structure. In the monotypic genus *Baldellia ranunculoides*, Lesser Water-plantain, which flowers in May, the inflorescence is umbellate, and in *Sagittaria sagittifolia*, Arrowhead, the flowers, which appear in July and August, are unisexual, the male ones having numerous stamens.

Fig. 84 Alismataceae, *Alisma plantago-aquatica*

A A portion of the inflorescence, a pyramidal panicle with whorls of branches bearing pedicelled flowers. Both the inflorescence-branches and the pedicels have bracts at their base.
Portion of inflorescence: 38 cm

B A flower-bud, showing the persistent, outer whorl of the perianth composed of 3 free sepals.
Bud: 2 × 2.75 mm

C The actinomorphic, hermaphrodite flower at anthesis. The deciduous, inner whorl of 3 free petals alternates with the sepals. Fifteen to thirty free carpels are arranged in a single whorl in the centre of the flower. The 6 stamens have filaments which broaden towards the base.
Flower: 10.5 mm in diameter Petal: 5 mm
Sepal: 2 mm

D Dorsal view of anther, showing point of attachment to the filament. The anthers are extrorse, 2-celled, and dehisce longitudinally.
Filament: 2 mm Anther: 0.4 × 0.5 mm

E L.S. of flower, showing 7 of the free carpels backed by the basal portion of a petal. At the side are 2 of the 3 sepals. The point of attachment of one of the stamens can also be seen. Each carpel consists of an ovary bearing a short, lateral style with a papillose stigma.
Ovary: 2 × 0.75 mm Style + stigma: 0.75–1 mm

F Detail of the apex of the style with its stigmatic papillae.
Papillae: 0.06 mm

G L.S. of a carpel still attached to a portion of the receptacle. The ovary has a single loculus containing a solitary ovule on a basal placenta. The remains of the style are visible on one side of the ovary.
Ovule: 1.25 × 0.75 mm

H L.S. of a mature fruit, showing the single seed with its horseshoe-shaped embryo. The fruit is dry and indehiscent forming an achene.
Achenes: 2 × 1–1.25 mm

Fig. 84

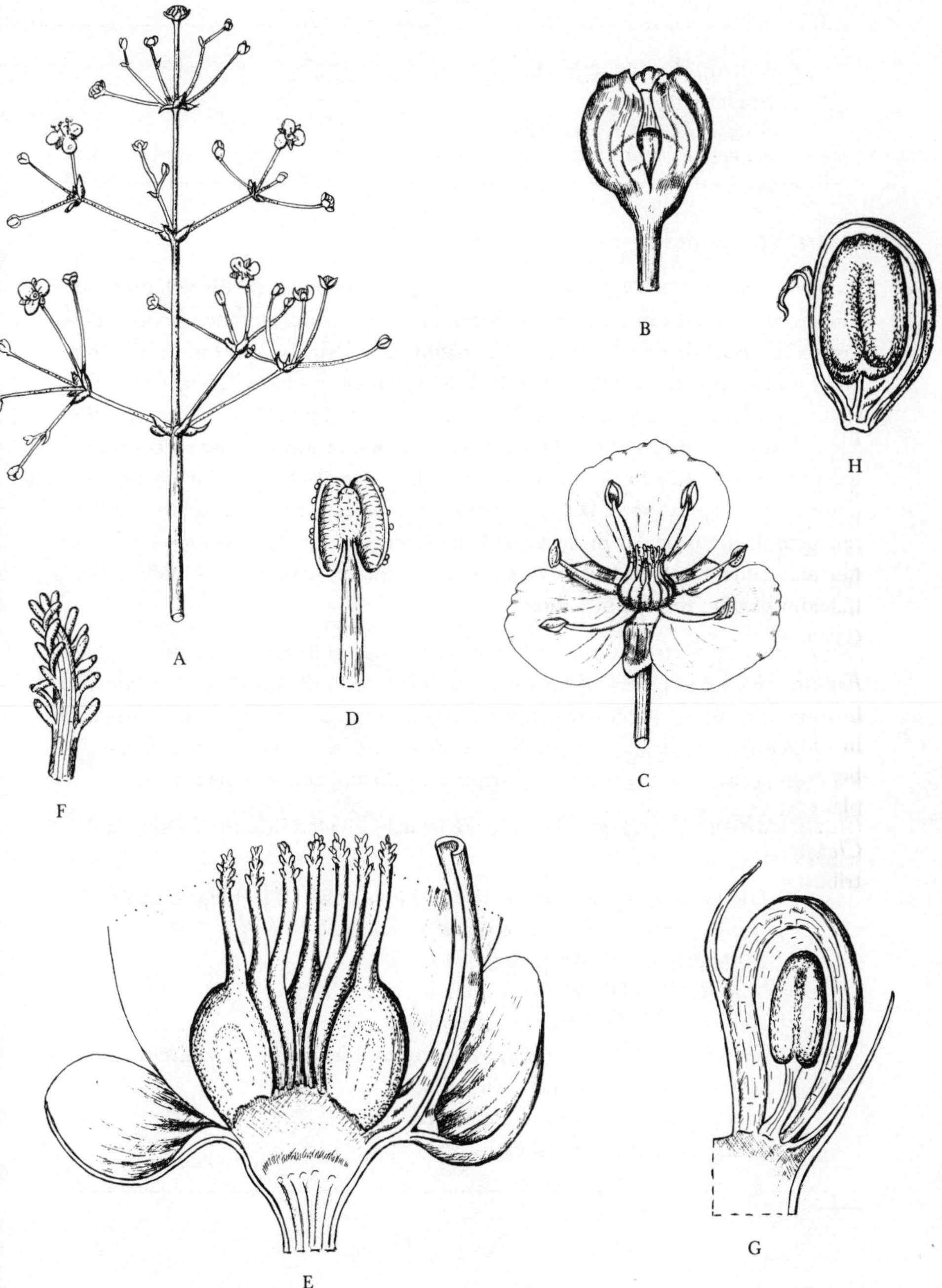

85 Commelinaceae R. Br.

Spiderwort family

38 genera and 500 species

Distribution. Mostly tropics and subtropics.

General characteristics. Herbs (occasionally twining) with jointed, more or less succulent stems and alternate, sheathing leaves. Inflorescence usually a cincinnus (cf. Boraginaceae). Flowers hermaphrodite, usually actinomorphic, commonly blue, sometimes subtended by a boat-shaped spathe or foliaceous bracts. Calyx of 3 free sepals. Corolla of 3 free, rarely united petals. Stamens typically 3 + 3, but frequently some reduced to staminodes, or absent; filaments often hairy; anthers basifixed, 2-celled, dehiscing by a longitudinal slit or (in *Dichorisandra*) by an apical pore. Ovary superior, of 3 united carpels, 3-locular, with a few orthotropous ovules in each loculus; placentation axile. Fruit usually a loculicidal capsule, rarely indehiscent; seed often arillate, with copious endosperm. Calcium oxalate present in tissues.

Economic and ornamental plants. The family is devoid of economic plants, but contains a few genera of horticultural interest. *Tradescantia* and *Commelina* include some hardy species which are suitable as border plants, and a larger number which require glasshouse conditions. *Zebrina pendula* is a popular house plant on account of the attractive striping on the upperside of its leaves.

Classification. The Commelinaceae are usually divided into the following 2 tribes:

1. Tradescantieae (flowers usually actinomorphic; 6 fertile stamens or 3 of one series more or less reduced or aborted).
   Tradescantia (60) America.
   Cyanotis (50) Old World tropics.
   Dichorisandra (35) tropical America.
2. Commelineae (flowers usually zygomorphic; usually 3 adjacent stamens only – from both series – functional).
   Commelina (230) tropics and subtropics.
   Aneilema (100) warm regions.

TRADESCANTIA × ANDERSONIANA Ludw. & Rohw.

Distribution. A hybrid assemblage of hardy perennials preferring rather moist soils. Many of the plants grown in European gardens as *T. virginiana* are really this hybrid, which is a cross between the true *T. virginiana* and other species of N. American origin, including *T. ohiensis* and *T. subaspera.*

Vegetative characteristics. An erect, fleshy, glabrous or slightly hairy perennial, from 30 cm to 1 m in height, with sheathing, linear-lanceolate leaves.

Floral formula. K3 C3 A3+3 $\underline{G}$(3)

Flower and inflorescence. The terminal, cymose inflorescence forms a cincinnus which is subtended by 1–3 leaf-like bracts (see Fig. 85.A). The actinomorphic, hermaphrodite flowers last only for a day, but as they open in succession, the plant has a long flowering period, extending from June to October. The colour of the flowers of the various cultivars ranges from blue and purple to white.

Pollination. Little seems to be known about the pollination of *T.* × *andersoniana*, but in the case of at least one of the parent species, *T. virginiana*, the flowers are protandrous. As they fade, the petals become pulpy and their surface is covered with a layer of liquid which attracts small flies. Plants grown in isolation are self-sterile and produce no seed, but when cultivated in association with others they are capable of producing viable seed which may result in the appearance of self-sown seedlings.

Alternative flowers for study. Any of the hardy members of the genus would be suitable alternatives, also some of the more tender species, e.g., *T. fluminensis*, which are usually grown for their foliage but will readily flower if not cut back. Comparison between genera will show some differences in floral structure, e.g., axillary inflorescences in *Rhoeo discolor*, gamopetalous flowers in *Zebrina pendula* and only 2 or 3 fertile stamens in *Commelina.*

Fig. 85 Commelinaceae, *Tradescantia* × *andersoniana*

A The inflorescence, a cincinnus, subtended by leaf-like bracts and sheathed by the foliage leaves. The flower of the day is fully open, and the large petals are spread out wide. The sepals, which are smaller, can be seen in the 2 drooping flower-buds.
Flowers: 30–35 mm in diameter Bud: 9 mm
Petals: 17–18 × 13–14 mm Sepals: 9–10 × 4–5 mm

B L.S. of lower portion of flower at early anthesis, showing the superior ovary bearing a style with a capitate stigma. Surrounding the ovary are 3 of the 6 stamens with long hairs on the filaments. The filaments will lengthen until they are level with or even exceed the stigma.
Young filament: 4 mm Mature filament: 9–10 mm
Anther: 1 × 1.75 mm Gynoecium: 11 mm

C A stamen detached from the flower. The basifixed anther is made up of 2 divergent cells which dehisce by longitudinal slits. The long hairs, which are attached to the filaments, are moniliform and, when viewed with a microscope, may be used for the study of protoplasm in circulation.
Filament hairs: *c.* 7 mm

D L.S. of the superior ovary, showing 2 of the 3 loculi with 2 orthotropous ovules in each loculus. The ovary is slightly hairy at its apex.
Ovary: 3 × 3 mm

E T.S. of ovary, showing the 3 loculi containing ovules attached to axile placentas.
T.S. of ovary: 2.5 mm

F The fruit at dehiscence surrounded by the 3 hairy, persistent sepals, one of which has been removed to reveal 2 of the 3 valves of the loculicidal capsule. The glandular hairs indicate that *T. subaspera* may be involved in the parentage of the plant illustrated, since *T. virginiana* and *T. ohiensis* have only eglandular hairs on the sepals.
Fruit: 6.5 × 4 mm Sepals: 11 × 5 mm

G A single seed with rugose surface. The seeds are somewhat flattened due to compression within the loculus.
Seed: 3 × 2 mm

Fig. 85

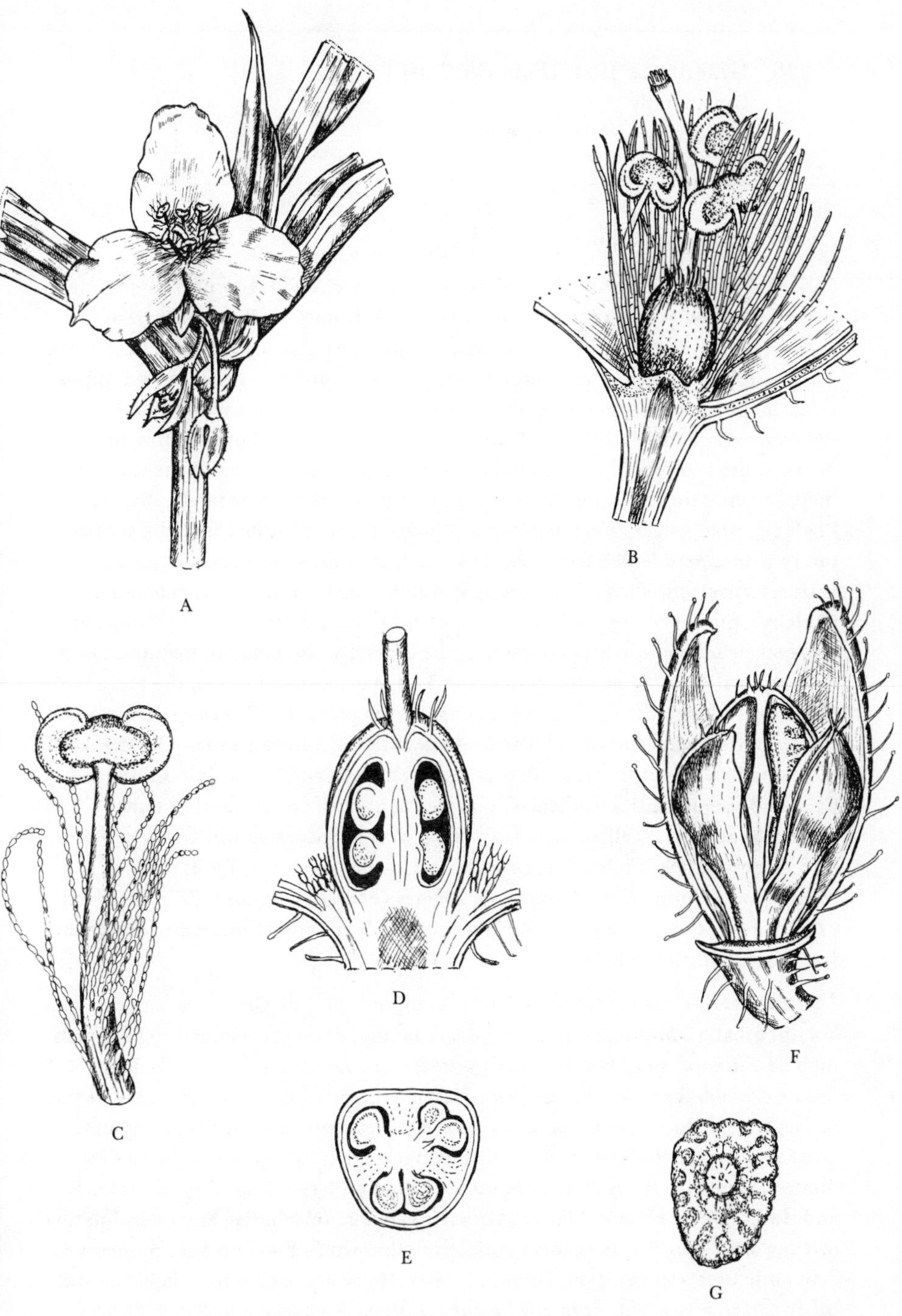

86 Gramineae Juss. (Poaceae Nash)

Grass family

620 genera and 10,000 species

Distribution. Cosmopolitan.

General characteristics. Annual or perennial herbs with fibrous roots, sometimes woody as in some of the Bamboos, forming loose to dense tufts or mats. Stems (culms) simple or branched, erect or prostrate, usually cylindrical, jointed, usually hollow except at the joints (nodes). Leaves alternate in 2 rows on opposite sides of the stem, originating at the nodes, often crowded at the base, and consisting of sheath, ligule and blade; sheath with margins free and overlapping or more or less united to form a tube round the stem; ligule (at junction of sheath and blade) membranous, sometimes reduced to a fringe of hairs or rarely absent; blade (lamina) usually long, narrow, and passing imperceptibly into the sheath, rarely with a petiole-like base, e.g., in some Bamboos; veins normally parallel. Inflorescence consisting of numerous spikelets which form spikes, racemes or panicles; spikelets composed of a series of alternating scales borne in 2 opposite rows on a short, jointed axis (rachilla), the 2 basal scales being termed the lower and upper glumes. Above the glumes are 1 to many pairs of scales, the outer scale of each pair being the lemma, the inner scale the palea, the 2 scales bearing a flower in their axil, the whole being termed a floret. Flowers usually hermaphrodite, small and inconspicuous, consisting of stamens and gynoecium subtended by 2 or 3 minute scales (lodicules) representing a reduced perianth. Stamens usually 3; anthers 2-celled, usually opening by a longitudinal slit. Ovary superior, of 3 united carpels, unilocular, containing 1 ovule; styles usually 2, rarely 1 (e.g., *Nardus*) or 3 (some of the Bamboos); stigmas generally plumose. Fruit 1-seeded, usually a caryopsis, rarely a nut or berry (some Bamboos). Endosperm abundant (see Fig. F. on page 442).

Economic and ornamental plants. Economically the Gramineae are probably of greater importance than any other family. Plants providing food for man include *Triticum* spp., Wheat, *Avena sativa*, Oats, *Hordeum vulgare*, Barley, *Secale cereale*, Rye, *Zea mays*, Maize (termed 'Corn' in the United States), *Oryza sativa*, Rice, *Panicum miliaceum*, Common Millet, *Sorghum vulgare*, Sorghum, and *Saccharum officinarum*, Sugar Cane. Many others are grown as fodder for domestic animals. *Phragmites communis*, Common Reed, is used for thatching and *Ammophila arenaria*, Marram Grass, is valuable for binding and consolidating drifting sand. The various genera collectively known as Bamboo have numerous economic uses, especially in Asian countries. Many grasses are used in lawns and other turfed areas, e.g., *Agrostis*, *Festuca*, *Lolium*, *Cynosurus* and *Poa*. Others grown for ornament in the garden or dried and used for floral decoration include

Cortaderia selloana, Pampas Grass, the hardy Bamboos, *Phalaris arundinacea* 'Picta', Ribbon Grass, *Lagurus ovatus*, Hare's-tail Grass, *Coix lacryma-jobi*, Job's Tears, *Setaria italica*, Foxtail Millet, and *Briza* spp., Quaking Grass.

Classification. The Gramineae form one of the largest families of flowering plants. Superficially they resemble the Cyperaceae, but are separated by the usually terete, hollow, jointed stems, distichous leaves and structure of the spikelets. They may be distantly related to the Commelinaceae. Between 50 and 60 tribes have been recognised, some of the more important being given below together with their principal genera:

1. Bambuseae.
   Arundinaria (150) warm regions.
2. Oryzeae.
   Oryza (25) tropics.
3. Stipeae.
   Stipa (300) tropical and temperate regions.
4. Arundineae.
   Arundo (12) tropical and temperate regions.
   Phragmites (3) cosmopolitan.
5. Bromeae.
   Bromus (50) temperate regions, tropical mountains.
6. Triticeae.
   Agropyron (100–150) temperate regions.
   Elymus (70) N. temperate region, S. America.
   Triticum (20) Europe, Mediterranean region, W. Asia.
   Hordeum (20) temperate regions.
7. Festuceae.
   Poa (300) cosmopolitan.
   Festuca (80) cosmopolitan.
   Briza (20) N. temperate region, S. America.
   Lolium (12) temperate Eurasia.
   Dactylis (5) temperate Eurasia.
8. Aveneae.
   Avena (70) temperate regions, tropical mountains.
   Holcus (9) Canary Islands, N. and S. Africa, temperate Eurasia.
   Arrhenatherum (6) Europe, Mediterranean region.
9. Agrostideae.
   Agrostis (150–200) cosmopolitan.
   Alopecurus (50) temperate Eurasia, S. America.
   Phleum (15) temperate Eurasia, America.
10. Meliceae.
    Melica (70) temperate regions.
11. Nardeae.
    Nardus (1) Europe, W. Asia, *N. stricta*.

12. Eragrostideae.
    Eragrostis (300) cosmopolitan.
13. Spartineae.
    Spartina (16) mainly temperate regions.
14. Paniceae.
    Panicum (500) tropical and warm temperate regions.
    Digitaria (380) warm regions.
    Paspalum (250) warm regions.
    Pennisetum (150) warm regions.
15. Andropogoneae.
    Andropogon (113) tropics and subtropics.
    Sorghum (60) tropics and subtropics.
16. Maydeae.
    Zea (1) tropical America, *Z. mays.*

ARRHENATHERUM ELATIUS (L.) J. & C. Presl
Tall or False Oat-grass

Distribution. Native in N. Africa, W. Asia and most of Europe including the British Isles, and introduced into N. America, Australia and New Zealand. It is a common constituent of rough grassland and is also found in hedgerows and on waste ground. It is a leafy grass which is palatable to cattle but does not withstand heavy grazing although it is most suitable for hay.

Vegetative characteristics. A tufted, usually glabrous perennial, with more or less erect stems 60–120 cm in height. The leaves are finely pointed, flat and scabrid, and the ligules short and membranous.

Floral formula. P2 A3 $\underline{G}$(3)

Flower and inflorescence. The inflorescence is a panicle with about 10 nodes from which spring clusters of 4–6 stalks bearing the pedicels of the spikelets. Each spikelet normally consists of 2 florets (occasionally 3 or 4), the upper one hermaphrodite and the lower male. The flowering period extends from June until September or even as late as November if the weather is suitable. The flowers usually open between 5 and 7 a.m., but some have been seen to open at 11 a.m. and others between 6 and 7 p.m. The observation of flowers at various stages may be made easier by spraying them carefully with a fixative as used by artists and hairdressers.

Pollination. Like other grasses, *A. elatius* is well adapted to wind-pollination, having pendulous stamens easily moved by currents of air, abundant pollen and feathery stigmas presenting a large surface to the wind-borne pollen from other plants. The flowers are also inconspicuous in colouring, another feature typical of those plants which rely on the wind as the chief pollinating agent rather than on insects or birds. The opening of the flowers is brought about by the lodicules which become swollen with sap and force apart the lemma and the palea. The filaments then rapidly elongate, causing the heavy anthers to be pushed over the sides and hang suspended, dehiscing as they fall. Not all the pollen, however, is shed at once. Some remains in the boat-shaped cavity, to be shaken out gradually by the wind. The flowers are homogamous, but the relative positions of stamens and stigmas, and the varying times of flower-opening, contribute to cross-pollination, though self-pollination can occur.

Alternative flowers for study. Any species of the closely related genus *Avena*, including *A. fatua*, Wild Oat, would be a suitable alternative. Floral variations within the Gramineae are many, and shape of inflorescence, arrangement of spikelets and number of florets in a spikelet are only a few of the differences that may be found. Comparison should be made not only with other native grasses but with those commonly grown as cereals, including the monoecious *Zea mays*, Maize. This monotypic genus has a stout, solid stem bearing large, broad leaves. The male inflorescence (tassel) is a terminal panicle, while the female inflorescences (cobs) arise from the leaf-axils lower down the stem. The cobs are enclosed by modified leaves (husks). Protruding from the ends of these are the numerous long styles (silks) which continue growing until pollination has taken place.

Fig. 86 Gramineae, *Arrhenatherum elatius*

A A single spikelet detached from the panicle. At its base are the 2 glumes, the outer one smaller than the inner (see C1 and C2). Above these can be seen the 2 awned lemmas. The lemma with the longer awn belongs to the male floret, while the one with the shorter awn belongs to the hermaphrodite floret.

B A spikelet at anthesis. The upper (hermaphrodite) floret is fully open and its 3 stamens are hanging down to shed their pollen. One of the 2 feathery stigmas protrudes just below the short-awned lemma that is backed by the upper glume. Two paleas can be seen in the centre, one from each floret. The rest of the lower (male) floret is hidden by the long-awned lemma that is backed by the lower glume.

C The 2 persistent glumes from the base of a spikelet. C1: The single-nerved, lower (outer) glume. C2: The 3-nerved, upper (inner) glume.
Lower glume: 4–5.5 x 1 mm Upper glume: 7.5–8 x 3 mm

D The lemma and palea of a male floret. D1: The 7-nerved lemma with its long, dorsal, geniculate awn. In contrast, the lemma of the hermaphrodite floret is awnless or only slightly awned and may have dorsal bristles (see B). D2: The palea, which has a small, hairy ridge at each side. The palea is angular and readily fits into the lemma, the 2 scales forming a protective chamber for the inner parts of the floret.
Long-awned lemma: 8 x 2 mm Awn: 14–17 mm
Short-awned lemma (not shown): 0–4.5 x 2 mm
Palea: 7 x 1 mm

E L.S. of a flowering spikelet, showing the 2 sessile florets borne on a short rachilla. The superior ovary of the upper (hermaphrodite) floret contains a single ovule in its loculus. The ovary is surmounted by 2 styles with feathery stigmas (one only shown). At its base can be seen one of the 2 lodicules which are considered to be the vestiges of a perianth. The 3 stamens are inserted on the rachilla between the ovary and the lodicules. The lower (male) floret exhibits an abortive ovary. Two of the 3 stamens are shown. The long, narrow structures in both florets are sections of the lower portions of the glumes, lemmas and paleas. At maturity, the part of the rachilla above the glumes breaks away, carrying with it both male and hermaphrodite florets (see Fig. G. on page 443).

F Detail of the 2-celled, basifixed anther which dehisces longitudinally, liberating a little of the pollen at a time from the boat-shaped cavity.
Filament: 5 mm Anther: 4 mm

G A hermaphrodite floret with stamens removed to reveal the superior ovary, terminated by 2 styles, and the 2 lodicules which become turgid at the base at the onset of anthesis and cause the flower to open. Below the lodicules is a portion of the lemma and behind the ovary is the lower part of the palea.
Ovary: 2 x 0.8 mm Styles: 3–4 mm
Lodicules: 1.5–2.25 mm

Fig. 86

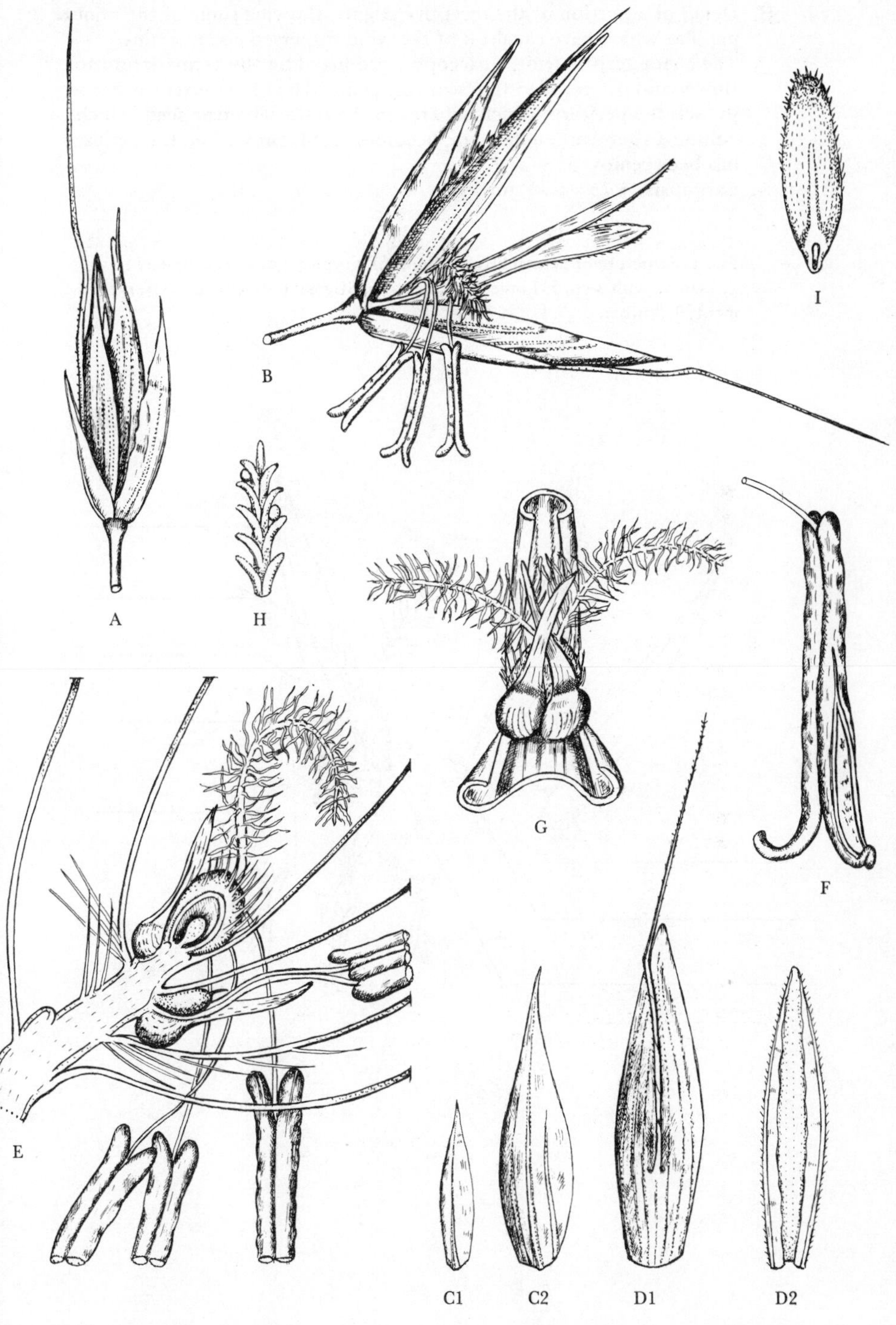

H Detail of a portion of the receptive stigma, showing some of the minute papillae which have caught 2 of the wind-dispersed pollen grains.

I The characteristic fruit, a caryopsis, produced by the hermaphrodite flower and often called the 'seed' or 'grain'. The hairy outer surface is, in fact, the pericarp, which adheres to the testa. The true seed, which contains abundant endosperm, becomes visible only when the pericarp has been removed.
Caryopsis: 4.75 × 1.75 mm

Fig. F. Structure of hermaphrodite floret of a typical grass: 1, Glume; 2, Lemma with awn; 3, Lodicule; 4, Palea; 5, Stigma; 6, Rachilla; 7, Ovary; 8, Filament; 9, Anther.

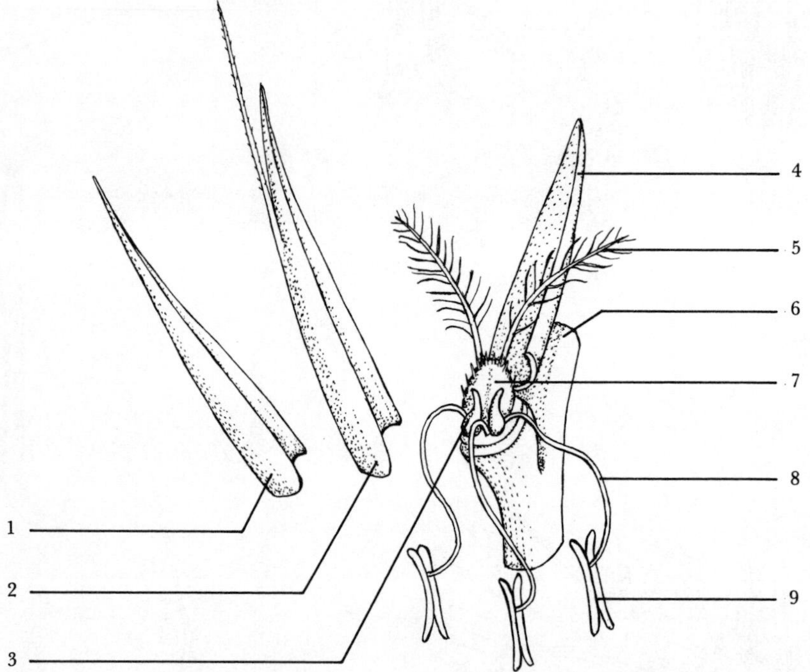

Fig. G. L.S. of flowering spikelet.
Hermaphrodite floret: 1, Glume; 2, Lemma; 3, Stigma; 4, Palea; 5, Lodicule; 6, Ovary; 7, Stamen.
Male floret: 8, Palea; 9, Aborted ovary; 10, Lodicule; 11, Lemma; 12, Glume; 13, Stamen.

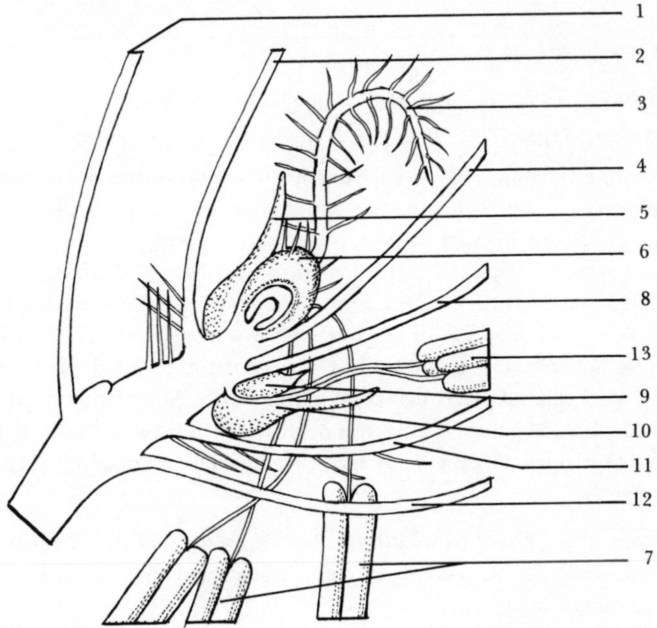

87 Juncaceae Juss.

Rush family

9 genera and 400 species

Distribution. Temperate and arctic regions, tropical mountains.

General characteristics. Plants (sometimes dioecious) of damp, cool places, usually with a creeping, sympodial rhizome, one joint of the sympodium appearing above ground each year as a leafy shoot. Leaves usually narrow, either flattened or circular in cross-section. Inflorescence usually a crowded mass of flowers in cymes of various types. Flowers usually hermaphrodite, actinomorphic, wind-pollinated. Perianth of 6 segments, in 2 whorls of 3, usually sepaloid. Stamens 6, in 2 whorls of 3 (or the inner whorl absent); anthers dehiscing laterally or introrse; pollen in tetrads. Ovary superior, of 3 united carpels, unilocular with 3 parietal placentas (basal in *Luzula*) or 3-locular with axile placentation, with 1 to many anatropous ovules in each loculus. Style simple, with 3 brush-like stigmas. Fruit a loculicidal capsule; embryo straight, in starchy endosperm.

Economic and ornamental plants. Various species of *Juncus*, Rush, are used commercially in the making of mats, baskets and chair-seats, and horticulturally in waterside planting.

Classification. The chief genera are:

Juncus (300) cosmopolitan.
Luzula (80) cosmopolitan, especially temperate Eurasia.

LUZULA CAMPESTRIS (L.) DC.

Field Wood-rush

Distribution. Almost cosmopolitan, and common throughout the British Isles in grassy places that are often drier than those normally associated with rushes.

Vegetative characteristics. A compact, tufted perennial with shortly creeping stolons. Leaves mostly radical, grass-like, with sheathing bases, and thinly clothed with long, colourless hairs (see Fig. 87.A).

Floral formula. P3+3 A3+3 $\underline{G}(3)$

Flower and inflorescence. The actinomorphic, hermaphrodite, chestnut-brown flowers are arranged in clusters that together form a panicle. The flowering period is from March until June.

Pollination. The absence of nectar or scent, the dull colour of the perianth, the prominent anthers and the protruding, papillose stigmas (see Fig. 87, C and E) together indicate adaptation to wind-pollination.

Alternative flowers for study. Any other member of the genus would be suitable, including the larger species *Luzula pilosa*, Hairy Wood-rush, which is widespread and locally frequent in woods and hedgebanks, and which flowers from April until June. Comparison should be made with species of the very large genus *Juncus*, of wide distribution in damp places. The more important floristic differences between the two genera are that *Juncus* has a many-seeded capsule (3-seeded in *Luzula*) and that the seed-coat is finely sculptured (smooth in *Luzula*). Reduction in the number of flowers is exhibited by *J. biglumis*, Two-flowered Rush, *J. triglumis*, Three-flowered Rush, and *J. trifidus*, Three-leaved Rush, which has a solitary terminal flower and up to 3 lateral flowers.

Fig. 87 Juncaceae, *Luzula campestris*

A The whole plant, showing the flowering stem arising from the tuft of leaves. The stem rarely exceeds 15 cm in height and bears a loose, cymose inflorescence consisting of 1 sessile and 3–6 stalked clusters of flowers, 3–12 in each cluster. The branches of the inflorescence are more or less curved and become reflexed in fruit. Note the sheathing bases of the leaves and the long hairs on their margins.
Flowering stem: 11 cm Leaves: up to 3.5 cm
Cluster of flowers (prior to anthesis): *c.* 8 × 8 mm

B A single, stalked flower at late bud stage, subtended by bracteoles (2 of the 3 are shown). The perianth-segments, in 2 whorls of 3, are chestnut-brown with a transparent margin. The 3 papillose stigmas protrude well beyond the perianth.

C The actinomorphic, hermaphrodite flower at anthesis. The superior ovary, with its single style terminated by the 3 stigmas, is surrounded by 6 free stamens situated opposite the 6 free perianth-segments.
Flower: *c.* 6 mm in diameter Perianth-segments: 3 × 0.75–1 mm
Ovary: 2.5 mm Style + stigmas: 3.5–4 mm

D L.S. of part of a flower. The bases of the filaments broaden out into a well-developed ring round the base of the ovary (see C). The position of the ovules is just visible through the thin ovary wall. Parts of the bracteoles can be seen below the perianth-segments. The prominent yellow anthers are basifixed and 2-celled, each cell dehiscing by a longitudinal slit to liberate the pollen. The pollen grains remain in tetrads.
Filaments: 0.3–0.9 mm Anthers: *c.* 1.75 mm

E Detail of a brush-like stigma showing the long papillae which catch the wind-borne pollen.
Papillae: *c.* 0.015 mm

F L.S. of the superior ovary prior to anthesis surrounded by part of the perianth. The ovary, which is formed from 3 united carpels, has a single loculus containing 3 ovules (2 only are shown). The anatropous ovules are attached to a central, basal placenta.
Ovary: 2.3 × 0.6 mm Ovule: 0.4 × 0.3 mm

G T.S. of the unilocular ovary with the 3 ovules.
Ovary: 0.6–0.7 mm in diameter

H The fruit at dehiscence, surrounded by the persistent perianth. The single-celled, loculicidal capsule dehisces by 3 valves to shed the 3 smooth, shiny seeds which have a white basal appendage up to half their length.
Valves of capsule: *c.* 2.5 × 1.25 mm Seeds: 1 × 0.75 mm

Fig. 87

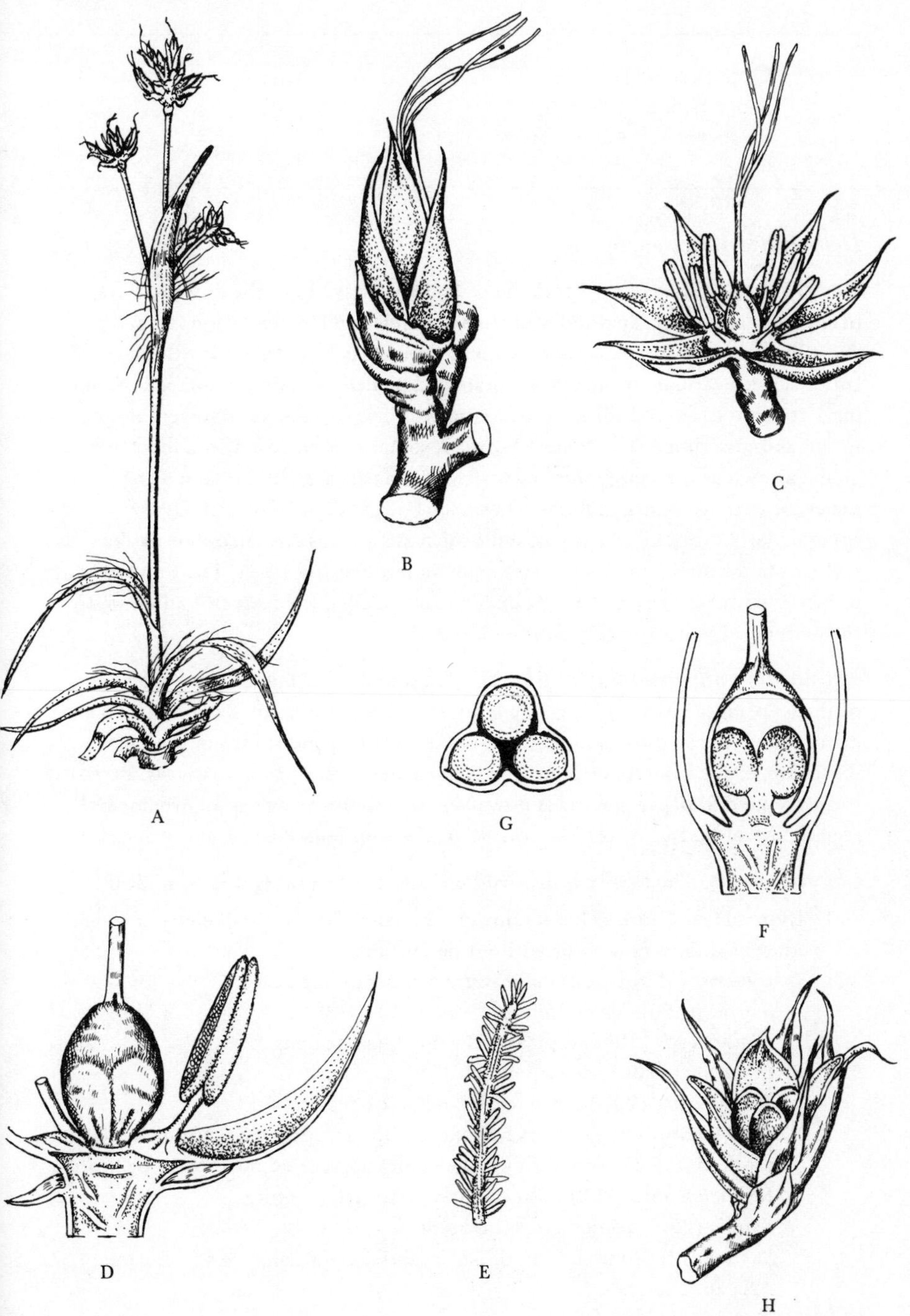

88 Cyperaceae Juss.

Sedge family

90 genera and 4000 species

Distribution. Cosmopolitan.

General characteristics. Perennial or annual, grass-like herbs of wet places, often with a creeping, sympodial rhizome. Culms solid, often trigonous, with sheathing leaves in 3 ranks. Sheaths usually entire, not articulated with blade. Inflorescence various, from a much-branched panicle to a simple spike, containing 1 to many-flowered spikelets of hermaphrodite or unisexual flowers arising in the axil of a glume. The female flower is sometimes enclosed by a modified glume known as the perigynium or utricle. Perianth of bristles (often 6) or scales, or entirely wanting. Stamens usually 1–3; anthers basifixed. Ovary superior, of 2 or 3 united carpels, unilocular, with 1 basal, anatropous ovule; style often deeply divided, branches equal in number to carpels. Fruit a trigonous or biconvex achene (except *Scirpodendron* and allies), the testa not adhering to the pericarp. Flowers usually wind-pollinated.

Economic and ornamental plants. The starchy rhizome of some members of the Cyperaceae is edible, e.g., *Cyperus esculentus*, Chufa or Tiger-nut, while many others are employed in mat and basket-making, and some are used for thatching, e.g., *Cladium mariscus*, the Great Fen-sedge. *Cyperus papyrus*, Papyrus, formerly used in paper-making is nowadays sometimes grown as an ornamental aquatic in hot-houses, as is *C. alternifolius*, the Umbrella Plant.

Classification. The family is divided here into the following 3 subfamilies:

I. Cyperoideae (flowers hermaphrodite in many-flowered spikelets, or single unisexual flowers with or without perianth).

Cyperus (550) tropical and warm temperate regions.
Scirpus (300) cosmopolitan.
Fimbristylis (300) tropical and subtropical regions.
Eleocharis (200) cosmopolitan.
Eriophorum (20) N. temperate and arctic region.

II. Rhynchosporoideae (flowers hermaphrodite or unisexual, with or without perianth, in few-flowered, spike-like cymes aggregated into spikes or heads).

Rhynchospora (200) cosmopolitan, especially tropics.
Scleria (200) tropics and subtropics.
Schoenus (100) mainly in S. Africa and Australasia; a few in Europe, Asia and America.
Mapania (80) tropics.

Cladium (50–60) tropical and temperate regions, especially Australia.
Scirpodendron (1) Indomalaysia, Australia and Polynesia; perhaps the most 'primitive' member of the family.

III. Caricoideae (flowers unisexual, naked, usually in many-flowered spikes; female flowers enclosed by a utricle persisting in fruit).

Carex (1500–2000) cosmopolitan, especially temperate regions.
Uncinia (35) mountains of Borneo, New Guinea, Australasia, Central and S. America, antarctic islands.

CAREX ACUTIFORMIS Ehrh.
Lesser Pond-sedge

Distribution. Commonly found growing on the banks of slow-flowing rivers and ponds in most of Europe (except the extreme north) and in temperate areas of Asia, N. and S. Africa, and N. America. In the British Isles it is scattered but locally abundant.

Vegetative characteristics. A glabrous perennial with creeping rhizomes, sharply trigonous stems and long, narrow leaves.

Floral formula. Male: P 0 A 3 Female: P 0 $\underline{G}$(3)

Flower and inflorescence. Each inflorescence is a spike of usually all male or all female flowers, the terminal and upper lateral spikes being male, the lower spikes female. However, as in many other species, a spike containing both male and female flowers may occur. The flowering period is June to July.

Pollination. As in other members of the Cyperaceae the main pollinating agent is the wind, although pollen-eating beetles may also transport pollen. Most British species of *Carex* are protandrous, but those with several male spikes may still be shedding pollen when the first stigmas in the female spikes become receptive.

Alternative flowers for study. Any other species of *Carex* would be suitable, but it should be noted that in some species the ovary is biconvex and has only 2 stigmas. This situation occurs in all British members of subgenus Vignea and in some members of subgenus Carex. Most species of *Carex* are in flower between May and July.

Fig. 88 Cyperaceae, *Carex acutiformis*

A The upper portion of the flowering shoot, showing one terminal and one lateral male spike, and below them 2 female spikes arising from the axils of the leaf-like bracts.
Male spikes: 10–45 mm Female spikes: 35–50 mm

B Female spike at a later stage of development. Compare with A. Note point of attachment of the spike in the axil of the bract.

C The male flower with the 3 young stamens subtended by a dark brown glume with a pale midrib.

D A mature male flower showing the elongated filaments and the longitudinal dehiscence of the 2-celled, basifixed anthers.
Filament (in young flower): 1 mm
(fully developed): 7 mm
Anther: 3.5 mm

E A single female flower showing the 3 stigmas protruding from the beaked perigynium (see F). The stigmas, which fall soon after the ovule has been fertilised, have small hair-like structures attached to them, called papillae.

F L.S. of fruit. The fruit consists of an inflated sac known as the perigynium or utricle which completely surrounds the ovary.

G T.S. of the ovary at an early stage. The solitary basal, erect and anatropous ovule is doubly protected, being enclosed both by the wall of the ovary and by the perigynium.
Ovary: 2 × 1 × 0.4 mm

H T.S. of fruit after fertilisation. The single seed is contained within the 3-sided ovary wall which later hardens to form a trigonous, indehiscent nut. In its development, the nut partly fills the perigynium.
T.S. of ovary after fertilisation: 2.5 × 2 mm
Ovule: 0.6 mm in diameter

I The mature fruit showing the beak at the apex of the perigynium, a useful diagnostic character.
Fruit: 3.5 × 2 mm

Fig. 88

89 Typhaceae Juss.

Reedmace family

1 genus and 15 species

Distribution. Temperate and tropical regions.

General characteristics. Monoecious, perennial herbs with thick rhizomes and erect stems. Leaves alternate, in 2 ranks, linear, parallel-veined, sessile. Inflorescence a dense spike, the smaller, upper portion consisting of usually yellow, male flowers and the lower, cylindrical part of brown, female flowers. Flowers surrounded by slender hairs or more or less forked scales (possibly the reduced perianth). Male flowers with 2–5 stamens; filaments free or connate; anthers linear-oblong, basifixed, the connective projecting beyond the anthers; pollen often in tetrads. Female flowers of 1 carpel; ovary on a stipe bearing slender hairs and containing 1 pendent ovule. Fruit an achene, bearing the persistent style and surrounded by the persistent threads or scales which aid in distribution. Seed with straight embryo and mealy endosperm.

Economic and ornamental plants. The rootstocks of some species of *Typha* are used locally as food, and the leaves in making mats and baskets. Horticulturally, the hardy species are of value as decorative, aquatic plants for pond and river margins.

Classification. The only genus in the family is

Typha (15) temperate and tropical regions.

TYPHA LATIFOLIA L.

Great Reedmace, Cat's-tail

Distribution. Native throughout much of the N. hemisphere, including the British Isles where it is a common and sometimes dominant feature of ponds, lakes, slow-flowing rivers and canals, especially on inorganic substrata or where there is silting and rapid decay of organic matter.

Vegetative characteristics. A robust, monoecious perennial with erect stems up to 1.5–2.5 m in height arising from a thick rhizome. The linear leaves, sheathing at the base and mostly radical, often overtop the inflorescence.

Floral formula. Male: P0 A3   Female: P0 $\underline{G}$1

Flower and inflorescence. In June and July the long stem projects high above the water surface and bears a prominent inflorescence in the form of a dense spike, often termed a spadix (see Fig. 89.A). The spike is in 2 parts, the upper portion, which is yellow in colour, containing the male or staminate flowers, while the lower, brown portion is composed of female or pistillate flowers. The 2 portions are usually contiguous.

Pollination. The male flowers produce a large amount of pollen which is carried by the wind to the stigmas of the female flowers. The question of whether the flowers are protandrous or protogynous appears to be unresolved at present.

Alternative flowers for study. T. latifolia is the most readily available species in the British Isles, though *T. angustifolia*, Lesser Reedmace, which is locally common and also flowers in June and July may be used as an alternative. *T. angustifolia* is distinguished from *T. latifolia* by its more slender inflorescence, and by a 1–9 cm space between the groups of male and female flowers. Also, the pistillate flowers of *T. angustifolia* are subtended by bracteoles which are absent in *T. latifolia.*

Fig. 89 Typhaceae, *Typha latifolia*

A The complete inflorescence, a dense, terminal spike. The female or pistillate flowers form a dark brown, cylindrical mass in the lower part of the inflorescence, while the yellow, male or staminate flowers are situated immediately above them. The group of male flowers tapers to a point, and in the drawing there is a slight abnormality in the form of a leaf-like structure about half way up.
Female inflorescences: 100–200 × 15–25 mm
Male inflorescences: 105–170 × *c.* 15 mm

B A portion of the main axis with 3 male flowers. Each flower has 3 stamens whose filaments are connate below, forming a common stalk. Surrounding the filaments are slender hairs which are considered by some authorities to be the reduced perianth.
Anthers: 2.25–2.5 mm Connate filaments: up to 10 mm
Hairs: up to 6 mm

C Detail of a single male flower, showing the 3 stamens on their common stalk. The stamens have linear-oblong anthers with the connective projecting slightly beyond them. The pollen is grouped into tetrads which are easily dispersed by the wind.

D A portion of the main axis with one of the female flowers that arise from short, lateral outgrowths. In the drawing the flower is shown erect although in nature it would project horizontally from the stem. The flower is borne on a slender stalk or gynophore and consists of a spindle-shaped ovary crowned by a long style which broadens at the apex into the stigma. From the base of the gynophore arise 30–50 long slender hairs.
Gynophore: 5 mm Ovary: 1–1.25 mm
Style: 5 mm Stigma: 2.25 mm
Hairs: 11–23 mm

E Detail of the base of the gynophore, showing the point of attachment of some of the numerous hairs. The gynophore and the hairs later lengthen so as to aid dispersal of the fruit by the wind.

F L.S. of the ovary at late anthesis. The unilocular ovary is formed from one carpel and contains a single, large, pendulous ovule. The base of the ovary tapers into the gynophore while the apex tapers into the style.

G The fruit at an early stage of development bearing the withering remains of the style. The fruit later develops into an achene.
Young fruit: 1 × 0.3 mm

H An abortive ovary detached from the gynophore of a sterile female flower. Such flowers are not uncommon and are variously shaped. A vestigial style terminates the abortive ovary.
Abortive ovary: 1 × 0.4 mm

I The seed removed from its protective covering. The funicle, which joins the seed to the wall of the ovary, can be seen at the upper end.
Seed: 1.25 mm

Fig. 89

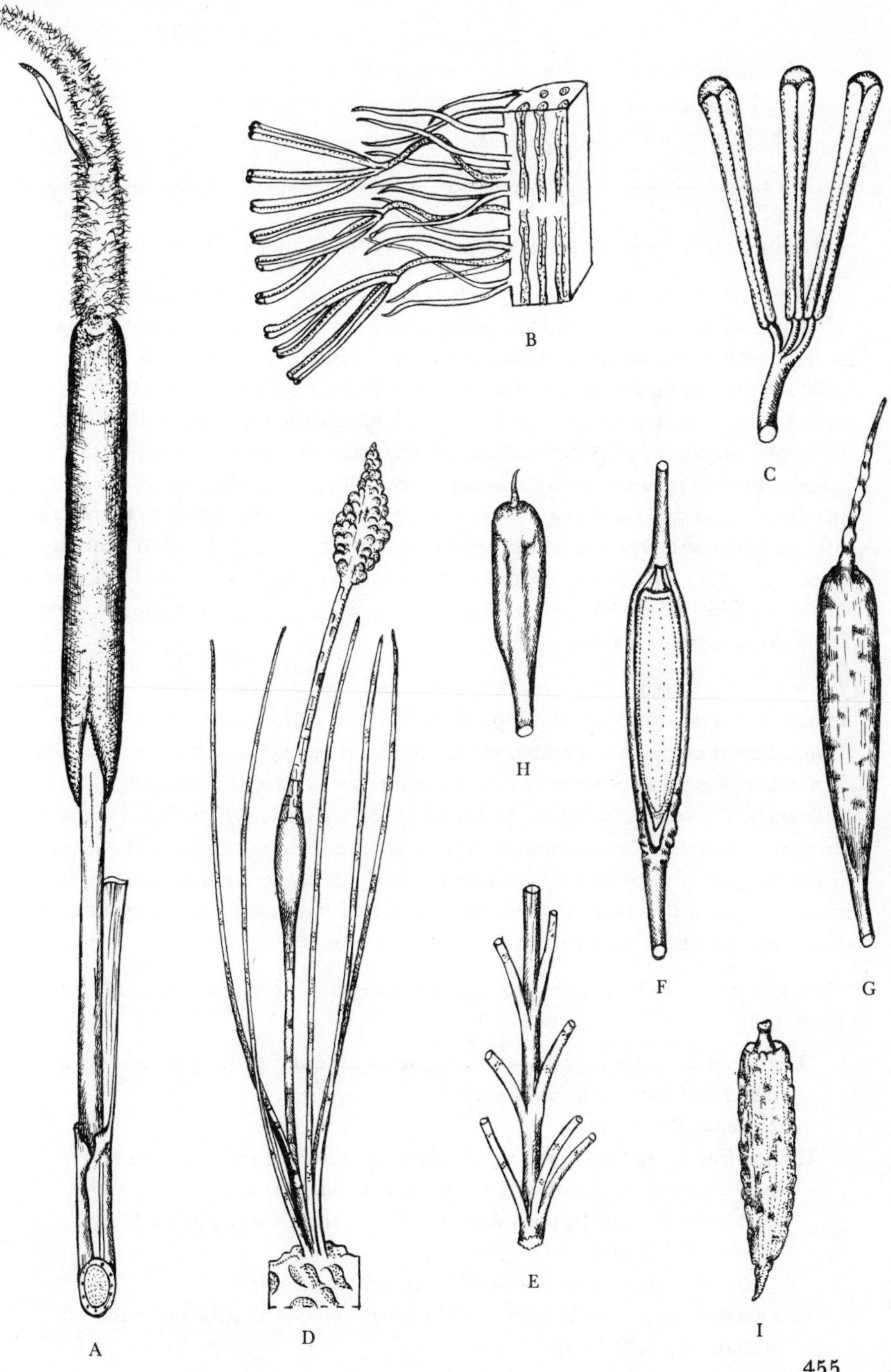

90 Bromeliaceae Juss.

Pineapple family

45 genera and 2000 species

Distribution. Chiefly tropical America, W. Indies; 2 species in W. Africa.

General characteristics. Mostly epiphytes, often xerophytic in habit. Stem usually reduced, with a rosette of fleshy leaves, channelled above, close-fitting at base, so that the whole plant forms a kind of funnel which fills with water, dead leaves, decaying animal matter, etc. The leaves have a thick cuticle, and their bases are covered with scaly hairs which enable the water to be absorbed. Inflorescence a head or panicle usually arising from the centre of the plant; coloured bracts present. Flowers usually hermaphrodite, actinomorphic. Calyx persistent, of 3 free sepals. Corolla of 3 free or united petals. Stamens 6, introrse, often epipetalous. Ovary inferior, half-inferior or superior, of 3 united carpels, 3-locular, with numerous anatropous ovules on axile placentas. Style 1, stigmas 3. Fruit a berry or capsule; seeds, in the latter case, very light or winged; embryo small, in mealy endosperm.

Economic and ornamental plants. The principal economic plant in this family is *Ananas comosus*, Pineapple, which is now cultivated in Hawaii and tropical areas of the Old World as well as in its native region of tropical America. A number of genera are grown as indoor plants for ornament, including *Tillandsia*, *Vriesea*, *Nidularium*, *Billbergia*, *Aechmea* and *Cryptanthus*. *Tillandsia usneoides*, Spanish Moss, is sometimes cultivated in glasshouses as a curiosity, as it has the general appearance of a lichen and the ability to grow pendent and rootless from any suitable object. Some species of *Puya* are fairly hardy and may be grown outside in milder parts of the British Isles.

Classification. The classification into 4 subfamilies is based on that of Harms (in Ref. 9).

I. Navioideae (leaves spinose-dentate; ovary superior, fruit a capsule; seeds naked, without wings or hairs).

Navia (60) Guyana.

II. Pitcairnioideae (leaves entire or spinose; ovary superior or half-inferior; fruit a capsule; seeds usually variously appendaged).

Pitcairnia (250) tropical America, W. Indies, 1 species in W. Africa.
Puya (120) Andes.
Dyckia (80) warm regions of S. America.

III. Tillandsioideae (leaves entire; ovary superior; fruit a capsule; seeds with plumose coma of hairs).

Tillandsia (500) warm regions of S. America, 1 species in W. Africa.
Vriesea (190) tropical America.

IV. Bromelioideae (leaves toothed or spinose; ovary inferior; fruit a berry; seeds naked).

Aechmea (150) W. Indies, S. America.
Billbergia (50) warm regions of America.
Bromelia (40) tropical America, W. Indies.
Ananas (5) tropical America.

BILLBERGIA NUTANS H. Wendl.

Distribution. Native in Brazil and introduced into the British Isles in 1868. It is a popular pot-plant for the house or glasshouse.

Vegetative characteristics. An epiphytic perennial that readily produces suckers. The numerous basal leaves are stiff, linear, greyish green and have small teeth on the margins.

Floral formula. K3 C3 A6 $\overline{G}(3)$

Flower and inflorescence. An arching peduncle, clothed with conspicuous rose-pink bracts bears 4–12 pendent, hermaphrodite, actinomorphic flowers on slender pedicels. Flowering may take place at any time between November and March.

Pollination. Humming-birds are considered to be the most likely pollinators of *Billbergia nutans*, although insects may also play a part. The birds would be attracted by the prominent pink bracts along the stem and by the pink and green flowers which are a source of nectar. At the base of the petals are scales (see Fig. 90.F) which form 6 tubular structures on their inner surface. Because of the pendulous habit of the flowers, nectar, which is secreted at the base of the style, flows along the channels formed by the scales, leaving the centre of the flower dry. The humming-bird would be able to insert its tongue into the flower and reach the nectar which drips from the end of the scales. As it hovered there, its beak or feathers would come into contact with the exserted stamens and would be dusted with pollen. This would be transferred to the equally prominent style of another flower. Although stamens and style are close together, self-pollination is prevented by the protandry of the flowers.

Alternative flowers for study. Billbergia nutans is a readily available species but any other member of the genus would be an acceptable substitute, including the popular hybrid *B.* × *windii*, a cross between *B. nutans* and *B. decora*, which has slightly larger flowers. Comparison with genera in other subfamilies will reveal differences in the position of the ovary and the form of the fruit, a superior ovary forming a capsular fruit and an inferior ovary a berry. *Ananas*, like *Billbergia*, has an inferior ovary, but here the whole inflorescence including floral axes, bracts and fruit become fleshy to form a succulent syncarp. Growth of the main stem continues beyond the inflorescence producing the characteristic apical tuft of leaves.

Fig. H. T.S. of *Billbergia nutans* flower just above ovary: 1, Sepal; 2, Petal with 2 basal scales; 3, Style; 4, Filament.

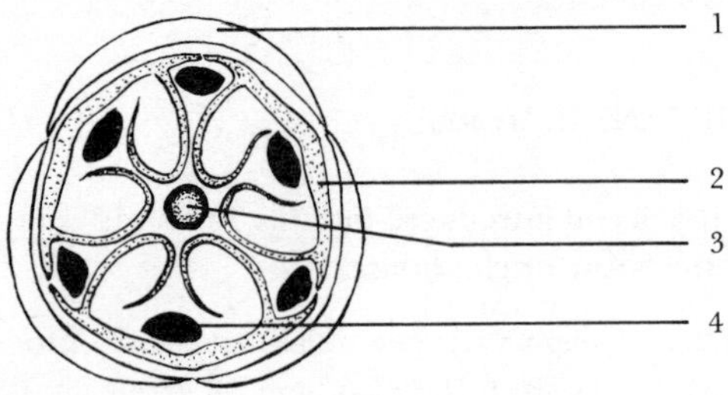

Fig. 90 Bromeliaceae, *Billbergia nutans*

A The actinomorphic, hermaphrodite flower with inner and outer perianth-segments clearly differentiated. The 3 sepals are salmon-pink with a purplish blue margin, and the 3 reflexed petals are yellowish green with a deep blue margin. The long style protrudes even beyond the 6 well-exserted stamens.
Flower: 45 mm Sepal: 15 mm Petal: 32 mm

B Detail of upper portion of stamen, showing the versatile anther which is introrse, 2-celled and dehisces longitudinally.
Filaments: 30–32 mm Anthers: 7 mm

C L.S. of lower portion of flower. The perianth is situated on a marginal ridge above the inferior ovary, while the style rises from the centre of a cup-shaped depression. The filaments (2 shown) are attached to the base of the petals. Two of the scales that function as nectar-channels can be seen (also see F).

D Detail of upper portion of style, terminated by 3 spirally twisted stigmas.
Stigma: 6 mm

E L.S. of ovary, showing the numerous anatropous ovules attached to axile placentas.
Ovary: 10 mm

F L.S. of the lower part of the perianth, exposing 3 of the tubular, basal scales. Two of the scales are shown in their entirety, but one has been cut longitudinally to show the structure. The bases of 2 of the filaments can be seen between the scales.
Scale: 5 mm

G T.S. of the 3-locular ovary with numerous ovules in each loculus. After fertilisation, the ovary may develop into a many-seeded berry, though this rarely forms in cultivation.
Ovary: 4 mm in diameter

Fig. 90

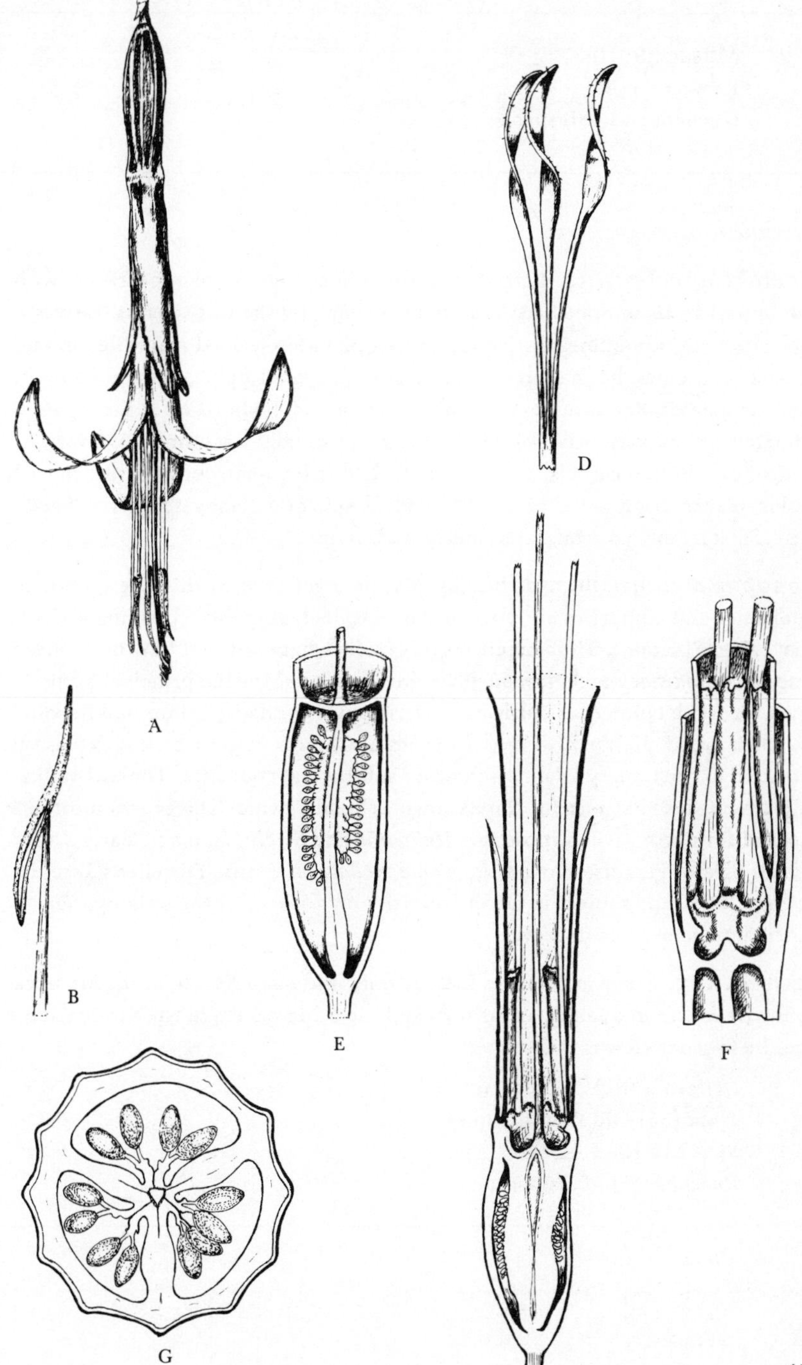

91 Musaceae Juss.

Banana family

6 genera and 130 species

Distribution. Tropics.

General characteristics. Large herbs, sometimes tree-like in appearance, with what appear to be unbranched aerial stems formed by the leaf-sheaths. Leaves large, alternate, sometimes distichous, entire, pinnately-veined and rolled in bud. Flowers in racemes, hermaphrodite or unisexual, zygomorphic, subtended by bracts or spathes. Perianth petaloid, of 6 segments in 2 whorls of 3, the sepals and petals free or variously united. Stamens 6 (1 usually a staminode). Ovary inferior, of 3 united carpels, 3-locular, with 1 to many anatropous ovules in each loculus; placentation axile. Fruit a 3-celled capsule or an elongated berry. Seeds often arillate; embryo straight, in mealy perisperm.

Economic and ornamental plants. *Musa* is a genus of world-wide economic importance and comprises all types of Bananas, including those cooking varieties often called Plantains. The nomenclature is somewhat confused but the Linnean names *M. paradisiaca* and *M. sapientum* are often used for the principal kinds grown for food. Numerous cultivars, differing in size, shape, colour and flavour, have been raised. *M. nana* is a dwarf species of Chinese origin which is extensively cultivated in the Canary Islands and other subtropical countries. The leaf-stalks of *M. textilis* provide a useful fibre known as Manila hemp. The genera most favoured for decorative purposes are *Heliconia* and *Strelitzia*, particularly *S. reginae*, Bird-of-Paradise Flower. *Ravenala madagascariensis*, Traveller's Tree, is used as a landscape subject in warm countries on account of the striking, fan-shaped arrangement of its leaves.

Classification. Some authorities include only *Musa* and *Ensete* in the Musaceae, placing *Heliconia* in a family of its own and the other genera in the Strelitziaceae. Here the broader view is maintained.

Heliconia (80) tropical America.
Musa (35) Old World tropics.
Strelitzia (5) S. Africa.
Ravenala (1) Madagascar.

STRELITZIA REGINAE Banks
Bird-of-Paradise Flower

Distribution. Native in the coastal region of the Cape Province of South Africa, where it grows along river banks and in open glades of the bush, and widely cultivated in gardens in warm climates. It was introduced into the British Isles in 1773 and is a striking plant for the warm glasshouse. It is best propagated by division, as seed germination is very slow.

Vegetative characteristics. A vigorous, herbaceous perennial, reaching a height of about 1.2 m in its native habitat and up to 2 m in cultivation. The dark green, oblong-lanceolate, leathery leaf-blades are borne on very long petioles which arise in 2 ranks from a large rhizome with numerous fleshy roots.

Floral formula. K3 C(3) A5 $\overline{G}$(3)

Flower and inflorescence. The long, rigid, cylindrical scape is terminated by a short, condensed flower-spike enclosed in a green bract tinged with pink (see Fig. 91.A). From this more or less horizontal, sheathing bract arise the orange and blue flowers. They usually appear in succession at any time from January to June, each flower lasting about a week, and the time of flowering may vary somewhat according to the prevailing conditions.

Pollination. The flowers of *Strelitzia* are adapted to pollination by birds, but they are quite different in structure from the flowers of *Columnea* (Gesneriaceae) and *Callistemon* (Myrtaceae) which are also ornithophilous. The bright orange sepals are held in a more or less erect position and are a conspicuous feature of the flower, contrasting with the blue petals. The 2 larger petals are united along their lower margins to form a trough in which lie the 5 stamens and the style, while the third and smaller petal is transformed into a nectary. The larger petals are lobed towards the base, making the structure arrow-shaped (see Fig. 91.C). In their native habitat, the flowers are visited by several species of sugar-birds and sun-birds. When a bird in search of nectar alights on the lobes, its weight causes the upper edges of the petals to part, exposing the anthers which deposit their sticky pollen on to the underside of the bird's body. The flowers are protandrous and in the first stage, when the anthers are shedding pollen, the stigma is immature, so self-pollination cannot take place. However, if the bird subsequently alights on older flowers, where the stamens have withered, the pollen is deposited on to the receptive stigma and cross-pollination is achieved.

Alternative flowers for study. S. reginae is the commonest species of *Strelitzia* found in cultivation, though *S. augusta*, a much larger plant with white flowers is occasionally grown and may be used as an alternative. If possible, comparison should be made with other genera, e.g., *Heliconia*, which has decorative foliage and brilliantly coloured bracts, and more particularly *Musa*, whose fruit is the Banana. The flowers of the latter genus are borne in terminal spikes. The flower-clusters towards the apex of the spike are, by abortion, functionally male, while the fruit-forming flowers are clustered towards the base of the inflorescence. In the wild the fruit is an elongated berry with few or many seeds depending on the species, but the fruit of the cultivated Banana is seedless.

Fig. 91 Musaceae, *Strelitzia reginae*

A The terminal flower-spike, supported on a rigid, cylindrical scape. In the illustration 3 flowers have emerged from the horizontal bract to form a 'crest'.

B A single, zygomorphic, hermaphrodite flower still attached to the scape. The bract and the 2 other flowers have been removed for clarity. The outer whorl of the perianth consists of 3 long-pointed, orange sepals, one being somewhat boat-shaped and more acuminate than the others. The inner whorl is made up of 3 blue petals, 2 of which are united to enclose the stamens and style. Below the perianth the long, inferior ovary resembling a pedicel arises from the cavity of the bract. Parts of the ovaries of the 2 accompanying flowers can be seen.
Lateral sepals: 114 × 26–30 mm Boat-shaped sepal: 100 × 10 mm

C A flower with the sepals removed. The 2 large petals lie parallel to each other and are united along their lower margins to form a sagittate structure that encloses the anthers and the style. The third, free petal, which is considerably shorter, acts as a nectary. Just below the short petal is the point of attachment of 2 of the sepals. The style, which lies in the groove formed by the 2 united petals, is terminated by a long, pointed stigma that protrudes well beyond the apex of the petals.
United petals: 44 × 20 mm Free petal: 17 × 10 mm
Stigma: 26 mm

D A portion of the groove formed by the 2 united petals, showing the pollen grains on 2 of the 5 linear anthers. The anthers, which dehisce longitudinally by means of slits, are attached to rigid filaments below the sagittate structure. The prominent wavy line indicates the edge of the groove.
Portion of groove illustrated: 4.5 × 2 mm
Anther: 0.4 mm wide

E L.S. of the inferior ovary, exposing one of the loculi containing numerous ovules. If fertilisation is successful the ovary develops into a 3-valved, loculicidal capsule. Inside the capsule are shiny, black seeds possessing a fluffy, bright orange aril which contains a large amount of fat attractive to birds. While eating the arils the birds scatter the seeds about and aid their dispersal. At the base of the ovary illustrated are the lower portions of 2 other flowers.
Ovary: 60 × 6 mm

F T.S. of the ovary formed from 3 united carpels, showing the anatropous ovules attached to axile placentas.
T.S. of ovary: 10 × 6 mm

Fig. 91

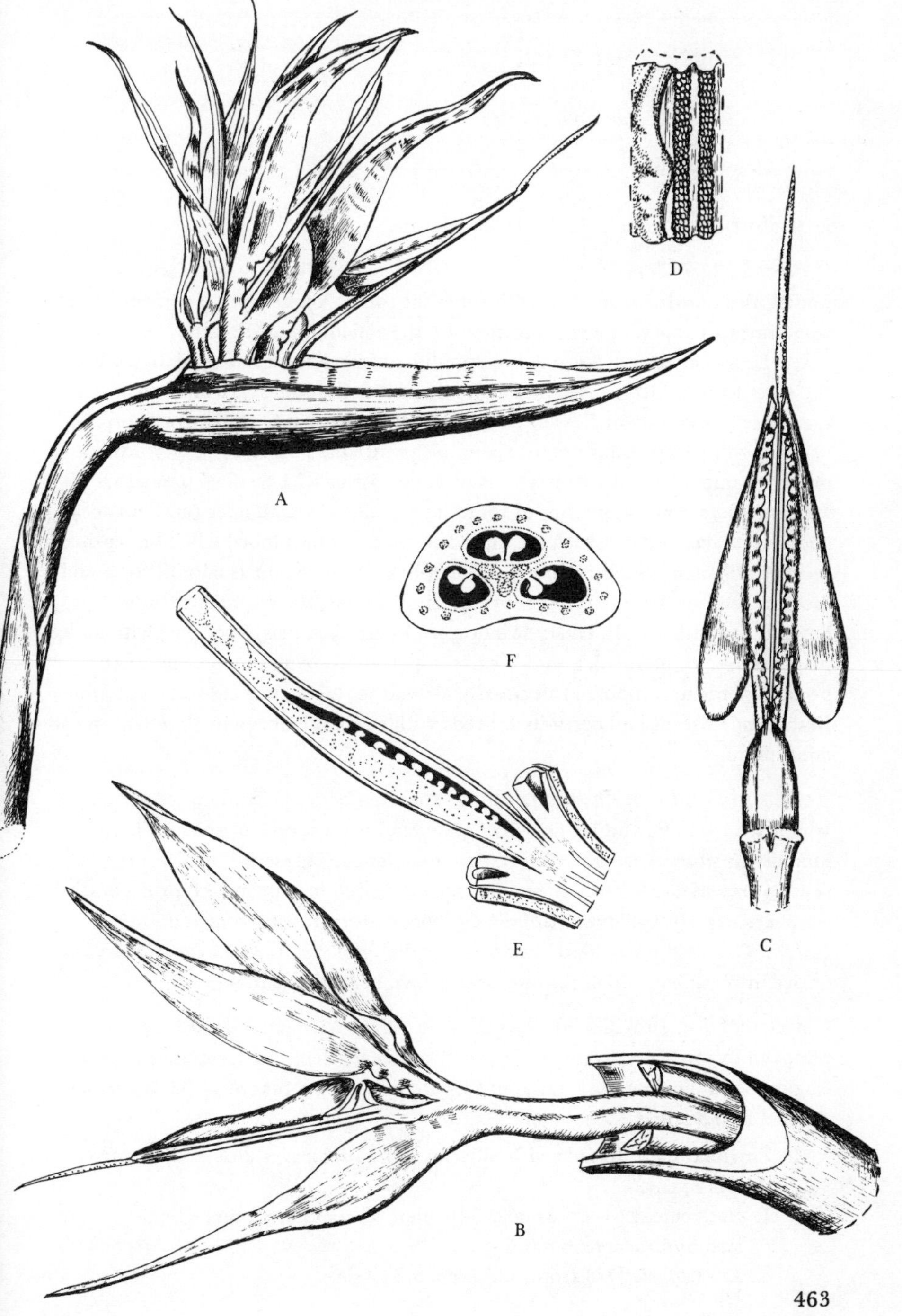

92 Zingiberaceae Lindl.

Ginger family

49 genera and 1000 species

Distribution. Tropics, chiefly Indomalaysia.

General characteristics. Perennial, often aromatic herbs, usually with sympodial, fleshy rhizomes, often with tuberous roots. Aerial stem, if present, short; sometimes an apparent stem is formed by the rolled-up leaf-sheaths (cf. Musaceae). Leaves alternate, 2-ranked or spirally arranged, with sheathing bases. At the top of the sheath is a characteristic ligule (cf. Gramineae). Inflorescence racemose or cymose, or flowers solitary. Flowers hermaphrodite, zygomorphic, usually with conspicuous bracts. Calyx of 3 united sepals. Corolla of 3 united petals. Stamens basically in 2 whorls of 3, consisting of 1 fertile, epipetalous stamen, and a varying number of sometimes petaloid staminodes (cf. Cannaceae and Marantaceae) 2 of which are united to form a 2 or 3-lobed labellum; anther 2-celled; filament usually slender and deeply grooved. Style usually filiform and more or less enveloped in the groove of the filament, the stigma protruding beyond the anther-cells. Ovary inferior, of 3 united carpels, 3-locular, with axile placentation, or unilocular with 3 parietal placentas; ovules numerous, anatropous or semi-anatropous. Fruit usually a 3-valved, loculicidal capsule, sometimes fleshy, indehiscent and berry-like. Seeds with straight embryo in abundant, mealy endosperm.

Economic and ornamental plants. The Zingiberaceae contain several plants which are aromatic and are grown commercially as a source of spices. These include *Zingiber officinale*, Ginger, *Curcuma longa*, Turmeric, and *Elettaria cardamomum*, Cardamom. A number are cultivated in glasshouses for decorative purposes, e.g., *Hedychium*, *Alpinia*, *Globba*, *Costus*, *Kaempferia* and *Brachychilum*. One of the few hardy members of the family is *Roscoea cautleoides*, native in W. China, which is sometimes grown in British gardens.

Classification. In Willis (5), the genus *Costus* and 3 small related genera are removed to form a separate family, the Costaceae. Here, the treatment follows Engler (4) in dividing the family into 2 subfamilies, the first of which is further divided into 3 tribes:

I. Zingiberoideae (plants with oil-cells, aromatic; leaves 2-ranked, sheaths always open).

   1. Hedychieae (ovary usually 3-locular, placentation axile; lateral staminodes large, petaloid).
      Kaempferia (70) tropical Africa, S.E. Asia.

Curcuma (60) Indomalaysia, China.
Hedychium (50) Madagascar, Indomalaysia, S.W. China.

2. Globbeae (ovary unilocular, placentation parietal; lateral staminodes large, petaloid).
Globba (50) S. China, Indomalaysia.
3. Zingibereae (ovary 3-locular, placentation axile; lateral staminodes usually small or absent, rarely petaloid).
Alpinia (250) warm regions of Asia, Polynesia.
Amomum (150) Old World tropics.
Zingiber (80–90) E. Asia, Indomalaysia, N. Australia.
Elettaria (7) Indomalaysia.

II. Costoideae (plants without oil-cells; leaves spirally arranged, sheaths closed, at least at first).
Costus (150) tropics.

HEDYCHIUM GARDNERANUM Roscoe
Kahili Ginger

Distribution. Native in N. India and introduced into Britain about 1820. Owing to its relative hardiness it may be grown outside as a subtropical bedding plant as well as indoors in large pots or planted out in glasshouse borders.

Vegetative characteristics. A strong-growing perennial with reed-like stems up to 2 m in height which arise from a fleshy rhizome. The leaves are 2-ranked, large, oblong-lanceolate and narrowed to a slender point.

Floral formula. K(3) C(3) A 1 $\overline{\text{G}}$(3)

Flower and inflorescence. The pale yellow, hermaphrodite flowers, each with one well-exserted red filament, are borne in terminal spikes from mid-June to August.

Pollination. The scent of the flowers and the conspicuous petals and petaloid staminodes indicate pollination by insects, while the long-exserted stamen and style suggest that hovering insects in particular are involved. Wing-scales of Lepidoptera have been found adhering to the stigmas of some species of *Hedychium* in their native habitat. Long-tongued butterflies, in search of the nectar secreted by the prominent gland near the base of the style (see Fig. 92.E), are likely to brush first the stigma and then the anther-cells with their wings as they approach the flower. In this way cross-pollination is carried out.

Alternative flowers for study. Any other species of *Hedychium* would be suitable as an alternative including *H. coronarium*, Garland Flower, which has fragrant, white flowers. Several other genera are also cultivated as ornamental plants, e.g., *Alpinia* which produces indehiscent dry or fleshy fruits in contrast to the 3-valved capsule of *Hedychium*. Comparison should be made with *Costus*, which is placed in the other subfamily, and also with the related families Cannaceae and Marantaceae (see Fig. J. on page 468).

Fig. 92 Zingiberaceae, *Hedychium gardneranum*

A A fully open, hermaphrodite, zygomorphic flower and one still in bud, surrounded by a sheathing bract that hides the 3-lobed, tubular calyces. The long, narrow, corolla-tube separates at its apex into 3 linear lobes. Within these are the petaloid staminodes, 2 of which are united to form a bilobed lip or labellum, the other 2 remain separate. The long, single filament, which arises from the top of the corolla-tube, is exserted well beyond the corolla-tube and is terminated by a single anther.
Calyx: 31 mm Corolla-tube: 55 × 2 mm
Corolla-lobes: 38 × 3–4 mm Labellum: 25–27 × 22–23 mm
Lateral staminodes: 34–38 × 5–8 mm
Exserted portion of filament: 53 mm

B The upper part of the stamen and the style. B1: The natural position. B2: The 2 anther-cells have been prised apart to show the slender style which lies along the deep ventral groove in the filament. The capitate stigma protrudes slightly beyond the anther-cells which dehisce longitudinally.
Anther: 8 mm

C T.S. of the filament just below the anther showing how the style is enveloped and held within the groove. The upper part of the filament is glabrous, but a thin line of hairs is present along the inside edges of the groove towards its base.
T.S. of filament: 1.3 × 1 mm

D A section of the upper part of the corolla-tube viewed from the adaxial side. Arising from the tube is the filament which envelops the style (shown as a small ring within the groove). At each side of the filament are the hairy, swollen bases of the 2 lateral staminodes. Below these are the basal portions of 2 lobes of the corolla. Part of the third corolla-lobe can be seen at one side of the filament.

E L.S. of the lower part of a flower and adjacent stem. At its base is the inferior ovary, one of whose 3 loculi has been exposed to reveal the numerous anatropous ovules on a well-developed placenta. Above the ovary is the lower portion of the corolla-tube surrounded by the tubular calyx. The slender style is situated slightly off-centre, and at its base is a prominent nectary. Strands of vascular bundles can be seen leading from the stem to the floral organs.
Ovary: 2.5 × 3 mm Nectary: 3 × 1 mm

F Detail of the capitate stigma, showing the glandular hairs on its receptive surface.
Stigma: 1 mm in diameter

G T.S. of the ovary formed from 3 fused carpels. Each of the 3 loculi contains numerous ovules attached to an axile placenta. After fertilisation the ovary develops into a globose capsule that dehisces loculicidally by 3 valves. The inner surface of the valves is bright orange and the round seeds are covered by a bright red aril.
T.S. of ovary: 3 × 3.5 mm

Fig. 92

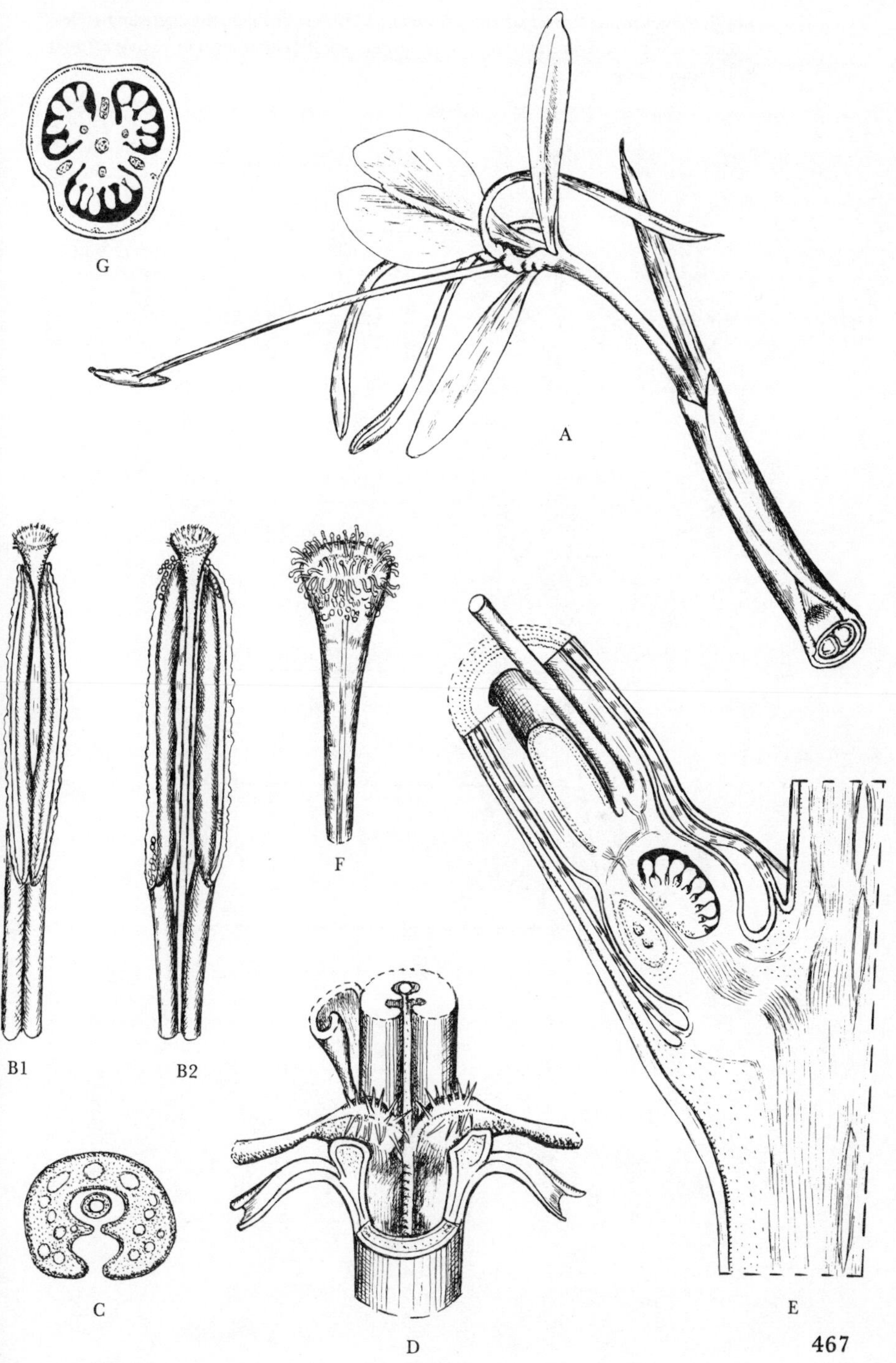

Fig. J. Structure of flower of *Hedychium*: 1, Anther; 2, Filament enclosing style; 3, Corolla-lobe; 4, Labellum; 5, Lateral staminode; 6, Corolla-tube; 7, Calyx-tube; 8. Inferior ovary.

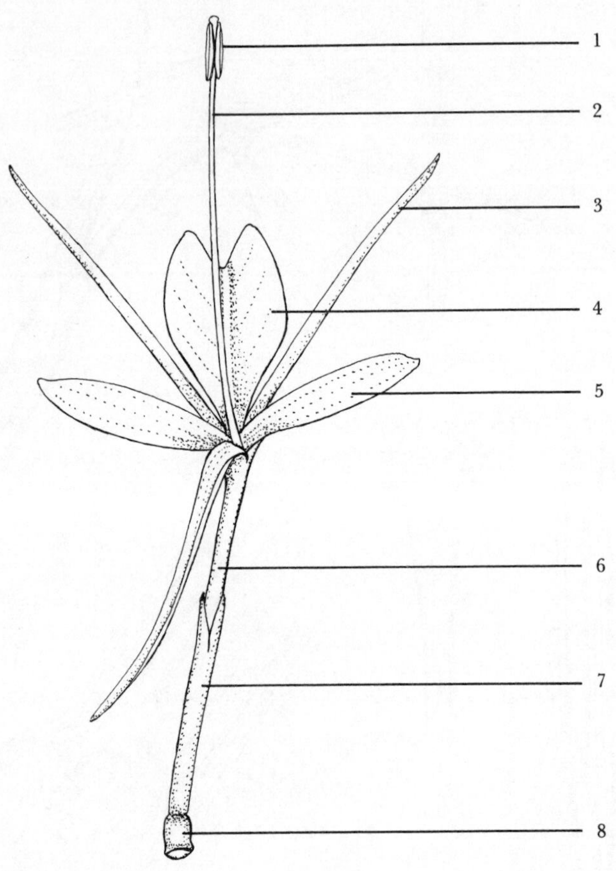

93 Cannaceae Juss.

Canna family

1 genus and 55 species

Distribution. America.

General characteristics. Perennial herbs, with a tuberous rootstock. Leaves cauline, usually oblong to broadly elliptical, pinnately veined with a prominent midrib, the petiole sheathing the stem. Inflorescence racemose, terminal, usually composed of 2-flowered cincinni. Flowers large and showy, hermaphrodite, asymmetric. Calyx of 3 free sepals. Corolla of 3 united petals. Stamens petaloid, 2–5, one bearing a single, fertile anther-cell on its edge, the others infertile staminodes; one of the latter is reflexed and is termed the lip or labellum, the other 2 (when present) are often referred to as the wings; when a fourth staminode is present, it stands behind the fertile stamen. In some species the only staminode present is the labellum (cf. Marantaceae and Zingiberaceae). Ovary inferior, of 3 united carpels, 3-locular, with 2 rows of anatropous ovules in each loculus; placentation axile; style petaloid. Fruit a warty capsule; seed with perisperm and straight embryo.

Economic and ornamental plants. Canna edulis, Queensland Arrowroot, is cultivated in the tropics for its tubers which are the source of an edible starch. Several other species are grown as decorative plants for the glasshouse or for summer bedding, particularly *C. indica*, and a number of hybrids of mixed parentage have been raised.

Classification. The family contains a single genus, which can be distinguished from the Zingiberaceae by the absence of a ligule and from the Marantaceae by the absence of a pulvinus.

CANNA × GENERALIS L.H. Bailey

Common Garden Canna

Distribution. C. × *generalis* is the name given to a group of hybrids of mixed parentage which have arisen in cultivation. These large-flowered cannas may be grown in glasshouses or outdoors during the summer months, preferably on light, rich soils, as tropical bedding plants or in a flower border. They are usually propagated by division of the rhizomes as they are somewhat difficult to raise from seed.

Vegetative characteristics. An erect perennial, with stout stems arising from a fleshy rootstock. The alternate leaves are long and broad, and have sheathing bases. Some cultivars exhibit a bronze or purple tint to the leaf, probably due to the genetic influence of *C. warscewiczii*, a species frequently grown in gardens (and therefore one of the likely parents) which has purplish or brown-purple foliage.

Floral formula. K3 C(3) A1 $\overline{\text{G}}$(3)

Flower and inflorescence. The large, zygomorphic, hermaphrodite flowers are borne terminally and more or less erect in a racemose inflorescence. The flowers are usually red, sometimes orange or yellow, and make a spectacular show from late July to the end of September, or even October, if the weather is mild.

Pollination. Self-pollination has been observed to be more frequent than cross-pollination though the latter may take place through the agency of various kinds of insects including honey-bees. The large, brightly coloured, petaloid stamen and staminodes are more conspicuous than the perianth, which would normally fulfil the function of attracting insects. In addition, nectar is secreted round the base of the style by glands situated in the septa of the ovary. The single-celled anther sheds its pollen while the flower is still in bud and the proximity of the stigma at this time may easily result in self-pollination. As the flower opens, visiting insects are able to transfer pollen from another flower to the stigma, and may also collect pollen which has fallen on to the inner parts of the flower.

Alternative flowers for study. *Canna* × *generalis* is frequently cultivated in parks and gardens but any species will act as a suitable alternative, including *C. indica*, Indian Shot, a commonly grown summer-flowering plant which has somewhat smaller flowers. If several species and hybrids are obtainable it will be found that there is variation in the number of staminodes present and also in the degree of erectness of the flower-parts. Comparison should be made with flowers in the closely related families Zingiberaceae and Marantaceae.

Fig. 93 Cannaceae, *Canna* × *generalis*

A A zygomorphic, hermaphrodite flower at anthesis together with one that is still in bud. The flowers, borne on short pedicels, occur in pairs forming a 2-flowered cincinnus. Each flower is subtended by a bract. The outer whorl of the perianth consists of 3 free, imbricate sepals, the inner of 3 basally united petals. In the flower illustrated there are 5 petaloid staminodes, with the petaloid stamen and style visible at the centre of the flower.
Sepals: 23 × 8 mm Petals: 50–60 × 3–4 mm
Staminodes: 50–70 × 20–50 mm Stamen: 50 × 30 mm

B L.S. of the centre of a flower showing the petaloid stamen and style surrounded by parts of the staminodes and perianth. The stamen can be recognised by the presence of the single anther-cell along its upper margin. Two of the 3 sepals can be seen at the base of the corolla-tube. All these flower-parts are situated above the inferior ovary which is borne on a short pedicel.
Ovary: 8 × 8 mm

C The single anther-cell, situated near the apex of the petaloid stamen, dehisces by a longitudinal slit to liberate the pollen, some of which is still adhering to its margin.
Anther-cell: 16 × 2 mm

D Parts of the petals, staminodes and stamen have been cut away to reveal the petaloid style with the stigma along its upper edge.
Style: 48–60 × 5–9 mm

E T.S. of the 3-locular ovary formed from 3 united carpels. Each loculus contains numerous anatropous ovules attached to an axile placenta. The surface of the ovary is strongly verrucose.
Ovary (at early anthesis): 8 mm in diameter

F The fruit before dehiscence, crowned by the persistent sepals. The capsule has a warty pericarp that disintegrates at maturity to release the large round seeds.
Fruit: 20 × 18 mm

Fig. 93

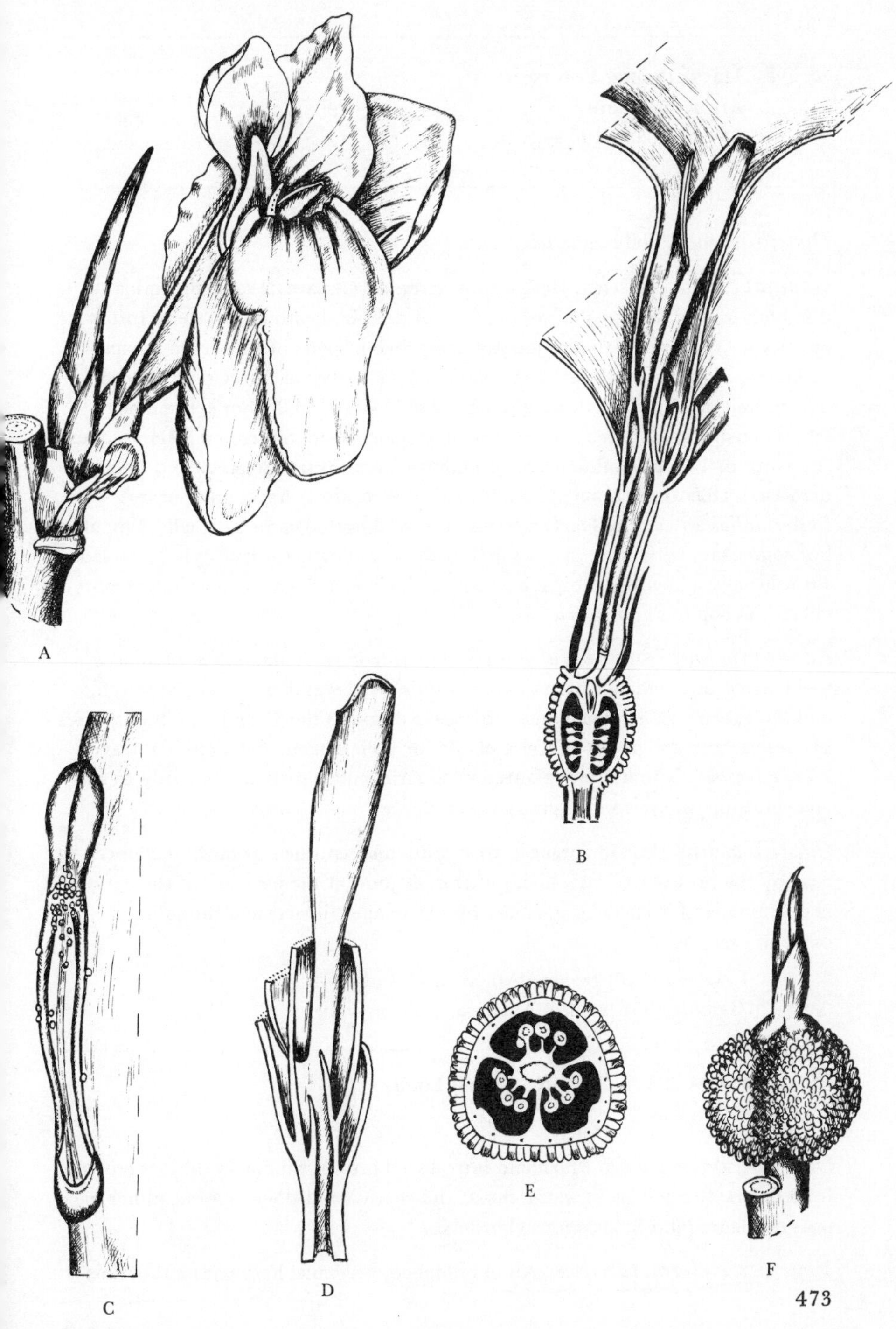

94 Marantaceae Petersen

Arrowroot family
30 genera and 400 species

Distribution. Chiefly America.

General characteristics. Herbaceous perennials. Leaves 2-ranked, lamina pinnately-veined, often asymmetric, base of petiole sheathing the stem. Inflorescence a spike or panicle, frequently composed of pairs of flowers in the axils of bracts. Flowers hermaphrodite, zygomorphic. Calyx of 3 free sepals. Corolla of 3 united petals. Stamens usually petaloid, basically in 2 whorls, the inner whorl consisting of 1 fertile stamen with a single-celled anther and 2 staminodes, the outer of 1 or 2 staminodes or sometimes absent (cf. Cannaceae and Zingiberaceae). One of the inner staminodes is hooded, the other is more or less fleshy or has a fleshy callus. Ovary inferior, of 3 united carpels, usually 3-locular, but sometimes with only one of the 3 ovules reaching maturity; style 1, curved. Fruit usually a loculicidal capsule, rarely indehiscent. Seed often arillate; embryo curved, in copious endosperm.

Economic and ornamental plants. The rhizomes of *Maranta arundinacea* yield arrowroot, which is prepared by grinding and washing to free the starch, and the tubers of *Calathea allouia* are used as food in the W. Indies. Other species of these genera are grown as house plants for their ornamental, coloured leaves. *Thalia dealbata* is sometimes planted in aquaria, and is sufficiently hardy to survive outdoors where winters are fairly mild.

Classification. The Marantaceae are readily distinguished from the Zingiberaceae by the presence of a swollen pulvinus or joint at the junction of the petiole and lamina, and from the Cannaceae by the smaller flowers and the solitary ovule in each loculus.

Calathea (150) tropical America, W. Indies.
Maranta (23) tropical America.

CALATHEA ZEBRINA (Sims) Lindl.

Zebra Plant

Distribution. Native in Brazil and introduced into Britain in 1815. It is grown for its attractive foliage in warm, moist and shady conditions indoors, either in pots or planted out in glasshouse borders.

Vegetative characteristics. An unbranched, perennial herb with a tuberous rootstock. The large, petiolate, sheathing leaves have the characteristic pulvinus just below the leaf-blade. The leaves are a velvety green, with alternating stripes of pale yellow-green and dark green extending outwards from the midrib. The under surface of the leaf is purplish at maturity.

Floral formula. K3 C(3) A1 $\overline{G}(3)$

Flower and inflorescence. The terminal, cone-like spike (see Fig. 94.A), which is borne on a strong stalk, is composed of spirally arranged bracts, each subtending 2 violet and white, zygomorphic, hermaphrodite flowers. The flowers appear during May, but the flower-spike continues to stand erect even after the flowers have withered.

Pollination. The flowers are pollinated by insects which forage for the nectar that is secreted into the corolla-tube by glands in the septa of the ovary. The lower staminode of the almost horizontal flower is hooded and envelops the style which is curved down at the end and terminates in a funnel-shaped stigma (see Fig. 94.E). Above the stigma is the single-celled anther which dehisces while the flower is still in bud, depositing pollen on the upper side of the style. When an insect alights on the hooded staminode, the style is suddenly released, causing the stigma to touch the insect's body and collect any pollen adhering to it. At the same time, fresh pollen, from the upper side of the style, falls on to the insect. When the insect flies away to another flower the process is repeated, resulting in cross-pollination. In the course of its movement, the style curves back, blocking the way to the nectar, so that each flower is only visited once.

Alternative flowers for study. A number of species of *Calathea* are in cultivation and may be used as alternatives. It should be noted that many flowers, including those of *C. zebrina*, are best dissected at early anthesis as later on they become crumpled and consequently more difficult to study. The genera *Thalia* and *Maranta* are distinguished from *Calathea* by the branched inflorescence and the single developed loculus, the other loculi aborting. Unfortunately flowers are rarely produced in cultivated plants of *Maranta*. Comparison should be made with the closely related families Cannaceae and Zingiberaceae.

Fig. 94 Marantaceae, *Calathea zebrina*

A The cone-like inflorescence, borne on a rigid stalk, is composed of a number of large, spirally arranged bracts. Each bract subtends 2 flowers which will later extend beyond the bracts.
Inflorescence: 35–40 × 40–45 mm

B One of a pair of flowers, showing the large bract beneath and a smaller one above, still attached to part of the main axis. The zygomorphic, hermaphrodite flowers are borne horizontally on very short pedicels, eventually protruding beyond the bracts and exposing the corolla-lobes and essential organs.
Lower bract: 20 × 18 mm Upper bract: 16 mm

C A young flower without the protecting bracts. The outer whorl of the perianth consists of 3 free sepals, the inner of 3 petals which are united in their lower part to form a tube. One of the petals or corolla-lobes has been removed. At the centre is the curved style with its funnel-shaped stigma surrounded by 3 petaloid staminodes. Directly behind the style is the callous staminode, so called from its rather fleshy structure. Next to it is the 3-lobed, hooded staminode. At the back is the largest petaloid staminode. The solitary stamen, bearing one anther, is attached to the callous staminode (see Fig. K). The corolla-tube will eventually extend beyond the sepals.
Flower: 25 mm Corolla-tube: 13 mm
Sepal: 18 mm

D L.S. of the lower part of a flower, showing the sepals and corolla-tube surrounding the base of the curved style, and below these the inferior ovary, with one loculus opened to reveal the solitary ovule attached at its base in an erect position.
Ovary: 1.5 mm Style + stigma: 20 mm

E Detail of the curved end of the style which broadens into a funnel-shaped stigma.
Stigma: 1.75 mm in diameter

F T.S. of the 3-locular ovary formed from 3 fused carpels. The 3 ovules, one in each loculus, are attached to the base of the ovary and the placentas are therefore not visible. After fertilisation, the ovary develops into a 3-valved, loculicidal capsule containing the 3 arillate seeds.
Ovary: 1.25 mm in diameter

Fig. 94

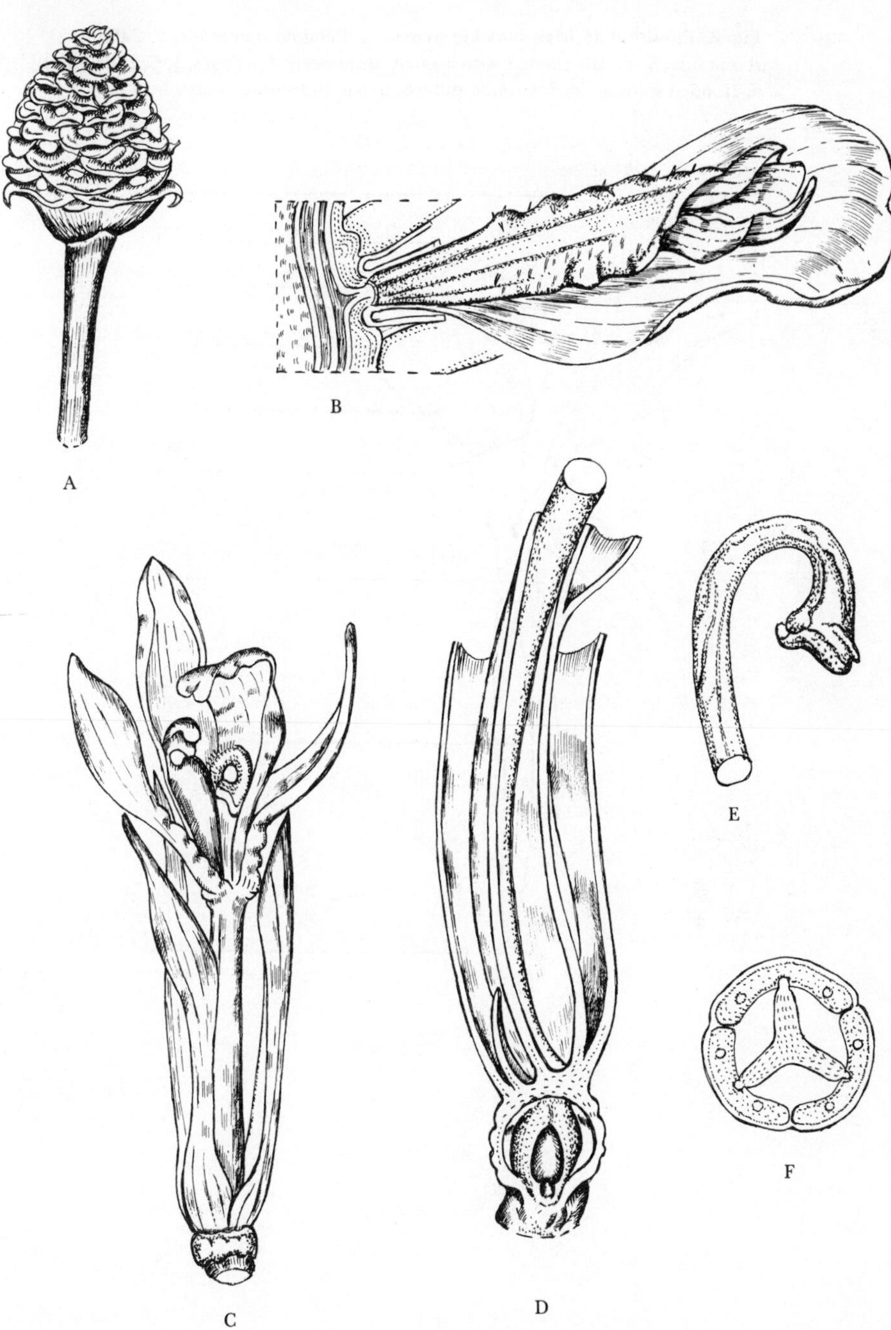

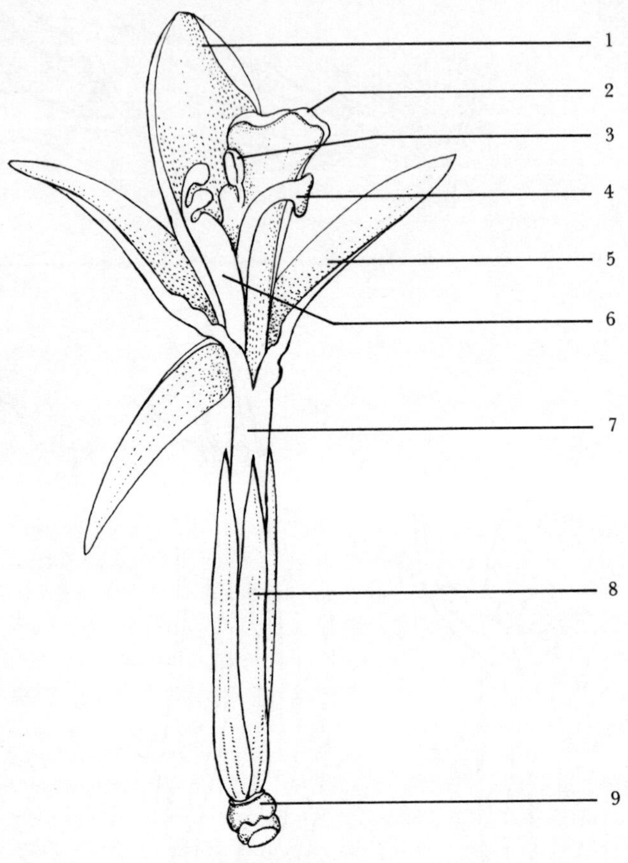

Fig. K. Flower at anthesis showing stamen: 1, Petaloid staminode; 2, Callous staminode; 3, Fertile stamen with a single anther-cell; 4, Stigma; 5, Corolla-lobe; 6, Hooded staminode; 7, Corolla-tube; 8, Sepal; 9, Inferior ovary.

95 Palmae Juss. (Arecaceae Schultz-Schultzenst.)

Palm family

217 genera and 2500 species

Distribution. Tropics and subtropics.

General characteristics. Shrubs, vines or trees, often monoecious or sometimes dioecious, with leaves forming a terminal cluster in the arborescent species, or alternate in the climbing and some shrubby species. Leaves palmate (fan palms) or pinnate (feather palms), usually large and with the base of the petiole often sheathing the stem. Inflorescence usually large and much-branched, often termed a spadix, and subtended by one or more bracts. Flowers usually unisexual, actinomorphic. Perianth of 6 free or united segments in 2 whorls of 3. Stamens 6, in 2 whorls of 3. Ovary superior, of 3 free or united carpels, 3-locular with a single anatropous ovule in each loculus, or sometimes unilocular. Fruit a berry or drupe. Endosperm abundant.

Economic and ornamental plants. Palms form an important part of the economy in tropical and subtropical regions, and in some cases their products are familiar in other parts of the world also. Some of the best known palms include *Phoenix dactylifera*, Date Palm, *Cocos nucifera*, Coconut Palm, *Elaeis guineensis*, Oil Palm, *Arenga saccharifera*, Sugar Palm, *Metroxylon sagu*, Sago Palm, *Copernicia cerifera*, Wax Palm, and *Raphia* spp. the source of Raffia. Many palms are locally important for their fibre, their stems are often used in building, and their leaves for thatching and for making baskets and mats. They are also widely planted for ornament in warm regions. In the British Isles the only species which is really hardy is *Trachycarpus fortunei*, Chusan Palm, from central China, though *Chamaerops humilis*, Dwarf Fan Palm, the only palm native to S. Europe, may succeed in the milder parts of the country. Several other species are frequently cultivated as pot plants for the house or conservatory, e.g., *Chamaedorea elegans*, *Microcoelum weddellianum*, *Howea belmoreana* and *H. forsterana.*

Classification. The Palmae are easily separable from other plant families, and there is fairly close agreement on the primary division of the family (using a variety of characters) into 6–9 groups. The chief genera are:

Raphia (30) tropical and S. Africa, Madagascar.
Phoenix (17) warm regions of Africa and Asia.
Metroxylon (15) S.E. Asia, Pacific islands.
Caryota (12) S.E. Asia, N.E. Australia.
Arenga (11) Indomalaysia.

Trachycarpus (8) Himalaya, E. Asia.
Cocos (1) probably native in tropical Asia or Polynesia, *C. nucifera.*

PHOENIX SYLVESTRIS (L.) Roxb.
Wild Date Palm

Distribution. Native in India and cultivated there for its sap which yields sugar, but often planted elsewhere as an ornamental tree. It was introduced into Britain in 1863 and is grown under glass in Botanic Gardens and more rarely in conservatories.

Vegetative characteristics. A fast-growing, dioecious tree with a robust, unbranched trunk reaching a height of 16 m. The leaves are pinnate and have a spiny stalk. The leaflets (pinnae) are greyish green, usually in 2–4 ranks, and often somewhat clustered.

Floral formula. Male: P3+3 A3+3 Female: P3+3 $\underline{G}$(3)

Flower and inflorescence. The large inflorescence consists of a branched panicle of numerous, unisexual, actinomorphic flowers which arises from amongst the crown of leaves. The panicle, which is subtended by a woody bract or spathe, is sometimes referred to as a spadix. In its native country flowering occurs at the beginning of the hot season (April) but under glasshouse conditions it appears to be more variable and may take place at any time between January and April.

Pollination. The large inflorescence composed of numerous, inconspicuous flowers indicates adaptation to wind-pollination.

Alternative flowers for study. The family exhibits interesting characteristics which deserve first-hand study, but this may be hampered by the difficulty of obtaining flowering or fruiting specimens. *P. dactylifera*, if available, would be an excellent alternative since the familiar fruits of this species are larger than those of *P. sylvestris. Trachycarpus fortunei* and *Chamaerops humilis*, already mentioned in the section on Ornamental Plants, are likely to be the most convenient sources of flowering material. The most easily obtainable fruit, apart from the Date, is probably the Coconut which is produced by the monoecious palm *Cocos nucifera.* This fruit, when purchased from a shop, is normally devoid of the thin, smooth outer layer and the fibrous middle layer (the coir of commerce) although part of the latter may remain attached as a tuft. What is often assumed to be the outer surface of the Coconut is the hard inner layer which protects the seed. The seed is composed of a thin brown testa covering a thick layer of edible, white endosperm (dried to form copra) but it is hollow at its centre, the cavity being partially filled by a whitish juice (coconut-milk). At the base of the Coconut are 3 pores or 'eyes' indicating that the ovary developed from 3 carpels. The whole fruit is buoyant and is adapted to dispersal by water.

Fig. 95 Palmae, *Phoenix sylvestris*

A Part of one of the flattened branches of a male inflorescence with the bases of flower-spikes. Each inflorescence-branch bears normally 7 or more groups of spikes, with 4–15 spikes in each group. Each spike bears a large number of small, actinomorphic flowers.
Inflorescence: 38 cm Flower-spikes: 10–20 cm

B A portion of the spike, showing one sessile, male flower at bud-break. The 3 fleshy, inner perianth-segments, which are subtended by 3 minute outer ones, are valvate.
Bud: 6 x 4 mm Outer perianth-segments: 1.5 mm

C A male flower at anthesis. The 3 inner perianth-segments have opened wide, exposing the 6 stamens to the wind. Flowers with 4 segments are sometimes present towards the base of the spike, and these tend to be larger than the normal flowers.
Male flowers: 7–8 mm in diameter
Inner perianth-segments: 5–6 x 3 mm

D L.S. of a male flower, showing 3 of the 6 stamens. The filaments, which are very short, bear anthers that are 2-celled and dehisce longitudinally.
Anthers: 3–3.25 mm

E A portion of the spike from a female inflorescence with one of the sessile, actinomorphic flowers. A scar on the side of the axis indicates the position of a discarded flower. The 3 fleshy, inner perianth-segments, subtended by 3 minute outer ones, differ from those of the male flower in being imbricate and in not opening fully. Three styles can be seen at the apex of the flower.
Female flowers: 4 x 3.5–4 mm

F L.S. of a female flower. On each side of the ovary are 2 overlapping sections of the imbricate, inner perianth-segments. The superior ovary is composed of 3 separate carpels (2 only shown), each loculus containing a single ovule with basal placentation. The receptacle shows a deep, basal cavity where it was removed from the flower-spike. The outer perianth-segments appear as an extension of the margin of the receptacle.
Carpel: 2 x 1.25 mm Ovule: 0.5 x 0.25 mm

G Two of the fruits (Wild Dates) attached to a portion of the spike. The scars denote the position of discarded fruits. The Date is, botanically, a one-seeded berry which has developed from one of the 3 carpels. The fruits of both *P. sylvestris* and the more familiar *P. dactylifera* become wrinkled with age.
Fruits: 17–18 x 10 mm

H The single seed ('stone'), with the basal funicle shown at the top owing to the pendulous nature of the fruit. In the genus *Phoenix*, the 'stone' is the hard endosperm of the seed and the membranous skin surrounding it is the thin, inner layer of the fruit. The seed is conspicuously grooved along one side, and on the other there is a small bulge, indicating the position of the embryo. Germination is apparently slow, but can be achieved in a sufficiently high temperature, provided that the seed has not been allowed to dry out.
Seed: 15 x 7.5 mm

Fig. 95

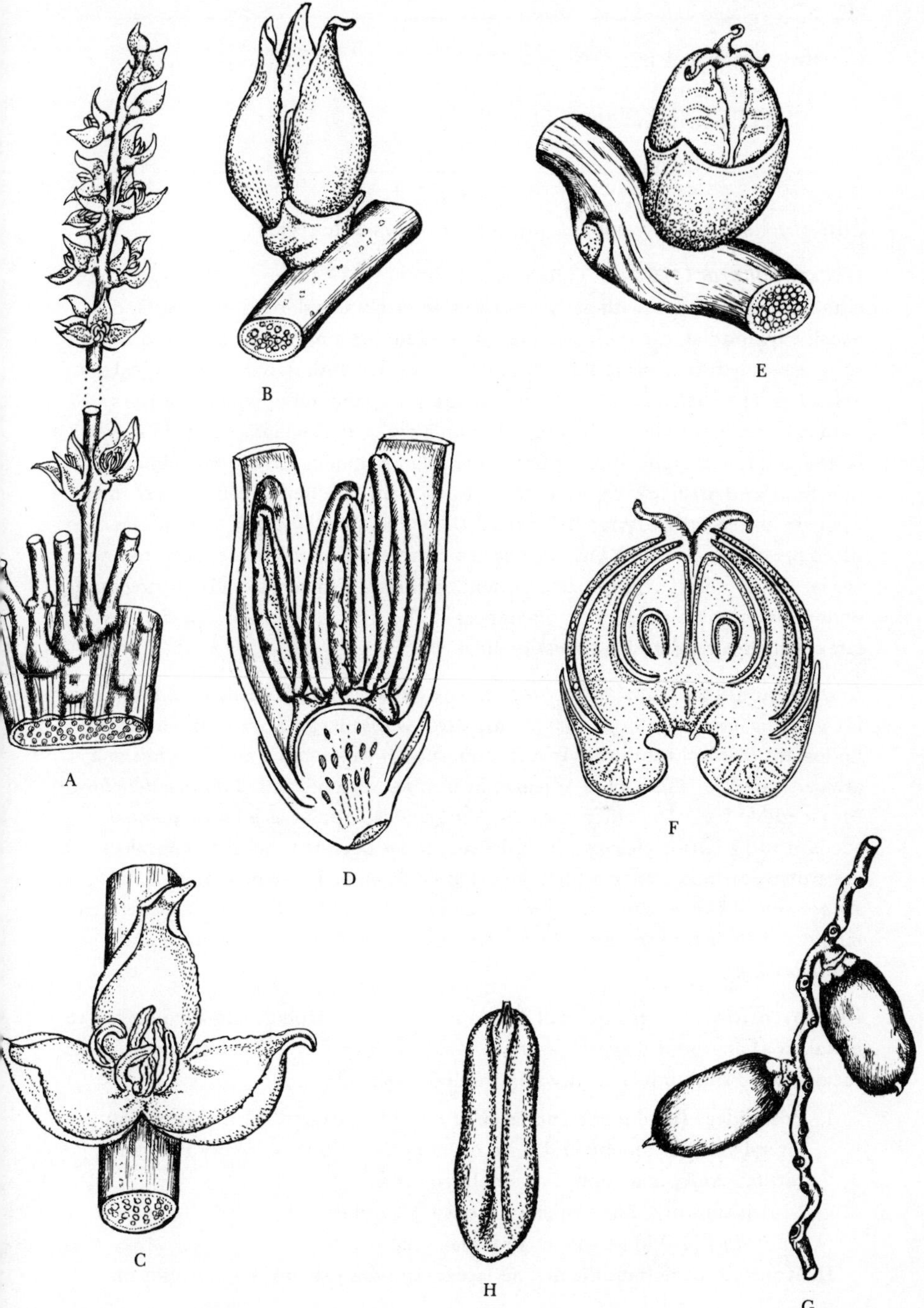

96 Araceae Juss.

Arum family

115 genera and 2000 species

Distribution. Mostly tropical, but some in temperate regions.

General characteristics. Tuberous or rhizomatous herbs or shrubs, sometimes climbing or epiphytic with aerial roots, more rarely marsh or water plants; usually sympodial, each joint of the sympodium beginning with one or more scale-leaves before bearing foliage leaves. Accessory buds often found in leaf-axils. Leaves of various types, often pinnately or palmately divided. Flowers without bracts, usually massed together on a cylindrical spadix enclosed in a large spathe; hermaphrodite, or unisexual (plants then usually monoecious, but dioecious in *Arisaema*). Perianth present or absent. Stamens usually fewer than 6, often united into a synandrium (e.g., *Colocasia*, *Spathicarpa*); staminodes often present and united. Ovary of varied structure, frequently reduced to one carpel. Fruit a berry. Endosperm present or absent. Flowers usually protogynous, with a disagreeable smell. Many members of the Araceae contain a poisonous latex, which is rendered harmless by heat.

Economic and ornamental plants. The chief economic plant in the Araceae is *Colocasia antiquorum* (*C. esculenta*), Taro, whose large, starchy tubers are an important tropical root crop. In addition, species of *Alocasia* and *Xanthosoma* are cultivated for their roots or leaves in tropical regions, and *Monstera deliciosa* for its edible fruit. In colder climates, *Monstera* is popular as a house plant on account of its curious leaves which develop holes between the ribs. Several genera are grown for their leaves, which are often variegated. These include *Caladium*, *Aglaonema*, *Philodendron*, *Syngonium* and *Dieffenbachia*. Others are cultivated mainly for their flowers, e.g., *Anthurium*, *Spathiphyllum*, *Sauromatum* and *Zantedeschia*.

Classification. The grouping of the Araceae is very difficult and account has to be taken of anatomical as well as morphological characters. The following classification into 8 subfamilies is based on that of Engler (4):

I. Pothoideae (land plants; no latex or raphides; leaves 2-ranked or spiral; lateral veins of second and third order netted; flowers usually hermaphrodite; ovules anatropous or amphitropous).

Anthurium (550) tropical America, W. Indies.

Pothos (75) Madagascar, Indomalaysia.

II. Monsteroideae (land plants; no latex; raphides present; lateral veins of third, fourth and sometimes second order netted; flowers hermaphrodite, usually naked; ovules anatropous or amphitropous).

Rhaphidophora (100) Indomalaysia, New Caledonia.

Monstera (50) tropical America, W. Indies.

Spathiphyllum (38) Central and tropical S. America, and Philippines to Solomon Islands.

Epipremnum (25) Indomalaysia.

III. Calloideae (land or marsh plants; latex present; flowers usually hermaphrodite; ovules anatropous or orthotropous; leaves never sagittate, usually not veined).

Calla (1) N. temperate and subarctic region, *C. palustris.*

Symplocarpus (1) N.E. Asia, Japan, Atlantic N. America, *S. foetidus.*

IV. Lasioideae (land or marsh plants; latex present; flowers hermaphrodite or unisexual; ovules anatropous or amphitropous; seed usually exalbuminous; leaves sagittate, often much lobed, not veined).

Amorphophallus (100) tropical Africa, Asia.

Dracontium (13) Mexico to tropical S. America.

V. Philodendroideae (land or marsh plants; latex present; flowers naked, unisexual; ovules anatropous or orthotropous; seed usually albuminous; leaves usually parallel-veined).

Philodendron (275) warm areas of America, W. Indies.

Zantedeschia (8 or 9) temperate and subtropical S. Africa, tropical Africa.

VI. Colocasioideae (land or marsh plants; latex present; flowers naked, unisexual; stamens in synandria; ovules orthotropous or anatropous; seed albuminous or not; leaves net-veined).

Alocasia (70) Indomalaysia.

Xanthosoma (45) Mexico to tropical S. America, W. Indies.

Colocasia (8) Indomalaysia, Polynesia.

VII. Aroideae (land or marsh plants; latex present; leaves various; net veined; stems mostly tuberous; flowers unisexual, usually naked; stamens free or in synandria; ovules anatropous or orthotropous; seed albuminous).

Arisaema (150) E. Africa, tropical Asia, Atlantic N. America to Mexico.

Arum (15) Europe, Mediterranean region.

Spathicarpa (7) tropical S. America.

VIII. Pistioideae (floating plants; no latex; flowers unisexual, naked; male flowers in a whorl, female solitary).

Pistia (1) tropics and subtropics, *P. stratiotes.*

ARUM MACULATUM L.
Lords-and-Ladies, Cuckoo-pint

Distribution. Generally distributed throughout the British Isles, though less common in Scotland. It is also native in Europe northwards to southern Sweden and in N. Africa. It prefers shady places such as hedgerows and woods, and can become a persistent weed in established garden shrubberies.

Vegetative characteristics. An erect, glabrous, perennial herb with tuberous rootstock and long-stalked, triangular-hastate leaves which are often blackish-spotted.

Floral formula. Male: P 0 A 2–4
Female: P 0 $\underline{G}$ 1

Flower and inflorescence. The monoecious inflorescence, which appears in April and May after the leaves, is a club-shaped structure known as the spadix. The upper portion is naked and dull purple in colour; the lower portion bears whorls of small, unisexual flowers, the male above, the female below. The spadix is subtended by the spathe, a large yellowish green bract sometimes spotted with purple.

Pollination. The highly specialised structure described above forms a pitfall trap and is directly connected with pollination. The plant is protogynous, and when the female flowers are in a receptive state, the spadix emits a smell (and incidentally generates heat) which attracts small flies. If the insects alight on the spadix or the inner surface of the spathe, both of which are smooth and slippery, they lose their grip and fall through the ring of bristles attached to the sterile male flowers to the bottom of the pit. Here they come into contact with the mature female flowers and transfer to them any pollen they may be carrying from another plant. After pollination, the stigmas soon wither and the fertile male flowers situated above ripen and shower the trapped insects with pollen. The bristles of the sterile male flowers then wither and the surface of the spadix itself becomes wrinkled enabling the flies to climb up out of their prison and fly off to act as agents in the cross-pollination of another plant (cf. Aristolochiaceae).

Alternative flowers for study. A number of Araceae are grown indoors for their flowers and are easily obtainable, but it should be noted that there is often considerable variation between one genus and another.

Fig. 96 Araceae, *Arum maculatum*

A The complete flowering shoot on a short stem. The enveloping lower portion of the spathe conceals the reproductive organs, while the upper part of the spadix is clearly visible against the open portion of the spathe.
Spathe: 17–25 cm × 6.5–10.5 cm

B Part of the spathe has been cut away to reveal the pitfall trap formed by the lower portions of the spathe and spadix. Just below the constricted part of the spathe are the sterile male flowers with long bristles which exclude the larger insects; then come the fertile male flowers, below these a small number of sterile female flowers, also provided with bristles, and at the base the fertile female flowers (see Fig. L). Both male and female flowers are devoid of perianth-segments.
Lower (flower-bearing) part of spadix: 21–26 mm
Upper part of spadix: 43–60 mm

C A single, fertile male flower, prior to dehiscence, with portion of spadix attached.
Male flower: *c.* 1 × 1 mm Bristle of sterile male flower: 4–6 mm

D A male flower, showing pollen emerging from a slit in the stamens.

E L.S. of portion of the staminate inflorescence, showing the paired nature of the stamens and the pollen grains inside them.

F A single, fertile female flower with sessile stigma.
Female flower (at anthesis): 3 × 2.5 mm
Stigma: *c.* 0.9 mm in diameter

G T.S. of unilocular ovary showing the ovules attached to the placenta.

H L.S. of ovary with portion of spadix attached. One ovule is still connected to the placenta, the other has been slightly displaced in sectioning.
Loculus of ovary: *c.* 0.5 mm

I A sterile female flower with its long, hooked bristle.
Swollen portion: 2 mm Bristle: 3–5 mm

J The infructescence, which has developed from the individual carpels. The scarlet, few-seeded fruits are partly enveloped by the remains of the withered spathe.
Infructescence: 30–40 × 18–19 mm
Single berry: 8 × 8 mm to 9 × 7 mm

Fig. 96

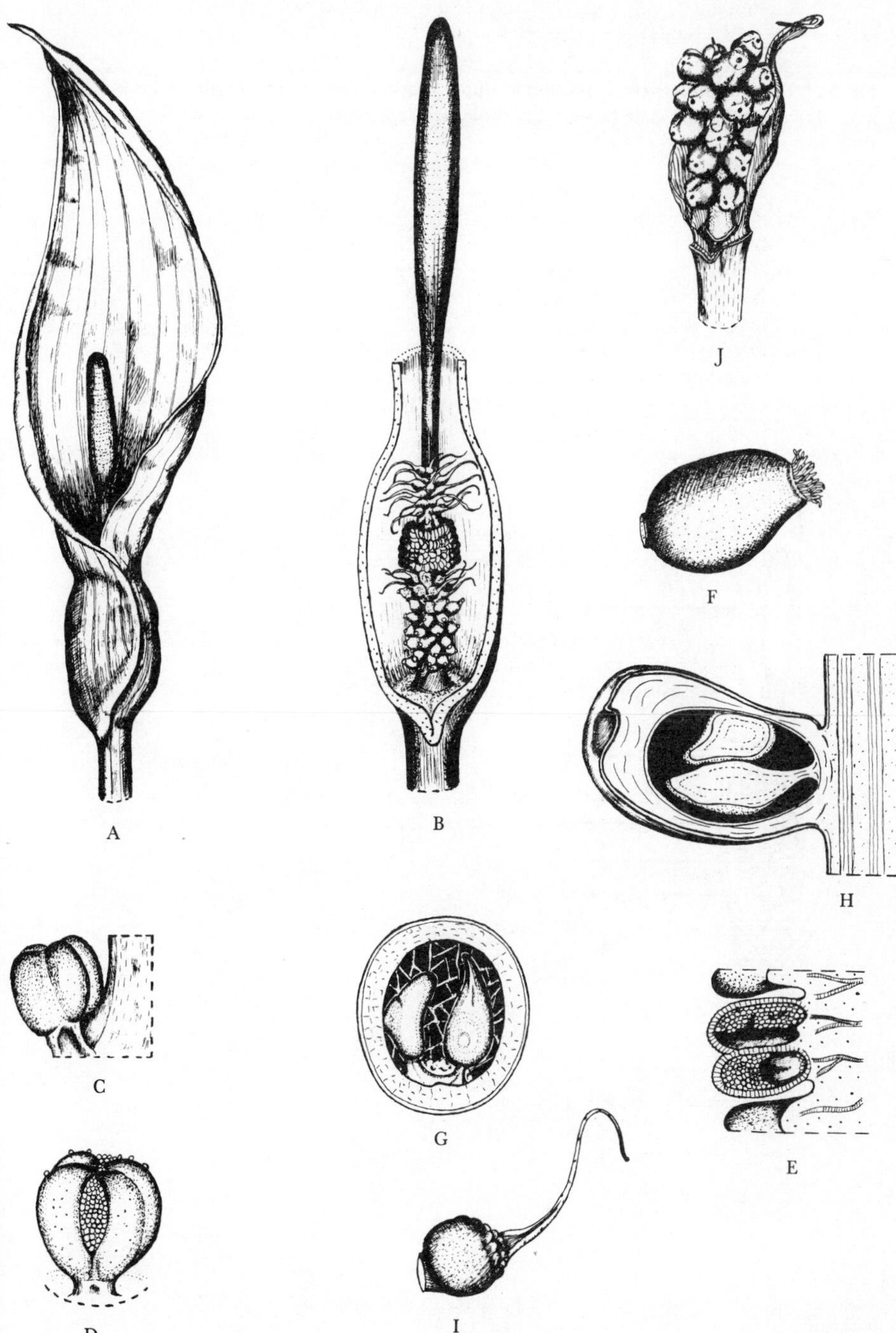

Fig. L. L.S. of inflorescence: 1, Spathe; 2, Upper portion of spadix; 3, Sterile male flowers; 4, Fertile male flowers; 5, Sterile female flowers; 6, Fertile female flowers.

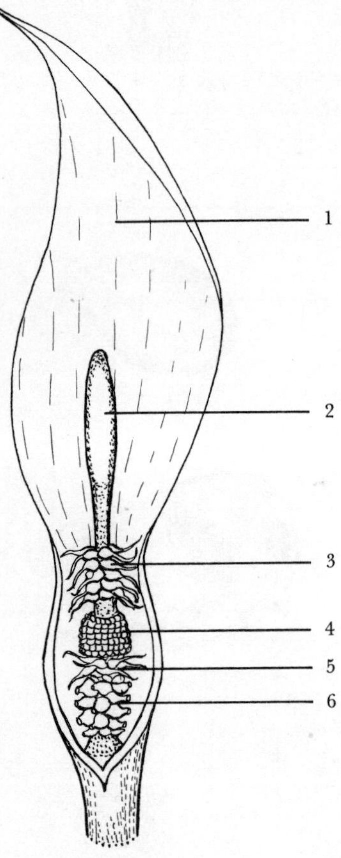

97 Liliaceae Juss.

Lily family

250 genera and 3700 species

Distribution. Cosmopolitan.

General characteristics. Mostly perennial herbs with sympodial rhizomes or bulbs, some woody plants (e.g., *Yucca, Dracaena*), succulent (e.g., *Aloe, Gasteria*) and climbers (e.g., *Smilax, Gloriosa*). *Ruscus* exhibits phylloclades. Inflorescence most commonly racemose, but cymose in the apparent umbels of *Allium, Agapanthus*, etc. Solitary terminal flowers occur in *Tulipa* etc. Flowers usually hermaphrodite, actinomorphic, 3-merous, hypogynous. Perianth-segments in 2 whorls of 3, free or united, petaloid or sometimes sepaloid. Stamens usually in 2 whorls of 3, with usually introrse anthers. Ovary superior, of 3 united carpels, usually 3-locular, with 2 rows of numerous, anatropous ovules on axile placentas in each loculus, rarely unilocular with parietal placentation. Fruit usually a capsule, sometimes a berry. Seed with abundant endosperm. Flowers usually insect-pollinated.

Economic and ornamental plants. The Liliaceae is one of the largest families of flowering plants but is of comparatively little economic value. *Phormium tenax*, New Zealand Flax, *Sansevieria*, Bowstring Hemp, and *Yucca* are commercially important as fibre-plants. *Xanthorrhoea* and *Dracaena* produce resins, and *Aloe, Colchicum, Urginea* and *Veratrum* are of medicinal value. Food plants include *Asparagus officinalis*, Asparagus, and various species of *Allium*, which are known to the vegetable gardener as Onions, Leeks, Shallots, Chives, Garlic, etc. By contrast, a considerable number of genera are highly decorative and of horticultural significance. Many of those listed in the next section are in this category.

Classification. The Liliaceae, divided here into 12 subfamilies, is considered in a broad sense (after Krause, in Ref. 9), but some authorities restrict it to certain subfamilies, raising the others to the rank of family. The latter view affects the whole of subfamilies IV, VII, XI and XII, and some genera in III, VIII and IX.

I. Melanthioideae (rhizome, or bulb covered with scale-leaves; inflorescence terminal; fruit never a berry).

Colchicum (65) Europe, Mediterranean region to central Asia and N. India.

Veratrum (25) N. temperate region.

Tofieldia (20) N. temperate region, S. America.

II. Herrerioideae (tuber, with climbing stem; leaves in tufts; flowers in racemes or panicles; fruit a septicidal capsule).

Herreria (8) S. America.

III. Asphodeloideae (usually a rhizome with radical leaves, rarely with leafy stem or a bulb; fruit usually a capsule).

Aloe (330) tropical and S. Africa, Madagascar, Arabia.
Haworthia (150) S. Africa.
Kniphofia (75) E. and S. Africa, Madagascar.
Gasteria (70) S. Africa.
Hemerocallis (20) temperate Eurasia.
Xanthorrhoea (15) Australia.
Asphodelus (12) Mediterranean region to Himalaya.
Hosta (10) China, Japan.
Phormium (2) New Zealand.

IV. Allioideae (bulb or short rhizome; inflorescence usually an umbel subtended by 2 bracts).

Allium (450) northern hemisphere.
Gagea (70) temperate Eurasia.
Agapanthus (5) S. Africa.

V. Lilioideae (bulb; stem leafy; inflorescence terminal and racemose; fruit a loculicidal capsule, except in *Calochortus*).

Tulipa (100) temperate Eurasia.
Fritillaria (85) N. temperate region.
Lilium (80) N. temperate region.

VI. Scilloideae (bulb; stem leafless; fruit a loculicidal capsule).

Ornithogalum (150) temperate regions of the Old World.
Scilla (80) temperate Eurasia, S. and tropical Africa.
Muscari (60) Europe, Mediterranean region, W. Asia.
Hyacinthus (30) Mediterranean region, Africa.
Endymion (10) W. Europe, W. Mediterranean region.

VII. Dracaenoideae (stem erect with leafy crown, except in *Astelia*; leaves sometimes leathery, never fleshy; fruit a berry or capsule).

Dracaena (150) warm regions of the Old World.
Yucca (40) S. United States, Mexico, W. Indies.

VIII. Asparagoideae (rhizome subterranean and sympodial; fruit a berry).

Asparagus (300) Old World.
Polygonatum (50) N. temperate region.
Trillium (30) W. Himalaya to Japan, N. America.
Ruscus (7) Madeira, W. and central Europe, Mediterranean region to Iran.
Convallaria (1) N. temperate region, *C. majalis*.

IX. Ophiopogonoideae (short rhizome, sometimes with suckers; leaves radical, narrow or lanceolate; fruit with thin pericarp; seeds 1–3 with fleshy testa).

Sansevieria (60) tropical and S. Africa, Madagascar, Arabia.
Ophiopogon (20) Himalaya to Japan and Philippines.

X. Aletridoideae (short rhizome; leaves radical, narrow or lanceolate; perianth-segments united; capsule loculicidal; seeds numerous, with thin testa).
Aletris (25) E. Asia, N. America.

XI. Luzuriagoideae (shrubs or undershrubs; inflorescence usually many-flowered, with scaly bract at base; fruit a berry with spherical seeds).
Luzuriaga (3) New Zealand, Peru to Tierra del Fuego.
Lapageria (1) Chile, *L. rosea.*

XII. Smilacoideae (climbing shrubs with net-veined leaves; flowers small in axillary umbels or racemes or terminal panicles; fruit a berry).
Smilax (350) tropics and subtropics.

ENDYMION NON-SCRIPTUS (L.) Garcke
Bluebell

Distribution. Common in woods and hedgerows throughout the British Isles. Native also in Belgium, the Netherlands, north, west and central France. Doubtfully native in the Iberian peninsula, N. Italy and N.W. Germany.

Vegetative characteristics. A glabrous herb with 5 or 6 linear leaves and a flowering scape arising from an ovoid bulb.

Floral formula. P3+3 A3+3 $\underline{G}(3)$

Flower and inflorescence. The violet-blue (rarely pink or white) flowers form a unilateral, racemose inflorescence from April to June. They are erect in bud but nodding when fully open. The tips of the perianth-segments are somewhat recurved, and the anthers are cream.

Pollination. Bumble-bees, hover-flies and other insects seek out the nectar secreted by glands in the ovary wall, an arrangement common to many members of the Liliaceae. The removal of nectar would normally leave the flower undamaged, but insects have been observed to bite through the base of the perianth-tube in order to reach the nectar. This 'illegitimate' behaviour, usually perpetrated by powerful but short-tongued bumble-bees occurs quite commonly in connection with tubular flowers.

Alternative flowers for study. *Endymion hispanicus*, Spanish Bluebell, which flowers in May, differs from *E. non-scriptus* in having blue anthers, and paler, more spreading perianth-segments which are not recurved. It is commonly grown in gardens and sometimes escapes and becomes naturalised. Comparison should be made with other spring-flowering genera in different tribes, e.g., *Tulipa* and *Convallaria.* Summer-flowering genera, commonly cultivated, include *Lilium*, *Hemerocallis* and *Yucca.* Many species of *Colchicum* flower in the autumn.

Fig. 97 Liliaceae, *Endymion non-scriptus*

A The unilateral raceme, with unopened buds erect and clustered near the apex while the fully opened flowers are drooping.
Inflorescence: 13 cm

B An individual flower with 2 bracts at the base of the curved pedicel. The 6 perianth-segments are in 2 whorls and are connate at their base.
Bracts: 10–12 mm Pedicels: 5–10 mm

C L.S. of flower exposing the superior ovary and 4 of the 6 stamens, whose filaments are adnate to the perianth-segments. The anthers are dorsifixed and versatile.
Perianth-segments: 13–17 × 4.5 mm

D A young inner stamen from a well-developed flower-bud. The filament is the same colour as the perianth-segment and is adnate to it. The cream-coloured anther is 2-celled, introrse and dehisces longitudinally. The anthers are attached to the perianth in 2 rows, the outer row to the middle of the segments and the inner row nearer the base.
Filament of outer stamen: 6 mm
Filament of inner stamen: 3 mm
Anther: 3.75–4 mm

E T.S. of the 3-locular ovary composed of 3 united carpels. The ovules are arranged in 2 rows in each loculus. Placentation is axile.
Ovary: 4 mm in diameter Ovules: 0.35–0.65 mm

F L.S. of ovary exposing the rows of ovules in 2 of the 3 loculi. The ovary is terminated by a long, simple style.
Ovary: 3.25–4 mm Style + stigma: 4–5 mm

G Detail of the receptive papillose stigma.

H The mature fruit, a 3-lobed, loculicidal capsule. The perianth-segments persist during fruit development.
Fruit: 11–16 × 10–14 mm

Fig. 97

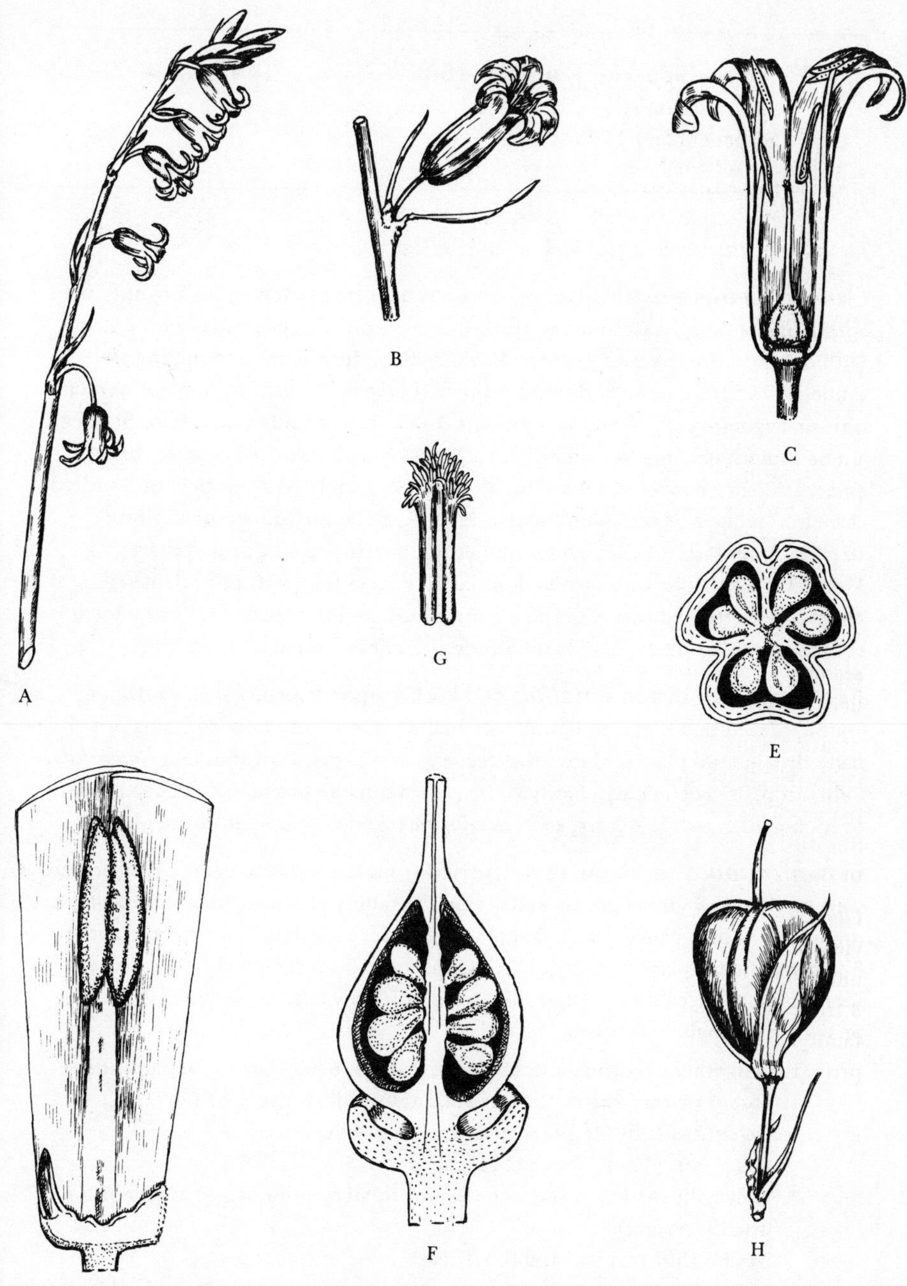

98 Amaryllidaceae Jaume St-Hil.

Daffodil family

85 genera and 1100 species

Distribution. Usually tropical or subtropical.

General characteristics. Usually xerophytic, often bulbous, leafing only in spring or the rainy season; many rhizomatous. Inflorescence usually on a scape, with one or more spathes, cymose, but often condensed and resembling an umbel or head; sometimes flowers solitary. Flowers hermaphrodite, actinomorphic or zygomorphic. Perianth-segments 3 + 3, free or united, petaloid. Stamens 3 + 3, sometimes some staminodial; anthers 2-celled, usually introrse, versatile or occasionally basifixed. Ovary usually inferior, rarely half-inferior, of 3 united carpels, 3-locular or rarely unilocular; style single, with 3-lobed or capitate stigma; placentation axile; ovules numerous, anatropous. In some genera (e.g., *Narcissus*) a conspicuous corona, looking like an extra whorl of perianth-segments, is situated between the normal whorl and the stamens. Fruit a loculicidal capsule or berry. Endosperm present; embryo small and straight.

Economic and ornamental plants. With economic plants such as *Allium*, Onion, placed in a separate family (see below), the family now contains principally ornamental plants. Many of these, e.g., *Narcissus*, Daffodil, and *Galanthus*, Snowdrop, are sufficiently hardy to be grown outside in the herbaceous border or rock-garden, while others, such as *Hippeastrum*, are popular house plants.

Classification. Hutchinson (15) departed from the hitherto generally accepted view of the Amaryllidaceae by restricting the family to those members having an umbellate inflorescence subtended by one or more spathes. More recently, the 3 tribes with a superior ovary (cf. Liliaceae) have been removed to form a separate family, the Alliaceae. The Amaryllidaceae, as it now appears in Willis (5), comprises the following 10 tribes:

1. Galantheae (corona absent, scape leafless, ovules many, perianth-tube absent or very short, flowers actinomorphic, solitary or few).
   Galanthus (20) Mediterranean region to Caucasus.
   Leucojum (12) S. Europe, Morocco.
2. Amaryllideae (as in Galantheae, but flowers more or less zygomorphic, usually several).
   Nerine (30) tropical and S. Africa.
   Brunsvigia (13) S. Africa.
   Amaryllis (1) S. Africa, *A. belladonna.*

3. Crineae (as in tribes 1 and 2, but perianth-tube present, inflorescence several-flowered).
   Crinum (100–110) tropics and subtropics.
   Cyrtanthus (47) tropical and S. Africa.
   Ammocharis (5) tropical and S. Africa.
   Vallota (1) S. Africa, *V. speciosa.*
4. Zephyrantheae (as in Crineae, but flowers solitary or paired).
   Zephyranthes (35–40) warm areas of America, W. Indies.
   Sternbergia (8) E. Mediterranean region to Caucasus.
5. Haemantheae (as in tribes 1–4, but ovules few).
   Haemanthus (50) tropical and S. Africa.
   Boophone (5) S. and E. Africa.
   Carpolyza (1) S. Africa, *C. tenella.*
6. Ixiolirieae (as in tribes 1–5, but scape leafy below, umbel subcompound).
   Ixiolirion (3) W. and central Asia.
7. Euchariteae (as in tribes 1–6, but conspicuous corona present, of expanded, often connate filaments).
   Hymenocallis (50) warm areas of America.
   Pancratium (15) Mediterranean region to tropical Asia and tropical Africa.
   Eucharis (10) tropical S. America.
8. Eustephieae (as in Euchariteae, but corona of small teeth present between filaments; perianth-lobes not spreading).
   Eustephia (6) Peru, Argentina.
   Phaedranassa (6) Andes.
9. Hippeastreae (as in Eustephieae, but corona of scales, and perianth-lobes spreading).
   Hippeastrum (75) tropical and subtropical America.
   Sprekelia (1) Mexico, *S. formosissima.*
10. Narcisseae (as in tribes 7–9, but corona distinct from filaments, annular or tubular or of separate scales).
   Narcissus (60) Europe, Mediterranean region, W. Asia.

GALANTHUS NIVALIS L.

Snowdrop

Distribution. A native of damp woods and stream-sides from southern Europe to the Caucasus, but commonly cultivated for ornament and widely naturalised elsewhere, making the northern limit of its native range uncertain.

Vegetative characteristics. A bulbous plant with the scape arising from between a pair of narrow, glaucous green leaves.

Floral formula. P3+3 A3+3 $\overline{G}(3)$

Flower and inflorescence. The solitary, pendulous, white flowers, with green, horseshoe-shaped markings on their inner perianth-segments, appear from January to March. The fruit becomes fully developed in June.

Pollination. The family is well adapted to insect pollination owing to the presence of conspicuous, often scented, flowers, and nectar-secreting glands. In *Galanthus*, the flower has 6 stamens which cluster round the longer style. Each of the anthers narrows upwards, ending in an apiculus which points outwards in the direction of the perianth. At maturity the anther-cells split open at their tips, forming a hollow cone. When an insect, such as a bee, searches for nectar and touches this cone, it causes pollen to fall upon itself. In the course of foraging trips the pollen is transferred to other flowers and cross-pollination is effected. If there is an absence of insect visits self-pollination is possible, as the cone of anthers becomes more lax, allowing the pollen to be shed directly on to the stigma below.

Alternative flowers for study. Any other species of *Galanthus* would be suitable, or any member of the closely related genus *Leucojum*, Snowflake, which is in flower between February and September according to species. Compare with the spring-flowering genus *Narcissus*, which has a conspicuous corona and with genera in other tribes mentioned above.

Fig. 98 Amaryllidaceae, *Galanthus nivalis*

A A solitary nodding flower, subtended by a membranous spathe. The 2 whorls of perianth-segments arise from the top of the prominent, green, inferior ovary. The outer, spreading segments are pure white. The shorter, inner segments are obovate, deeply emarginate, and have a green marking round the sinus.

B L.S. of flower. The actinomorphic flower arises from a spathe formed of 2 leafy, green bracteoles which are at first joined at their margins by 2 strips of membrane. One of these splits to allow the flower and pedicel to emerge, but the other remains intact, shrinking in width and drawing the 2 bracteoles together. The 6 perianth-segments are in 2 whorls, the segments of the outer whorl being about twice as long as those of the inner. One segment from each whorl has been removed. The ovary is clearly inferior.
Perianth-segments (outer): 19 mm
(inner): 11 mm
Spathe: 30 mm Ovary: 7–8 mm

C One of the 6 stamens inserted opposite a perianth-segment. The anthers are 2-celled and introrse. In most genera of the family they dehisce by vertical slits, but in *Galanthus* and *Leucojum* the anthers split open at the apex so that their tips form a hollow cone.
Anther: 6.25 mm Filament: 2 mm

D L.S. of the inferior ovary, showing the numerous ovules attached to the central axis. The single style is terminated by a capitate stigma.

E T.S. of the 3-locular ovary, showing some of the ovules attached to the axile placenta.
Ovary: 7 × 3.25 mm

F A developing fruit with the persistent style. The fruit is a loculicidal capsule. The seeds are provided with a horn-like elaiosome which attracts ants by its oil content, causing these insects to assist in the dispersal of the seeds.
Fruit: 10 × 5 mm Style: *c.* 8 mm

Fig. 98

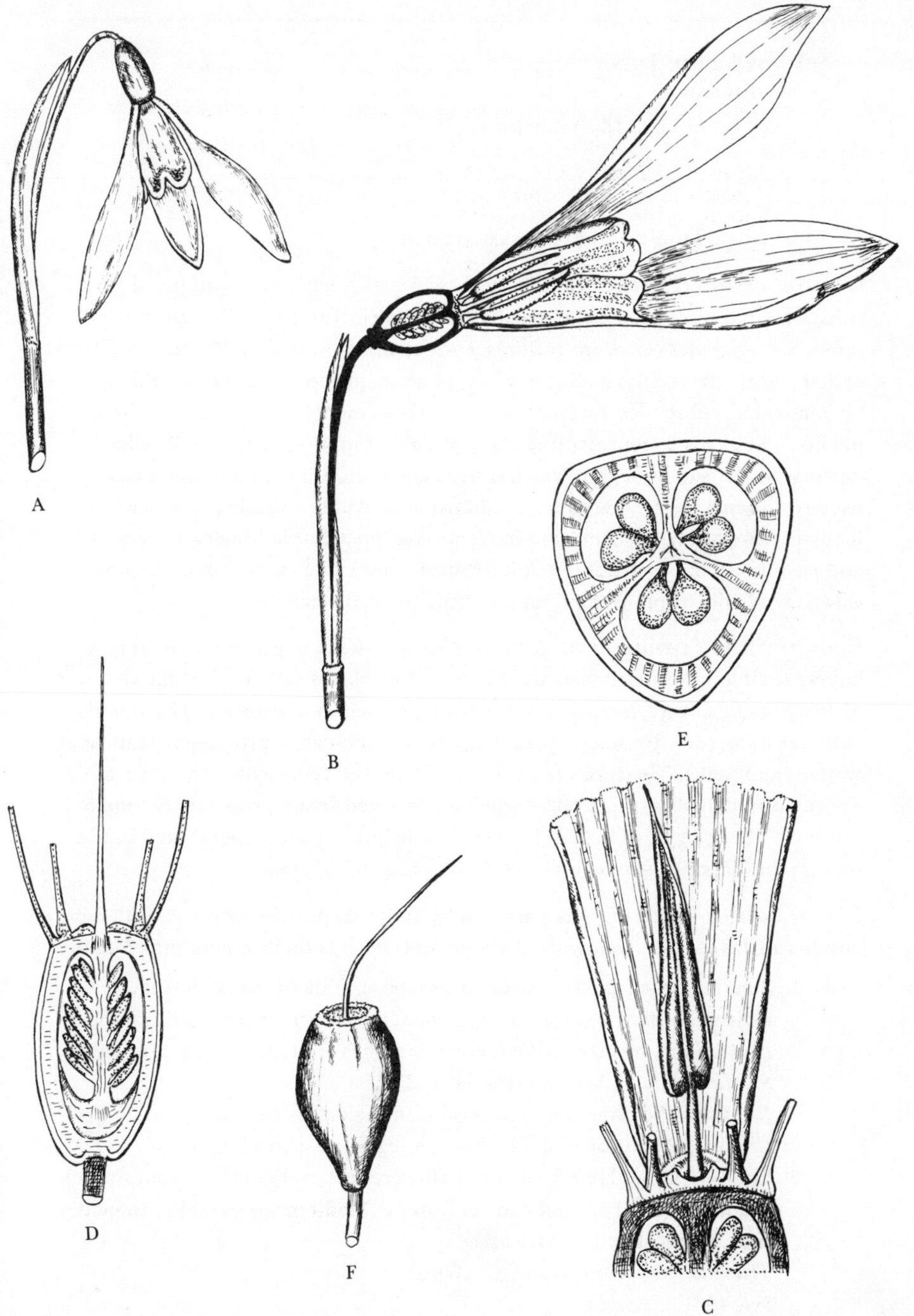

99 Iridaceae Juss.

Iris family

70 genera and 1500 species

Distribution. Tropical and temperate regions.

General characteristics. Chiefly perennial herbs, with roots produced from rhizome, bulb or corm. Leaves usually linear to ensiform, usually equitant in 2 ranks, with parallel venation. Inflorescence terminal, usually cymose, but flowers solitary in the tribe Sisyrinchieae. Flowers hermaphrodite, actinomorphic or zygomorphic, subtended by spathes. Perianth-segments 3 + 3, free or united, petaloid. Stamens 3 (representing the remaining outer whorl), with 2-celled, extrorse anthers dehiscing by vertical slits. Ovary inferior, of 3 united carpels, usually 3-locular, with few to many anatropous ovules on axile placentas (unilocular with parietal placentation in *Hermodactylus*); style usually 3-branched, and frequently more or less petaloid. Fruit a loculicidal capsule dehiscing by 3 valves; seed with copious, hard endosperm; embryo small.

Economic and ornamental plants. Crocus sativus is grown commercially, largely in Spain, for its stigmas which are the source of saffron, and the rhizome of *Iris germanica* var. *florentina*, Orris Root, is used in perfumery. The family contains numerous decorative plants, many of which are a prominent feature of gardens and parks. The genus *Iris* is particularly well represented by species, hybrids and cultivars, the group known as 'Bearded Irises' providing the main display. Other genera of horticultural value include all those mentioned in the section on Classification below, also *Crocosmia* and *Sparaxis*.

Classification. The family is classified by some authorities into a considerable number of tribes, but, following Diels (in Ref. 9), it is divided here into 3 only:

1. Sisyrinchieae (spathes terminal or lateral, stalked, rarely sessile; flowers solitary or more often several developed centrifugally round a central one, mostly stalked; style-branches alternating with stamens).
   Sisyrinchium (100) America, W. Indies.
   Romulea (90) Europe, Mediterranean region, S. Africa.
   Crocus (75) Europe, Mediterranean region to central Asia.
2. Ixieae (spathes lateral, sessile, 1-flowered; flowers often zygomorphic).
   Gladiolus (300) W. and central Europe, Mediterranean region to central Asia, tropical and S. Africa.
   Tritonia (55) tropical and S. Africa.
   Ixia (45) S. Africa.
   Freesia (20) S. Africa.

3. Irideae (flowers numerous, usually actinomorphic, in terminal or lateral, usually stalked spathes; style-branches opposite the stamens; leaves equitant).
Iris (300) N. temperate region.
Moraea (100) tropical and S. Africa.
Tigridia (12) Mexico to Chile.

IRIS PSEUDACORUS L.
Yellow Flag

Distribution. Native in W. Asia, N. Africa and most of Europe including the British Isles. It is commonly found in marshes, swamps, ditches and at the edges of rivers and ponds.

Vegetative characteristics. An erect, perennial herb reaching a height of 40–150 cm, with a well-developed, fleshy rhizome. The ensiform leaves are about as tall as the scape.

Floral formula. P3+3 A3 $\overline{G}(3)$

Flower and inflorescence. Groups of 2 or 3 large, showy, yellow flowers are borne at the top of a tall, stout scape from May until July. Each group is subtended by a green spathe with a papery margin.

Pollination. Bumble-bees and long-tongued flies are attracted to the flowers by the bright yellow perianth-segments and by the nectar that is secreted at their base. Functionally, the *Iris* flower is composed of 3 separate units, each containing an inner and an outer perianth-segment ('standard' and 'fall' respectively) which are united towards the base into a short tube, divided into 2 by the filament. The anther, borne by the filament, is protected from above by a branch of the petaloid style, which has a small, flap-like stigma situated just below the bifid tip. On visiting a flower, the insect lands on one of the 'falls', which is marked with nectar-guides in the form of converging lines. As it moves forward, the insect bends back the stigma, transferring pollen which it has brought from another flower. Further on, it brushes against the anther, which dusts its body with fresh pollen. After it has extracted nectar, the insect creeps out backwards. Self-pollination cannot normally occur since the insect's body only touches the non-receptive side of the stigma as it leaves the flower. Some insects have been observed to enter or leave at the side of the floral units, by creeping in or out between the style-branch and the 'fall'. In this way they may obtain nectar without contributing to the pollination of the flower.

Alternative flowers for study. Numerous species and hybrids of the genus *Iris* are in cultivation and would be suitable alternatives to *I. pseudacorus.* The commonly grown 'Bearded Irises', which are available in many colours, and are hybrids of a complex parentage, are so called because of the line of dense hairs at the base of the 'falls'. Floral differences between genera include zygomorphic flowers in *Gladiolus* and *Freesia*, and usually solitary flowers in the genus *Crocus*. In contrast to *Iris*, which has the style-branches opposite the stamens, *Crocus* and *Sisyrinchium* have them alternate.

Fig. 99 Iridaceae, *Iris pseudacorus*

A The hermaphrodite, actinomorphic flower. The yellow perianth is made up of 3 outer segments and 3 inner, united into a short tube above the inferior ovary. The outer segments ('falls') are narrow at their base but expand into a broad blade which hangs down. The inner segments ('standards') are more or less erect and smaller, but also have a well-defined limb and claw. Arising from the centre are the 3 petaloid style-branches which, for clarity, have been separated from the outer perianth-segments though in nature they lie close above them.
Outer perianth-segment: 55 × 35 mm
Inner perianth-segment: 25 × 5 mm Petaloid style-branch: 40 × 5 mm

B B1: L.S. of part of a flower with the perianth-segments separated to show the stamens inserted at the base of the outer segments. Two of the 3 inner segments are shown joined near the base with the outer segments. The inferior ovary, which contains numerous anatropous ovules attached to axile placentas, supports a stylar column which divides into 3 petaloid branches and protects the stamens from above.
Ovary: 20–30 × 6–7 mm
B2: The tip of a style-branch, showing the non-receptive side of the stigma, a small flap situated on the underside, just below the bifid apex ('crest'). The receptive surface appears when the flap is bent back.
Apex of style: 7 mm wide Lobe: 5 × 2.5 mm
Stigma: 1.5 × 1.5–1.75 mm

C A stamen, situated opposite one of the outer perianth-segments and flanked by 2 of the inner perianth-segments.

D Detail of an extrorse anther which is 2-celled and dehisces longitudinally.
Filament: 10 mm Anther: 15 mm

E T.S. of ovary, showing the 3 loculi and the ovules attached to axile placentas.

F The mature fruit, a capsule which dehisces loculicidally by 3 valves to liberate the brown, flattened seeds which are buoyant and readily dispersed by water.
Fruit (before dehiscence): 45–50 × 15 mm Seed: 10 × 8 mm

Fig. 99

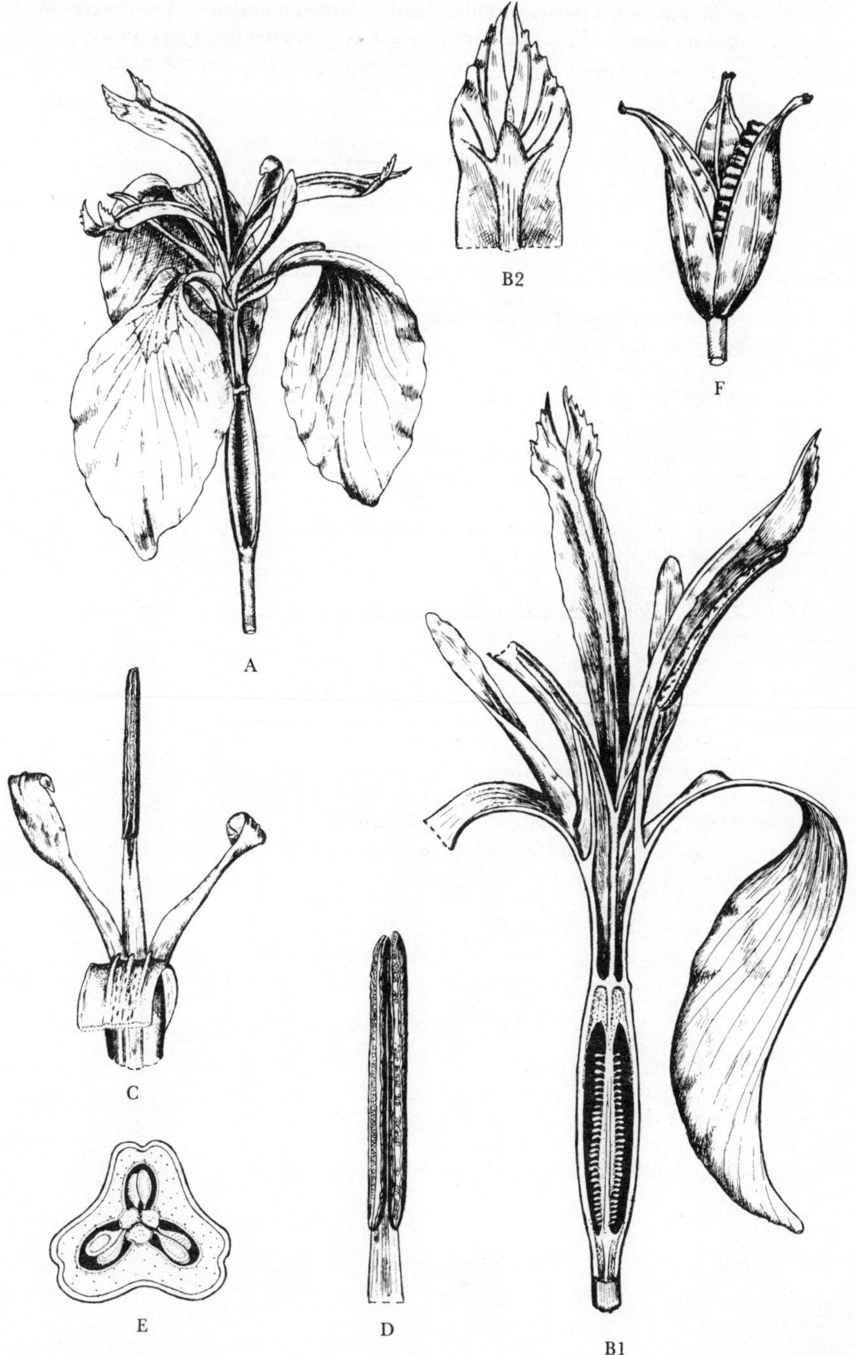

Fig. M. L.S. of *Iris* flower with floral parts in natural position: 1, Inner perianth-segment ('standard'); 2, Branch of petaloid style; 3, Bifid tip of style ('crest'); 4, Stigma; 5, Anther; 6, Outer perianth-segment ('fall'); 7, Perianth-tube; 8, Stylar column; 9, Inferior ovary; 10, Spathe; 11, Scape.

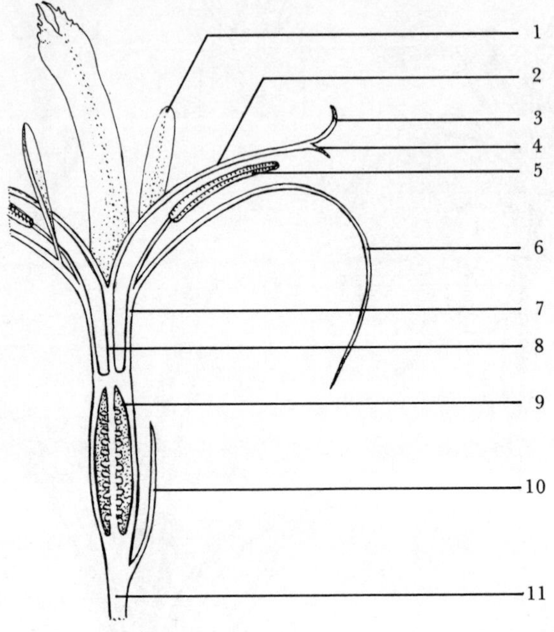

100 Orchidaceae Juss.

Orchid family

735 genera and 17,000 species

Distribution. Cosmopolitan.

General characteristics. Perennial herbs, generally terrestrial (particularly in temperate regions) or epiphytic (mainly in the tropics), occasionally saprophytic (e.g., *Neottia*). Mycorrhiza often present. Sometimes climbing (e.g., *Vanilla*). Terrestrial species often tuberous or rhizomatous, the epiphytic ones often with fleshy pseudobulbs consisting of one or more thickened stem-internodes. Stems leafy or scapose. Leaves simple, usually alternate, often distichous, sometimes reduced to scales. Inflorescence racemose or flowers solitary. Flowers usually hermaphrodite, zygomorphic, and extraordinarily diverse in size, shape and colour. Perianth typically of 6 segments in 2 whorls of 3, the outer whorl (sepals) often petaloid, the inner petaloid and with a highly modified posterior petal, the labellum, which is often projected basally into a spur. In many species the ovary twists through 180 degrees, causing the labellum to come round to the anterior side of the flower and providing a landing-place for pollinating insects. Stamens usually 1, more rarely 2 or 3. Pollen grains usually aggregated into pollinia (cf. Asclepiadaceae). Ovary inferior, of 3 united carpels, usually unilocular, with numerous ovules on the 3 parietal placentas. Style, stigmas and stamens variously adnate into a single, highly complex structure, the column. Stigmas basically 3, but only 2 fertile, the third represented by a sterile outgrowth, the rostellum. Fruit a capsule, containing a very large number of exceedingly small and light seeds, well suited to wind distribution.

Economic and ornamental plants. The climbing genus *Vanilla*, especially the Central American species *V. planifolia*, is of economic importance and is now cultivated in many tropical countries for its pods, which are the source of the flavouring used in confectionery. The starchy tubers of certain genera, e.g., *Orchis*, *Dactylorhiza* and *Eulophia*, are dried to form salep, which is used in parts of Europe and Asia for culinary and medicinal purposes. Many other members of the Orchidaceae are of horticultural value. Some terrestrial European orchids are grown in alpine collections, but it is the epiphytic tropical orchids, often with large, brightly coloured flowers of intriguing structure, which have attracted the greatest attention. The genus *Cattleya* is probably the most widely cultivated and numerous cultivars have been raised. Other genera commonly grown in glasshouses include *Paphiopedilum*, *Oncidium*, *Cymbidium*, *Dendrobium*, *Miltonia*, *Odontoglossum*, *Phalaenopsis* and *Epidendrum*. Orchid genera can be crossed easily and many bigeneric and plurigeneric hybrids are in existence.

Classification. While the characters of the Orchidaceae separate them clearly from other families, subdivision of the family has proved difficult and many schemes have been devised. The following classification, given in Willis (5), divides the family into 3 subfamilies and 6 tribes, those of the Orchidoideae being separated principally on anther and pollinia characters.

I. Apostasioideae (flowers more or less actinomorphic; labellum shallow; stamens 2 or 3; pollen in separate grains).

1. Apostasieae.
   Apostasia (10) tropical Asia, Australia.

II. Cypripedioideae (flowers zygomorphic; labellum deeply saccate; stamens 2, staminode usually shield-like; pollen as tetrads in a sticky fluid).

2. Cypripedieae.
   Cypripedium (50) N. temperate region.
   Paphiopedilum (50) tropical Asia to Solomon Islands.

III. Orchidoideae (flowers zygomorphic; stamen 1, staminode absent; pollen aggregated into pollinia).

3. Orchideae (pollinia granular; viscidium present; base of anther firmly attached to column).
   Platanthera (200) temperate and tropical Eurasia, N. Africa, N. and Central America.
   Orchis (35) Madeira, temperate Eurasia to India and S.W. China.
   Ophrys (30) Europe, W. Asia, N. Africa.
   Dactylorhiza (30) Atlantic Islands, N. Africa, temperate Eurasia, Alaska.
4. Neottieae (pollinia mealy; viscidium present; anther deciduous, apex lightly attached to column).
   Listera (30) N. temperate region.
   Spiranthes (25) cosmopolitan.
   Epipactis (24) N. hemisphere.
   Neottia (9) temperate Eurasia.
5. Epidendreae (pollinia waxy; viscidium absent or poorly developed; anther deciduous, attached by apex).
   Dendrobium (1400) Asia, Australasia, Polynesia.
   Epidendrum (400) tropical America.
   Vanilla (90) tropics and subtropics.
   Cattleya (60) Central and S. America, W. Indies.
6. Vandeae (pollinia horny or waxy; viscidium present; anther deciduous, attached by apex).
   Oncidium (350) Florida to temperate S. America, W. Indies.
   Angraecum (220) tropical and S. Africa.
   Odontoglossum (200) Mexico to tropical S. America.
   Vanda (60) China, Indomalaysia.

Fig. N. Structure of *Cypripedium* flower: 1, Dorsal sepal; 2, Lateral petal; 3, Column (with central staminode and 2 lateral anthers); 4, Labellum; 5, United lateral sepals; 6, Ovary.

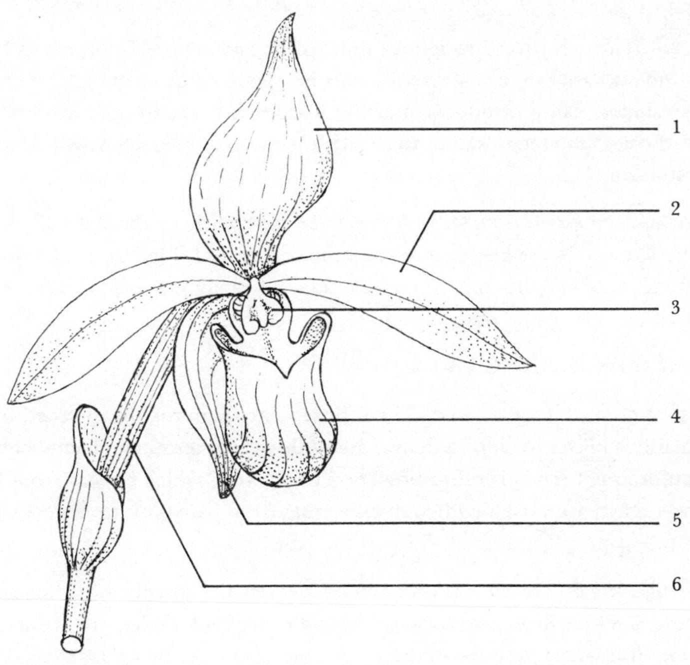

Fig. P. L.S. of *Cypripedium* flower: 1, Staminode; 2, Anther; 3, Stigma.

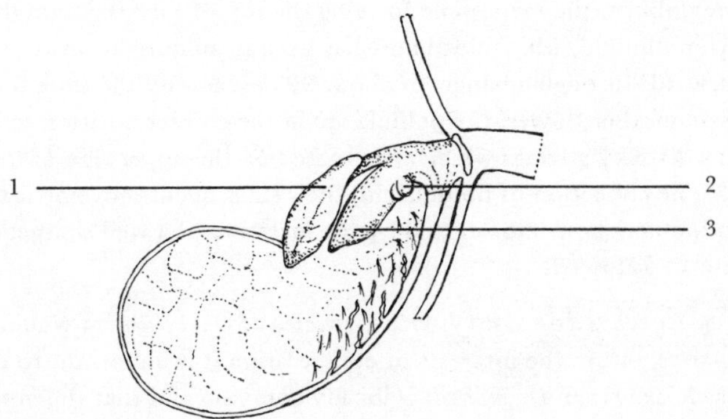

DACTYLORHIZA FUCHSII (Druce) Soó
Common Spotted Orchid

Distribution. Native throughout most of N. and central Europe and N. Asia. It is found on base-rich or calcareous soils in a wide range of habitats including grassy slopes, damp meadows, marshes, base-rich fens and open woods. Three or 4 subspecies are recognised in the British Isles, ssp. *fuchsii* having the widest distribution.

Vegetative characteristics. A perennial herb with a palmately divided root-tuber. The leaves are keeled and sometimes slightly folded, obovate-oblong to elliptical, occasionally unmarked but usually heavily marked with more or less transversely elongated dark blotches.

Floral formula. K3 C3 A1 $\overline{G}(3)$

Flower and inflorescence. The inflorescence is borne on an erect, unbranched stem and is in the form of a dense, many-flowered, terminal raceme composed of zygomorphic, hermaphrodite flowers. The flowers, which appear from May until early August, are variable in colour, ranging from pale pink to mauve or white, with red or purplish dots or lines on the labellum.

Pollination. D. fuchsii is pollinated by a variety of insects including solitary, humble and hive bees, and by syrphid and other flies, which are attracted by the raceme of brightly coloured flowers. A visiting insect alights on the large horizontal petal (labellum) which is prolonged backwards into a narrow spur. As the insect inserts its proboscis into the spur to obtain the liquid within its fleshy walls (it contains no free nectar) it touches the bursicle, the pouch-like base of the rostellum. This ruptures along the front, exposing the 2 sticky discs known as viscidia, one or both of which will touch and adhere firmly to the head of the insect. Each viscidium is attached to a group of pollen grains (pollinium) by elastic threads, collectively termed the caudicle. As the insect flies away bearing one or more pollinia, the membrane forming the top of the viscidium dries out, causing each pollinium, which until now has been in an upright position, to be directed forwards through an angle of about 90 degrees. By the time the insect has arrived at another flower, the pollinia are in the correct position to be transferred to the 2 sticky stigmas which are situated on the upper side of the entrance to the spur. The collection of pollinia and their subsequent movement to a horizontal position may be demonstrated by the insertion of a well-sharpened pencil into the spur of a flower.

Alternative flowers for study. Other species of *Dactylorhiza* would be suitable alternatives, but in the interests of conservation it is important to check that any species chosen (even *D. fuchsii*) is locally common and that the minimum number of flowers is collected. The equally widespread *D. maculata* ssp. *ericetorum*, Heath Spotted Orchid, is very similar to *D. fuchsii* and it is sometimes difficult to distinguish them. *D. maculata* ssp. *ericetorum*, however, grows on acid, peaty soils in contrast to *D. fuchsii.* The flowering period of these two orchids is very similar and hybrids have been found to occur in certain areas. Hybridisation has in fact been recorded between many of the species in this genus. Comparison should be made with other tribes in the same subfamily and also with *Cypripedium* and *Paphiopedilum* which, amongst other characters, are distinguished in having a deeply saccate labellum and 2 anthers (see Figs. N. and P. on page 508).

Fig. Q. L.S. of *Dactylorhiza* flower: 1, Pollinium; 2, Rostellum; 3, Stigma.

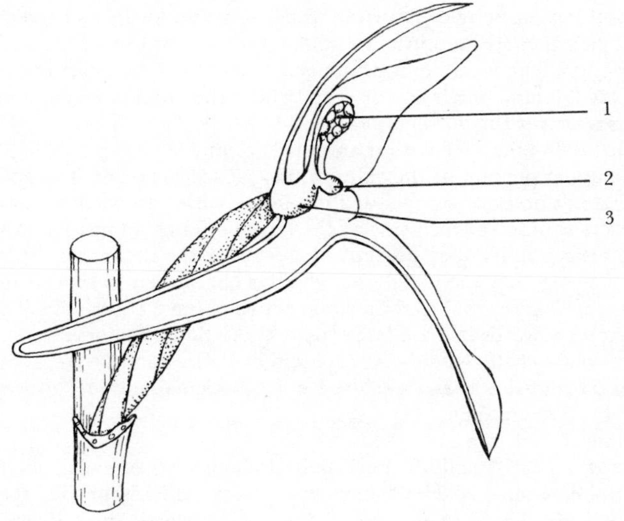

Fig. 100 Orchidaceae, *Dactylorhiza fuchsii*

A The inflorescence, a dense raceme of zygomorphic, hermaphrodite flowers which arise from the axils of leaf-like bracts. The flowers at the base of the inflorescence open first.
Inflorescence: 5 cm

B Anterior view of flower. The 3 petaloid sepals and the 3 petals arise from the top of the inferior ovary. One of the petals is considerably enlarged and modified to form the 3-lobed labellum. The other, smaller petals overlap slightly and help to protect the reproductive organs which combine to form the highly specialised and characteristic structure at the centre of the flower known as the column.
Dorsal sepal: 7 x 2.5 mm Lateral sepals: 7–8 x 3 mm
Labellum: 7–9 x 9–10 mm Lateral petals: 5 x 2 mm

C Lateral view of a flower with part of the perianth removed. The flower arises from the axil of a bract, and during growth the inferior ovary twists in a half-circle (shown by the ribs on the surface of the ovary) bringing the labellum into a suitable position for the insect visitors to alight on. The labellum is prolonged at its base into a narrow spur. One of the 2 lateral petals can be seen behind the column and arching over these parts is the dorsal sepal.
Spur: 4–6 mm Ovary (at anthesis): 8 mm

D The upper portion of the column, showing the stamen. The pollen masses or pollinia are enclosed in thecae, which are membranous containers situated at each side of the large, arching connective. At maturity, each theca splits open (as shown) to expose the pollinium. At the base of the stamen is the rostellum, which is considered to represent a third stigma. This is prolonged upwards between the 2 thecae. Each theca narrows towards its base into a pouch-like flap, the bursicle, which protects from air the viscid disc or viscidium. The concave areas to the left and right of the pair of viscidia are the 2 stigmas and just above these are the 2 staminodes.
Stamen: 2 x 1 mm

E One of a pair of pollinia. Each pollinium consists of a club-shaped mass of pollen-grains, which when viewed under a microscope are seen to be joined together in small, compact masses by elastic threads. The lower portions of the threads form a stalk, the caudicle, which connects the pollinium to the viscidium.
Pollinium: 1.5 mm

F T.S. of the inferior ovary, which consists of a single loculus containing numerous ovules attached to 3 parietal placentas.
Ovary: 1.25 mm in diameter

G L.S. of the ovary, showing 2 of the 3 placentas bearing the numerous ovules. The ovary develops into a capsule which at maturity splits into 6 parts. Three of these bear the placentas with the seeds attached, the others are merely strips of pericarp. All 6 remain joined at their tips as well as at the base. In dry weather the capsule contracts, causing the slits to open and allowing the minute seeds to escape and be dispersed by the wind. But in damp weather the dry pericarp absorbs moisture and lengthens, closing the slits and preventing seed-dispersal.
Ovary (at late anthesis): 11 x 4 mm

Fig. 100

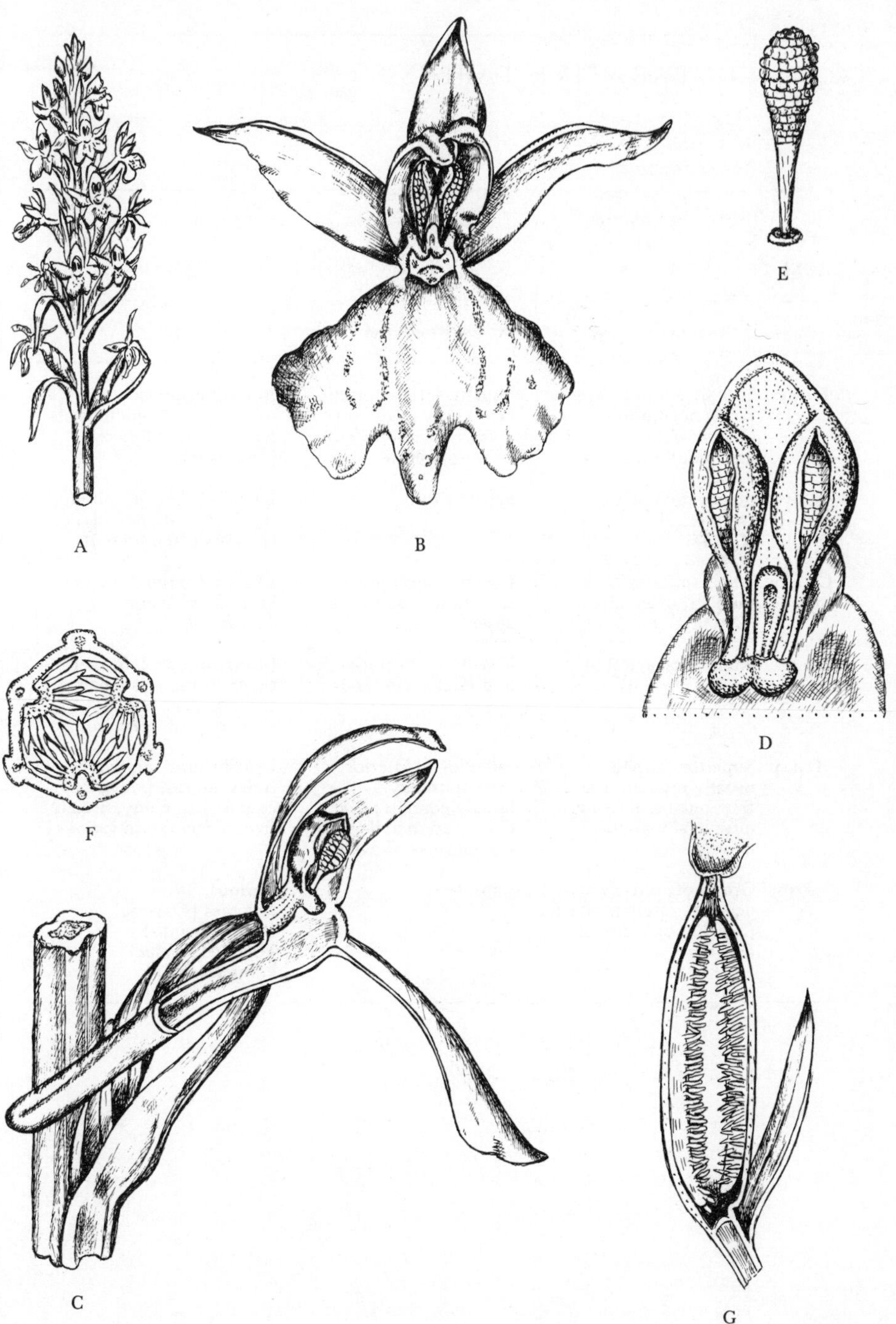

COMPARATIVE TABLES

TABLE 1

	5. Ranunculaceae	36. Saxifragaceae	38. Rosaceae
Flower	Usually actinomorphic, rarely zygomorphic (*Delphinium*); entomophilous	Usually actinomorphic, very rarely zygomorphic (*Saxifraga stolonifera*); entomophilous	Actinomorphic; usually entomophilous, rarely anemophilous (*Poterium*)
Calyx	Sepals variable in number, often petaloid (*Caltha*)	Sepals 4 or 5	Usually 5 free or united sepals, 4 in *Alchemilla*; epicalyx often present
Corolla	Petals variable in number, often absent (*Caltha*)	Usually 4 or 5 petals, sometimes connate or absent	Usually 5 petals, absent in *Alchemilla* and *Sanguisorba*
Stamens	Numerous, spiral or cyclic	Usually in 2 whorls of 5, obdiplostemonous	Numerous, or 2, 3, or 4 times as many as sepals and whorled; only 4(–5) stamens in *Alchemilla*
Ovary	Superior, carpels usually numerous and free; one basal ovule or several marginal ovules	Superior to inferior; carpels usually 2, basally connate; placentation axile with several rows of ovules	Usually superior, sometimes inferior (*Malus*); carpels 1 to many, usually free; placentation various
Fruit	Group of follicles or achenes, rarely a capsule (*Nigella*) or berry (*Actaea*)	Capsule	Various: Achenes (*Fragaria*) Drupe (*Prunus*) Follicle (*Spiraea*) Pome (*Malus*)

TABLE 2

	70. Boraginaceae	72. Labiatae	74. Scrophulariaceae
Flower	Usually actinomorphic, rarely zygomorphic (*Echium*)	Zygomorphic	Zygomorphic, sometimes nearly actinomorphic (*Verbascum*)
Calyx	Five united sepals, sometimes only basally connate (*Lithospermum*)	Five united sepals, sometimes 2-lipped	Usually 5 united sepals (4 in *Veronica*)
Corolla	Five united petals	Five united petals, usually 2-lipped	Usually 5 united petals (4 in *Veronica*), often 2-lipped
Stamens	Five, epipetalous	Usually 4, didynamous and epipetalous, sometimes 2	Usually 4, didynamous, sometimes 2, rarely 5 (*Verbascum*), epipetalous
Ovary	Superior; 2 united carpels becoming 4 at maturity, placentation axile	Superior; 2 united carpels becoming 4 at maturity; placentation axile	Superior; 2 united carpels; placentation axile
Fruit	Usually 4 achenes (nutlets) separating at maturity, sometimes a drupe	Usually 4 achenes (nutlets) separating at maturity, sometimes a drupe	Usually a capsule, sometimes a berry or drupe

TABLE 3

	86. Gramineae	87. Juncaceae	88. Cyperaceae
Inflor-escence	Spike, raceme or panicle	Cymose panicle or head	Various, from a simple spike to a much-branched panicle
Perianth-segments	Usually 2 or 3 scales (lodicules)	Six, in 2 whorls of 3, usually sepaloid	Often 6 bristles or scales, or absent
Stamens	Usually 3	Usually 6, in 2 whorls of 3, or the inner whorl absent	Usually 1–3
Style	Usually 2, 1 in *Nardus*, 3 in some Bambuseae	Simple, with 3 brush-like stigmas	Often deeply divided into 2 or 3 branches
Fruit	Usually a caryopsis, rarely a nut or berry	Loculicidal capsule	Trigonous or biconvex achene
Seeds	One	Three to many	One
Stem	Jointed, usually cylindrical, often with pith only at the nodes	Usually without joints, cylindrical, pith often present but not solid	Not jointed, usually 3-angled and with solid pith
Leaves	Basal, or, if cauline, 2-ranked, usually linear to lanceolate and flat; sheath usually open	Mostly in basal tufts, usually linear, grass-like and flat, or terete	In basal tufts, or, if cauline, 3-ranked, usually grass-like; sheath usually closed

TABLE 4

	97. Liliaceae	98. Amaryllidaceae	99. Iridaceae
Inflor-escence	Usually racemose, sometimes umbellate (*Allium*) or flowers solitary (*Tulipa*)	Cymose, often umbellate, sometimes flowers solitary (*Galanthus*)	Cymose, flowers sometimes solitary in tribe Sisyrinchieae
Flower	Actinomorphic	Actinomorphic or zygomorphic	Actinomorphic or zygomorphic
Perianth-segments	Three + 3, free or united, petaloid or sometimes sepaloid	Three + 3, free or united, petaloid	Three + 3, free or united, petaloid
Stamens	Usually 3 + 3, anthers introrse	Three + 3, sometimes some staminodial, anthers usually introrse	Three (inner whorl suppressed), anthers extrorse
Ovary	Superior; carpels 3; united; placentation usually axile	Inferior; carpels 3, united; placentation axile	Inferior; carpels 3, united; placentation usually axile
Fruit	Loculicidal or septicidal capsule, or berry	Usually a loculicidal capsule, sometimes a berry (*Clivia*)	Loculicidal capsule
Storage organ	Rhizome, bulb or fleshy root	Bulb or rhizome	Rhizome, bulb or corm

APPROXIMATE FLOWERING TIMES: DICOTYLEDONS

Family	Species		Jan.	Feb.	Mar.	Apr.	May	Jun.	Jul.	Aug.	Sep.	Oct.	Nov.	Dec.
1. Magnoliaceae	Magnolia × soulangeana					—	—							
2. Lauraceae	Laurus nobilis					—	—							
3. Piperaceae	Peperomia incana	GH	····	—	····	····	····	····	····	····	····	····	····	····
4. Aristolochiaceae	Aristolochia clematitis							—	—	—	—			
5.1. Ranunculaceae	Delphinium ambiguum							—	—					
5.2. Ranunculaceae	Ranunculus repens						—	—	—	—	····			
6. Berberidaceae	Berberis darwinii					—	—	—	····	····	····	····	····	
7. Papaveraceae	Papaver rhoeas							—	—	—				
8. Hamamelidaceae	Hamamelis mollis		—	—	—									—
9. Fagaceae	Fagus sylvatica					—	—							
10. Betulaceae	Betula pendula					—	—							
11. Corylaceae	Corylus avellana		—	—	—	—								
12. Cactaceae	Schlumbergera × buckleyi	GH	—	—	····								····	—
13. Caryophyllaceae	Cerastium tomentosum						—	—	—	—	—			
14. Portulacaceae	Calandrinia grandiflora								—	—				

APPROXIMATE FLOWERING TIMES: **DICOTYLEDONS** (continued)

Family	Species	Jan.	Feb.	Mar.	Apr.	May	Jun.	Jul.	Aug.	Sep.	Oct.	Nov.	Dec.
15. Chenopodiaceae	Spinacia oleracea						—	—	—				
16. Polygonaceae	Polygonum persicaria						—	—	—	—	—		
17. Plumbaginaceae	Armeria maritima				—	—	—	—	—	—	—		
18. Theaceae	Camellia × williamsii	—	—	—	—	—						—	—
19. Guttiferae	Hypericum perforatum						—	—	—	—			
20. Tiliaceae	Tilia platyphyllos						—	—					
21. Malvaceae	Malva sylvestris						—	—	—	—			
22. Ulmaceae	Ulmus procera		—	—									
23. Moraceae	Morus nigra					—							
24. Urticaceae	Urtica dioica						—	—	—				
25. Violaceae	Viola riviniana				—	—	—		…	…	…		
26. Passifloraceae	Passiflora caerulea						—	—	—	—			
27. Cistaceae	Helianthemum nummularium						—	—	—	—			
28. Begoniaceae	Begonia semperflorens						—	—	—	—	…	…	
29. Cucurbitaceae	Cucurbita pepo						—	—	…	…			
30. Salicaceae	Salix caprea			—	—								

GH, glasshouse plant; ———, main flowering period; . . . . . . , occasional extension of flowering period.

Note that the flowering times shown are based upon the authors' own observations and on information obtained from standard botanical and horticultural works.

APPROXIMATE FLOWERING TIMES: DICOTYLEDONS (continued)

Family	Species		Jan.	Feb.	Mar.	Apr.	May	Jun.	Jul.	Aug.	Sep.	Oct.	Nov.	Dec.
31. Cruciferae	Arabis caucasica													
32. Resedaceae	Reseda lutea													
33. Ericaceae	Erica herbacea													
34. Primulaceae	Primula vulgaris													
35. Crassulaceae	Kalanchoe tubiflora	GH												
36. Saxifragaceae	Saxifraga granulata													
37. Grossulariaceae	Ribes uva-crispa													
38.1. Rosaceae	Filipendula ulmaria													
38.2. Rosaceae	Rosa canina													
38.3. Rosaceae	Malus × domestica													
38.4. Rosaceae	Prunus spinosa													
39.1. Leguminosae	Acacia armata	GH												
39.2. Leguminosae	Cercis siliquastrum													
39.3. Leguminosae	Vicia faba													
40. Lythraceae	Lythrum salicaria													
41. Thymelaeaceae	Daphne laureola													

APPROXIMATE FLOWERING TIMES: **DICOTYLEDONS** (continued)

Family	Species	Jan.	Feb.	Mar.	Apr.	May	Jun.	Jul.	Aug.	Sep.	Oct.	Nov.	Dec.
42. Myrtaceae	Callistemon citrinus GH												
43. Onagraceae	Fuchsia magellanica												
44. Cornaceae	Cornus sanguinea												
45. Loranthaceae	Viscum album												
46. Celastraceae	Euonymus europaeus												
47. Buxaceae	Buxus sempervirens												
48. Euphorbiaceae	Euphorbia helioscopia												
49. Rhamnaceae	Ceanothus thyrsiflorus												
50. Vitaceae	Vitis vinifera												
51. Hippocastanaceae	Aesculus hippocastanum												
52. Aceraceae	Acer pseudoplatanus												
53. Anacardiaceae	Cotinus coggygria												
54. Rutaceae	Choisya ternata												
55. Juglandaceae	Juglans regia												
56. Linaceae	Linum perenne												
57. Geraniaceae	Geranium pratense												

GH, glasshouse plant; ———, main flowering period; . . . . . . occasional extension of flowering period.

APPROXIMATE FLOWERING TIMES: **DICOTYLEDONS** (continued)

Family	Species		Jan.	Feb.	Mar.	Apr.	May	Jun.	Jul.	Aug.	Sep.	Oct.	Nov.	Dec.
58. Oxalidaceae	Oxalis acetosella					—	—	....	....	....				
59. Polygalaceae	Polygala vulgaris						—	—	—	—	—			
60. Araliaceae	Hedera helix										—	—	—	....
61. Umbelliferae	Heracleum sphondylium							—	—	—	—	....		
62. Loganiaceae	Buddleia davidii							—	—	—	—	—		
63. Gentianaceae	Gentiana septemfida									—	—			
64. Apocynaceae	Vinca minor				—	—	—							
65. Asclepiadaceae	Asclepias curassavica	GH						—	—	—	—	—		
66. Oleaceae	Syringa vulgaris						—							
67. Solanaceae	Solanum dulcamara							—	—	—	—			
68. Convolvulaceae	Convolvulus arvensis							—	—	—	—			
69. Polemoniaceae	Polemonium caeruleum							—	—	....				
70. Boraginaceae	Symphytum officinale						—	—	—					
71. Verbenaceae	Verbena rigida								—	—	....			
72. Labiatae	Lamium album						—	—	—	—	—	—	—	—
73. Plantaginaceae	Plantago lanceolata					—	—	—	—	—				

APPROXIMATE FLOWERING TIMES: DICOTYLEDONS (continued)

Family	Species		Jan.	Feb.	Mar.	Apr.	May	Jun.	Jul.	Aug.	Sep.	Oct.	Nov.	Dec.
74.1. Scrophulariaceae	Antirrhinum majus													
74.2. Scrophulariaceae	Verbascum thapsus													
74.3. Scrophulariaceae	Veronica persica													
75.1. Gesneriaceae	Saintpaulia ionantha	GH												
75.2. Gesneriaceae	Columnea × banksii	GH												
76. Bignoniaceae	Catalpa bignonioides													
77. Acanthaceae	Acanthus mollis													
78. Campanulaceae	Campanula rotundifolia													
79. Rubiaceae	Galium verum													
80.1. Caprifoliaceae	Viburnum opulus													
80.2. Caprifoliaceae	Lonicera periclymenum													
81. Valerianaceae	Centranthus ruber													
82. Dipsacaceae	Knautia arvensis													
83.1. Compositae	Doronicum plantagineum													
83.2. Compositae	Taraxacum officinale													

GH, glasshouse plant; ———, main flowering period; . . . . . . , occasional extension of flowering period.

APPROXIMATE FLOWERING TIMES: MONOCOTYLEDONS

Family	Species		Jan.	Feb.	Mar.	Apr.	May	Jun.	Jul.	Aug.	Sep.	Oct.	Nov.	Dec.
84. Alismataceae	Alisma plantago-aquatica							—	—	—				
85. Commelinaceae	Tradescantia × andersoniana							—	—	—	—	—		
86. Gramineae	Arrhenatherum elatius							—	—	—	—	…	…	
87. Juncaceae	Luzula campestris				—	—	—	—						
88. Cyperaceae	Carex acutiformis							—	—					
89. Typhaceae	Typha latifolia							—	—					
90. Bromeliaceae	Billbergia nutans	GH	—	—	—								—	—
91. Musaceae	Strelitzia reginae	GH	—	—	—	—	—	—						
92. Zingiberaceae	Hedychium gardneranum	GH						—	—	—				
93. Cannaceae	Canna × generalis								—	—	—	…		
94. Marantaceae	Calathea zebrina	GH					—							
95. Palmae	Phoenix sylvestris	GH	—	—	—	—								
96. Araceae	Arum maculatum					—	—							
97. Liliaceae	Endymion non-scriptus					—	—	—						
98. Amaryllidaceae	Galanthus nivalis		—	—	—									
99. Iridaceae	Iris pseudacorus						—	—	—					
100. Orchidaceae	Dactylorhiza fuchsii						—	—	—	—				

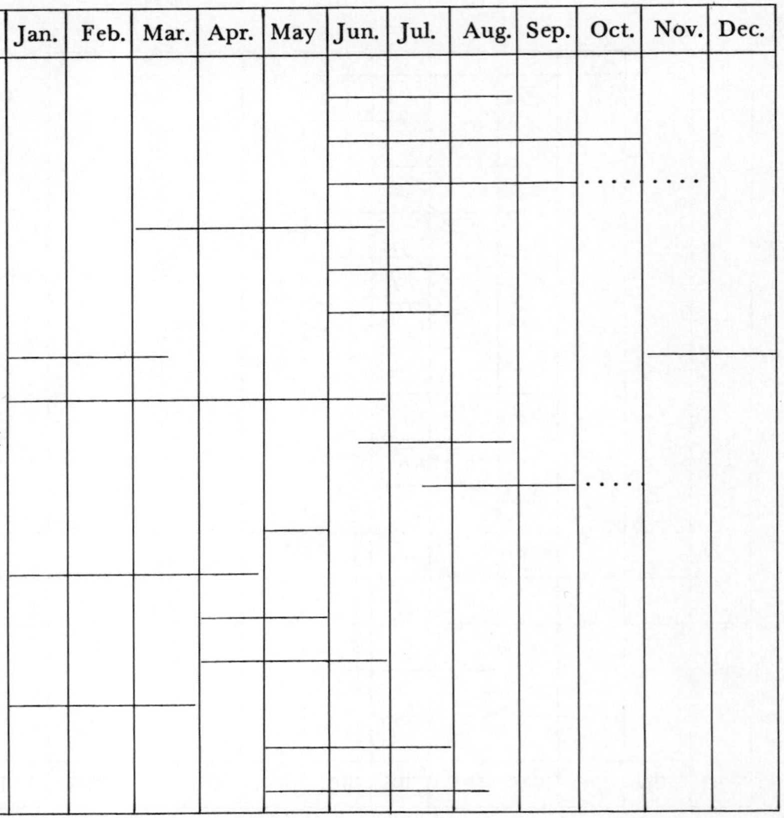

GH, glasshouse plant; ———, main flowering period; . . . . . . , occasional extension of flowering period.

GLOSSARY

Figs. I–IX are to be found within the Introduction, while Figs. X–XIV follow the Glossary.

abaxial The side of an organ away from the axis, dorsal.
abortion Non-formation or incompletion of a part.
abortive Imperfectly developed.
accrescent Increasing in size with age, as the calyx of some plants after flowering.
achene A small, dry, one-seeded, indehiscent fruit.
actinomorphic Regular, divisible into equal halves in 2 or more planes.
acuminate Narrowing gradually to a point.
acute Sharply pointed.
acyclic Arranged spirally rather than in whorls.
adaxial The side of an organ towards the axis, ventral.
adnate United (with a different part).
adventitious Produced from an unusual place, as buds from the stem (instead of the leaf-axils) or roots from a rhizome (instead of an existing root).
aestivation The arrangement of the calyx or corolla in the bud.
agamospermy The production of seeds by asexual means.
aggregate Collected together.
ala A wing or lateral petal in some of the Leguminosae.
alate Winged.
albumen Nutritive material stored within the seed.
albuminous Possessing albumen.
alternate (leaves) Placed singly at different heights on the stem or axis.
amphitropous (ovule) Curved, so that both ends are brought near to each other (Fig. VI, p. 13).
amplexicaul (leaves) Clasping the stem, but not completely encircling it (Fig. XII, p. 548).
anatropous (ovule) With the body inverted so that it lies alongside the funicle (Fig. VI, p. 13).
androecium The male sex organs (stamens) collectively.
androgynophore A stalk bearing both androecium and gynoecium, as in many Passifloraceae.
androphore A stalk bearing the androecium, as in some Tiliaceae.
anemophilous Depending on the wind to convey pollen for fertilisation.
angiosperm A plant having its seeds enclosed in an ovary.
annual A plant that completes its life-cycle within one year.
annular Ring-like.

annulus Ring.
anterior Front, away from the axis.
anther The part of the stamen that produces pollen (Fig. III, p. 7).
anther-cell (theca) One of the pollen sacs that comprise the anther; at first there are usually 4 (2 in each lobe) but before anthesis the tissue separating each pair disintegrates and the anther then appears 2-celled (Fig. V, p. 11).
anthesis Flowering.
antipetalous (stamens) The same number as, and opposite to the corolla-segments (Fig. IV, p. 8).
antisepalous (stamens) The same number as, and opposite to the calyx-segments (Fig. IV, p. 8).
apetalous Without petals.
apex (plur. **apices**) The tip of an organ; adj., **apical.**
apiculus A short, sharp point.
apocarpous Having free carpels (Fig. IX, p. 16).
apomixis Reproduction, including vegetative propagation, which does not involve sexual processes, often used in the narrower sense of agamospermy; adj. **apomictic.**
apopetalous With free petals.
aposepalous With free sepals.
appendage An attached subsidiary part.
arborescent Tree-like in growth or general appearance.
areole One of the small, spine-bearing areas on the stem of a cactus.
aril An outgrowth of the funicle, forming an appendage or outer covering on a seed.
arillate With an aril.
arilloid Resembling an aril.
arista An awn or stiff bristle.
aristate With an arista.
articulated Jointed.
ascending Directed upwards at an oblique angle.
asepalous Without sepals.
asexual Non-sexual.
atypical Not conforming to type.
auricle A small lobe or ear-shaped appendage; adj. **auriculate.**
awn A bristle-like appendage, especially occurring on the glumes or lemmas of some grasses.
axil The angle formed by the upper side of a leaf and the stem.
axile placentation A type of placentation in which the ovules are borne on placentas on the central axis of an ovary having 2 or more loculi (Fig. VII, p. 14).
axillary In the axil.
axis (plur. **axes**) An imaginary line, round which the organs are developed.
baccate Berry-like.
basal At the base of an organ.
basifixed Attached to the base (Fig. V, p. 11).

bast Fibrous tissues serving for support.
beak A pointed projection.
berry A fleshy, indehiscent fruit with the seed or seeds immersed in pulp.
biconvex Convex on both sides, lenticular.
biennial A plant that completes its life-cycle in 2 years, flowering in its second year.
bifid Divided to about half-way into 2 parts.
bilabiate With 2 lips.
bilocular With 2 loculi.
bilobed With 2 lobes.
bipinnate Pinnate, with the primary leaflets again pinnate (Fig. XI, p. 547).
biserrate Serrate, with the teeth themselves serrate.
bisexual Having both stamens and carpels in the same flower.
biternate Consisting of 3 parts, each part again divided into 3 (Fig. XI, p. 547).
bivalved With 2 valves.
blade The expanded part of a leaf.
bract A much-reduced leaf, especially the small or scale-like leaves associated with a flower or flower-cluster.
bracteate Bearing bracts.
bracteolate Bearing bracteoles.
bracteole A small bract.
bulb A usually underground organ, consisting of a short disc-like stem bearing fleshy scale-leaves and one or more buds, often enclosed in protective scales; adj., **bulbous.**
bulbil A small bulb, often arising from the axil of a leaf or the inflorescence.
bursicle The pouch-like base of the rostellum in some Orchidaceae.
caducous Falling off early.
caespitose Tufted.
calcicole Growing on soils containing lime.
calcifuge Growing on lime-free soils.
callus A hard or tough swollen area.
calyculus The epicalyx, appearing as a small rim or fringe in some of the Loranthaceae.
calyx (plur. **calyces**) The outer perianth, composed of free or united sepals.
campanulate Bell-shaped (Fig. II, p. 4).
campylotropous (ovule) With the body curved so that it appears to be attached by its side to the funicle (Fig. VI, p. 13).
capitate Pin-headed, as the stigma of a Primrose (Fig. IX, p. 16); growing in heads, as the flowers of Compositae.
capitulum A head of sessile or almost sessile flowers (Fig. XIV, p. 553).
capsule A dry, dehiscent fruit formed from a syncarpous ovary (adj., **capsular**).
carina A keel, as that formed by the 2 united lower petals in some of the Leguminosae.
carpel The unit of which the gynoecium is composed; in an apocarpous gynoecium the individual carpels are separate, in a syncarpous one the carpels are assumed to be joined in various ways (Fig. IX, p. 16).

carpophore The stalk to which the carpels are attached, as in Geraniaceae and Umbelliferae.
cartilaginous Gristly.
caruncle A protuberance near the hilum of a seed.
carunculoid Resembling a caruncle.
caryopsis (plur. **caryopses**) The fruit of grasses, a one-seeded indehiscent fruit with the pericarp united to the seed.
catkin An often pendulous inflorescence of unisexual, apetalous flowers.
caudicle The group of threads connecting a pollinium with the rostellum.
cauline Borne on the stem.
chalaza The basal portion of the nucellus of an ovule (Fig. VI, p. 13).
chasmogamy The opening of the perianth at the time of flowering; adj., **chasmogamous.**
chromosomes Thread-like structures which normally occur in 2 sets (diploid) in the nucleus of all vegetative cells and carry the genes or units of heredity; the sexual reproductive cells usually contain one set (haploid).
ciliate Fringed with hairs.
cincinnus A monochasial cyme, in which successive flowers are on alternate sides of the axis (Fig. XIV, p. 550).
circumscissile Dehiscing by a line round the fruit or anther, the top coming off as a lid.
clavate Club-shaped, thickened towards the apex.
claw The narrowed base of some petals.
cleistogamy Self-fertilisation within the unopened flower, as in some Violaceae; adj., **cleistogamous.**
clone A group of plants that have arisen by vegetative reproduction from a single parent and which are therefore genetically identical.
coalescence Growing together; adj., **coalescent.**
collateral Standing side by side.
columella The persistent central axis round which the carpels of some fruits are arranged.
column The combination of stamens and styles into a single structure, as in Asclepiadaceae and Orchidaceae.
coma The tuft of hairs at the end of some seeds.
commissure The face by which 2 carpels adhere, as in Umbelliferae.
compound Composed of 2 or more similar elements.
concave Appearing as if hollowed out.
conduplicate Folded together lengthwise.
confluent Merging together.
conifer A plant bearing cones, as the Pine.
connate United (with a similar part) (Fig. XII, p. 548).
connective The part of the stamen that joins the 2 pairs of anther-cells (Fig. V, p. 11).
connivent Converging.
contiguous Touching or adjoining.
contorted Twisted (Fig. I, p. 3).
convex Having a rounded surface.

convolute Rolled together lengthwise (Fig. I, p. 3).
cordate Heart-shaped (Fig. X, p. 544).
coriaceous Leathery.
corm A swollen underground stem, somewhat bulb-like in appearance, but solid and not composed of fleshy scale-leaves.
corolla The inner perianth, composed of free or united petals.
corona A structure occurring between the stamens and the corolla, as in Asclepiadaceae and Amaryllidaceae (Fig. II, p. 5).
corpusculum The clip connecting the 2 bands (retinacula) which are attached to the pollinia in Asclepiadaceae.
corymb A racemose inflorescence with pedicels of different lengths, causing the flower-cluster to be flat-topped (Fig. XIV, p. 552); adj., **corymbose.**
cotyledon One of the first leaves of the embryo.
coumarin An aromatic substance, the smell of newly mown hay, especially of the grass *Anthoxanthum odoratum.*
crenate With rounded teeth (Fig. XIII, p. 549).
crest An elevation or ridge on the summit of an organ.
cross-pollination The transfer of pollen from one plant to another.
cucullus A corona-hood in the Asclepiadaceae.
culm The hollow stem of grasses.
cultivar A horticultural variety that has originated and persisted under cultivation, not necessarily referable to a botanical species.
cuneate Wedge-shaped.
cupule A cup-shaped structure composed of coalescent bracts, as in the Fagaceae; adj., **cupular.**
cuticle The waxy covering secreted by the cells of the epidermis.
cyathium The inflorescence of the genus *Euphorbia.*
cyclic Arranged in whorls.
cyme A determinate inflorescence (Fig. XIV, pp. 550, 551); adj., **cymose.**
cymule A small cyme or portion of one.
cystolith A crystal or deposit, usually of calcium carbonate, within a cell.
deciduous (leaves) Falling in autumn,
(other organs) falling before the majority of the associated organs.
decumbent Lying along the ground, but with the apex ascending.
decurrent Running down, as when the base of a leaf is prolonged down the stem as a wing (Fig. XII, p. 548).
decussate In opposite pairs, each pair at right angles to the next.
dehiscence The process of splitting open at maturity.
dehiscent Splitting open along definite lines.
deltoid Triangular (Fig. X, p. 544).
dentate Toothed (Fig. XIII, p. 549).
denticulate Minutely toothed.
determinate (inflorescence) One in which the terminal flower opens first and prevents further growth of the axis.
diadelphous With 2 groups of stamens, as in some Leguminosae (Fig. IV, p. 9).
dichasial cyme (dichasium) A cyme with lateral branches on both sides of the main axis (Fig. XIV, p. 551).

dichotomous Forked with 2, usually equal, branches.

dicotyledon A plant having 2 seed-leaves; adj., **dicotyledonous.**

didynamous Having one pair of stamens longer than the other pair, as in most Labiatae (Fig. IV, p. 9).

digitate Resembling the fingers of a hand.

dimerous (2-merous) Having the parts of the flower in twos.

dimorphic Occurring in 2 forms.

dioecious With male and female flowers on separate plants.

diploid Having 2 sets of chromosomes in each cell.

diplostemonous Having the stamens in 2 whorls, the inner whorl opposite the petals and the outer whorl opposite the sepals.

disc A development of the receptacle within the perianth, often bearing nectar-glands; the central part of the capitulum in Compositae.

discoid Resembling a disc.

distichous Arranged in 2 vertical ranks.

distinct Separate.

dithecous With 2 anther-cells.

divergent Spreading away from each other.

divided (leaves) With lobing extending to the midrib or the base.

dorsal The side of an organ away from the axis, abaxial.

dorsifixed Attached at the back (Fig. V, p. 11).

double (flowers) Having more than the usual number of petals, usually at the expense of the stamens.

drupaceous Resembling a drupe.

drupe A fleshy fruit with the seed enclosed by a stony endocarp, as in the genus *Prunus.*

drupelet One of the small drupes which together form an aggregate fruit such as the Blackberry.

ebracteolate Without bracteoles.

eglandular Without glands.

elaiosome An appendage on the seeds of some plants which contains oily substances attractive to ants.

ellipsoidal Elliptical in outline with a 3-dimensional body.

elliptic(al) In the form of an ellipse (Fig. X, p. 544).

emarginate Having a notch in the margin at the apex.

embryo The rudimentary plant within the seed (Fig. VI, p. 13).

endocarp The innermost layer of the pericarp.

endosperm The albumen of a seed, particularly that deposited within the embryo sac.

ensiform Sword-shaped (Fig. X, p. 544).

entire Without division of toothing of any kind (Fig. XIII, p. 549).

entomophilous Depending on insects to convey pollen for fertilisation.

epicalyx (plur. **epicalyces**) A whorl of sepal-like organs just below the true sepals.

epicarp The outermost layer of the pericarp.

epidermis The true cellular skin or covering of a plant; adj., **epidermal.**

epigynous With the sepals, petals and stamens inserted near the top of the ovary (Fig. VIII, p. 15).

epipetalous Borne upon the petals (Fig. IV, p. 8).

epiphyte A plant which grows on another plant but does not derive nourishment from it; adj., **epiphytic.**

equitant Overlapping in 2 ranks, as in the leaves of *Iris.*

ericoid Resembling *Erica.*

exalbuminous Of seeds, lacking endosperm.

excentric Not having its axis placed centrally.

exserted Protruding.

exstipulate Without stipules.

extrafloral Outside the flower, as nectaries situated on the petiole.

extrastaminal Situated outside the whorl of stamens.

extrorse (anthers) Facing and opening outwards, away from the centre of the flower.

falcate Sickle-shaped (Fig. X, p. 544).

fall One of the 3 outer perianth-segments in *Iris.*

family A group of genera resembling one another in general appearance and technical characters.

fascicle A close cluster or bundle; adj., **fasciculate.**

fertile Able to reproduce sexually.

fertilisation The effect of pollen, deposited on the stigma or stigmatic surface, causing the change of flower into fruit and ovule into seed.

fibrous roots Adventitious roots arising from the base of the stem, as in grasses.

filament A thread, particularly the stalk of an anther (Fig. III, p. 7 and Fig. V, p. 11).

filiform Thread-like.

fimbriate Fringed, having the margin cut into long, slender lobes.

floral Pertaining to the flower.

floret A small flower as in, e.g., Compositae and Gramineae.

flower The structure in Angiosperms concerned with sexual reproduction.

foliaceous Leaf-like.

follicle A dry fruit formed from a single carpel, containing more than one seed, and splitting open along the suture.

free Not united with similar parts.

free-central placentation A type of placentation in which the ovules are borne on placentas on a free, central column within a unilocular ovary (Fig. VII, p. 14).

fruit A matured ovary with its enclosed seeds and sometimes with attached external structures.

funicle The stalk connecting an ovule to its placenta (Fig. VI, p. 13).

funnelform Funnel-shaped (Fig. II, p. 4).

fused Completely united.

fusiform Spindle-shaped.

fusion Complete union of parts.

gamopetalous With united petals.

gamosepalous With united sepals.

genes Hereditary units carried in chromosomes.

genetics The study of variation and heredity.

geniculate Abruptly bent.
genus (plur. **genera**) A group consisting of one or more similar species.
glabrous Without hairs.
gland A secretory organ.
glandular Possessing glands.
glaucous With a waxy, greyish-blue bloom.
globose Spherical.
glochid A barbed hair or bristle.
glomerule A condensed cyme of almost sessile flowers; a compact cluster.
glume One of the pair of bracts at the base of a spikelet in Gramineae; the single bract subtending the flower in Cyperaceae.
gymnosperm A plant having its seed naked, i.e., not enclosed in an ovary, as a Conifer.
gynobasic (style) Arising from below the ovary and between the carpels, as in Boraginaceae and Labiatae.
gynoecium The female sex organs (carpels) collectively.
gynomonoecious Having female and hermaphrodite flowers on the same plant.
gynophore The stalk bearing a carpel or gynoecium.
gynostegium The staminal crown in Asclepiadaceae.
gynostemium The column formed by the androecium and gynoecium combined, as in Aristolochiaceae.
half-inferior (ovary) Having the lower part embedded in the pedicel and the upper part exposed, as in *Saxifraga granulata.*
halophytic Growing on saline soils.
haploid Having a single set of chromosomes in each cell.
haplostemonous With a single series of stamens in one whorl.
hastate Arrowhead-shaped, but with the basal lobes directed outwards (Fig. X, p. 544).
haustorium A sucker of parasitic plants.
head A short, dense spike of flowers, the capitulum in Compositae.
herb A non-woody plant, or one that is woody only at the base; adj., **herbaceous.**
hermaphrodite Bisexual, having both stamens and carpels in the same flower.
heterostyly Variation in the length of the style (and stamens) in different flowers within a species, as in Primulaceae and Lythraceae; adj., **heterostylous.**
hexamerous (6-merous) Having the parts of the flower in sixes.
hilum The scar left on a seed where it was previously attached to the funicle.
hirsute Covered in rough, coarse hairs.
hispid Having stiff, bristly hairs.
homogamous With only one kind of flower.
honey-leaf A nectary, as in many Ranunculaceae.
hybrid A plant resulting from a cross between 2 or more genetically-unlike plants.
hygrophilous Requiring abundant moisture.
hygroscopic Extending or shrinking according to changes in moisture content.
hymenopterous (insects) Having membranous wings.
hypanthium A cup-shaped enlargement of the receptacle or the bases of the floral parts.

hypocotyl The axis of an embryo below the cotyledons.
hypocrateriform Salverform (Fig. II, p. 4).
hypogeal Growing or remaining below ground.
hypogynous With the sepals, petals and stamens attached to the receptacle or axis below the ovary (Fig. VIII, p. 15).
imbricate Overlapping (Fig. I, p. 3).
imbricate-ascending (aestivation) Having the vexillum within the lateral petals. (Fig. I, p. 3).
imbricate-descending (aestivation) Having the vexillum outside the lateral petals, vexillary (Fig. I, p. 3).
imparipinnate Pinnate, with a terminal leaflet (Fig. XI, p. 547).
incised Cut deeply and sharply into narrow, angular divisions (Fig. XIII, p. 549).
indefinite (flower-parts) Of a large enough number to make a precise count difficult.
indehiscent (fruit) Remaining closed at maturity.
indeterminate (inflorescence) One in which the lower or outer flowers open first, and the axis continues growth.
induplicate Rolled or folded inwards.
inferior Below, as when the ovary appears embedded in the pedicel below the other floral parts (Fig. VIII, p. 15).
infertile Not fertile.
inflexed Abruptly bent inward.
inflorescence The arrangement of the flowers on the floral axis; a flower-cluster.
infructescence A cluster of fruits, derived from an inflorescence.
inserted Growing from another organ.
integument The outer covering of an ovule, usually 2 being present in Angiosperms, which later hardens to become the testa, or seed-coat (Fig. VI, p. 13).
internode The part of the stem between 2 nodes.
intrastaminal Within the stamens.
introduced Brought in from another region.
introrse (anthers) Facing and opening inwards, towards the centre of the flower.
involucel The involucre of a partial umbel in Umbelliferae.
involucre A ring of bracts surrounding the head of flowers in Compositae or subtending the umbel in Umbelliferae; adj., **involucral.**
jaculator A hook-like outgrowth from the stalk of a seed which aids its dispersal, as in Acanthaceae.
keel (carina) The 2 united lower petals in some flowers of the Leguminosae.
labellum A lip, especially the highly modified third petal in Orchidaceae.
labiate Lipped (Fig. II, p. 4), or a member of the Labiatae.
laciniate Cut into narrow lobes.
lamina Leaf-blade.
lanceolate Lance-shaped (Fig. X, p. 544).
lateral On or at the side.
latex Milky juice.
leaf A lateral outgrowth from the stem, usually consisting of a stalk (petiole) and a flattened blade (lamina).

leaflet A leaf-like segment of a compound leaf (Fig. X, p. 546).
legume The 2-valved fruit formed from a single carpel in the Leguminosae.
lemma The lower of the 2 bracts enclosing a grass flower.
lenticular Lens-shaped, biconvex.
liane A woody, tropical climber.
ligulate Strap-shaped or tongue-shaped (Fig. II, p. 4).
ligule A strap-shaped body, such as the limb of the ray-florets in Compositae; the scarious projection from the top of the leaf-sheath in grasses (Fig. XII, p. 548).
limb The expanded part of a gamopetalous corolla, as distinct from the tube.
limen The rim at the base of the androgynophore in *Passiflora.*
linear Long and narrow, with parallel sides (Fig. X, p. 544).
lip One of the 2 divisions of a bilabiate corolla or calyx; the labellum in Orchidaceae.
lobe Any division of an organ, especially if the part is rounded.
lobed Having one or more lobes.
lobulate Having small lobes.
loculicidal (capsule) Dehiscing through the midrib of the carpels.
loculus Compartment of an ovary or anther.
lodicule One of usually 2 minute scales in a grass flower, generally considered to be a vestigial perianth.
lomentum A legume constricted into a series of one-seeded portions.
lyrate Pinnatifid, with the terminal lobe rounded and much larger than the others (Fig. X, p. 544).
marginal placentation A type of placentation in which the ovules are borne on a placenta situated along the ventral suture of a single carpel (Fig. VII, p. 14).
membranous Thin and semi-transparent like a membrane.
mericarp A portion of a fruit which splits away as a perfect fruit, as one of the 2 carpels forming the fruits of most Umbelliferae.
mesocarp The middle layer of the pericarp.
micropyle The opening in the integuments of an ovule, through which the pollen-tube grows after pollination (Fig. VI, p. 13); adj., **micropylar.**
microspecies Species founded on minute differences and used mostly for apomictic plants, as in the genera *Taraxacum* and *Hieracium.*
midrib The middle and principal vein of a leaf.
monadelphous Having the stamens united in one group by the fusion of their filaments, as in some of the Leguminosae and Malvaceae (Fig. IV, p. 9).
moniliform Like a string of beads.
monochasial cyme (monochasium) A cyme with lateral branching on only one side of the main axis (Fig. XIV, p. 550).
monocotyledon A plant having one seed-leaf; adj., **monocotyledonous.**
monoecious With male and female flowers on the same plant.
monopodial With a simple main stem or axis, growing by apical extension and bearing lateral branches.
monothecous With one anther-cell.
monotypic Having only one representative, as a genus with one species.

mucilaginous Slimy.
mucronate Terminated abruptly by a short, straight point.
multicellular Many-celled.
multilocular With many loculi.
mycorrhiza The association of fungi and the roots of plants to their mutual advantage.
myrmecophilous (plants) Providing food and sometimes shelter for ants and in return deriving benefit from their presence, as in some Rubiaceae.
naked (flowers) Lacking a perianth.
naturalised Established thoroughly after introduction from another region.
nectar A sugar solution, attracting insects or birds to flowers for the purpose of pollination.
nectar-guide A line or other marking leading insects to the nectary.
nectariferous Producing nectar.
nectary A gland which produces nectar.
node The point on a stem where one or more leaves are borne.
nucellus The mass of cells in the middle of the ovule which contain the embryo (Fig. VI, p. 13).
nut A dry, one-seeded, indehiscent fruit with a woody pericarp.
nutlet A small nut, sometimes applied to an achene or part of a schizocarp.
obcordate Cordate, but having the notch at the apex, and the attachment at the point.
obdiplostemonous Having the stamens in 2 whorls, the inner whorl opposite the sepals and the outer whorl opposite the petals (Fig. IV, p. 8).
oblanceolate Lanceolate, but attached at the narrower end (Fig. X, p. 545).
oblong Longer than broad, with nearly parallel sides (Fig. X, p. 545).
obovate Ovate, but attached at the narrower end (Fig. X, p. 545).
obtuse Blunt or rounded at the end.
ochrea A tubular sheath, formed by the fusion of 2 stipules at the nodes of many Polygonaceae.
ochreate Provided with ochreae (Fig. XII, p. 548).
offset A short runner, producing a new plant at its tip, as *Sempervivum.*
operculum The lid of a circumscissile fruit; the membranous cover of the nectar-secreting ring in *Passiflora.*
opposite (leaves) Two at a node, one on each side of the stem.
orbicular Circular (Fig. X, p. 545).
ornithophilous Depending on birds to convey pollen for fertilisation.
orthotropous (ovule) Borne on a straight funicle (Fig. VI, p. 13).
oval Broadly elliptical (Fig. X, p. 545).
ovary The lower part of a carpel (or carpels) which contains the ovules (Fig. III, p. 7 and Fig. VIII, p. 15).
ovate Shaped like a longitudinal section of a hen's egg (Fig. X, p. 545).
ovoid Egg-shaped.
ovule A structure in the ovary which, after fertilisation, develops into a seed.
palate The swollen part of the lower lip of a gamopetalous corolla which almost or entirely closes the throat, as in *Antirrhinum.*
palea The upper of the 2 bracts enclosing a grass flower.

pali The shorter filaments in the corona of the Passifloraceae.
palmate (leaves) Divided to the base into separate leaflets, all the leaflets arising from the apex of the petiole (Fig. XI, p. 547).
panicle A much-branched inflorescence (Fig. XIV, p. 552); adj., **paniculate.**
papilla A minute protuberance.
papillose Bearing papillae.
pappus The specialised calyx of hairs or scales occurring mainly in the Compositae.
parasite A plant which lives on another plant and derives nourishment from it; adj., **parasitic.**
parenchyma Tissue composed of thin-walled cells with intercellular spaces.
parietal placentation A type of placentation in which the ovules are borne on placentas situated on the inner surface of the ovary-wall or on outgrowths from it (Fig. VII, p. 14).
paripinnate Having an equal number of leaflets and lacking the terminal one (Fig. XI, p. 547).
parted Cut, but not quite to the base.
parthenocarpy The formation of fruit without preliminary fertilisation and usually without development of seeds; adj., **parthenocarpic.**
patent Spreading.
pectinate Pinnatifid with narrow segments set close like the teeth of a comb.
pedicel The stalk of a single flower (Fig. III, p. 7); adj., **pedicillate.**
peduncle The stalk of an inflorescence.
peltate (leaves) Having the blade attached by its lower surface to the stalk, instead of by its margin (Fig. X, p. 545).
pendent, pendulous Hanging down.
pentamerous (5-merous) Having the parts of the flower in fives.
pepo The unilocular, many-seeded fruit in Cucurbitaceae.
perennating Surviving from one season to another.
perennial A plant that lives for more than 2 years, flowering in its second and subsequent years.
perfoliate (leaves) Completely encircling the stem (Fig. XII, p. 548).
perianth The outer, non-reproductive organs of a flower, often differentiated into calyx and corolla.
pericarp The fruit-wall.
perigynium The utricle which encloses the female flower in some Cyperaceae.
perigynous With the sepals, petals and stamens inserted around the ovary on a concave structure developed from the receptacle (Fig. VIII, p. 15).
perisperm The albumen of a seed formed outside the embryo sac.
persistent Remaining attached.
personate (corolla) Bilabiate with a prominent palate (Fig. II, p. 4).
petal A single segment of the corolla (Fig. III, p. 7).
petaloid Petal-like.
petiolate Having a leaf-stalk.
petiole Leaf-stalk.
phylloclade A branch taking on the form and functions of a leaf, as in some Euphorbiaceae and *Ruscus* in Liliaceae.

phyllode A petiole taking on the form and functions of a leaf, as in *Acacia armata.*

pilose Softly hairy.

pin-eyed A dimorphic flower, as the Primrose, which has a long style and short stamens.

pinna A primary division or leaflet of a pinnate leaf.

pinnate Having separate leaflets along each side of a common stalk (Fig. XI, p. 547).

pinnatifid Pinnately divided, but not to the rachis.

pinnatisect Pinnately divided down to the rachis.

pistillate (flowers) Having only female organs.

placenta The part of the ovary to which the ovules are attached; adj., **placental.**

placentation The arrangement of placentas in an ovary.

plantlet A small plant, as those formed on the margins of the leaves in some species of *Kalanchoe.*

plica A fold.

plicate Folded, usually lengthwise.

plumose Feather-like.

plumule The rudimentary shoot of an embryo.

pod A dry, many-seeded, dehiscent fruit, particularly the legume in Leguminosae.

pollen The small grains which contain the male reproductive cells of the flower.

pollination The placing of pollen on the stigma or stigmatic surface.

pollinium A pollen-mass, as in Asclepiadaceae and Orchidaceae.

polygamo-monoecious Polygamous, but in the main monoecious.

polygamous Bearing both unisexual and hermaphrodite flowers on the same plant.

polypetalous With a corolla of separate petals.

polyploid Having more than 2 sets of chromosomes in each cell.

polysepalous With a calyx of separate sepals.

pome A fruit consisting of a core, formed by several united carpels, enclosed within the firm, fleshy receptacle, e.g., Apple.

pore A small aperture.

poricidal (anthers) Opening by pores (Fig. V, p. 11).

posterior Back, towards the axis.

posticous On the posterior side, next the axis.

prickle A sharp-pointed outgrowth from the superficial tissues of the stem, as in the Rose.

proboscis The elongated part of the mouth of some insects.

procumbent Lying along the ground.

prostrate Lying flat.

protandrous Having stamens which mature and shed their pollen before the stigmas of the same flower become receptive.

protogynous Having stigmas which become receptive before the stamens of the same flower mature and shed their pollen.

protoplasm The viscous, semi-transparent substance which forms the basis of all living plant-cells.

pseudobulb The thickened and bulb-like stem of some orchids.

pseudocarp A structure comprising the mature ovary combined with some other part of the plant, as the 'hip' in *Rosa canina*.
pubescence Hairiness.
pubescent Covered in soft hairs.
pulvinus An enlarged portion of the petiole, as its base in *Mimosa pudica* or at its junction with the lamina in Marantaceae.
punctate Marked with dots, depressions, or translucent glands.
punctiform In the form of a point or dot.
pungent Ending in a stiff, sharp point.
pyxis (plur. **pyxides**) A capsule with circumscissile dehiscence.
quincuncial Partially imbricated of 5 parts, 2 being exterior, 2 interior, and the fifth having one margin exterior and the other interior (Fig. I, p. 3).
raceme An indeterminate inflorescence with pedicillate flowers (Fig. XIV, p. 552); adj., **racemose.**
rachilla A secondary axis in the inflorescence of grasses.
rachis (plur. **rachides**) The axis of a compound leaf or an inflorescence.
radical (leaves) Arising directly from the rootstock.
radicle The rudimentary root of an embryo.
radii The longer filaments in the corona of the Passifloraceae.
raphides Needle-shaped crystals found in the cells of some plants.
ray A primary branch of an umbel.
raylet A smaller, secondary branch of an umbel.
receptacle The part of the axis which bears the flower parts, the torus (Fig. III, p. 7).
recurved Curved backward.
reflexed Bent backward abruptly.
reniform Kidney-shaped (Fig. X, p. 545).
replum The central partition in the fruits of Cruciferae.
resin A sticky, sometimes fragrant substance, insoluble in water, which is secreted by special cells in some plants for protective purposes; adj., **resinous.**
reticulate Marked with a network pattern.
retinaculum The band connecting a pollinium to the corpusculum in Asclepiadaceae.
revolute Rolled back from the margin or apex.
rhipidium A monochasial cyme, in which the lateral branches are developed alternately, resulting in a fan-shaped inflorescence (Fig. XIV, p. 550).
rhizomatous Resembling or possessing a rhizome.
rhizome A root-like stem, lying horizontally on or situated under the ground, bearing buds or shoots and adventitious roots.
rhomboid More or less diamond-shaped (Fig. X, p. 546).
rootstock A frequently subterranean stem or rhizome.
rosette A radiating cluster of leaves, usually at ground level.
rostellum The beak-like extension of the stigma in some Orchidaceae.
rotate (corolla) Gamopetalous, with a short tube and spreading lobes (Fig. II, p. 5).
rugose Wrinkled.

ruminate Looking as though chewed.
runcinate With sharply cut divisions directed backward towards the base (Fig. X, p. 546).
runner A prostrate shoot, rooting at the nodes and giving rise to new plants.
sac A small pouch or bag-shaped structure; adj., **saccate.**
sagittate Arrowhead-shaped (Fig. X, p. 546).
salverform (corolla) Having a slender tube expanding abruptly into a flat or saucer-shaped limb (Fig. II, p. 4).
samara A dry indehiscent, winged fruit.
samaroid Resembling a samara.
sap The juice of a plant.
saprophyte A plant which lives upon dead organic matter; adj., **saprophytic.**
scabrid Rough.
scale A reduced leaf, usually membranous, and often found covering buds and bulbs.
scandent Climbing.
scape A leafless stalk, bearing one or more flowers, which arises from the ground.
scapose Having a scape.
scarious Thin, dry and membranous.
schizocarp A fruit which splits into one-seeded portions (mericarps).
scion A young shoot which is inserted into a rooted stock in grafting.
scorpioid cyme A monochasial cyme, in which the branching is always on the same side, resulting in a coiled inflorescence (Fig. XIV, p. 550).
secund Directed towards one side.
seed A unit of sexual reproduction developed from the fertilised ovule.
segment One of the divisions of an organ.
sepal A single segment of the calyx (Fig. III, p. 7).
sepaloid Sepal-like.
septicidal (capsule) Dehiscing through the septa or carpel margins.
septum A partition; adj., **septate.**
serrate With a saw-toothed margin (Fig. XIII, p. 549).
serrulate Minutely serrate.
sessile Not stalked.
sheath A tubular covering.
shoot A developed bud or young green stem.
shrub A perennial, woody plant, generally smaller than a tree and having several main stems instead of a single trunk.
silicula The capsular fruit of the Cruciferae when less than 3 times as long as broad.
siliqua The capsular fruit of the Cruciferae when at least 3 times as long as broad.
simple Of one piece.
sinuate (leaves) Having the blade flat but with the margin winding strongly inward and outward (Fig. XIII, p. 549).
sinus The recess between 2 lobes, e.g., of a leaf.
solitary Borne singly.
spadix (plur. **spadices**) A spike with a fleshy axis, as in Araceae.
spathe A large bract enclosing an inflorescence, usually a spadix.

spathulate Spatula-shaped (Fig. X, p. 546).
species (plur. **species**) A group of closely related, mutually fertile individuals, showing constant differences from allied groups, the basic unit of classification.
spicate Spike-like.
spike An indeterminate inflorescence with sessile flowers (Fig. XIV, p. 552).
spikelet A unit of the inflorescence in grasses, consisting of one or more flowers subtended by a common pair of glumes.
spine A sharp, woody or hardened outgrowth from a leaf, sometimes representing the entire leaf.
spinose Spiny (Fig. XIII, p. 549).
spur A slender, hollow extension, often nectariferous, of the calyx or corolla (Fig. II, p. 5).
stamen One of the male sex organs, usually consisting of anther, connective and filament (Fig. III, p. 7); adj., **staminal.**
staminate (flowers) Having only male organs.
staminode A sterile stamen; adj., **staminodial.**
standard The upper petal in some flowers of the Leguminosae, or one of the 3 inner perianth-segments in *Iris.*
stellate Star-shaped.
stem The main supporting axis of a plant.
sterile Unable to reproduce sexually.
stigma The apex of the style, usually enlarged, on which the pollen grains alight and germinate (Fig. III, p. 7 and Fig. IX, p. 16); adj., **stigmatic.**
stipe The stalk of a stalked carpel or gynoecium.
stipulate Having stipules (Fig. XII, p. 548.
stipule A leafy outgrowth, often paired, arising at the base of the petiole.
stock The rooted stem into which the scion is inserted when grafted.
stolon A lateral stem growing horizontally at ground level, rooting at the nodes and producing new plants from its buds, as in the Strawberry.
stoloniferous Bearing stolons.
stoma (plur. **stomata**) A breathing pore in the epidermis of a leaf or young stem.
stone The hard endocarp of a drupe.
style The often elongated apical part of a carpel or gynoecium bearing the stigma at its tip (Fig. III, p. 7 and Fig. IX, p. 16); adj., **stylar.**
stylopodium The enlargement at the base of the styles in Umbelliferae.
subapical Almost at the apex.
subbasal Almost at the base.
subcompound More or less compound.
subcordate More or less heart-shaped.
subcuneate More or less wedge-shaped.
subequal Almost equal.
subfamily A subdivision of a family, usually found desirable in large families with many components.
subgenus A subdivision of a genus.
subglabrous Almost without hairs.
subglobose Nearly globular.

suborbicular Nearly circular.
subsessile Almost devoid of a stalk.
subshrub A low shrub, sometimes with partially herbaceous stems; an undershrub.
subspecies A subdivision of a species, often used for a geographically or ecologically distinct group of plants.
subtend To stand below and close to, to extend under.
subulate Awl-shaped (Fig. X, p. 546).
subvalvate Almost valvate.
succulent Fleshy and juicy.
sucker A shoot of subterranean origin.
suffrutescent Woody only at the base of the stem.
suffruticose Woody in the lower part of the stem.
sulcate Grooved or furrowed.
superior Above, as when the ovary is situated above the other floral parts on the receptacle (Fig. VIII, p. 15).
surmounted Capped or crowned.
suture A seam or line of joining.
syconium A multiple hollow fruit, as the Fig.
sympodial With the main stem or axis ceasing to elongate but growth being continued by the lateral branches.
synandrium An androecium coherent by the anthers, as in some Araceae.
syncarp An aggregate fruit, often fleshy, as the Mulberry.
syncarpous Having united carpels (Fig. IX, p. 16).
syngenesious With anthers united into a tube, but filaments free, as in Compositae (Fig. IV, p. 9).
taproot A strongly developed main root which grows vertically downwards bearing lateral roots much smaller than itself.
taxon A classificatory unit of any rank, e.g., Daisy, *Bellis perennis* (species); *Bellis* (genus); Compositae (family).
tendril A twining, thread-like structure produced from the stem or leaf which enables a plant to hold its position securely (Fig. X, p. 546).
tepal One of the petals or sepals of a flower in which all the perianth-segments closely resemble each other.
terete Slender and cylindrical, usually tapering, but more or less circular in any cross-section.
terminal At the apex or end.
ternate In threes.
testa The outer coat of the seed, formed by the hardened integuments.
tetrad (pollen) A group of 4 cells which remain together until they are almost or quite mature.
tetradynamous Having 4 long stamens and 2 short, as in Cruciferae (Fig. IV, p. 9).
tetramerous (4-merous) Having the parts of the flower in fours.
theca Pollen sac of an anther (Fig. V, p. 11).
thorn A sharp-pointed branch.
throat The opening into a corolla or calyx which has united segments.

thrum-eyed A dimorphic flower, as the Primrose, which has long stamens and a short style.

thyrse A mixed inflorescence in which the main axis is indeterminate and the secondary and ultimate axes determinate or cymose.

tomentose Densely covered in soft hairs.

torus The receptacle of a flower.

translator The clip (corpusculum) and bands (retinacula) which connect the pair of pollinia in Asclepiadaceae.

translucent Allowing light to pass through, but not transparent.

tree A perennial, woody plant with an evident trunk.

tribe A subdivision, usually of a large subfamily, but sometimes a family may be divided directly into tribes.

trichotomous Forked, with 3 usually equal branches.

trifid Divided to about half-way into 3 parts.

trifoliate Three-leaved.

trifoliolate With 3 leaflets (Fig. XI, p. 547).

trigonous Three-angled.

trilobate (trilobed) With 3 lobes.

trimerous (3-merous) Having the parts of the flower in threes.

trimorphic Occurring in 3 forms.

triploid Having 3 sets of chromosomes in each cell.

truncate Appearing as if cut off at the base or apex.

tuber An underground stem or root, swollen with reserves of food, as the Potato.

tubercle A small tuber.

tuberous Resembling or producing tubers.

tubular Cylindrical and hollow (Fig. II, p. 5).

umbel A usually flat-topped inflorescence in which the pedicels arise from the same point on the peduncle, as in most Umbelliferae (Fig. XIV, p. 553); adj., **umbellate.**

umbelliform In the shape of an umbel.

undershrub A low shrub, sometimes with partially herbaceous stems, a subshrub.

undulate (leaves) With a wavy margin curving up and down (Fig. XIII, p. 549).

unguiculate Narrowed at the base into a claw.

unicellular One-celled.

unilateral One-sided.

unilocular With a single loculus.

unisexual (flowers) Having only male or female organs.

united Joined.

urceolate Urn-shaped (Fig. II, p. 5).

utricle A bladder-like structure, especially the membranous sac in the fruit of *Carex.*

vaginate Sheathed (Fig. XII, p. 548).

valvate (flower) When the parts of the corolla or calyx meet exactly without overlapping (Fig. I, p. 3);
(anther or fruit) opening by valves (Fig. V, p. 11).

valve One of the pieces into which a capsule naturally separates at maturity.

variegated Having 2 or more colours in the leaves.

variety A rank subordinate to species used to designate a group of plants varying in flower-colour, habit or some other way.
vascular bundle A strand of tissue which conducts water and nutrients.
vein A strand of vascular tissue in a leaf or other flat organ.
venation The arrangement of veins as in a leaf.
ventral The side of an organ towards the axis, adaxial.
verrucose Warty.
versatile (anther) Attached at the middle and turning freely on its filament (Fig. V, p. 11).
verticillaster A false whorl, as in Labiatae.
verticillate Whorled.
vestigial Imperfect development of an organ which was fully developed in some ancestral form.
vexillary (aestivation) Having the vexillum folded over the other petals, imbricate-descending (Fig. I, p. 3).
vexillum The standard petal in the Leguminosae.
viscid (viscous) Sticky.
viscidium The sticky disc at the base of the caudicle in Orchidaceae.
viscin A sticky substance surrounding the seeds in Loranthaceae.
vitta An oil-canal in the fruits of Umbelliferae.
whorl The arrangement of organs in a circle round the axis; adj., **whorled.**
wing A lateral petal in some of the Leguminosae; a flat, often dry and membranous extension to an organ.
xerophyte A plant which can live in very dry conditions, as in a desert; adj., **xerophytic.**
zygomorphic Divisible into equal halves in one plane only.

LATIN PLURALS

The following examples show how the plural is formed for most of the Latin terms in the glossary. Less common plurals, where required, are given after the terms concerned.

locul-us	locul-i
al-a	al-ae
haustori-um	haustori-a

Fig. X. Leaf shapes – simple leaves and compound leaf with tendrils.

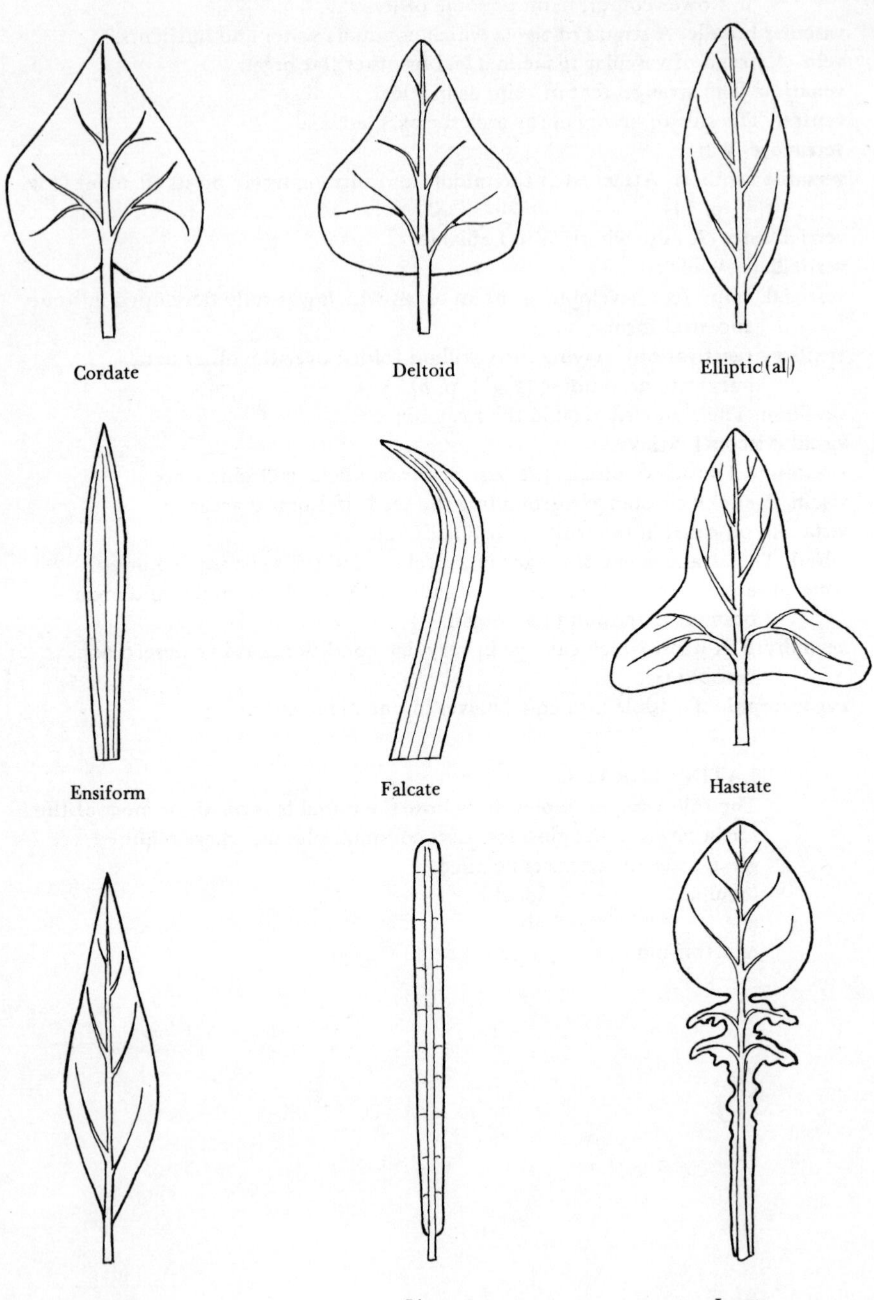

Fig. X. *continued*

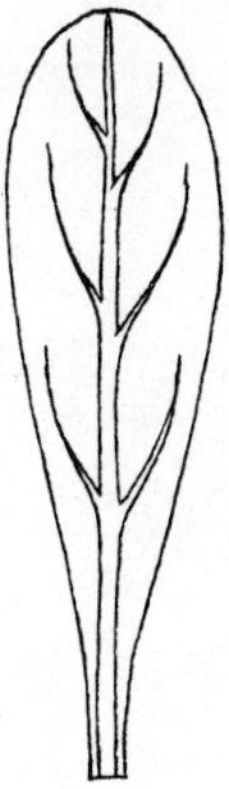

Oblanceolate

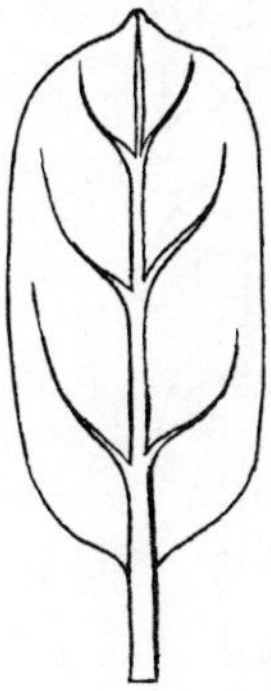

Oblong

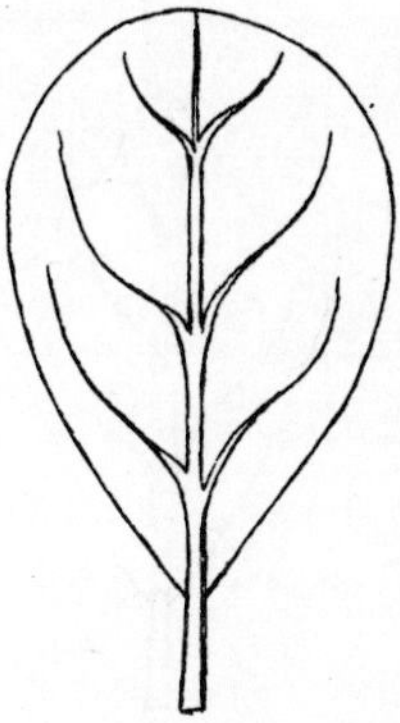

Obovate

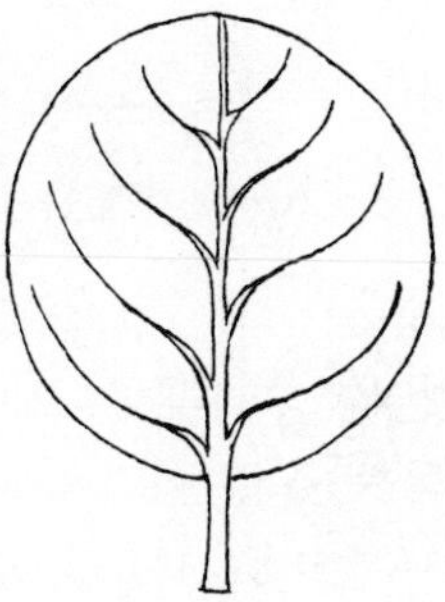

Orbicular

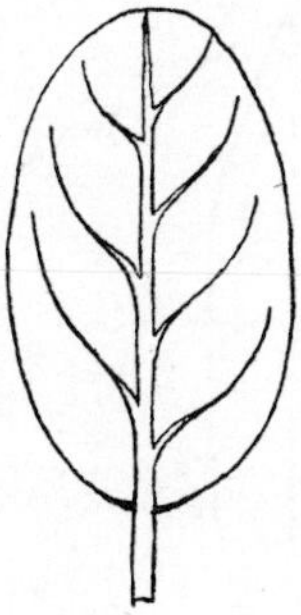

Oval

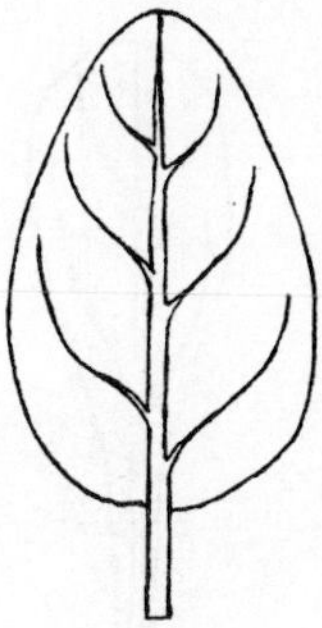

Ovate

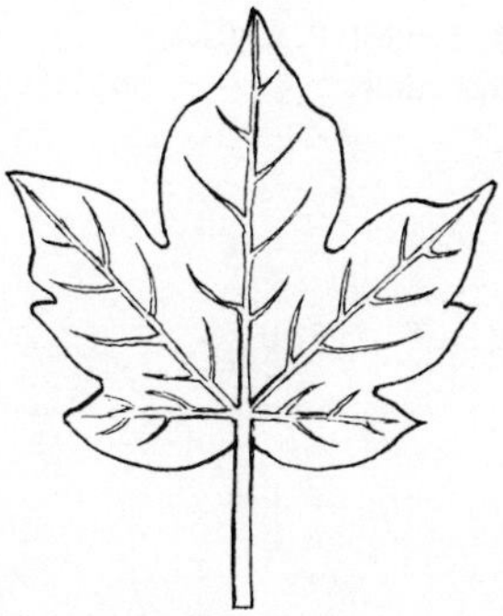

Palmately-lobed

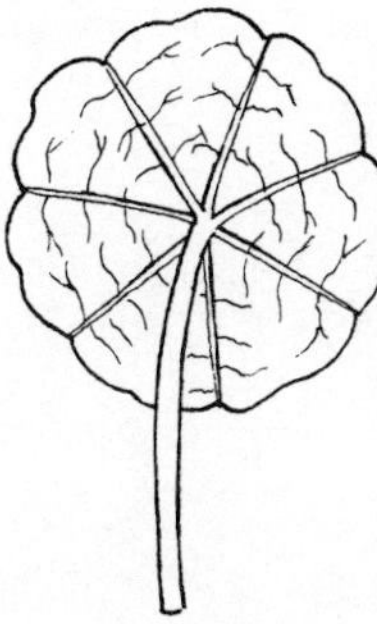

Peltate

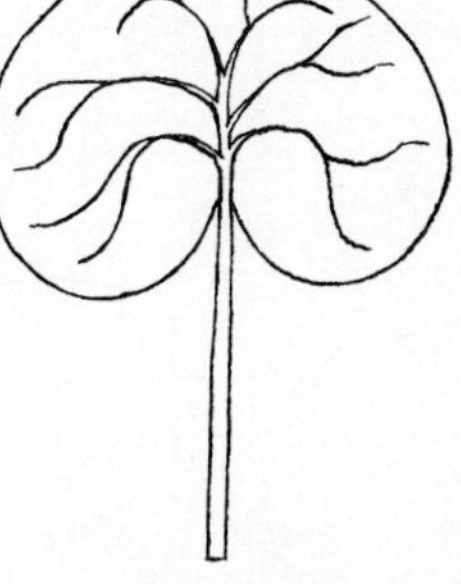

Reniform

Fig. X. *continued*

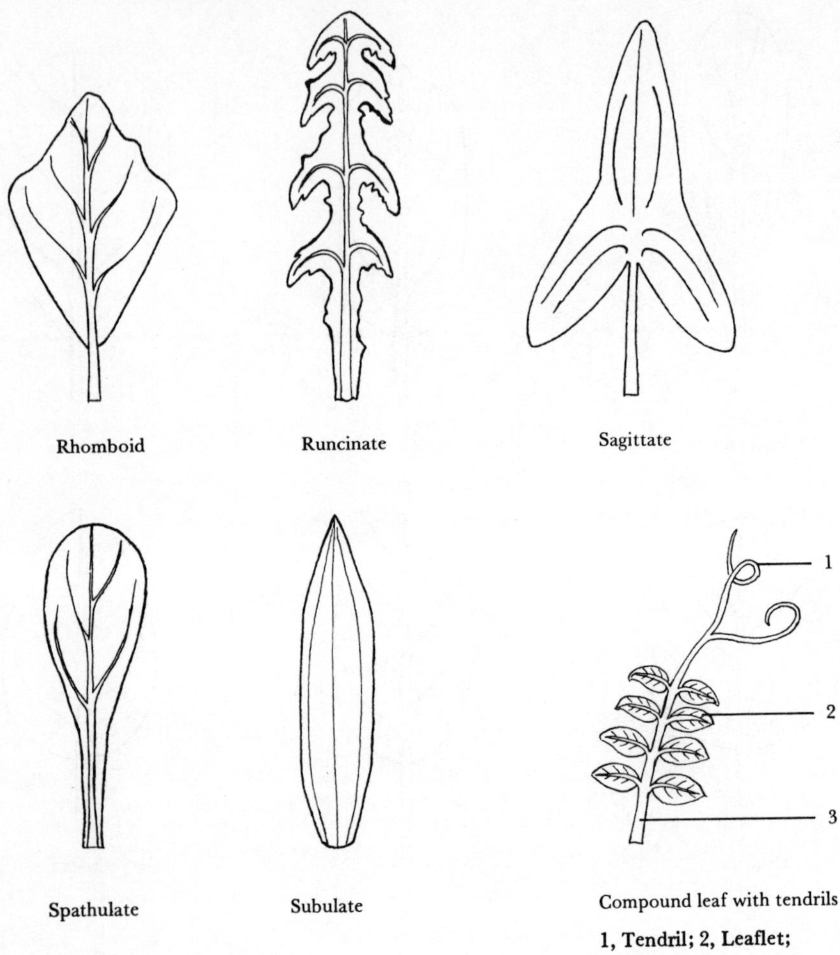

Rhomboid

Runcinate

Sagittate

Spathulate

Subulate

Compound leaf with tendrils

1, Tendril; 2, Leaflet; 3, Petiole.

Fig. XI. Leaf shapes – compound leaves.

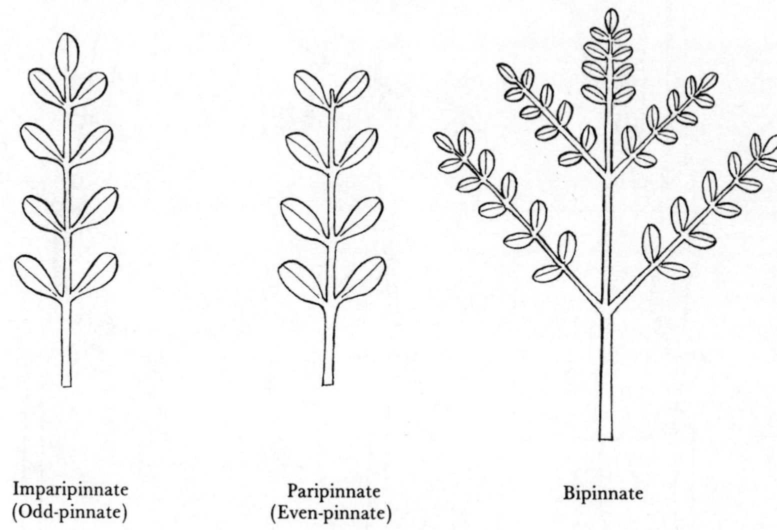

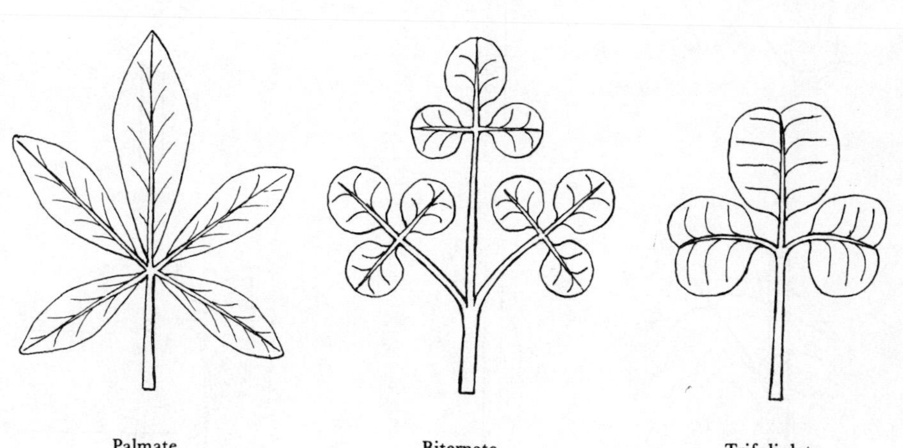

Fig. XII. Petioles and points of attachment: 1, Leaf-blade; 2, Ligule; 3, Leaf-sheath.

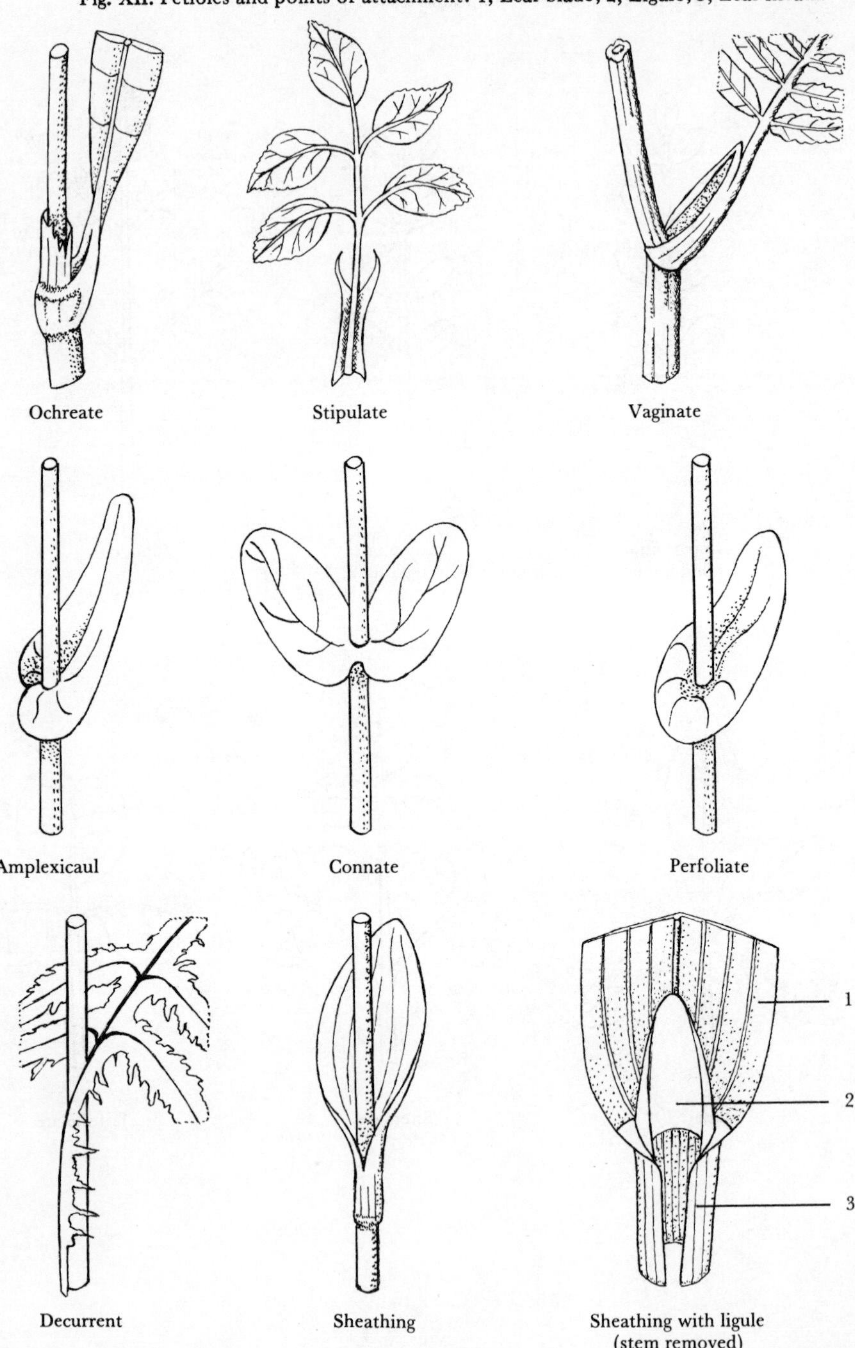

Fig. XIII. Leaf margins.

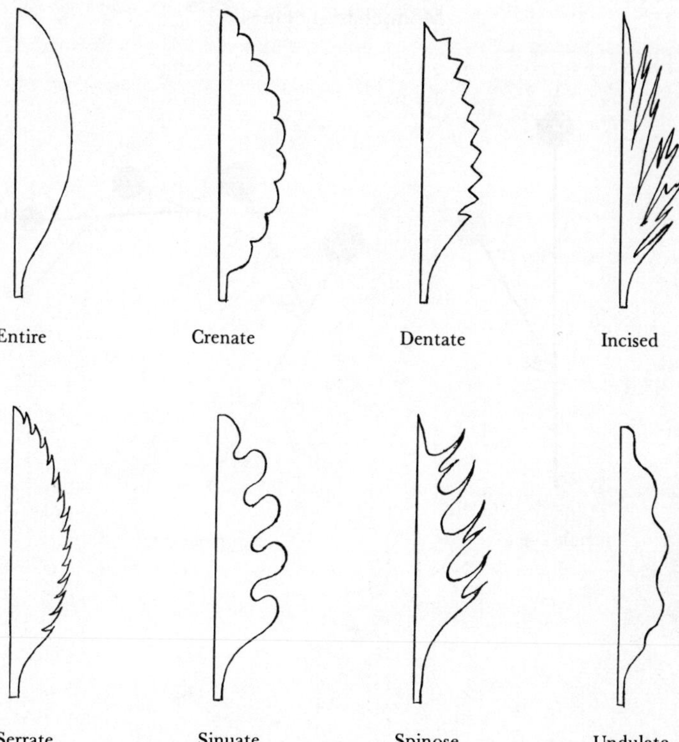

Fig. XIV. Inflorescences.

Monochasial cymes

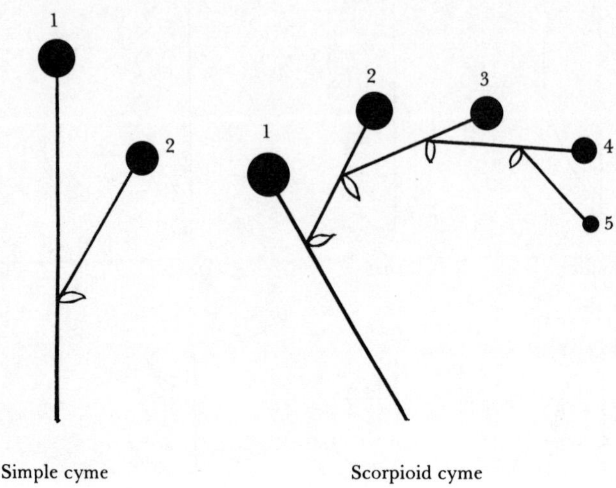

Simple cyme

Scorpioid cyme

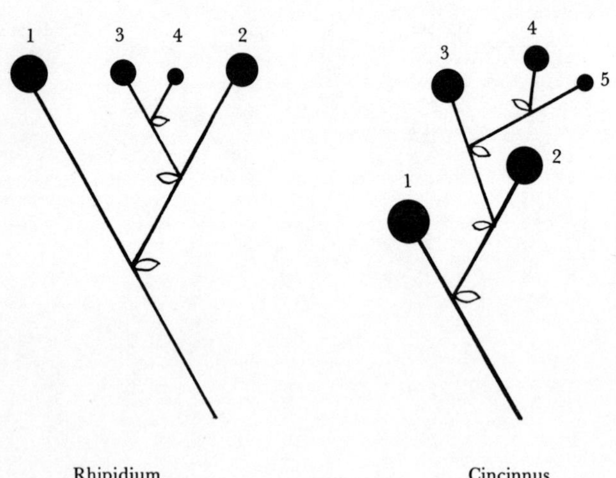

Rhipidium

Cincinnus

Dichasial cymes

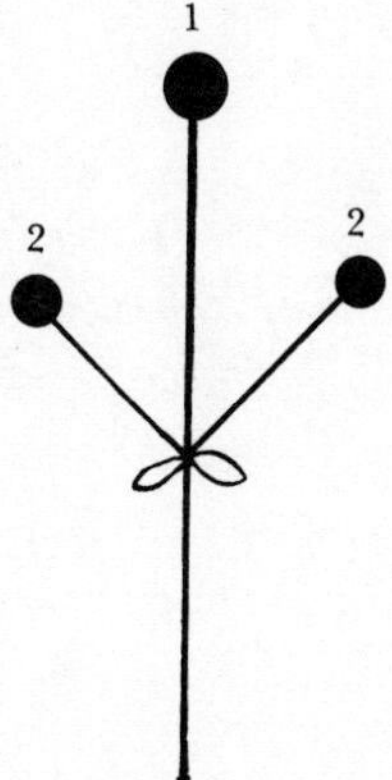

Simple cyme

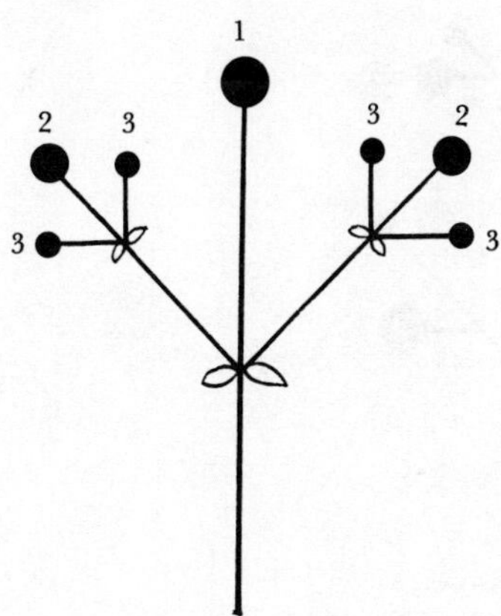

Compound cyme

Fig. XIV. *continued.* Inflorescences.

Racemose inflorescences

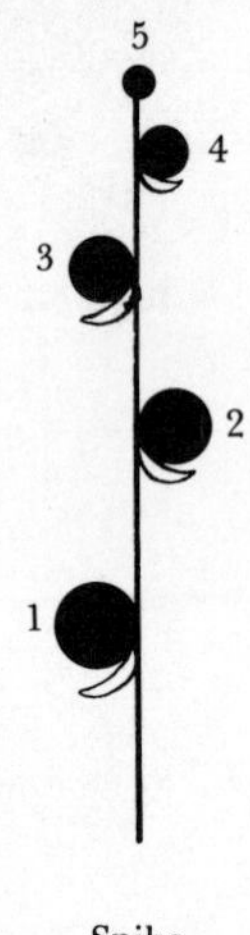

Spike

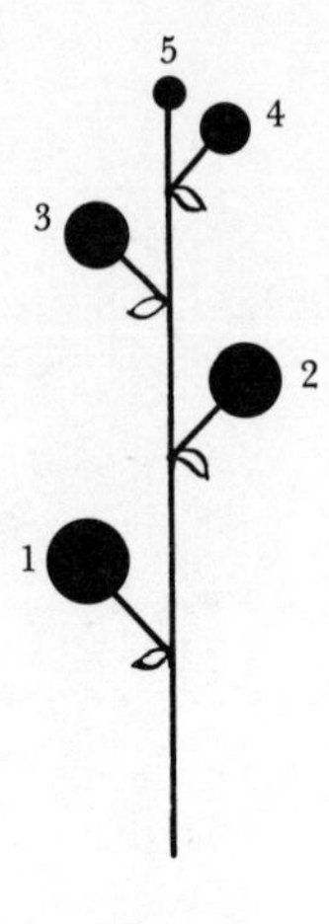

Raceme

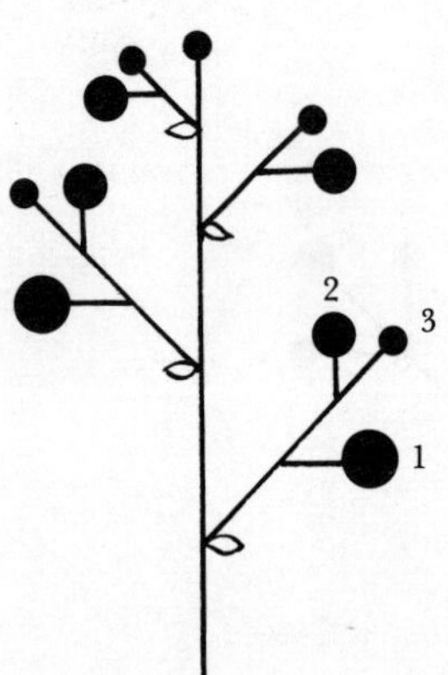

Panicle

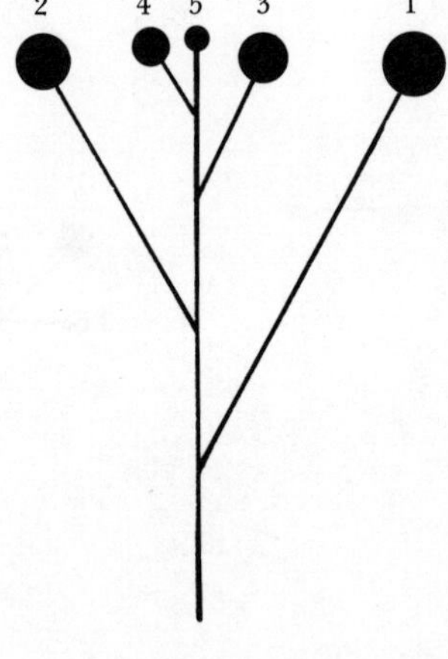

Corymb

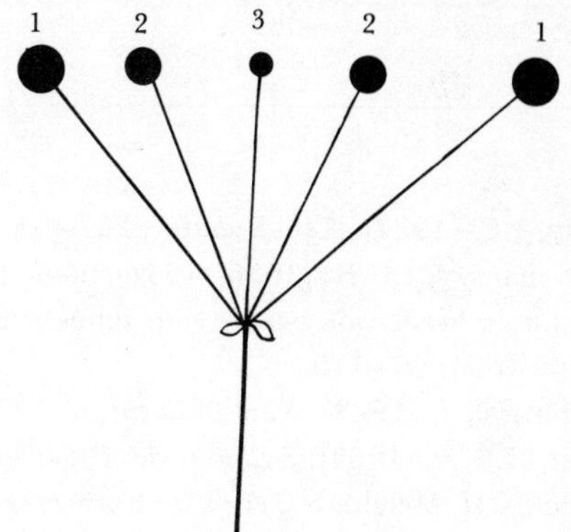

Simple umbel

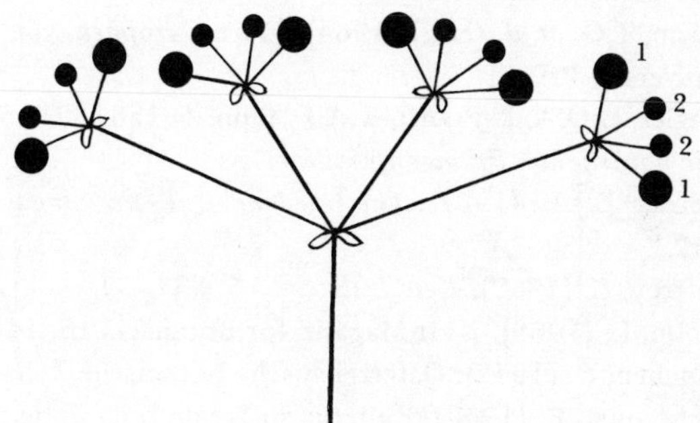

Compound umbel

Capitulum

REFERENCES

(1) Dandy, J.E. (1927). *Kew Bulletin*, 257–64.

(2) Kostermans, A.J.G.H. (1936–1938). *Mededeelingen van het botanisch museum en herbarium van de rijks universiteit te Utrecht*, 25, 12–50; 42, 500–604; 46–119.

(3) Hutchinson, J. (1923). *Kew Bulletin*, 65–89.

(4) Engler, H.G.A. (1964). *Syllabus der Pflanzenfamilien*, vol. 2, 12th edn., revised by H. Melchior. Gebrüder Borntraeger, Berlin.

(5) Willis, J.C. (1973). A Dictionary of the Flowering Plants and Ferns, 8th edn., revised by H.K. Airy Shaw. Cambridge University Press.

(6) Forman, L.L. (1964). *Kew Bulletin*, 17, 381–96.

(7) Britton, N.L. & Rose, J.N. (1919–1923). *The Cactaceae*, vols. 1–4. The Carnegie Institution of Washington.

(8) Tutin, T.G. *et al.* (eds.) (1964). *Flora Europaea*, vol. 1. Cambridge University Press.

(9) Engler, H.G.A. & Prantl, K.A.E. (edn. 1, 1887–1915; edn. 2, 1924–). *Die natürlichen Pflanzenfamilien.*

(10) Corner, E.J.H. (1962). *The Gardens' Bulletin, Singapore*, 20, 187–252.

(11) Jeffrey, C. (1962). *Kew Bulletin*, 15, 337–71.

(12) Holm, L. (1969). Nytt Magasin for Botanikk, 16, 147–50.

(13) Janchen, E. (1942). Österreichische botanische Zeitschrift, 91, 1–28.

(14) Verdcourt, B. (1958). Bulletin du Jardin botanique de l'État à Bruxelles, 28, 209–90.

(15) Hutchinson, J. (1934). *The families of Flowering Plants*, 1st edn., vol. 2. London.

INDEX OF FAMILIES AND GENERA

Botanical names of the 100 families in **bold** type